Advancing Maths for AQA

CORE MATHS 2

Sam Boardman, Tony Clough, David Evans and Maureen Nield

Series editors
Sam Boardman Roger Williamson Ted Graham David Pearson

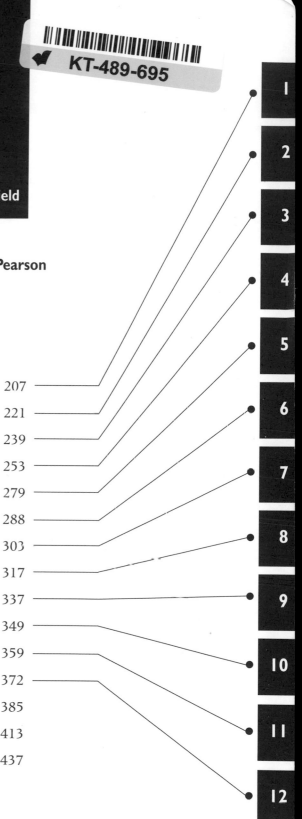

Core Maths 2

1
2
3
4
5
6
7
8
9
10
11
12

Heinemann

Heinemann is an imprint of Pearson Education Limited,
a company incorporated in England and Wales, having
its registered office at Edinburgh Gate, Harlow, Essex,
CM20 2JE. Registered company number: 872828

Heinemann is a registered trademark of
Pearson Education Limited

First published 2004

10
10 9

British Library Cataloguing in Publication Data is available
from the British Library on request.

ISBN 978 0 435 51330 6

Edited by Alex Sharpe, Standard Eight Limited
Typeset and illustrated by Tech-Set Limited, Gateshead, Tyne & Wear.
Original illustrations © Harcourt Education Limited, 2004
Cover design by Miller, Craig and Cocking Ltd
Printed in China (CTPS/09)

Acknowledgements
The publishers' and authors' thanks are due to the AQA for permission to reproduce questions from past examination papers.

The answers have been provided by the authors and are not the responsibility of the examining board.

Every effort has been made to contact copyright holders of material reproduced in this book. Any omissions will be rectified in subsequent printings if notice is given to the publishers.

About this book

This book is one in a series of textbooks designed to provide you with exceptional preparation for AQA's 2004 Mathematics Specification. The series authors are all senior members of the examining team and have prepared the textbooks specifically to support you in studying this course.

Finding your way around

The following are there to help you find your way around when you are studying and revising:

- **edge marks** (shown on the front page) – these help you to get to the right chapter quickly;
- **contents list** – this identifies the individual sections dealing with key syllabus concepts so that you can go straight to the areas that you are looking for;
- **index** – a number in bold type indicates where to find the main entry for that topic.

Key points

Key points are not only summarised at the end of each chapter but are also boxed and highlighted within the text like this:

> An equation of the form $ax + b = 0$, where a and b are constants, is said to be a **linear equation** with variable x.

Exercises and exam questions

Worked examples and carefully graded questions familiarise you with the syllabus and bring you up to exam standard. Each book contains:

- Worked examples and Worked exam questions to show you how to tackle typical questions; Examiner's tips will also provide guidance;
- Graded exercises, gradually increasing in difficulty up to exam-level questions, which are marked by an [A];
- Test-yourself sections for each chapter so that you can check your understanding of the key aspects of that chapter and identify any sections that you should review;
- Answers to the questions are included at the end of the book.

Contents C1

1 Advancing from GCSE maths: algebra review

2 Surds

3 Coordinate geometry of straight lines

Contents C2

1 Indices

2 Further differentiation

3 Further integration and the trapezium rule

4 Basic trigonometry

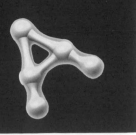

CHAPTER 1

C1: Advancing from GCSE maths: algebra review

Learning objectives

After studying this chapter, you should be able to:

- solve simple linear equations
- solve simultaneous linear equations
- solve linear inequalities
- use function notation.

1.1 Advancing from GCSE

In your GCSE course you will have begun to realise the importance of algebra in solving mathematical problems.

For instance, suppose you had a pet rabbit and you wanted to make a rectangular enclosure so it had the greatest possible area to graze. If the total perimeter fencing you had available was 12 m, what is the largest area you could enclose?

By trial and error you will soon realise that a square with each side 3 m encloses the greatest area of 9 m^2.

Suppose, however, that you could make use of a wall in your garden so you only needed to fence three of the sides to make a rectangular enclosure. What are the dimensions of the rectangle now that would give you the greatest possible area for grazing?

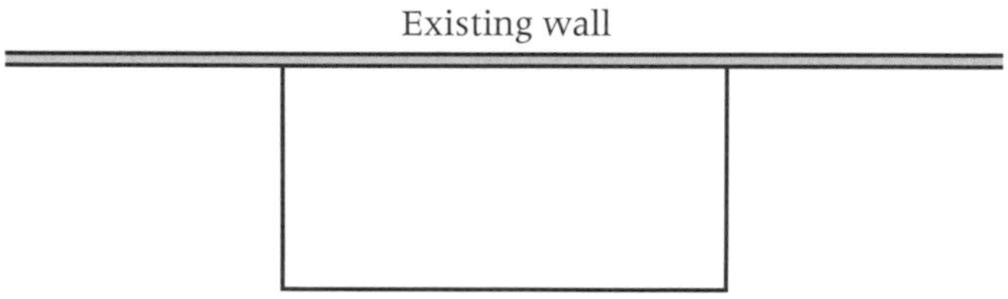

You could try lots of possibilities, but you really need algebra to solve this problem. After completing chapter 4 you should be able to solve this problem algebraically, and you may be surprised that the greatest area does not arise from making a square!

Here is another problem that you should enjoy thinking about, and once again, the answer may not be what you first expect.

Imagine you have a piece of rope that is tied around the earth along the equator to form a circle. Now imagine that an extra metre of rope is added to the circumference and the new circle is raised above the equator on little sticks so that it forms a new circle concentric with the equator. Can you visualise the situation?

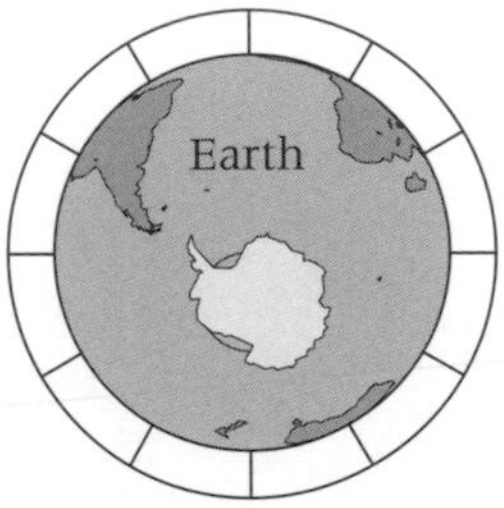

The question is: what size of creature could now crawl under the rope?
An ant? A worm? A mouse? A cat? A dog? A pig? A horse?

The solution to this and the previous problem are given at the end of the chapter.

If you have solved this problem, you will see the usefulness of algebra, even though it only involves solving a linear equation.

In this chapter you will need to revise your basic algebra and you will start by reviewing linear equations.

1.2 Solving linear equations

An equation of the form $ax + b = 0$, where a and b are constants, is said to be a **linear equation** with **variable** x.

The method of solving linear equations is to collect all the terms involving x on one side of the equation and everything else on the other side.

Worked example 1.1

Solve each of the following equations:

(a) $5x - 3 = 2x + 15$,

(b) $3x - 5 + 7x + 3 = 5x - 11 + 3x - 7$,

(c) $5(x + 3) + 4(2x - 3) = 2(2x + 15)$.

Solution

(a) $5x - 3 = 2x + 15$
$\Rightarrow 5x - 2x = 15 + 3$
$\Rightarrow 3x = 18$
$\Rightarrow x = 6$

Collect all the terms involving x onto one side of the equation.

(b) $3x - 5 + 7x + 3 = 5x - 11 + 3x - 7$
$\Rightarrow 3x - 5x + 7x - 3x = 5 - 11 - 3 - 7$
$\Rightarrow 2x = -16$
$\Rightarrow x = -8$

(c) $5(x + 3) + 4(2x - 3) = 2(2x + 15)$
$\Rightarrow 5x + 15 + 8x - 12 = 4x + 30$
$\Rightarrow 5x + 8x - 4x = 30 - 15 + 12$
$\Rightarrow 9x = 27$
$\Rightarrow x = 3$

You can check the answer is correct by substituting back into the original equation:
right-hand side $= 2 \times 21 = 42$
left-hand side $= 5 \times 6 + 4 \times 3 = 42$ ✓

Note. The logical proof symbol $\Rightarrow$ meaning 'implies' has been used at each step. Beware of using a 'trailing equal sign' as you proceed from one line to the next. In each of the solutions above, the argument is valid if you work back from the last line to the first, and so we could have used the symbol $\Leftrightarrow$ in place of $\Rightarrow$.

> Sometimes the linear equations are disguised because they involve fractions. A good strategy is to multiply every term by the lowest common multiple (LCM) of the denominators.

Worked example 1.2

Solve the equation $\dfrac{x - 5}{4} - \dfrac{4 - x}{3} = 5$

Solution

$$\frac{x - 5}{4} - \frac{4 - x}{3} = 5$$

Multiply through by 12.

$$\Rightarrow \frac{12(x - 5)}{4} - \frac{12(4 - x)}{3} = 12 \times 5$$

$\Rightarrow 3(x - 5) - 4(4 - x) = 60$
$\Rightarrow 3x - 15 - 16 + 4x = 60$
$\Rightarrow 3x + 4x = 60 + 15 + 16$
$\Rightarrow 7x = 91$
$\Rightarrow x = 13$

Notice the need to introduce brackets around the terms being multiplied. A common error is to leave these out and then the minus sign causes problems.

An historical problem

The earliest book on algebra using a mathematical notation to represent equations was called *Arithmetica* and was written by a Greek, Diophantus, around 250 AD in Alexandria. An interesting problem posed in a collection called the *Greek Anthology* allows us to calculate the age he lived if we form and solve an algebraic equation.

Worked example 1.3

'Diophantus passed one-sixth of his life as a boy, one-twelfth in youth, and one seventh as a bachelor. Five years after his marriage he was granted a son who died four years before his father, at half the measure of his father's life.' How long did Diophantus live?

Solution

Let the age when Diophantus died be x years.

$\frac{x}{6} + \frac{x}{12} + \frac{x}{7} + 5$ is the age when his son was born,

so if we add the son's age when he died, namely $\frac{x}{2}$, this must give us the age of Diophantus at that time, which must be $x - 4$. Hence, the equation to allow us to solve the problem is

$$\frac{x}{6} + \frac{x}{12} + \frac{x}{7} + 5 + \frac{x}{2} = x - 4$$

$\Rightarrow 14x + 7x + 12x + 420 + 42x = 84x - 336$

Multiply throughout by 84 to clear the fractions.

$\Rightarrow 75x + 756 = 84x$

$\Rightarrow 756 = 9x \qquad \Rightarrow x = 84$

So Diophantus lived to the age of 84.

Worked example 1.4

Solve the equation $\frac{7}{x+1} = \frac{8}{x-2}$.

Solution

$$\frac{7}{x+1} = \frac{8}{x-2}$$

Multiply through by $(x + 1)(x - 2)$

$$\Rightarrow \frac{7\cancel{(x+1)}(x-2)}{\cancel{(x+1)}} = \frac{8(x+1)\cancel{(x-2)}}{\cancel{(x-2)}}$$

Notice that we can cancel the algebraic factors $(x + 1)$ and $(x - 2)$ but only when they are at the top and bottom of the same expression.

$\Rightarrow 7(x-2) = 8(x+1)$

$\Rightarrow 7x - 14 = 8x + 8$

$\Rightarrow 7x - 8x = 8 + 14$

$\Rightarrow -x = 22$

$\Rightarrow x = -22$

You could have made use of the fact that $\frac{a}{b} = \frac{c}{d} \Rightarrow a \times d = c \times b$ (sometimes known as cross-multiplication) and obtained this equation directly from the original equation.

EXERCISE 1A

Solve each of the equations in questions **1–12**.

1 $3x - 4 = 2(7 - x)$.

2 $4 - 3x + 7x - 8 = 3x + 5 - x$.

3 $5x - 17 + 3x = 3(5 - 2x)$.

4 $15 - 3x + 7(x - 2) = 3(x - 4)$.

5 $\dfrac{7x-1}{5} = \dfrac{3(5+2x)}{3}$.

6 $\dfrac{7x-8}{3} = \dfrac{3x+5}{2}$.

7 $\dfrac{3x-1}{2} - \dfrac{2(4+3x)}{13} = 2$.

8 $\dfrac{2x+3}{5} - \dfrac{5x+2}{8} + 1 = 0$.

9 $\dfrac{5}{2x+1} = \dfrac{4}{x+2}$.

10 $\dfrac{3}{2x-5} = \dfrac{7}{3-5x}$.

11 $\dfrac{10}{3x+1} = \dfrac{1}{2x-5}$.

12 $\dfrac{5}{3x-7} = \dfrac{2}{4-3x}$.

13 A quadrilateral has three of its sides of lengths x, $2x - 5$ and $3x + 8$ cm. The fourth side has a length equal to one-third of the sum of the other three sides. Find the value of x if the total perimeter is 45 cm.

14 I am thinking of my age. When I multiply it by 7, subtract 3 and divide by 8 and then take away one-third of my age I get an answer of 11. What is my age?

15 I have a number of sweets in a bag. If I removed three of them and divided the number remaining by 4, I would have two fewer than if I removed 7 and divided the number remaining by 3. How many sweets do I have?

1.3 Solving simultaneous linear equations

We shall be concerned primarily with two equations in two variables such as x and y, and this idea will be used extensively in the next chapter to find points of intersection of straight lines.

In the equation $3x - 5y = 7$, the coefficient of x is 3 and the coefficient of y is -5.

The first example of solving simultaneous linear equations demonstrates the method of **elimination** where the coefficients of one of the variables are made equal and then the two equations are added or subtracted in order to eliminate it.

Worked example 1.5

Solve the simultaneous equations

$$2x + 3y = 13$$
$$7x - 5y = -1$$

Solution

$2x + 3y = 13$ [A]
$7x - 5y = -1$ [B]

> We label the equations *A* and *B*, say, so that we can refer to the equations in our working.

Make the coefficients of y the same: multiply [A] by 5
multiply [B] by 3

$10x + 15y = 65$ [C]
$21x - 15y = -3$ [D]

> We could have chosen to multiply [A] by 7 and [B] by 2 and obtained
> $14x + 21y = 91$
> $14x - 10y = -2$
> We would then subtract the two equations to give $31y = 93$, and hence $y = 3$.

Adding equations [C] + [D] gives

$$31x = 62$$
$$\Rightarrow x = 2$$

Substitute $x = 2$ into equation [A]

$$4 + 3y = 13$$
$$\Rightarrow y = 3.$$

The solution is $x = 2$, $y = 3$.

> Another method of solving simultaneous linear equations involves rearranging one of the equations and substituting into the other. This technique is called the method of **substitution**.

Worked example 1.6

Solve the simultaneous equations

$$7x + 2y = 11$$
$$4x + y = 7$$

Solution

$7x + 2y = 11$ [A]
$4x + y = 7$ [B]

> Rearrange [B] to make y the subject.

$$y = 7 - 4x$$

Substitute $(7 - 4x)$ for y in equation [A]

$$\Rightarrow 7x + 2(7 - 4x) = 11$$
$$\Rightarrow 7x + 14 - 8x = 11$$
$$\Rightarrow x = 3. \quad \Rightarrow y = 7 - 4(3) = -5$$

The solution is $x = 3$, $y = -5$.

EXERCISE 1B

Solve the following pairs of simultaneous equations.

1 $5x + 2y = 11$, $4x + 3y = 13$.

2 $3x - 2y = 8$, $7x + 3y = 11$.

3 $4x - 7y = 13$, $3x - 5y = 9$.

4 $x - 7y = -11$, $3x + 4y = -8$.

5 $3x + 2y = 10$, $4x + 7y = 12$.

6 $5x - 8y = -3$, $3x + 2y = 1$.

7 $3x + 2y + 5 = 0$, $4x + 7y + 3 = 0$.

8 $5x + 4y + 5 = 3x - 7y$, $4(x + 3y) + 3 = 5x$.

9 $x + 4(y - 1) = 3(x - 3) + 2y$, $3(x - y) = 5(x - 1)$.

10 $\dfrac{x + 2y}{x - 3} = 2$, $\dfrac{x - 7}{2y + 3} = 5$.

1.4 Linear inequalities

We solve linear inequalities in very much the same way as linear equations. However, instead of the = sign, we use one of the following signs:

$>$ greater than	$<$ less than
$\geqslant$ greater than or equal to	$\leqslant$ less than or equal to

The solution, rather than having a single value, is in the form of an interval of values such as $x > 2$, for example.

In a single inequality, it is customary to present the final result with the variable on the left-hand side so that a solution

$2 < x$

is usually rewritten as

$x > 2$.

Worked example 1.7

Solve the inequality $3(x + 2) > x - 1$.

Solution

$3(x + 2) > x - 1$

$\Rightarrow 3x + 6 > x - 1$

$\Rightarrow 3x - x > -1 - 6$

$\Rightarrow \quad 2x > -7$

$\Rightarrow \quad x > -3\frac{1}{2}$

The answer means that any value of x greater than $-3\frac{1}{2}$ is part of the solution interval.

It is always good to check that the solution is correct. Choose a value of x greater than $-3\frac{1}{2}$, such as -2 for example, and test to see that it satisfies the original inequality.

Left-hand side $= 3(x + 2) = 3(-2 + 2) = 0$
Right-hand side $= x - 1 = -2 - 1 = -3$

Is $0 > -3$? Yes, so we have confidence in our solution.

> **An important difference between solving equations and inequalities:**
>
> Whenever we multiply or divide an inequality by a **negative** number we must **reverse** the inequality sign.

So, for example, if we have $-3y > -6$ then $y < 2$.

You can verify this with a value for y such as 1 since $-3 > -6$ and $1 < 2$.

Worked example 1.8

Solve the inequality $2(3 - 4x) \leqslant 5 - 2(3x + 5)$

Solution

$2(3 - 4x) \leqslant 5 - 2(3x + 5)$

$\Rightarrow \quad 6 - 8x \leqslant 5 - 6x - 10$

$\Rightarrow 6x - 8x \leqslant 5 - 6 - 10$

$\Rightarrow \quad -2x \leqslant -11$

$\Rightarrow \quad x \geqslant 5\frac{1}{2}$

Dividing by -2 so we need to reverse the inequality sign.

Note. On the third line of the solution we could have collected the terms involving x on the right instead of the left-hand side and the solution would have been written as:

$$6 - 5 + 10 \leqslant 8x - 6x$$
$$\Rightarrow \quad 11 \leqslant 2x$$
$$\Rightarrow \quad 5\tfrac{1}{2} \leqslant x \Rightarrow x \geqslant 5\tfrac{1}{2}$$

EXERCISE 1C

Solve the inequalities in questions **1–8**.

1 $3(x - 3) < 2x + 7$.

2 $2(3x - 5) \leqslant 8x - 3$.

3 $7x + 9 > 5x + 3(x - 1)$.

4 $7(2x + 1) > -3 - 5x$.

5 $\dfrac{7x - 1}{5} > \dfrac{2(5 + 2x)}{3}$.

6 $\dfrac{7x - 8}{3} \geqslant \dfrac{3x + 5}{2}$.

7 $\dfrac{4(x - 2)}{3} > \dfrac{5(3 - x)}{2}$.

8 $\dfrac{3x - 1}{5} - \dfrac{x + 1}{2} \leqslant 3$.

9 A rectangular plot of land has dimensions $2x + 3$ metres by $5x - 4$ metres. Explain why $x > \dfrac{4}{5}$.

Given that the perimeter is less than 40 m, find another inequality satisfied by x.

10 A piece of string has length x cm. A second piece needs to be double the length of the first plus an extra 5 cm. A third piece is obtained by doubling the length of the second piece and subtracting 20 cm. The total length of the three pieces must be less than 1 m. Find two inequalities satisfied by x.

1.5 Function notation

Imagine you have a machine into which you feed various numbers and for each number input the machine gives a particular output, but always following a single rule.

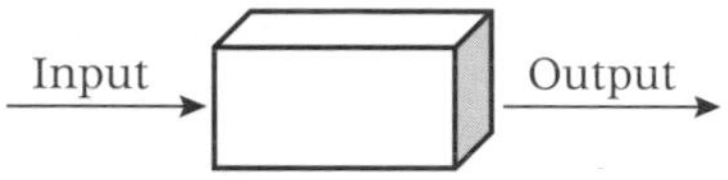

For instance, suppose you have the following situation.

Input	Output
1	4
8	88
−2	−2
5	?

Could you predict what output you get when the input is 5?

This is not easy. However if you know that the machine is programmed to *'square the number you input and add the result to three times the number input'* you will have no problem in discovering the output for 5.

A shorthand way to describe the **function** of this machine is to use the symbol f and to write

$$f(x) = x^2 + 3x$$

You read f(x) as 'f of x'.

Hence, $f(8) = 8^2 + (3 \times 8) = 64 + 24 = 88$ and
$f(5) = 5^2 + (3 \times 5) = 25 + 15 = 40$

You read this as 'f of 5 is equal to 40'.

The output of this is therefore 40 when the input is 5.

Worked example 1.9

A function f is defined by $f(x) = 2x^2 - 5$. Find the values of f(3) and f(−4).

Any letter can be used to describe a function. For instance, the function g may be defined as $g(x) = 3x + 7$. In words it has the effect of 'multiplying by three then adding 7'.

Solution

You need to realise that $2x^2$ means 'square x then multiply your answer by 2'.

Hence, $f(3) = 2 \times 3^2 - 5 = 2 \times 9 - 5 = 18 - 5 = 13$
and $f(-4) = 2 \times (-4)^2 - 5 = 2 \times 16 - 5 = 32 - 5 = 27$.

A function f has a defining rule such as $f(x) = x^2 + 3$ to enable you to find its value for different values of x.

EXERCISE 1D

1 Describe the effect of each of the following functions in words:

(a) $f(x) = 5x + 2$, **(b)** $g(x) = 3x - 4$, **(c)** $h(x) = x^2 + 7$

Although any letter can be used for a function it is most common to use the letter f since it is the first letter of 'function'. Once f has been used, it is common to use g, then h, etc.

2 The function f is defined by $f(x) = 4x - 5$. Find:

(a) $f(3)$, **(b)** $f(0)$, **(c)** $f(1)$, **(d)** $f(-2)$.

3 The function g is defined by $g(x) = 3x^2 - 1$. Find:

(a) $g(1)$, **(b)** $g(0)$, **(c)** $g(-1)$, **(d)** $g(-2)$.

4 The function h is defined by $h(x) = (3x)^2 - 1$. Find:

(a) $h(1)$, **(b)** $h(0)$, **(c)** $h(-1)$, **(d)** $h(-2)$.

5 Find the values of: **(i)** $f(2)$, **(ii)** $f(0)$, **(iii)** $f(-1)$, **(iv)** $f(-3)$ for each of the following functions:

(a) $f(x) = (x + 1)^2$,

(b) $f(x) = x^3 + 1$,

(c) $f(x) = 4 - 3x^2$,

(d) $f(x) = \dfrac{1}{x + 2}$.

The topic of functions will be covered in much greater depth in the A2 course.
For the AS examination you will only need to be familiar with function notation as illustrated in this exercise.

1.6 Use of calculators

At GCSE you will already be used to taking one examination in which no calculators are permitted, whereas a calculator can be used in the other exam.

In a similar way, you will **not** be allowed to use any type of calculator for the C1 examination. This means that exam questions will not be set involving complicated arithmetic. Answers will sometimes be required in surd form (see chapter 2) where you will need to leave an answer in the form $\sqrt{2}$, for example, rather than a calculator approximation.

However, one of the main reasons for excluding calculators in the C1 exam is to allow the testing of simple graph sketching which most modern calculators can do very easily.

You are encouraged to use modern technology: computer software, applets from the Internet and graphics calculators.

Even though you cannot use a calculator in the C1 examination, you can use one in every other mathematics examination and its use is positively encouraged whilst learning the C1 material.

With a graphics calculator you could check that you have sketched graphs correctly and become familiar with its special features from your first weeks of your AS course.

Solution to rabbit enclosure problem

The maximum area is achieved by having two pieces perpendicular to the wall each of length 3 m and one piece parallel to the wall of length 6 m. The area is then 18 m^2. Note that if you made a square by using three pieces of length 4 m, the area would only be 16 m^2.

Solution to rope around the equator problem

Assume that all lengths are in metres.
Let the radius of the earth be R.
The circumference is then $2\pi R$.

Once you add a metre of rope, the new circle has circumference $2\pi R + 1$.

Let the length of the little sticks be x so that the radius of the new circle round the earth has radius $R + x$.
The circumference of the new circle is $2\pi(R + x)$.

You can now form an equation by equating the two results:

$$\begin{aligned} 2\pi R + 1 &= 2\pi(R + x) \\ &= 2\pi R + 2\pi x \end{aligned}$$

The equation reduces to $1 = 2\pi x$, hence,

$$x = \frac{1}{2\pi} \approx 0.159 \ldots$$

So the length of each stick is approximately 0.159 m or about 16 cm. Certainly a cat could get under the rope and maybe even some dogs!

(*Be honest – how many of you thought it would only allow an ant or a worm to squeeze through?*)

Key point summary

1	A linear equation is of the form $ax + b = 0$.	*p2*
2	To solve a linear equation, collect all the terms involving x on one side of the equation.	*p2*
3	If equations involve fractions try multiplying through by the LCM of the denominators.	*p3*
4	Simultaneous linear equations can be solved by elimination or by substitution.	*pp5, 6*
5	Inequalities can be handled like equations, but remember to reverse the inequality sign whenever you multiply or divide by a negative number.	*p8*
6	A function f has a defining rule such as $f(x) = x^2 + 3$ to enable you to find its value for different values of x.	*p9*

Test yourself	What to review
1 Solve the equation $3x - 2 = 5 - 2x$.	*Section 1.2*
2 Solve the simultaneous equations: $3x + 7y = 15$ $5x - 2y = -16$	*Section 1.3*
3 Solve the inequality $5(2x - 1) > 3(1 + 5x)$.	*Section 1.4*
4 The function f is defined by $f(x) = 4x^2 - 7$. Find the values of: **(a)** f(3), **(b)** f(−1).	*Section 1.5*

Test yourself ANSWERS

1 $x = \frac{7}{5}$.

2 $x = -2, y = 3$.

3 $x < -\frac{8}{5}$.

4 **(a)** 29; **(b)** −3.

CHAPTER 2

C1: Surds

Learning objectives

After studying this chapter, you should be able to:

- distinguish between rational and irrational numbers
- understand what is meant by a surd
- simplify expressions involving surds
- perform arithmetic involving surds.

2.1 Special sets of numbers

When you learned to count, you started to use the numbers 1, 2, 3, 4, … which are known as natural numbers and denoted by the symbol $\mathbb{N}$.

As the need arose to use zero and negative numbers the set of integers, denoted by $\mathbb{Z} = \{\ldots -3, -2, -1, 0, 1, 2, 3, 4, \ldots\}$, was constructed.

Some numbers such as fractions cannot be written as a whole number. Those which can be written in the form $\frac{a}{b}$, where a and b are integers and b is not equal to zero, are called **rational** numbers and are denoted by $\mathbb{Q}$, since they can be written as the quotient of two integers.

Are these sensible or reasonable numbers?

The term rational is used because these numbers can be written as the **ratio** of two integers.

When rational numbers are written as decimals they either terminate such as $\frac{3}{4} = 0.75$ or $\frac{71}{1000} = 0.071$ or they have a sequence of recurring digits such as $\frac{5}{11} = 0.45454545\ldots$ or $\frac{5}{7} = 0.714285714285714\ldots$.

However, numbers that cannot be written as a fraction in the form $\frac{a}{b}$, where a and b are integers, are called **irrational**.

The decimal representation of an irrational number neither terminates nor has a recurring pattern of digits, no matter how many decimal places we write down.

Examples include $\sqrt{3} = 1.73205\ldots$ and $\sqrt[3]{5} = 1.709975947\ldots$ and $\pi = 3.14159\ldots$ which each has a non-repeating pattern of digits in its decimal representation.

The set of rationals combined with the set of irrationals gives us the set of **real numbers** and is denoted by $\mathbb{R}$.

2.2 Surds

In mathematics, we often arrive at answers that contain root signs (they may be square roots, cube roots, etc.).

Remember that although the equation $x^2 = 9$ has solutions $x = \pm 3$, the symbol $\sqrt{}$ means the **positive** square root so that $\sqrt{9} = 3$.

We will find that some of these numbers with a root sign are easy to deal with since they have an exact decimal representation.

For instance $\sqrt{16} = 4$, $\sqrt[3]{8} = 2$, $\sqrt{11.56} = 3.4$, $\sqrt[5]{\frac{1}{32}} = 0.5$.

This is because each of these numbers is **rational**.

> Expressions with root signs involving irrational numbers such as $\sqrt{7} - 2$ or $\sqrt[3]{5}$ are called **surds**.

EXERCISE 2A

State whether each of the following is a rational or irrational number.

1 $\sqrt{7}$. **2** $\sqrt{9}$. **3** $\sqrt{\frac{5}{16}}$. **4** $\sqrt{\frac{50}{72}}$. **5** $\sqrt[3]{64}$.

6 $\sqrt[3]{\frac{8}{27}}$. **7** $2 - \sqrt[3]{6}$. **8** $3 + 2^{\frac{1}{5}}$. **9** $7 + \sqrt{4}$. **10** π^2.

Sometimes, you will be required to give answers to problems in surd form since these answers are **exact**. If you use your calculator to get a decimal form, it can only give the answer to a certain number of decimal places and so can only be an approximation.

For example, $\sqrt{3}$ is exact, whereas 1.732050808 is the full calculator display, but is still only an approximation to $\sqrt{3}$.

Order of a surd

The **order** of a surd is determined by the root symbol. For example:

In general, you will be handling surds of order 2.

$\sqrt{3}$ is a surd of order 2 (sometimes called a quadratic surd).
$\sqrt[3]{5}$ is a surd of order 3.
$\sqrt[n]{x}$ is a surd of order n.

A rational quantity may be expressed in the form of a root of any required order. For example, $3 = \sqrt{9} = \sqrt[3]{27} = \sqrt[4]{81}$, etc.

2

Worked example 2.1

Write the following surds in ascending size: $\sqrt[4]{9}$, $\sqrt[3]{5}$, $\sqrt[6]{26}$.

Solution

It is difficult to see which is the biggest at a glance since they are all of a different order.

Since we have the fourth root, the third root and the sixth root of various numbers, we can raise each of the terms to the power 12, because the lowest common multiple of 4, 3 and 6 is 12.

Let $a = \sqrt[4]{9}$, $b = \sqrt[3]{5}$ and $c = \sqrt[6]{26}$.

Hence, $a^4 = 9 \Rightarrow a^{12} = 9^3 = 729$

$b^3 = 5 \Rightarrow b^{12} = 5^4 = 625$

and $c^6 = 26 \Rightarrow c^{12} = 26^2 = 676$.

We can now see that $b < c < a$ and hence rearrange the surds in ascending size as $\sqrt[3]{5}$, $\sqrt[6]{26}$, $\sqrt[4]{9}$.

2.3 Simplest form of surds

When simplifying surds, we try to make the number under the root sign as small as possible.

Worked example 2.2

Simplify the following as far as possible.

1 $\sqrt{18}$.

2 $\sqrt[3]{48}$.

3 $\sqrt{216}$.

Hint. Look for a square number that divides exactly into the number under the square root sign (or a perfect cube if the surd is of order 3, etc.).

Solution

1 $\sqrt{18}$

9 is a square number that is also a factor of 18.

$$\sqrt{18} = \sqrt{9 \times 2} = \sqrt{9} \times \sqrt{2} = 3\sqrt{2}$$

$$\Rightarrow \sqrt{18} = 3\sqrt{2}$$

2 $\sqrt[3]{48}$

8 is a perfect cube which is also a factor of 48.

$$\sqrt[3]{48} = \sqrt[3]{8 \times 6} = \sqrt[3]{8} \times \sqrt[3]{6} = 2\sqrt[3]{6}$$

$$\Rightarrow \sqrt[3]{48} = 2\sqrt[3]{6}$$

3 $\sqrt{216}$

4 is a square number that is a factor of 216.

$$\sqrt{216} = \sqrt{4 \times 54} = 2\sqrt{54}$$

$$2\sqrt{54} = 2\sqrt{9 \times 6} = 2\sqrt{9} \times \sqrt{6} = 6\sqrt{6}$$

$$\Rightarrow \sqrt{216} = 6\sqrt{6}$$

but $\sqrt{54}$ can itself be simplified.

2.4 Manipulating square roots

There are some simple rules which apply to all positive numbers and they will help us when working with square roots. The first rule generalises the idea demonstrated in the last worked example.

Rules

1 $\sqrt{ab} = \sqrt{a} \times \sqrt{b}$

2 $\sqrt{\dfrac{a}{b}} = \dfrac{\sqrt{a}}{\sqrt{b}}$

Worked example 2.3

Simplify the following.

1 $\sqrt{112}$ **2** $\dfrac{\sqrt{63}}{3}$. **3** $\sqrt{\dfrac{25}{9}}$. **4** $\sqrt{12} \times \sqrt{21}$.

Solution

1 $\sqrt{112} = \sqrt{16 \times 7} = \sqrt{16} \times \sqrt{7} = 4\sqrt{7}$

2 $\dfrac{\sqrt{63}}{3} = \dfrac{\sqrt{9 \times 7}}{3} = \dfrac{3 \times \sqrt{7}}{3} = \sqrt{7}$

3 $\sqrt{\dfrac{25}{9}} = \dfrac{\sqrt{25}}{\sqrt{9}} = \dfrac{5}{3}$

4 $\sqrt{12} \times \sqrt{21} = \sqrt{3} \times \sqrt{4} \times \sqrt{3} \times \sqrt{7}$
$= \sqrt{4} \times (\sqrt{3} \times \sqrt{3}) \times \sqrt{7}$
$= 2 \times 3 \times \sqrt{7} = 6\sqrt{7}$

EXERCISE 2B

Simplify each of the surd expressions **1–18**.

1 $\sqrt{8}$. **2** $\sqrt{12}$. **3** $\sqrt{20}$.

4 $\sqrt{75}$. **5** $\sqrt{52}$. **6** $\sqrt{120}$.

7 $\sqrt{245}$. **8** $\sqrt{252}$. **9** $\sqrt{192}$.

10 $\sqrt{1000}$. **11** $\dfrac{\sqrt{27}}{3}$. **12** $\dfrac{\sqrt{32}}{4}$.

13 $\dfrac{\sqrt{125}}{\sqrt{5}}$. **14** $\dfrac{\sqrt{448}}{4}$. **15** $\sqrt{35} \times \sqrt{7}$.

16 $\sqrt{75} \times \sqrt{27}$. **17** $\sqrt{20} \times \sqrt{15} \times \sqrt{6}$. **18** $\dfrac{\sqrt{48} \times \sqrt{14}}{\sqrt{56} \times \sqrt{18}}$.

19 Express each of the following in the form $k\sqrt{p}$ where k is an integer and p is a prime number.

(a) $\sqrt{72} - \sqrt{8}$,

(b) $5\sqrt{28} - \sqrt{63}$,

(c) $2\sqrt{147} - 5\sqrt{48} + \sqrt{75}$.

20 Arrange the following surds in ascending size: $\sqrt[3]{3}$, $\sqrt[4]{5}$, $\sqrt[6]{10}$.

2.5 Use in geometry

One of the main areas in which you will need to work with surds is in the use of Pythagoras' theorem in geometry. Often you need the exact value of a particular length.

Worked example 2.4

The hypotenuse of a right-angled triangle has length 18 cm and one of the other sides has length 6 cm. Find the length of the remaining side.

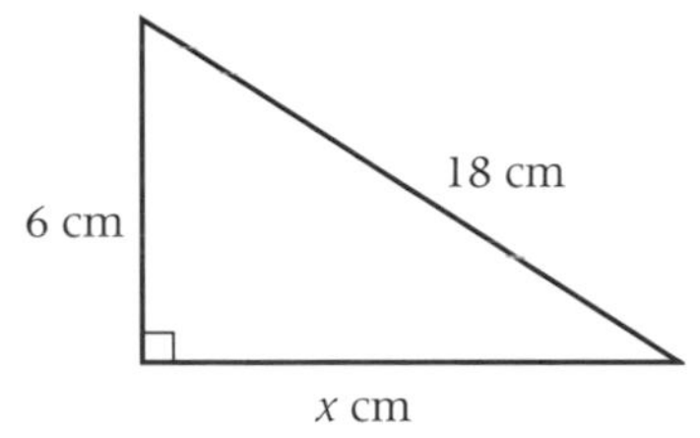

Solution

Let the remaining side have length x cm.
By Pythagoras' theorem

$$x^2 + 6^2 = 18^2 \text{ so that } x^2 = 324 - 36 = 288$$

$$\Rightarrow \quad x = \sqrt{144} \times \sqrt{2} = 12\sqrt{2}$$

$288 = 144 \times 2$

So the remaining side has length $12\sqrt{2}$ cm.

2.6 Like and unlike surds

'Like' surds have the same irrational factor.
For example $3\sqrt{5}$, $-6\sqrt{5}$, $13\sqrt{5}$ are 'like' surds since they have the same number under the root sign.

'Unlike' surds have different irrational factors.
For instance, $7\sqrt{3}$, $2\sqrt{6}$, $-4\sqrt{11}$ are 'unlike' surds since they have different numbers under the root sign.

Student health warning!

It is well worth noting that:

$\sqrt{a+b} \neq \sqrt{a} + \sqrt{b}$ and $\sqrt{a-b} \neq \sqrt{a} - \sqrt{b}$

For example, consider

$\sqrt{9+16} = \sqrt{25} = 5$

However,

$\sqrt{9} + \sqrt{16} = 3 + 4 = 7.$

This, of course, is not the same and demonstrates that we cannot split up $\sqrt{9+16}$ *as* $\sqrt{9} + \sqrt{16}$. This is a common mistake and must be guarded against.

2.7 Adding and subtracting surds

We can add or subtract surds as long as they are 'like' surds.

'Like' surds can be collected.
'Unlike' surds cannot be collected.

Worked example 2.5

Simplify each of the following as much as possible.

(a) $14\sqrt{5} - 6\sqrt{5}$, **(b)** $\sqrt{75} + 2\sqrt{27} - 5\sqrt{108}$,

(c) $4\sqrt{3} + 9\sqrt{6}$, **(d)** $5\sqrt{5} + 4\sqrt{3} - 2\sqrt{5} + 7\sqrt{3}$.

Solution

(a) $14\sqrt{5} - 6\sqrt{5} = 8\sqrt{5}$;

There are 14 lots of $\sqrt{5}$ minus 6 lots of $\sqrt{5}$ giving 8 lots of $\sqrt{5}$.

(b) $\sqrt{75} + 2\sqrt{27} - 5\sqrt{108}$

We first need to simplify these surds as far as possible.

$= (\sqrt{25} \times \sqrt{3}) + (2 \times \sqrt{9} \times \sqrt{3}) - (5 \times \sqrt{36} \times \sqrt{3})$

$= 5\sqrt{3} + 2 \times 3\sqrt{3} - 5 \times 6\sqrt{3}$

$= 5\sqrt{3} + 6\sqrt{3} - 30\sqrt{3}$

These are all 'like' surds and, therefore, can be collected.

$= -19\sqrt{3}$

(c) $4\sqrt{3} + 9\sqrt{6}$;

$\sqrt{3}$ and $\sqrt{6}$ are 'unlike' surds and, therefore, cannot be collected so this expression cannot be simplified any further.

(d) $5\sqrt{5} + 4\sqrt{3} - 2\sqrt{5} + 7\sqrt{3}$

We can collect all the terms with $\sqrt{5}$ and then separately collect the terms with $\sqrt{3}$

$= 5\sqrt{5} - 2\sqrt{5} + 4\sqrt{3} + 7\sqrt{3}$

$= 3\sqrt{5} + 11\sqrt{3}$

Note that we cannot combine these any further since $\sqrt{5}$ and $\sqrt{3}$ are 'unlike' surds.

2

EXERCISE 2C

Simplify each of the expressions **1–10** as far as possible.

1 $\sqrt{7} - 3\sqrt{7}$. **2** $\sqrt{12} + \sqrt{3}$.

3 $\sqrt{20} - 3\sqrt{5}$. **4** $\sqrt{72} + \sqrt{12}$.

5 $\sqrt{8} + \sqrt{18}$. **6** $\sqrt{12} + \sqrt{108} - \sqrt{27}$.

7 $\sqrt{20} - \sqrt{245} + 3\sqrt{5}$. **8** $4\sqrt{6} - 3\sqrt{24} + \sqrt{150}$.

9 $15\sqrt{45} - 3\sqrt{20} - \sqrt{180}$. **10** $3\sqrt{175} + 6\sqrt{18} - 3\sqrt{28} + 4\sqrt{72}$.

11 The two shorter sides of a right-angled triangle have lengths 3 m and 6 m. Find the length of the hypotenuse in the form $a\sqrt{b}$ metres, where a and b are prime numbers.

12 The hypotenuse of a right-angled triangle has length 24 cm and another one of the sides has length 18 cm. Find the exact length of the third side.

13 Find the length of the remaining side in each of the right-angled triangles below, giving your answers as simply as possible in surd form.

(a)

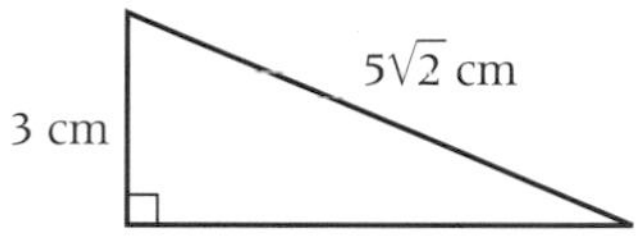

(b)

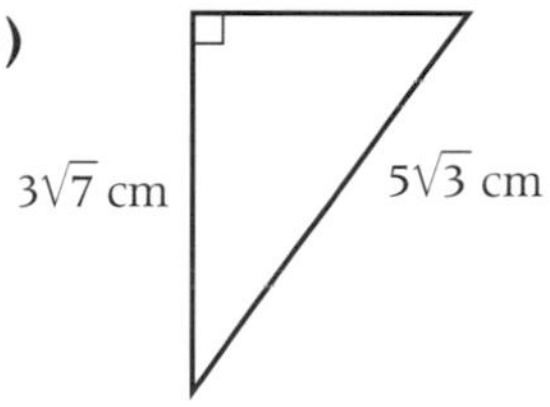

2.8 Multiplying surds

We multiply the rational factors then the irrational factors and then simplify (if possible).

Worked example 2.6

Simplify each of the following:

1 $3\sqrt{2} \times 5\sqrt{3}$. **2** $2\sqrt{3} \times 3\sqrt{3}$. **3** $5\sqrt{6} \times 3\sqrt{3}$.

Solution

1 $3\sqrt{2} \times 5\sqrt{3} = 3 \times 5 \times \sqrt{2} \times \sqrt{3} = 15\sqrt{6}$

2 $2\sqrt{3} \times 3\sqrt{3} = 6\sqrt{9} = 6 \times 3 = 18$

You should be able to recognise that $\sqrt{3} \times \sqrt{3} = 3$ and in general $\sqrt{n} \times \sqrt{n} = n$.

3 $5\sqrt{6} \times 3\sqrt{3} = 15\sqrt{18}$
$= 15 \times \sqrt{9} \times \sqrt{2}$
$= 45\sqrt{2}$

An expression such as $\sqrt{3} + \sqrt{5}$ is called a **compound surd**.

To multiply compound surds, we use the same idea as multiplying out brackets in algebra.

Recall that

$(a + b)(c + d) = ac + bc + bd + ad.$

Worked example 2.7

Simplify each of the following.

1 $(\sqrt{3} + \sqrt{5})(\sqrt{2} + \sqrt{3})$.

2 $(3 + \sqrt{2})(5 - \sqrt{7})$.

3 $(4\sqrt{2} - 3\sqrt{5})(3\sqrt{2} - 2\sqrt{5})$.

4 $(3 + \sqrt{5})^2$.

Solution

1 $(\sqrt{3} + \sqrt{5})(\sqrt{2} + \sqrt{3}) = \sqrt{6} + \sqrt{10} + \sqrt{9} + \sqrt{15}$
$= \sqrt{6} + \sqrt{10} + 3 + \sqrt{15}$

2 $(3 + \sqrt{2})(5 - \sqrt{7}) = 15 - 3\sqrt{7} + 5\sqrt{2} - \sqrt{14}$

3 $(4\sqrt{2} - 3\sqrt{5})(3\sqrt{2} - 2\sqrt{5}) = 12\sqrt{4} - 8\sqrt{10} - 9\sqrt{10} + 6\sqrt{25}$
$= 24 - 8\sqrt{10} - 9\sqrt{10} + 30$
$= 54 - 17\sqrt{10}$

4 $(3 + \sqrt{5})^2 = (3 + \sqrt{5})(3 + \sqrt{5})$
$= 9 + 3\sqrt{5} + 3\sqrt{5} + 5$
$= 14 + 6\sqrt{5}$

EXERCISE 2D

Multiply out the following and simplify the answers as far as possible.

1 $5\sqrt{3} \times 2\sqrt{6}$.

2 $6\sqrt{10} \times 5\sqrt{2}$.

3 $\sqrt{5}(\sqrt{6} - 2)$.

4 $\sqrt{8}(\sqrt{2} + \sqrt{8})$.

5 $2\sqrt{5}(3 + 6\sqrt{5})$.

6 $(3 + \sqrt{2})(7 - \sqrt{3})$.

7 $(\sqrt{5} - \sqrt{3})(\sqrt{5} + \sqrt{3})$.

8 $(3\sqrt{6} - \sqrt{7})(\sqrt{6} + 2\sqrt{7})$.

9 $(\sqrt{10} - 2)(\sqrt{10} + 2)$.

10 $(4\sqrt{11} - 2\sqrt{3})(3\sqrt{11} + 2)$.

2.9 Rationalising the denominator

Whenever we have a fraction in which the denominator is a surd, we can rewrite the fraction so that the denominator no longer contains a surd.

This is done by multiplying the top and bottom of the fraction by the same number so that the final answer has a **rational number in the denominator**.

Since $\sqrt{3} \times \sqrt{3} = 3$, we could simplify $\frac{2}{\sqrt{3}}$ by writing

$$\frac{2}{\sqrt{3}} = \frac{2}{\sqrt{3}} \times \frac{\sqrt{3}}{\sqrt{3}} = \frac{2\sqrt{3}}{3}.$$

Since $(2 + \sqrt{3})(2 - \sqrt{3}) = 2^2 - (\sqrt{3})^2 = 1$, we could simplify

$\frac{4}{2+\sqrt{3}}$ by writing $\frac{4}{2+\sqrt{3}} \times \frac{2-\sqrt{3}}{2-\sqrt{3}} = 4(2 - \sqrt{3})$.

Since $(p - q)(p + q) = p^2 - q^2$.

This process is called '**rationalising the denominator**'. The method is as follows.

> If the denominator is of the form $\sqrt{a}$ then multiply the top and bottom by $\sqrt{a}$.
>
> If the denominator is of the form $a + \sqrt{b}$ then multiply top and bottom by $a - \sqrt{b}$.
>
> If the denominator is of the form $a - \sqrt{b}$ then multiply top and bottom by $a + \sqrt{b}$.

Worked example 2.8

Rationalise the denominators of the following:

1 $\frac{4}{\sqrt{7}}$.   **2** $\frac{3}{3-\sqrt{6}}$.   **3** $\frac{1}{\sqrt{11}+\sqrt{7}}$.   **4** $\frac{\sqrt{6}+3\sqrt{2}}{\sqrt{5}+2\sqrt{3}}$.

Solution

1 $\frac{4}{\sqrt{7}}$.

Multiply the top and bottom by $\sqrt{7}$.

$$\frac{4}{\sqrt{7}} \times \left(\frac{\sqrt{7}}{\sqrt{7}}\right) = \frac{4\sqrt{7}}{\sqrt{49}} = \frac{4\sqrt{7}}{7}$$

We now have a rational number in the denominator.

$$\Rightarrow \quad \frac{4}{\sqrt{7}} = \frac{4\sqrt{7}}{7}$$

2 $\frac{3}{3-\sqrt{6}}$.

$$\frac{3}{(3-\sqrt{6})} \times \frac{(3+\sqrt{6})}{(3+\sqrt{6})} = \frac{3(3+\sqrt{6})}{9-6} = \frac{3(3+\sqrt{6})}{3} = 3 + \sqrt{6}$$

Multiply the top and bottom by $(3 + \sqrt{6})$.

$$\Rightarrow \quad \frac{3}{3-\sqrt{6}} = 3 + \sqrt{6}$$

3 $\dfrac{1}{\sqrt{11}+\sqrt{7}}$.

Multiply the top and bottom by $(\sqrt{11}-\sqrt{7})$.

$$\frac{1}{(\sqrt{11}+\sqrt{7})} \times \frac{(\sqrt{11}-\sqrt{7})}{(\sqrt{11}-\sqrt{7})} = \frac{\sqrt{11}-\sqrt{7}}{\sqrt{121}-\sqrt{49}} = \frac{\sqrt{11}-\sqrt{7}}{4}$$

$$\Rightarrow \frac{1}{\sqrt{11}+\sqrt{7}} = \frac{\sqrt{11}-\sqrt{7}}{4}$$

Note that the number that we multiply the top and bottom by in order to rationalise the denominator is sometimes called the **conjugate**.

The conjugate of $\sqrt{11}+\sqrt{7}$ is $\sqrt{11}-\sqrt{7}$.

The conjugate of $8-3\sqrt{5}$ is $8+3\sqrt{5}$.

4 $\dfrac{\sqrt{6}+3\sqrt{2}}{\sqrt{5}+2\sqrt{3}}$.

$$\frac{(\sqrt{6}+3\sqrt{2})}{(\sqrt{5}+2\sqrt{3})} \times \frac{(\sqrt{5}-2\sqrt{3})}{(\sqrt{5}-2\sqrt{3})} = \frac{\sqrt{30}-2\sqrt{18}+3\sqrt{10}-6\sqrt{6}}{\sqrt{25}-4\sqrt{9}}$$

$$= \frac{\sqrt{30}-6\sqrt{2}+3\sqrt{10}-6\sqrt{6}}{-7}$$

Multiply top and bottom by the conjugate $(\sqrt{5}-2\sqrt{3})$.

$$\Rightarrow \frac{\sqrt{6}+3\sqrt{2}}{\sqrt{5}+2\sqrt{3}} = \frac{\sqrt{30}-6\sqrt{2}+3\sqrt{10}-6\sqrt{6}}{-7}$$

$$= \frac{6\sqrt{2}+6\sqrt{6}-\sqrt{30}-3\sqrt{10}}{7}$$

Although this answer is perfectly correct, it is customary to try to leave a positive denominator, where possible.

EXERCISE 2E

Rationalise the denominators of the following:

1 $\dfrac{1}{\sqrt{10}}$.

2 $\dfrac{3}{\sqrt{2}}$.

3 $\dfrac{\sqrt{7}}{2\sqrt{5}}$.

4 $\dfrac{1}{\sqrt{2}-1}$.

5 $\dfrac{3}{\sqrt{21}-3}$.

6 $\dfrac{2}{\sqrt{5}-\sqrt{2}}$.

7 $\dfrac{5}{\sqrt{14}+2}$.

8 $\dfrac{\sqrt{6}}{\sqrt{6}+\sqrt{3}}$.

9 $\dfrac{\sqrt{7}+5}{\sqrt{7}+\sqrt{5}}$.

10 $\dfrac{2\sqrt{3}+\sqrt{7}}{5\sqrt{3}+\sqrt{7}}$.

11 $\dfrac{\sqrt{13}+\sqrt{5}}{\sqrt{13}-\sqrt{5}}$.

12 $\dfrac{10\sqrt{5}}{2\sqrt{15}-\sqrt{5}}$.

13 $\dfrac{4\sqrt{2}+\sqrt{5}}{4\sqrt{2}-\sqrt{5}}$.

14 $\dfrac{2\sqrt{2}+4\sqrt{7}}{5\sqrt{2}-3\sqrt{7}}$.

2

2.10 Equations and inequalities involving surds

An equation or inequality may look more complicated because it involves surds. The same techniques are used as in chapter 1. However, it may be necessary to rationalise the denominator if this is in surd form.

Worked example 2.9

Solve the equation $\sqrt{5} + 3x = 2\sqrt{5}x - 7$.

Solution

$$\sqrt{5} + 3x = 2\sqrt{5}x - 7$$
$$\Rightarrow \sqrt{5} + 7 = 2\sqrt{5}x - 3x = (2\sqrt{5} - 3)x$$
$$\Rightarrow \frac{\sqrt{5}+7}{2\sqrt{5}-3} = x$$

Although this answer is now correct it is better if we rationalise the denominator.

$$\Rightarrow x = \frac{\sqrt{5}+7}{2\sqrt{5}-3} \times \frac{2\sqrt{5}+3}{2\sqrt{5}+3}$$
$$\Rightarrow x = \frac{10 + 14\sqrt{5} + 3\sqrt{5} + 21}{20 - 9} = \frac{31 + 17\sqrt{5}}{11}$$

Worked example 2.10

Solve the inequality $\sqrt{3} - 2x > 4\sqrt{3}x - 5$

Solution

$$\sqrt{3} - 2x > 4\sqrt{3}x - 5$$
$$\Rightarrow \sqrt{3} + 5 > 4\sqrt{3}x + 2x$$
$$\Rightarrow \sqrt{3} + 5 > (4\sqrt{3} + 2)x$$
$$\Rightarrow \frac{\sqrt{3}+5}{4\sqrt{3}+2} > x$$

Although this answer is again correct, it is better if you rationalise the denominator. In fact, in an examination, you may well be required to do so.

$$\Rightarrow x < \frac{\sqrt{3}+5}{4\sqrt{3}+2} \times \frac{4\sqrt{3}-2}{4\sqrt{3}-2}$$
$$\Rightarrow x < \frac{12 - 2\sqrt{3} + 20\sqrt{3} - 10}{48 - 4}$$
$$\Rightarrow x < \frac{2 + 18\sqrt{3}}{44} \text{ or } x < \frac{1 + 9\sqrt{3}}{22}$$

EXERCISE 2F

Solve each of the following:

1 $3x - 5 = 7\sqrt{3}x + 11$.

2 $2\sqrt{5}x + 5 = \sqrt{5} - x$.

3 $7\sqrt{3} - 3x = 5\sqrt{3}x - 6$.

4 $6\sqrt{7} - 2x = 5 + 3\sqrt{7}x$.

5 $2\sqrt{2} - 3x = \sqrt{2}x - 4$.

6 $4\sqrt{3} + 2x = 7 + 5\sqrt{3}x$.

7 $3x + 4 > 5\sqrt{3}x + 6$.

8 $3\sqrt{5}x - 1 < \sqrt{5} - 2x$.

9 $3\sqrt{7} - 5x < 8 + 2\sqrt{7}x$.

10 $3\sqrt{2} - 5x > 4\sqrt{2}x - 7$.

Worked examination question 2.11

Express $\dfrac{4 + 5\sqrt{3}}{2 + 7\sqrt{3}}$ in the form $p + q\sqrt{3}$ where p and q are rational numbers.

Solution

$$\frac{4 + 5\sqrt{3}}{2 + 7\sqrt{3}} \times \frac{2 - 7\sqrt{3}}{2 - 7\sqrt{3}}$$

We need to rationalise the denominator and do this by multiplying top and bottom by $2 - 7\sqrt{3}$.

$$= \frac{8 - 28\sqrt{3} + 10\sqrt{3} - 35(\sqrt{3})^2}{4 - 49(\sqrt{3})^2}$$

$$= \frac{8 - 105 - 18\sqrt{3}}{4 - 147}$$

A common mistake is to think that $(\sqrt{3})^2$ is 9.

$$= \frac{-97 - 18\sqrt{3}}{-143}$$

$$= \frac{97}{143} + \frac{18\sqrt{3}}{143}$$

which is in the given form $p = \dfrac{97}{143}$ and $q = \dfrac{18}{143}$.

This would now score full marks, but it is good to state the actual values of p and q.

MIXED EXERCISE

1 Simplify the following as far as possible:

(a) $\sqrt{75}$,

(b) $\sqrt{300}$,

(c) $\sqrt{128}$,

(d) $\sqrt{28} - \sqrt{175} + \sqrt{63}$,

(e) $\sqrt{128} + \sqrt{98} - \sqrt{32}$,

(f) $\sqrt{24} - \sqrt{216} + \sqrt{150}$.

2 Expand the following and simplify where possible.

(a) $\sqrt{3}(5 - 2\sqrt{3})$,

(b) $\sqrt{5}(3 + \sqrt{45})$,

(c) $\sqrt{2}(\sqrt{50} - \sqrt{18})$,

(d) $(\sqrt{3} - 2)(\sqrt{3} + 7)$,

(e) $(\sqrt{5} - 7)(3\sqrt{5} + 4)$,

(f) $(3 + 2\sqrt{3})^2$.

3 Rationalise the denominator, simplifying where possible:

(a) $\dfrac{7\sqrt{3}}{\sqrt{5}}$,

(b) $\dfrac{1}{\sqrt{3} - 1}$,

(c) $\dfrac{\sqrt{5} + 1}{\sqrt{5} - 1}$,

(d) $\dfrac{1}{\sqrt{13} - \sqrt{11}}$,

(e) $\dfrac{11}{3\sqrt{3}+7}$, (f) $\dfrac{5\sqrt{3}-1}{3+2\sqrt{3}}$,

(g) $\dfrac{5-3\sqrt{2}}{2\sqrt{2}+3}$, (h) $\dfrac{\sqrt{7}}{\sqrt{2}(\sqrt{14}-\sqrt{7})}$.

4 Express each of the following in the form $p + q\sqrt{2}$ where p and q are rational:

(a) $(3-\sqrt{2})^2$, **(b)** $\dfrac{1}{(3-\sqrt{2})^2}$. [A]

5 **(a)** Write down:

(i) a rational number which lies between 4 and 5,
(ii) an irrational number which lies between 4 and 5.

(b) A student says, 'When you multiply two irrational numbers together the answer is always an irrational number'.

Simplify $(2+\sqrt{3})(2-\sqrt{3})$ and comment on the student's statement. [A]

6 Express each of the following in the form $p + q\sqrt{7}$ where p and q are rational numbers:

(a) $(2+3\sqrt{7})(5-2\sqrt{7})$, **(b)** $\dfrac{(5+\sqrt{7})}{(3-\sqrt{7})}$. [A]

7 **(a)** Express each of the following in the form $k\sqrt{5}$:

(i) $\sqrt{45}$, **(ii)** $\dfrac{20}{\sqrt{5}}$.

(b) Hence write $\sqrt{45} + \dfrac{20}{\sqrt{5}}$ in the form $n\sqrt{5}$, where n is an integer. [A]

8 **(a)** Express $(\sqrt{7}+1)^2$ in the form $a + b\sqrt{7}$, where a and b are integers.

(b) Hence express $\dfrac{(\sqrt{7}+1)^2}{(\sqrt{7}+2)}$ in the form $p + q\sqrt{7}$, where p and q are rational numbers. [A]

9 Express each of the following in the form $p + q\sqrt{3}$:

(a) $(2+\sqrt{3})(5-2\sqrt{3})$,

(b) $\dfrac{26}{4-\sqrt{3}}$. [A]

10 **(a)** Express $\dfrac{\sqrt{2}+1}{\sqrt{2}-1}$ in the form $a\sqrt{2} + b$, where a and b are integers.

(b) Solve the inequality $\sqrt{2}(x-\sqrt{2}) < x + 2\sqrt{2}$. [A]

Key point summary

1 A rational number is one which can be written in the form $\frac{a}{b}$, where a and b are integers. *p13*

2 A real number that is not rational is called irrational. Its decimal representation neither terminates nor has a recurring pattern of digits. *p13*

3 A surd is an irrational number containing a root sign. *p14*

4 $\sqrt{ab} = \sqrt{a} \times \sqrt{b}$ *p16*

5 $\sqrt{\frac{a}{b}} = \frac{\sqrt{a}}{\sqrt{b}}$ *p16*

6 'Like' surds can be collected. 'Unlike' surds cannot be collected. *p18*

7 To rationalise the denominator of the form $\sqrt{a}$, multiply top and bottom by $\sqrt{a}$. *p21*

8 To rationalise the denominator of the form $a + \sqrt{b}$, multiply top and bottom by $a - \sqrt{b}$. *p21*

9 To rationalise the denominator of the form $a - \sqrt{b}$, multiply top and bottom by $a + \sqrt{b}$. *p21*

Test yourself	What to review
1 State whether each of the following is rational or irrational: **(a)** $\sqrt[3]{8}$, **(b)** $\frac{1}{\sqrt{3}}$, **(c)** $\frac{2+\sqrt{3}}{5}$, **(d)** $\frac{\sqrt{12}}{\sqrt{3}}$.	*Section 2.1*
2 Simplify each of the surd expressions below: **(a)** $\sqrt{27} \times \sqrt{6}$, **(b)** $\frac{\sqrt{90}}{\sqrt{15}}$, **(c)** $\frac{\sqrt{3} \times \sqrt{15}}{6}$.	*Sections 2.3 and 2.4*
3 Simplify each of the following surd expressions as far as possible. **(a)** $\sqrt{27} - 2\sqrt{3}$, **(b)** $\sqrt{20} + \sqrt{45}$, **(c)** $\frac{\sqrt{13} - \sqrt{7}}{2}$.	*Section 2.7*
4 Find the length of the hypotenuse of a right-angled triangle if the two smaller sides have lengths $3\sqrt{3}$ cm and $3\sqrt{5}$ cm.	*Section 2.5*
5 Rationalise the denominator of the expression: $\frac{9\sqrt{6} - 8}{\sqrt{6} - 1}$.	*Section 2.9*

Test yourself ANSWERS

1 **(a)** rational; **(b)** irrational; **(c)** irrational; **(d)** rational.

2 **(a)** $9\sqrt{2}$; **(b)** $\sqrt{6}$; **(c)** $\frac{\sqrt{5}}{2}$.

3 **(a)** $\sqrt{3}$; **(b)** $5\sqrt{5}$; **(c)** cannot simplify.

4 $6\sqrt{2}$ cm.

5 $\frac{46 + \sqrt{6}}{5}$.

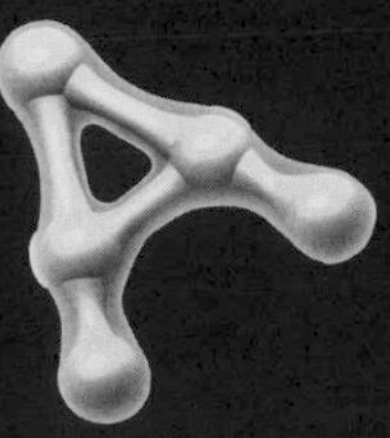

CHAPTER 3

C1: Coordinate geometry of straight lines

Learning objectives

After studying this chapter, you should be able to:

- use the language of coordinate geometry
- find the distance between two given points
- find the coordinates of the mid-point of a line segment joining two given points
- find, use and interpret the gradient of a line segment
- know the relationship between the gradients for parallel and for perpendicular lines
- find the equations of straight lines given (a) the gradient and y-intercept, (b) the gradient and a point, and (c) two points
- verify, given their coordinates, that points lie on a line
- find the coordinates of a point of intersection of two lines
- find the fourth vertex of a parallelogram given the other three.

Coordinate geometry is the use of algebraic methods to study the geometry of straight lines and curves. In this chapter you will consider only straight lines. Curves will be studied later.

3.1 Cartesian coordinates

In your GCSE course you plotted points in two dimensions. This section revises that work and builds up the terminology required.

Named after René Descartes (1596–1650), a French mathematician.

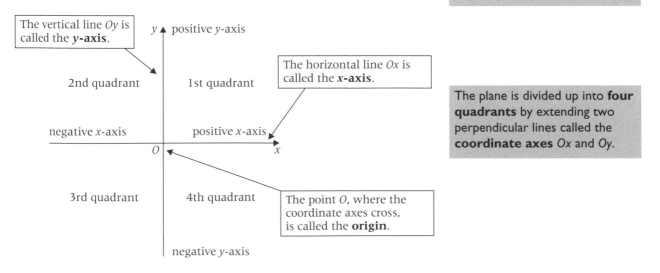

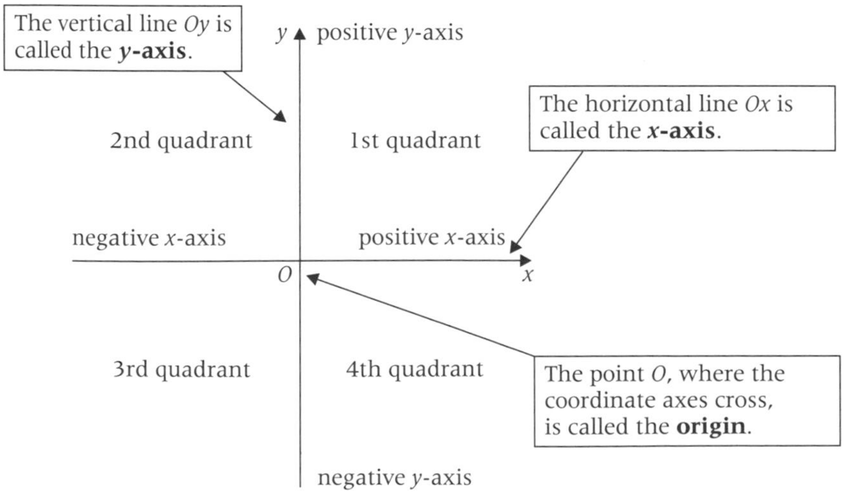
The vertical line Oy is called the y-axis.
y
positive y-axis
The horizontal line Ox is called the x-axis.
2nd quadrant
1st quadrant
negative x-axis
positive x-axis
O
x
3rd quadrant
4th quadrant
The point O, where the coordinate axes cross, is called the origin.
negative y-axis

The plane is divided up into **four quadrants** by extending two perpendicular lines called the **coordinate axes** Ox and Oy.

In the diagram below, the point P lies in the first quadrant.

Its distance from the y-axis is a.
Its distance from the x-axis is b.

> We say that:
> the **x-coordinate** of P is a
> the **y-coordinate** of P is b.
> We write 'P is the point (a, b)' or simply 'the point $P(a, b)$'.

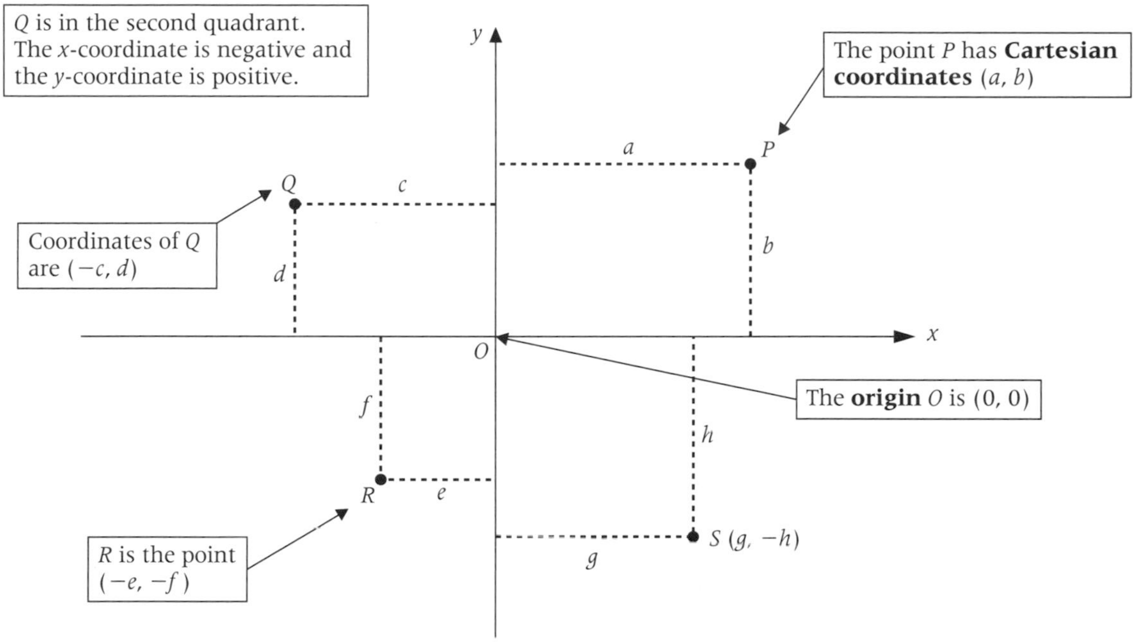

The diagram below shows part of the Cartesian grid, where L is the point $(-2, 3)$.

> The x-coordinate of L is -2 and the y-coordinate of L is 3.

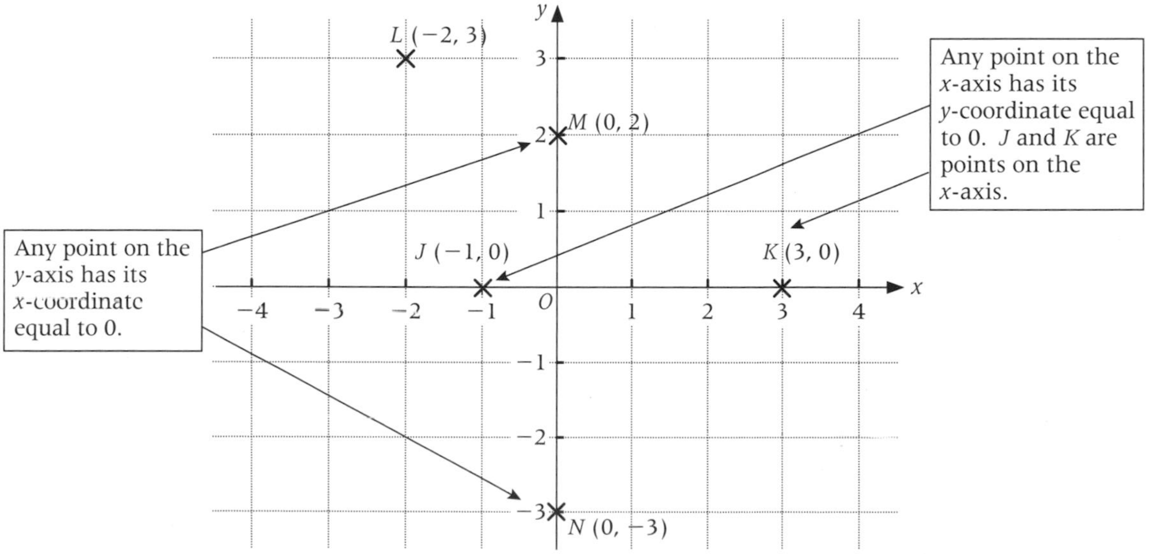

Worked example 3.1

1 Draw coordinate axes Ox and Oy and plot the points

$A(1, 2)$, $B(-3, 3)$, $C(1, -4)$ and $D(-2, -4)$.

2 The line joining the points C and D crosses the y-axis at the point E. Write down the coordinates of E.

3 The line joining the points A and C crosses the x-axis at the point F. Write down the coordinates of F.

Solution

1

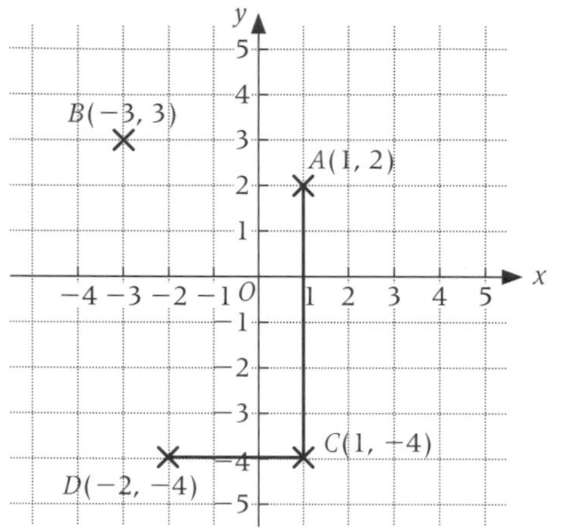

2 E is on the y-axis so its x-coordinate is 0. $E(0, -4)$.

3 F is on the x-axis so its y-coordinate is 0. $F(1, 0)$.

From the diagram in Worked example 3.1, you can see that the distance between the points $C(1, -4)$ and $D(-2, -4)$ is 3 units and the distance between the points $A(1, 2)$ and $C(1, -4)$ is 6 units, but how do you find the distance between points which do not lie on the same horizontal line or on the same vertical line?

3.2 The distance between two points

The following example shows how you can use Pythagoras' theorem which you studied as part of your GCSE course, in order to find the distance between two points.

Worked example 3.2

Find the distance AB where A is the point $(1, 2)$ and B is the point $(4, 4)$.

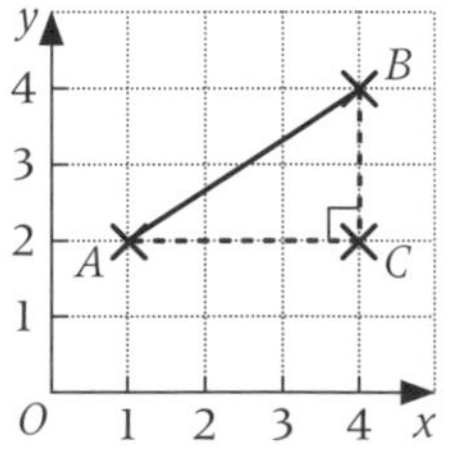

Solution

First plot the points, join them with a line and make a right-angled triangle ABC.

The distance $AC = 4 - 1 = 3$.
The distance $BC = 4 - 2 = 2$.

Using Pythagoras' theorem $AB^2 = 3^2 + 2^2$.

$$= 9 + 4 = 13$$

$$AB = \sqrt{13}$$

$AB \approx 3.6$ (to two significant figures) but since a calculator is not allowed in the examination for this unit you should leave your answers in exact forms. Where possible surd answers should be simplified.

We can generalise the method of the previous example to find a formula for the distance between the points $P(x_1, y_1)$ and $Q(x_2, y_2)$.

The distance $PR = x_2 - x_1$.
The distance $QR = y_2 - y_1$.

Using Pythagoras' theorem

$$PQ^2 = (x_2 - x_1)^2 + (y_2 - y_1)^2$$
$$PQ = \sqrt{[(x_2 - x_1)^2 + (y_2 - y_1)^2]}.$$

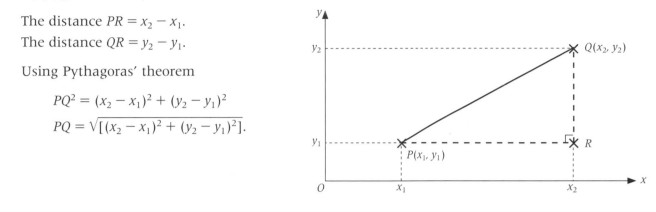

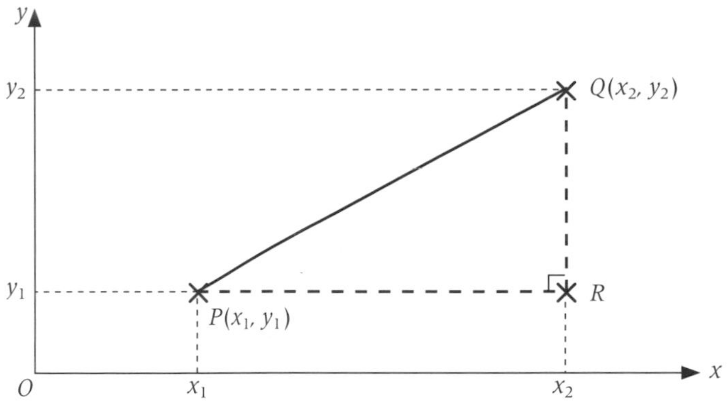

The distance between the points (x_1, y_1) and (x_2, y_2) is
$\sqrt{(x_2 - x_1)^2 + (y_2 - y_1)^2}$.

Worked example 3.3

The point R has coordinates $(-1, 2)$ and the point S has coordinates $(5, -6)$.

(a) Find the distance RS.

(b) The point T has coordinates $(0, 9)$. Show that RT has length $k\sqrt{2}$, where k is an integer.

An integer is a whole number.

Solution

(a) For points R and S:
the difference between the x-coordinates is $5 - (-1) = 6$
the difference between the y-coordinates is $-6 - 2 = -8$.

$$RS = \sqrt{(6)^2 + (-8)^2} = \sqrt{36 + 64} = \sqrt{100} = 10$$

If you draw a diagram you will probably find yourself dealing with distances rather than the difference in coordinates in order to avoid negative numbers. This is not wrong.

Of course you could have considered the differences as $-1 - 5 = -6$ and $2 - (-6) = 8$ and obtained the same answer.

(b) $RT = \sqrt{(1)^2 + (7)^2} = \sqrt{1 + 49} = \sqrt{50}$
$= \sqrt{25 \times 2}$
$= \sqrt{25} \times \sqrt{2}$
$= 5\sqrt{2}$

Used $\sqrt{a \times b} = \sqrt{a} \times \sqrt{b}$. See section 2.4.

Worked example 3.4

The distance MN is 5, where M is the point $(4, -2)$ and N is the point $(a, 2a)$. Find the two possible values of the constant a.

You may prefer to draw a diagram but it is also good to learn to work algebraically.

Solution

The difference between the x-coordinates is $a - 4$.
The difference between the y-coordinates is $2a - (-2) = 2a + 2$.

$$MN^2 = (a-4)^2 + (2a+2)^2$$
$$5^2 = (a-4)^2 + (2a+2)^2$$
$$25 = a^2 - 8a + 16 + 4a^2 + 8a + 4$$
$$25 = 5a^2 + 20$$
$$\Rightarrow a^2 = 1$$
$$\Rightarrow a = \pm 1.$$

Using $(x+y)^2 = x^2 + 2xy + y^2$.

Notice that this equation does have **two** solutions.

EXERCISE 3A

1 Find the lengths of the line segments joining:

(a) $(0, 0)$ and $(3, 4)$, **(b)** $(1, 2)$ and $(5, 3)$,

(c) $(0, 4)$ and $(5, 1)$, **(d)** $(-3, 1)$ and $(-1, 6)$,

(e) $(4, -2)$ and $(3, 0)$, **(f)** $(3, -2)$ and $(6, 1)$,

(g) $(-2, 7)$ and $(3, -1)$, **(h)** $(-2, 0)$ and $(6, -3)$,

(i) $(-1.5, 0)$ and $(3.5, 0)$ **(j)** $(2.5, 4)$ and $(1, 6)$,

(k) $(8, 0)$ and $(2, 2.5)$, **(l)** $(-3.5, 2)$ and $(4, -8)$.

The identity $(3k)^2 + (4k)^2 \equiv (5k)^2$ may be useful in **(j)**, **(k)** and **(l)**.

2 Calculate the lengths of the sides of the triangle ABC and hence determine whether or not the triangle is right-angled:

(a) $A(0, 0)$ $B(0, 6)$ $C(4, 3)$,

(b) $A(3, 0)$ $B(1, 8)$ $C(-7, 6)$,

(c) $A(1, 2)$ $B(3, 4)$ $C(0, 7)$.

From question **2** onwards you are advised to draw a diagram.

In longer questions, the results found in one part often help in the next part.

3 The vertices of a triangle are $A(1, 5)$, $B(0, -2)$ and $C(4, 2)$. By writing each of the lengths of the sides as a multiple of $\sqrt{2}$, show that the sum of the lengths of two of the sides is three times the length of the third side.

4 The distance between the two points $A(6, 2p)$ and $B(p, -3)$ is $5\sqrt{5}$. Find the possible values of p.

5 The vertices of a triangle are $P(1, 3)$, $Q(-2, 0)$ and $R(4, 0)$.

(a) Find the lengths of the sides of triangle PQR.

(b) Show that angle $QPR = 90°$.

(c) The line of symmetry of triangle PQR meets the x-axis at point S. Write down the coordinates of S.

(d) The point T is such that $PQTR$ is a square. Find the coordinates of T.

3.3 The coordinates of the mid-point of a line segment joining two known points

From the diagram you can see that the mid-point of the line segment joining (0, 0) to (6, 0) is (3, 0) or

$$\left(\frac{0+6}{2}, 0\right),$$

the mid-point of the line segment joining (0, 0) to (0, −4) is (0, −2) or

$$\left(0, \frac{0+(-4)}{2}\right),$$

and the mid-point of the line segment joining (0, −4) to (6, 0) is (3, −2) or

$$\left(\frac{0+6}{2}, \frac{(-4)+0}{2}\right).$$

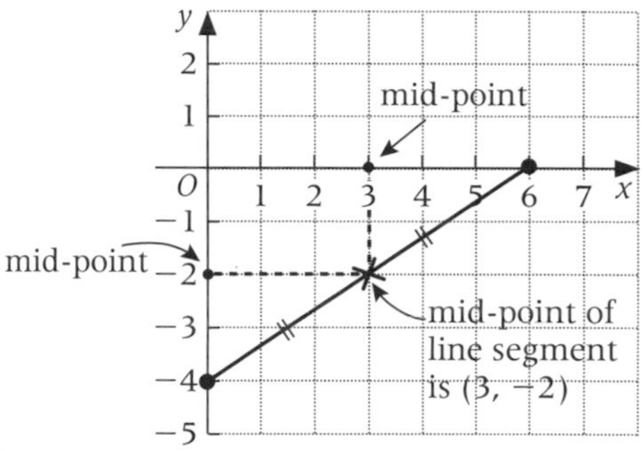

Going from $P(1, 2)$ to $Q(5, 8)$ you move 4 units horizontally and then 6 vertically.

If M is the mid-point of PQ then the journey is halved so to go from $P(1, 2)$ to M you move 2 (half of 4) horizontally and then 3 (half of 6) vertically.

So M is the point $(1 + 2, 2 + 3)$ or $M(3, 5)$ or $M\left(\frac{1+5}{2}, \frac{2+8}{2}\right)$.

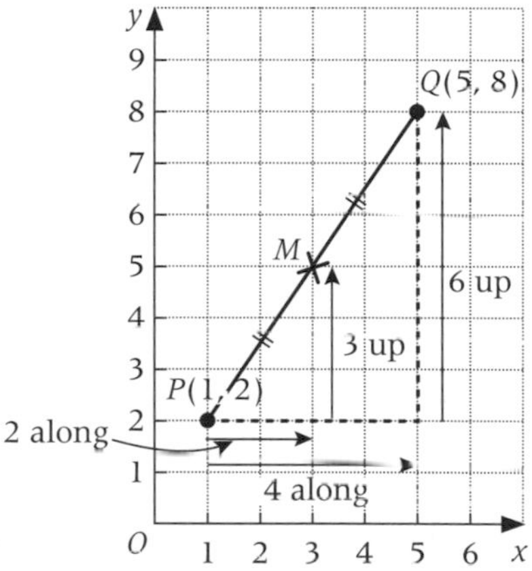

Check that M to Q is also 2 along and then 3 up.

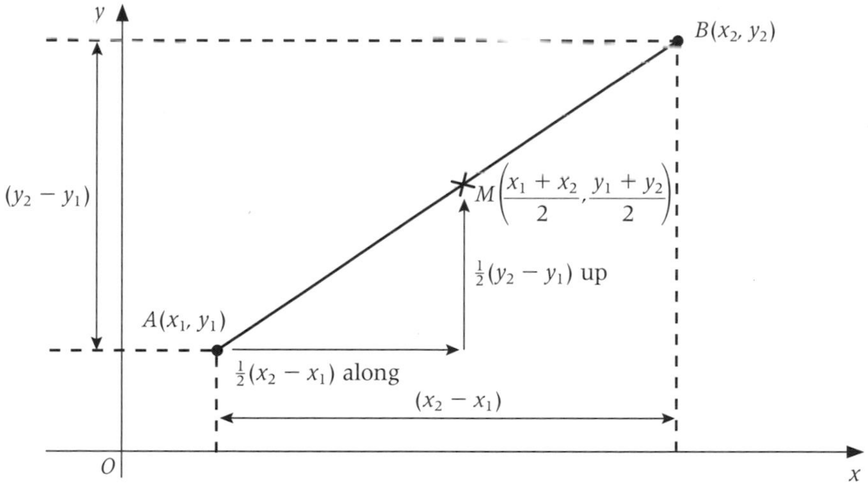

Note that $x_1 + \frac{1}{2}(x_2 - x_1) = \frac{1}{2}(x_1 + x_2)$ which is the x-coordinate of M.

> In general, the coordinates of the mid-point of the line segment joining (x_1, y_1) and (x_2, y_2) are
> $$\left(\frac{x_1+x_2}{2}, \frac{y_1+y_2}{2}\right).$$

Worked example 3.5

M is the mid-point of the line segment joining $A(1, -2)$ and $B(3, 5)$.

(a) Find the coordinates of M.

(b) M is also the mid-point of the line segment CD where $C(1, 4)$. Find the coordinates of D.

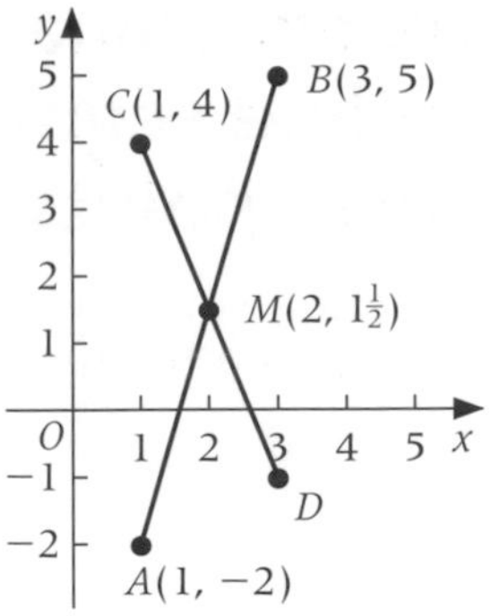

Solution

(a) Using $\left(\frac{x_1 + x_2}{2}, \frac{y_1 + y_2}{2}\right)$, M is the point $\left(\frac{1+3}{2}, \frac{-2+5}{2}\right)$ so $M(2, 1\frac{1}{2})$.

(b) Let D be the point (a, b) then the mid-point of CD is

$$\left(\frac{a+1}{2}, \frac{b+4}{2}\right) = (2, 1\tfrac{1}{2})$$ — coordinates of M

For this to be true $\frac{a+1}{2} = 2$ and $\frac{b+4}{2} = \frac{3}{2} \Rightarrow a = 3$ and $b = -1$.

The coordinates of D are $(3, -1)$.

Or, using the idea of 'journeys': C to M is 1 across and $2\frac{1}{2}$ down. So M to D is also 1 across and $2\frac{1}{2}$ down giving $D(2 + 1, 1\frac{1}{2} - 2\frac{1}{2})$ or $D(3, -1)$.

EXERCISE 3B

1 Find the coordinates of the mid-point of the line segments joining:

(a) $(3, 2)$ and $(7, 2)$,
(b) $(1, -2)$ and $(1, 3)$,
(c) $(0, 3)$ and $(6, 1)$,
(d) $(-3, 3)$ and $(-1, 6)$,
(e) $(4, -2)$ and $(3, 6)$,
(f) $(-3, -2)$ and $(-6, 1)$,
(g) $(-2, 5)$ and $(2, -1)$,
(h) $(-2, 5)$ and $(6, -3)$,
(i) $(-1.5, 6)$ and $(3.5, 0)$
(j) $(-3.5, 2)$ and $(4, -1)$.

2 M is the mid-point of the straight line segment PQ. Find the coordinates of Q for each of the cases:

(a) $P(2, 2)$, $M(3, 4)$,
(b) $P(2, 1)$, $M(3, 3)$,
(c) $P(2, 3)$, $M(1, 5)$,
(d) $P(-2, -5)$, $M(3, 0)$,
(e) $P(-2, 4)$, $M(1, 2\frac{1}{2})$,
(f) $P(-1, -3)$, $M(2\frac{1}{2}, -4\frac{1}{2})$.

3 The mid-point of AB, where $A(3, -1)$ and $B(4, 5)$, is also the mid-point of CD, where $C(0, 1)$.

(a) Find the coordinates of D.

(b) Show that $AC = BD$.

4 The mid-point of the line segment joining $A(-1, 3)$ and $B(5, -1)$ is D. The point C has coordinates $(4, 4)$. Show that CD is perpendicular to AB.

3.4 The gradient of a straight line joining two known points

The gradient of a straight line is a measure of how steep it is.

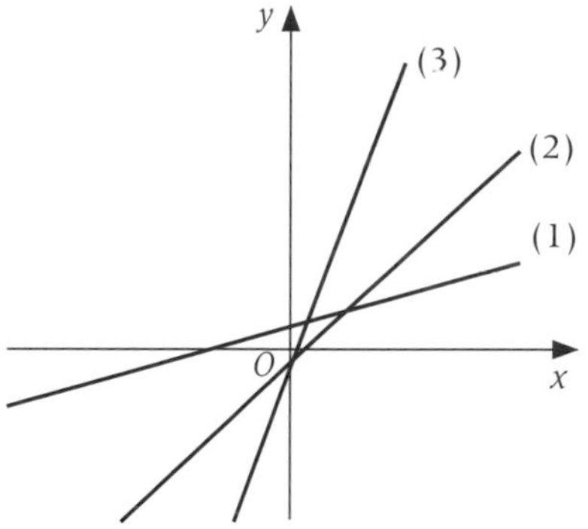

(a)

These three lines slope upwards from left to right. They have gradients which are positive. Line (2) is steeper than line (1) so the gradient of (2) is greater than the gradient of (1).

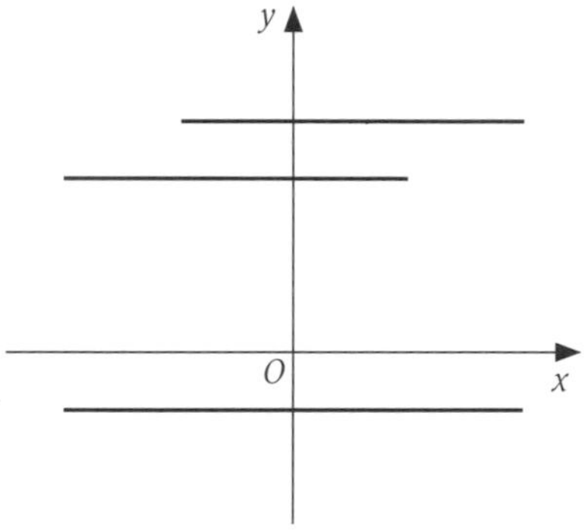

(b)

These three lines are all parallel to the x-axis. They are not sloping. Horizontal lines have gradient = 0.

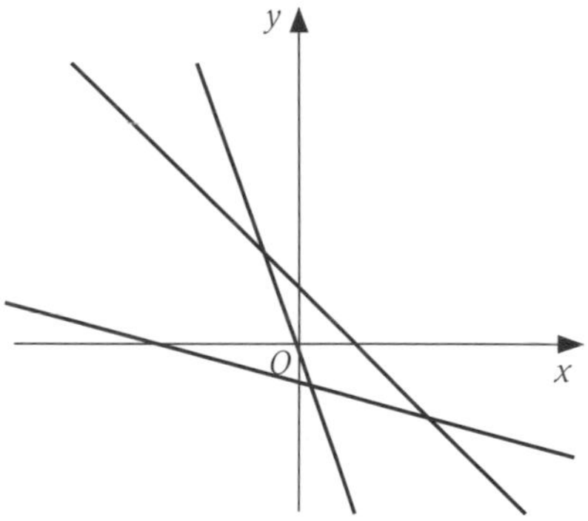

(c)

These three lines slope downwards from left to right. They have gradients which are negative.

The gradient of the line joining two points $= \dfrac{\text{change in } y\text{-coordinate}}{\text{change in } x\text{-coordinate}}$.

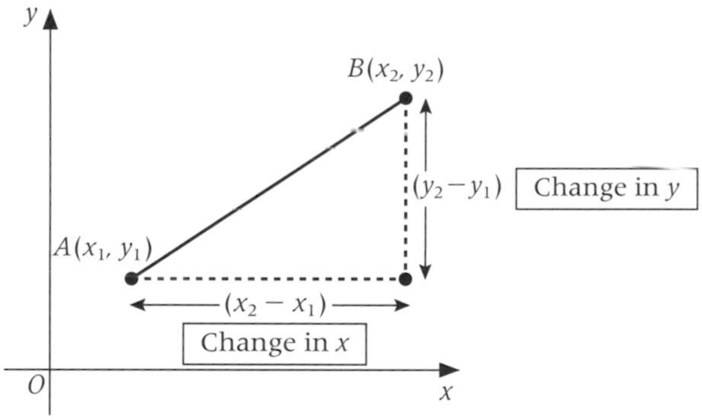

The gradient of the line joining the two points $A(x_1, y_1)$ and $B(x_2, y_2)$ is

$$\frac{y_2 - y_1}{x_2 - x_1}.$$

Lines which are equally steep are parallel.

Parallel lines have equal gradients.

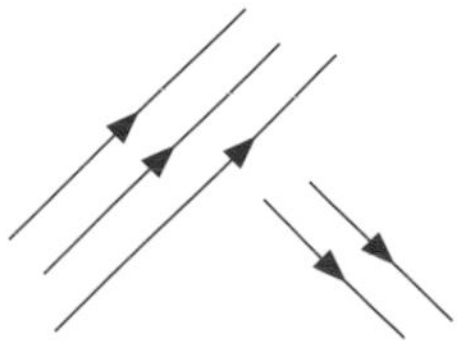

Worked example 3.6

$O(0, 0)$, $P(3, 6)$, $Q(0, 5)$ and $R(-2, 1)$ are four points.

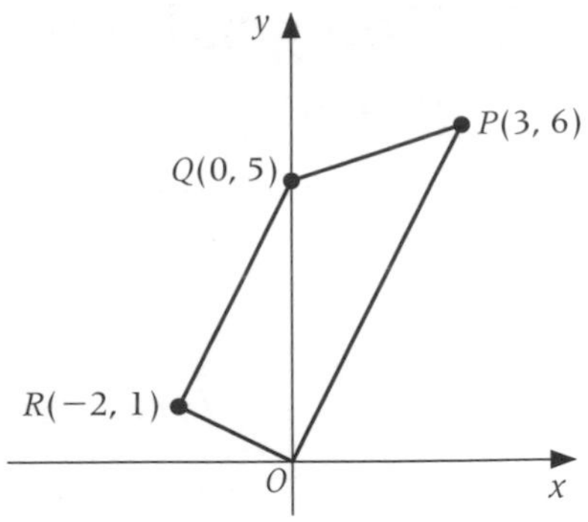

(a) Find the gradient of the line segment **(i)** OP, **(ii)** RQ.

(b) Find the gradient of the line segment **(i)** OR, **(ii)** PQ.

(c) What can you deduce from your answers?

Solution

A diagram is very helpful. You should try to position the points roughly in the correct place without plotting the points on graph paper, then you might see properties such as parallel lines and possible right angles.

(a) **(i)** Gradient of $OP = \dfrac{6-0}{3-0} = 2$;

(ii) Gradient of $RQ = \dfrac{5-1}{0-(-2)} = 2$.

(b) **(i)** Gradient of $OR = \dfrac{1-0}{-2-0} = -\dfrac{1}{2}$;

(ii) Gradient of $PQ = \dfrac{6-5}{3-0} = \dfrac{1}{3}$.

(c) The lines OP and RQ have gradients which are equal so they are parallel. Lines OR and PQ are not parallel since their gradients are not equal.

So we can deduce that the quadrilateral $OPQR$ is a trapezium.

EXERCISE 3C

1 By finding the gradients of the lines AB and CD determine if the lines are parallel.

(a) $A(2, 3)$ $B(3, 5)$ $C(0, 1)$ $D(1, 3)$,

(b) $A(3, 2)$ $B(5, 1)$ $C(-4, -3)$ $D(-2, -2)$,

(c) $A(-4, 5)$ $B(4, 5)$ $C(-1, -2)$ $D(0, -2)$,

(d) $A(-6, -3)$ $B(1, -2)$ $C(3\frac{1}{2}, 0)$ $D(7, \frac{1}{2})$.

2 By finding the gradients of the lines AB and BC show that $A(-2, 3)$, $B(2, 2)$ and $C(6, 1)$ are collinear points.

'Collinear' means 'in a straight line'.

3 $A(1, -3)$, $B(4, -2)$ and $C(6\frac{1}{2}, 0)$ are the vertices of triangle ABC.

(a) Find the gradient of each side of the triangle.

(b) Which side of the triangle is parallel to OP where O is the origin and P is the point $(-11, -6)$?

3.5 The gradients of perpendicular lines

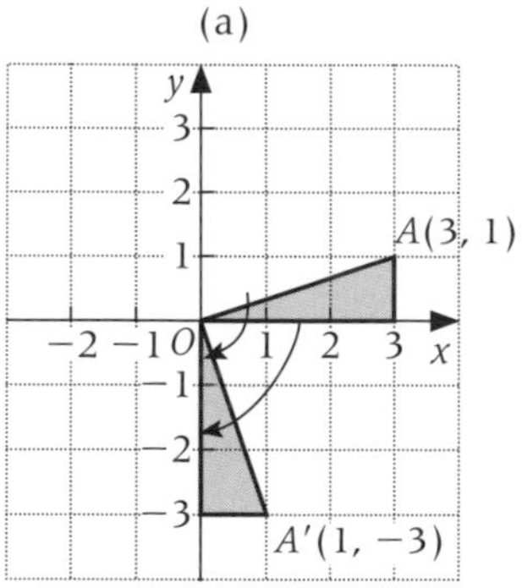

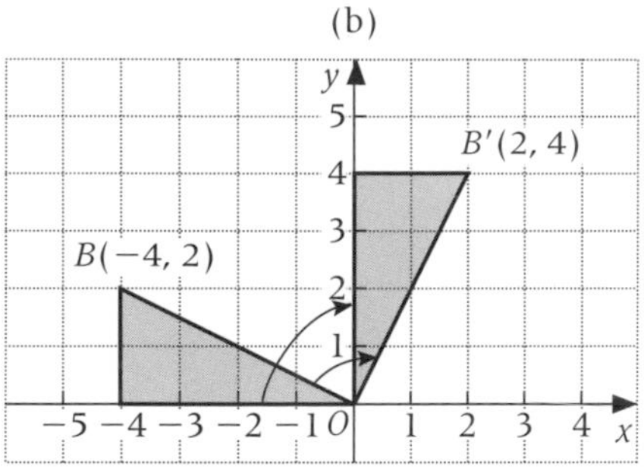

Rotate the shaded triangles clockwise through 90° as shown, keeping O fixed.

Line $OA \rightarrow$ line OA'
$A(3, 1) \rightarrow A'(1, -3)$

Gradient of $OA = \frac{1}{3}$

Gradient of $OA' = \frac{-3}{1}$

OA is perpendicular to OA'

Gradient of OA × Gradient OA'

$= \frac{1}{3} \times \frac{-3}{1} = -1$

Line $OB \rightarrow$ line OB'
$B(-4, 2) \rightarrow B'(2, 4)$

Gradient of $OB = -\frac{2}{4}$

Gradient of $OB' = \frac{4}{2}$

OB is perpendicular to OB'

Gradient of OB × Gradient OB'

$= -\frac{2}{4} \times \frac{4}{2} = -1$

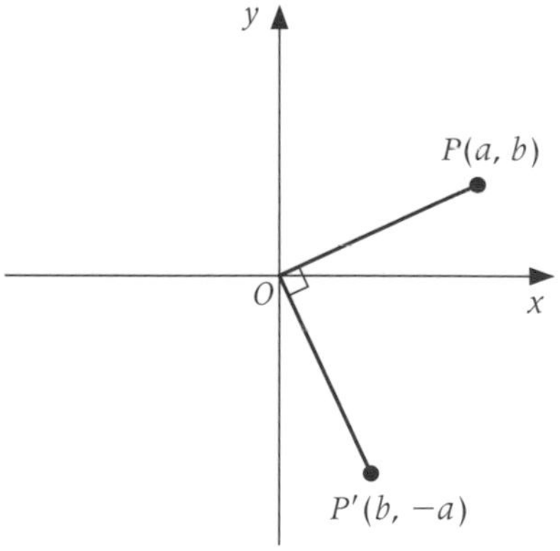

In general

gradient of $OP = \frac{b}{a}$ gradient of $OP' = -\frac{a}{b}$

Lines are perpendicular if the product of their gradients is −1.

> Lines with gradients m_1 and m_2
> – are parallel if $m_1 = m_2$,
> – are perpendicular if $m_1 \times m_2 = -1$.

Worked example 3.7

Find the gradient of a line which is perpendicular to the line joining $A(1, 3)$ and $B(4, 5)$.

Solution

Gradient of $AB = \dfrac{5-3}{4-1} = \dfrac{2}{3}$.

Let m_2 be the gradient of any line perpendicular to AB then

$\dfrac{2}{3} \times m_2 = -1 \Rightarrow m_2 = -\dfrac{3}{2}$.

The gradient of any line perpendicular to AB is $-\dfrac{3}{2}$.

EXERCISE 3D

1 Write down the gradient of lines perpendicular to a line with gradient:

(a) $\dfrac{2}{5}$, **(b)** $\dfrac{-1}{3}$, **(c)** 4, **(d)** $-3\frac{1}{2}$, **(e)** $2\frac{1}{4}$.

2 Two vertices of a rectangle $ABCD$ are $A(-2, 3)$ and $B(4, 1)$. Find:

(a) the gradient of DC,

(b) the gradient of BC.

3 $A(1, 2)$, $B(3, 6)$ and $C(7, 4)$ are the three vertices of a triangle.

(a) Show that ABC is a right-angled isosceles triangle.

(b) D is the point $(5, 0)$. Show that BD is perpendicular to AC.

(c) Explain why $ABCD$ is a square.

(d) Find the area of $ABCD$.

3.6 The y = mx + c form of the equation of a straight line

Consider any straight line which crosses the y-axis at the point $A(0, c)$. We say that **c is the y-intercept**.
Let $P(x, y)$ be any other point on the line.
If the gradient of the line is m then

$$\frac{y-c}{x-0} = m$$

so $y - c = mx$ or $y = mx + c$.

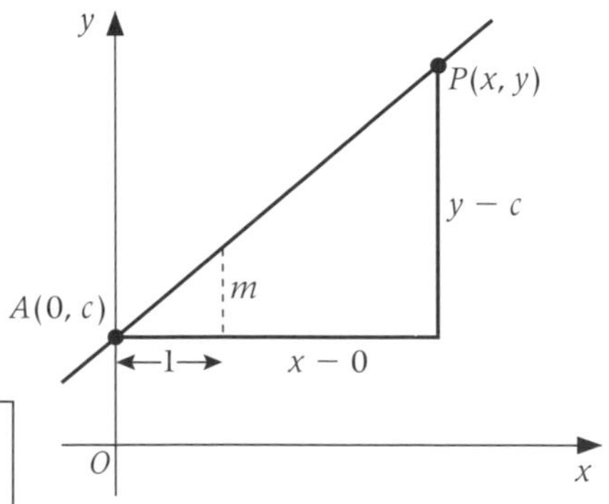

> $y = mx + c$ is the equation of a straight line with gradient m and y-intercept c.

> $ax + by + c = 0$ is the general equation of a line. It has gradient $-\dfrac{a}{b}$ and y-intercept $-\dfrac{c}{b}$.

Worked example 3.8

A straight line has gradient 2 and crosses the y-axis at the point $(0, -4)$. Write down the equation of the line.

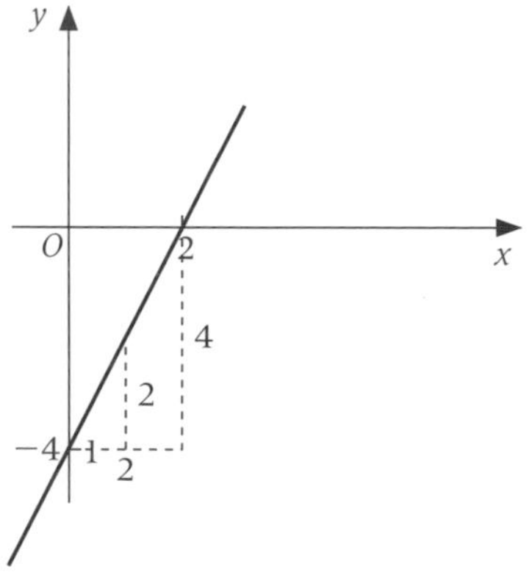

Solution

The line crosses the y-axis at $(0, -4)$ so the y-intercept is -4, so $c = -4$. The gradient of the line is 2, so $m = 2$.
Using $y = mx + c$ you get $y = 2x + (-4)$.
The equation of the line is $y = 2x - 4$.

Worked example 3.9

The general equation of a straight line is $Ax + By + C = 0$.
Find the gradient of the line, and the y-intercept.

Solution

You need to rearrange the equation $Ax + By + C = 0$ into the form $y = mx + c$.

You can write $Ax + By + C = 0$ as $By = -Ax - C$

$$\text{or } y = -\frac{A}{B}x - \frac{C}{B}.$$

Compare with $y = mx + c$,

you see that $m = -\frac{A}{B}$ and $c = -\frac{C}{B}$,

so $Ax + By + C = 0$ is the equation of a line

with gradient $-\frac{A}{B}$ and y-intercept $-\frac{C}{B}$.

EXERCISE 3E

1 Find, in the form $ax + by + c = 0$, the equation of the line which has:

(a) gradient 2 and y-intercept -3,

(b) gradient $-\frac{2}{3}$ and y-intercept 2,

(c) gradient $-\frac{1}{2}$ and y-intercept -3.

2 Find the gradient and y-intercept for the line with equation:

(a) $y = 2 + 3x$,

(b) $2y = 4x - 5$,

(c) $4y - 7 = 2x$,

(d) $2x + 3y = 8$,

(e) $8 - 5x + 4y = 0$,

(f) $0.5y = 4x - 3$,

(g) $5y - 3x = -2$,

(h) $4 - 3x = 2y$,

(i) $-2.5y + 5x = 3$,

(j) $2y = 4$.

3

3.7 The $y - y_1 = m(x - x_1)$ form of the equation of a straight line

Consider any line which passes through the known point $A(x_1, y_1)$ and let $P(x, y)$ be any other point on the line.

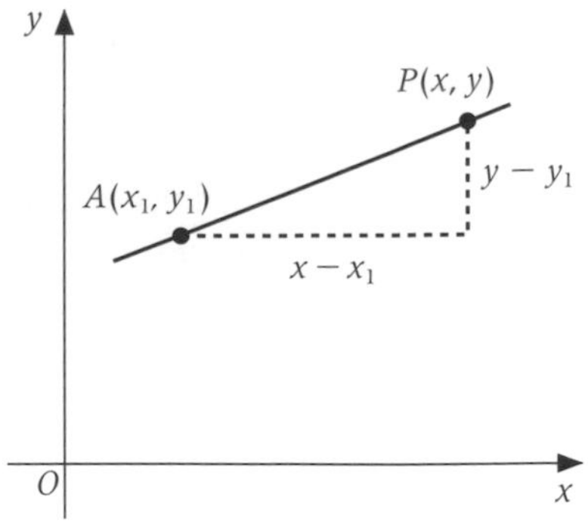

If m is the gradient of the line AP, then $\dfrac{y - y_1}{x - x_1} = m$

or $y - y_1 = m(x - x_1)$.

> The equation of the straight line which passes through the point (x_1, y_1) and has gradient m is
>
> $y - y_1 = m(x - x_1)$.

Worked example 3.10

Find the equation of the straight line which is parallel to the line $y = 4x - 1$ and passes through the point $(3, 2)$.

Solution

The gradient of the line $y = 4x - 1$ is 4,

Compare $y = 4x - 1$
↑
with $y = mx + c$

so the gradient of any line parallel to $y = 4x - 1$ is also 4.

We need the line with gradient 4 and through the point $(3, 2)$,

so its equation is, using $y - y_1 = m(x - x_1)$,

Parallel lines have equal gradients.

$$y - 2 = 4(x - 3)$$
$$\text{or } y = 4x - 10.$$

Worked example 3.11

(a) Find a Cartesian equation for the perpendicular bisector of the line joining $A(2, 1)$ and $B(4, -5)$.

(b) This perpendicular bisector cuts the coordinate axes at C and D. Show that $CD = 1.5 \times AB$.

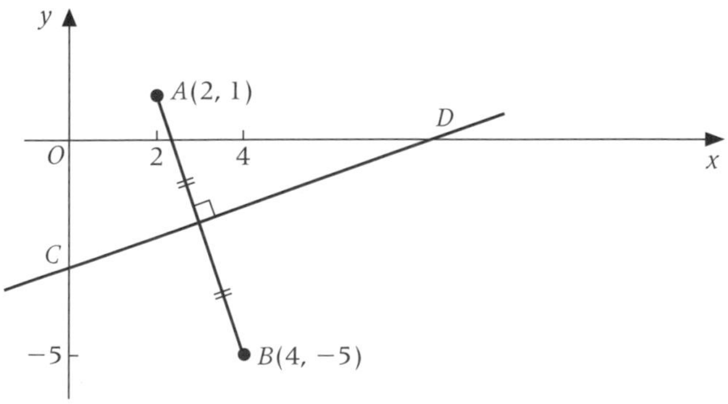

Solution

First, we draw a rough sketch.

(a) The gradient of $AB = \dfrac{-5 - 1}{4 - 2} = -3,$

so the gradient of the perpendicular is $\dfrac{1}{3}$.

$m_1 \times m_2 = -1$

The mid-point of AB is $\left(\dfrac{2+4}{2}, \dfrac{1+(-5)}{2}\right) = (3, -2)$.

Using $\left(\dfrac{x_1 + x_2}{2}, \dfrac{y_1 + y_2}{2}\right)$.

The perpendicular bisector is a straight line which passes through the point $(3, -2)$ and has gradient $\dfrac{1}{3}$, so its equation is

$$y - (-2) = \frac{1}{3}(x - 3) \text{ or } y = \frac{1}{3}x - 3.$$

Using $y - y_1 = m(x - x_1)$.

(b) Let C be the point where the line $y = \dfrac{1}{3}x - 3$ cuts the y-axis.

When $x = 0$, $y = \dfrac{1}{3} \times 0 - 3 = -3$, so we have $C(0, -3)$.

Any point on the y-axis has x-coordinate $= 0$.

Let D be the point where the line $y = \dfrac{1}{3}x - 3$ cuts the x-axis.

When $y = 0$, $0 = \dfrac{1}{3}x - 3$

Any point on the x-axis has y-coordinate $= 0$.

$\Rightarrow x = 9$, so we have $D(9, 0)$

Distance $CD = \sqrt{(9 - 0)^2 + (0 - (-3))^2} = \sqrt{81 + 9} = \sqrt{90}$

Using $\sqrt{(x_2 - x_1)^2 + (y_2 - y_1)^2}$.

Distance $AB = \sqrt{(4 - 2)^2 + (-5 - 1)^2} = \sqrt{4 + 36} = \sqrt{40}$

$$\frac{CD}{AB} = \frac{\sqrt{90}}{\sqrt{40}} = \sqrt{\frac{90}{40}} = \sqrt{\frac{9}{4}} = \frac{\sqrt{9}}{\sqrt{4}} = \frac{3}{2} = 1.5$$

so $CD = 1.5 \times AB$.

Using $\sqrt{\dfrac{a}{b}} = \dfrac{\sqrt{a}}{\sqrt{b}}$.
See section 2.4.

EXERCISE 3F

Because a straight line equation can be arranged in a variety of ways you are usually asked to find *an* equation rather than *the* equation.
All correct equivalents would score full marks unless you have been asked specifically for a particular form of the equation.

1 Find an equation for the straight line with gradient 2 and which passes through the point $(1, 6)$.

2 Find a Cartesian equation for the straight line which has gradient $-\dfrac{1}{3}$ and which passes through $(6, 0)$.

3 Find an equation of the straight line passing through $(-1, 2)$ which is parallel to the line with equation $2y = x + 4$.

4 Find an equation of the straight line that is parallel to $3x - 2y - 4 = 0$ and which passes through $(1, 3)$.

5 Find an equation of the straight line which passes through the origin and is perpendicular to the line $y = \frac{1}{2}x + 3$.

6 Find the y-intercept of the straight line which passes through the point $(-2, 2)$ and is perpendicular to the line $3y = 2x + 1$.

7 Find a Cartesian equation for the perpendicular bisector of the line joining $A(2, 3)$ and $B(0, 6)$.

8 The vertex, A, of a rectangle $ABCD$, has coordinates $(2, 1)$. The equation of BC is $y = -\frac{1}{2}x + 3$. Find, in the form $y = mx + c$, the equation of:

(a) AD,

(b) AB.

9 Given the points $A(0, 3)$, $B(5, 4)$, $C(4, -1)$ and $E(2, 1)$:

(a) show that BE is the perpendicular bisector of AC,

(b) find the coordinates of the point D so that $ABCD$ is a rhombus,

(c) find an equation for the straight line through D and A.

Hint for (b). The diagonals of a rhombus are perpendicular and bisect each other.

10 The perpendicular bisector of the line joining $A(0, 1)$ and $C(4, -7)$ intersects the x-axis at B and the y-axis at D. Find the area of the quadrilateral $ABCD$.

11 Show that the equation of any line parallel to $ax + by + c = 0$ is of the form $ax + by + k = 0$.

3.8 The equation of a straight line passing through two given points

The equation of the straight line which passes through the points (x_1, y_1) and (x_2, y_2) is

$$\frac{y - y_1}{y_2 - y_1} = \frac{x - x_1}{x_2 - x_1}$$

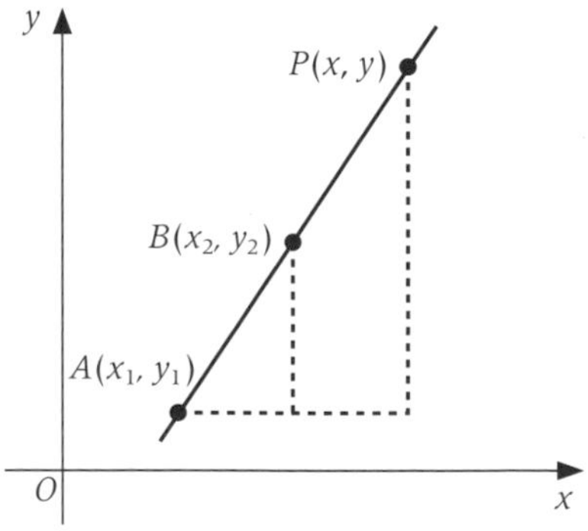

The derivation of this equation is similar to previous ones and is left as an exercise to the reader.

Worked example 3.12

Find a Cartesian equation of the straight line which passes through the points $(2, 3)$ and $(-1, 0)$.

Solution

If you take $(x_1, y_1) = (2, 3)$ and $(x_2, y_2) = (-1, 0)$ and substitute into the general equation

$$\frac{y - y_1}{y_2 - y_1} = \frac{x - x_1}{x_2 - x_1},$$

you get

$$\frac{y - 3}{0 - 3} = \frac{x - 2}{-1 - 2},$$

or $\quad y - 3 = x - 2$ which leads to $y = x + 1$.

Note that you can check that your line passes through the points (2, 3) and (−1, 0) by seeing if the points satisfy the equation $y = x + 1$.

$y = x + 1$

Checking for (2, 3)

LHS = 3 RHS = 2 + 1 = 3 ✓

Checking for (−1, 0)

LHS = 0 RHS = −1 + 1 = 0 ✓

3.9 The coordinates of the point of intersection of two lines

In this section you will need to solve simultaneous equations. If you need further practice, see section 1.3 of chapter 3.

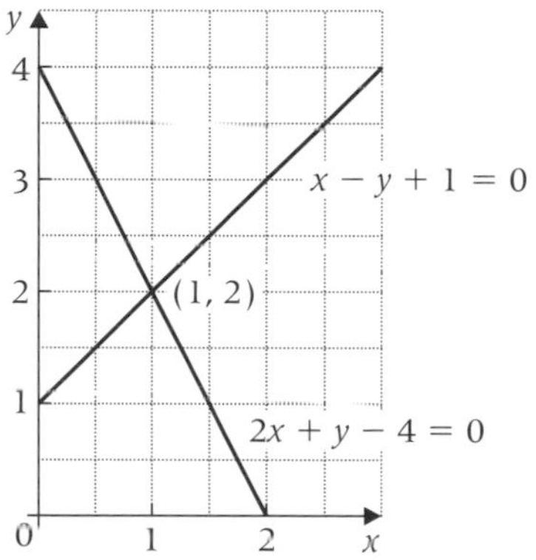

A point lies on a line if the coordinates of the point satisfy the equation of the line.

When two lines intersect, the point of intersection lies on both lines. The coordinates of the point of intersection must satisfy the equations of both lines. The equations of the lines must be satisfied simultaneously.

Given accurately drawn graphs of the two intersecting straight lines with equations $ax + by + c = 0$ and $Ax + By + C = 0$, the coordinates of the point of intersection can be read off. These coordinates give the solution of the simultaneous equations $ax + by + c = 0$ and $Ax + By + C = 0$.

From the diagram, you can clearly read off (1, 2) as the point of intersection of the lines $x - y + 1 = 0$ and $2x + y - 4 = 0$. So the solution to the simultaneous equations $x - y + 1 = 0$ and $2x + y - 4 = 0$ is $x = 1, y = 2$.

To find the coordinates of the point of intersection of the two lines with equations $ax + by + c = 0$ and $Ax + By + C = 0$, you solve the two linear equations simultaneously.

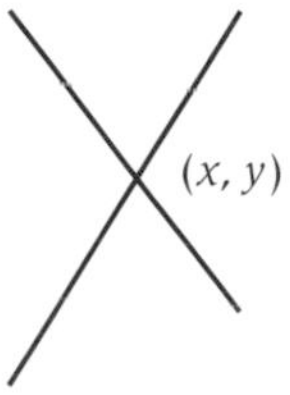

Worked example 3.13

Find the coordinates of the point of intersection of the straight lines with equations $y = x + 2$ and $y = 4x - 1$.

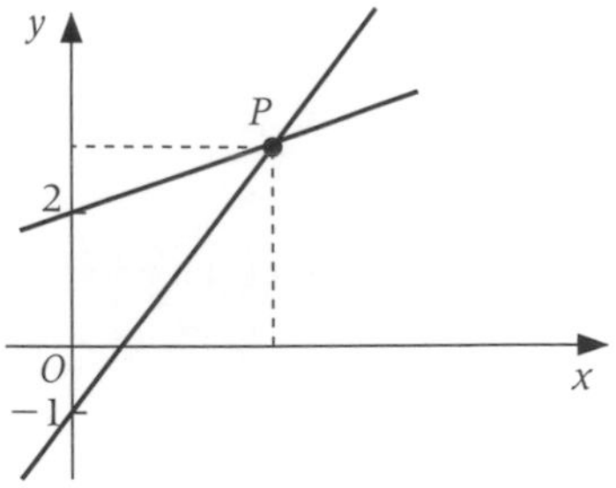

Solution

At the point of intersection P, $y = x + 2$ and $y = 4x - 1$.
So eliminating y gives $4x - 1 = x + 2$.
Rearranging gives $4x - x = 2 + 1$
$\Rightarrow 3x = 3 \Rightarrow x = 1$.

The x-coordinate of P is 1.

To find the y-coordinate of P, put $x = 1$ into $y = x + 2 \Rightarrow y = 3$
The point of intersection is $(1, 3)$.

Checking that $(1, 3)$ lies on the line $y = x + 2$;
LHS = 3 RHS = 1 + 2 = 3 ✓

Checking that $(1, 3)$ lies on the line $y = 4x - 1$;
LHS = 3 RHS = 4 − 1 = 3 ✓

Worked example 3.14

The straight lines with equations $5x + 3y = 7$ and $3x - 7y = 13$ intersect at the point R. Find the coordinates of R.

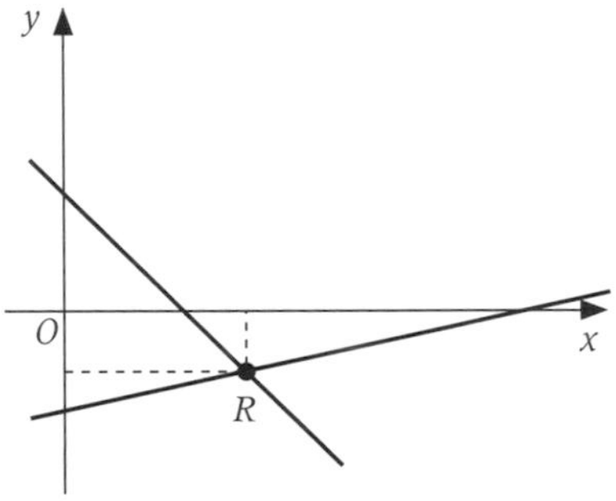

Solution

To find the point of intersection R, you need to solve the simultaneous equations

$$5x + 3y = 7 \quad [A]$$
$$3x - 7y = 13 \quad [B]$$

Multiply equation [A] by 7 and equation [B] by 3.

$$35x + 21y = 49 \quad [C]$$
$$9x - 21y = 39 \quad [D]$$

Adding [C] + [D].

$$44x + 0 = 88$$
$$\Rightarrow x = 2.$$

Substituting $x = 2$ in [A] gives $10 + 3y = 7$
$\Rightarrow 3y = -3 \qquad \Rightarrow y = -1$.

The coordinates of the point R are $(2, -1)$.

Checking in [A]
LHS = 10 + (−3) = 7 = RHS ✓;
and in [B]
LHS = 6 − 7(−1) = 13 ✓

Hint. Checking that the coordinates satisfy each line equation is advisable, especially if the result is being used in later parts of an examination question.
It can usually be done in your head rather than on paper.

EXERCISE 3G

1 Verify that (2, 5) lies on the line with equation $y = 3x - 1$.

2 Which of the following points lie on the line with equation $3x + 2y = 6$:

(a) (3, 0), **(b)** (2, 0), **(c)** (4, −3)

(d) (−2, 6), **(e)** (0, 2)?

3 Find the coordinates of the points where the following lines intersect the x-axis:

(a) $y = x + 4$, **(b)** $y = 2x - 6$,

(c) $2x + 3y + 6 = 0$, **(d)** $3x - 4y + 12 = 0$.

4 The point $(k, 2k)$ lies on the line with equation $2x + 3y - 6 = 0$. Find the value of k.

5 Show that the point (−4, 8) lies on the line passing through the points (1, 3) and (7, −3).

6 (a) Find the equation of the line AB where A is the point (−3, 7) and B is the point (5, −1).

(b) The point $(k, 3)$ lies on the line AB. Find the value of the constant k.

7 $A(-5, 2)$, $B(-2, 3)$, $C(-2, -1)$ and $D(-4, -2)$ are the vertices of the quadrilateral $ABCD$.

(a) Find the equation of the diagonal BD.

(b) Determine whether or not the mid-point of AC lies on the diagonal BD.

8 Find the coordinates of the point of intersection of these pairs of straight lines:

(a) $y = 2x + 7$ and $y = x + 1$,

(b) $3y + x = 7$ and $2y - x = 3$,

(c) $5x + 2y = 16$ and $3x + 2y = 8$,

(d) $y = 8x$ and $y = 40 + 3x$,

(e) $y = -7$ and $5y = -x - 1$,

(f) $y - 3x = 3$ and $2y - 5x = 9$,

(g) $4y + 9x = 8$ and $5y + 6x = 3$,

(h) $8y = 3x - 11$ and $2x - 5y = 6$.

9 Point A has coordinates $\left(\frac{11}{2}, -1\right)$, point B has coordinates $\left(-3, \frac{61}{60}\right)$ and point C has coordinates $\left(-\frac{19}{6}, \frac{3}{5}\right)$. The straight line AC has equation $12x + 65y - 1 = 0$ and the straight line BC has equation $60y = 150x + 511$. Write down the solution of the simultaneous equations $12x + 65y - 1 = 0$ and $60y = 150x + 511$.

10 The straight line $23x + 47y + 105 = 0$ passes through the point of intersection of the two straight lines $y = x$ and $7x - 5y = -3$. Write down the solution of the simultaneous equations $23x + 47y + 105 = 0$ and $7x - 5y = -3$.

Worked example 3.15

ABCD is a parallelogram in which the coordinates of *A*, *B* and *C* are $(1, 2)$, $(7, -1)$ and $(-1, -2)$, respectively.

(a) Find the equations of *AD* and *CD*.

(b) Find the coordinates of *D*.

(c) Prove that angle $BAC = 90°$.

(d) Calculate the area of the parallelogram.

(e) Show that the length of the perpendicular from *A* to *BC* is $\frac{6\sqrt{65}}{13}$.

Solution

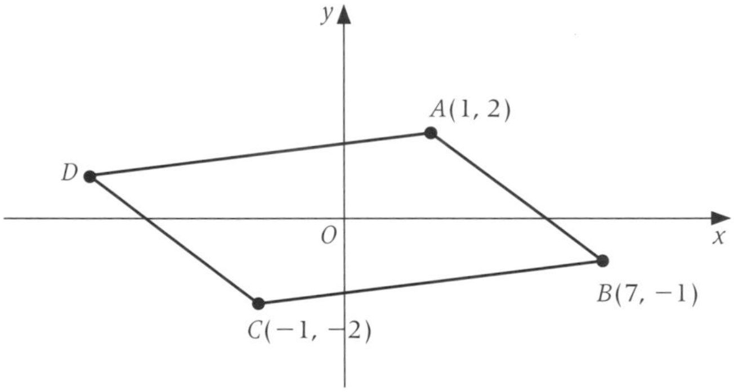

Always start with a good sketch; this helps to spot obvious errors like wrong signs for gradients or wrong quadrants for points. Note that when it says the parallelogram *ABCD*, the points must be connected in that order which determines where *D* must be.

(a) *AD* is parallel to *BC* and *CD* is parallel to *BA*.

Opposite sides of a parallelogram are equal and parallel.

Gradient of $BC = \frac{-1-(-2)}{7-(-1)} = \frac{1}{8} \Rightarrow$ gradient of AD $= \frac{1}{8}$

Using $\frac{y_2 - y_1}{x_2 - x_1}$.

AD is a line through $(1, 2)$ and has gradient $\frac{1}{8}$ so its equation is $y - 2 = \frac{1}{8}(x - 1)$ or $8y - x = 15$.

Using $y - y_1 = m(x - x_1)$.

Gradient of $BA = \frac{-1-2}{7-1} = -\frac{3}{6} = -\frac{1}{2}$

$\Rightarrow$ gradient of $CD = -\frac{1}{2}$.

You could check the signs of the two gradients using your sketch.

CD is a line through $(-1, -2)$ and has gradient $-\frac{1}{2}$ so its equation is $y - (-2) = \frac{-1}{2}[x - (-1)]$ or $2y + x = -5$.

(b) D is the point of intersection of AD and CD.

Solving $8y - x = 15$

and $2y + x = -5$ simultaneously,

adding gives $10y + 0 = 10 \Rightarrow y = 1$.

Substitution in the second equation gives $2(1) + x = -5$
$\Rightarrow x = -7$, i.e. $D(-7, 1)$

> Alternatively, since BA is equal and parallel to CD, the journey from $B \rightarrow A$ (6 left then 3 up) is the same as from C to D.
> $D(-1 - 6, -2 + 3)$ or
> $D(-7, 1)$.

> From the sketch you know D is in the 2nd quadrant. Checking in both equations $8 - (-7) = 15$ ✓
> $2 + (-7) = -5$ ✓

3

(c) Gradient of $AC = \dfrac{2 - (-2)}{1 - (-1)} = \dfrac{4}{2} = 2.$

From earlier work, gradient of $BA = -\dfrac{1}{2}$.

Gradient of $AC \times$ gradient of $BA = 2 \times -\dfrac{1}{2} = -1,$

so AC is perpendicular to BA and angle $BAC = 90°$.

> Angle $BAC = 90°$ if AC and BA are perpendicular.
> Aim to show that $m_1 \times m_2 = -1$.

> **Hint.** In exam questions look for possible links between the parts.

(d) Since angle $BAC = 90°$ you can use:

Area of parallelogram $ABCD$ $= \text{base} \times \text{height}$
$= AB \times AC$.

$$AB = \sqrt{(7-1)^2 + (-1-2)^2} = \sqrt{36+9} = \sqrt{45}$$

$$AC = \sqrt{(-1-1)^2 + (-2-2)^2} = \sqrt{4+16} = \sqrt{20}.$$

> $\sqrt{(x_2 - x_1)^2 + (y_2 - y_1)^2}$
> Keep your answers in surd form.

So area of parallelogram $ABCD = \sqrt{45} \times \sqrt{20} = \sqrt{900}$
$= \sqrt{9} \times \sqrt{100}$
$= 30$ square units

(e) Using the base of the parallelogram as CB and letting h = length of the perpendicular from A to BC,

$\Rightarrow$ area of parallelogram $ABCD = CB \times h,$

$$30 = \sqrt{(7 - (-1))^2 + (-1 - (-2)))^2} \times h,$$

$$30 = \sqrt{64 + 1} \times h, \quad \Rightarrow h = \frac{30}{\sqrt{65}} = \frac{30\sqrt{65}}{\sqrt{65}\sqrt{65}}$$

$$= \frac{30\sqrt{65}}{65} = \frac{6\sqrt{65}}{13}$$

> **Hint.** In exam questions look for links between the parts:
> Length of perpendicular from A to BC is required. We have just found the area of the parallelogram $ABCD$. Can the two be linked?

MIXED EXERCISE

1 The point A has coordinates $(2, 3)$ and O is the origin.

(a) Write down the gradient of OA and hence find the equation of the line OA.

(b) Show that the line which has equation $4x + 6y = 13$:

(i) is perpendicular to OA,

(ii) passes through the mid-point of OA. [A]

2 The line AB has equation $5x - 2y = 7$. The point A has coordinates $(1, -1)$ and the point B has coordinates $(3, k)$.

(a) (i) Find the value of k.
(ii) Find the gradient of AB.

(b) Find an equation for the line through A which is perpendicular to AB.

(c) The point C has coordinates $(-6, -2)$. Show that AC has length $p\sqrt{2}$, stating the value of p. [A]

3 The point P has coordinates $(1, 10)$ and the point Q has coordinates $(4, 4)$.

(a) Show that the length of PQ is $3\sqrt{5}$.

(b) (i) Find the equation of the perpendicular bisector of PQ.
(ii) This perpendicular bisector intersects the x-axis at the point A. Find the coordinates of A.

4 The point A has coordinates $(3, -5)$ and the point B has coordinates $(1, 1)$.

(a) (i) Find the gradient of AB.
(ii) Show that the equation of the line AB can be written in the form $rx + y = s$, where r and s are positive integers.

(b) The mid-point of AB is M and the line MC is perpendicular to AB.
(i) Find the coordinates of M.
(ii) Find the gradient of the line MC.
(iii) Given that C has coordinates $(5, p)$, find the value of the constant p. [A]

5 The points A and B have coordinates $(13, 5)$ and $(9, 2)$, respectively.

(a) (i) Find the gradient of AB.
(ii) Find an equation for the line AB.

(b) The point C has coordinates $(2, 3)$ and the point X lies on AB so that XC is perpendicular to AB.
(i) Show that the equation of the line XC can be written in the form $4x + 3y = 17$.
(ii) Calculate the coordinates of X. [A]

6 The equation of the line AB is $5x - 3y = 26$.

(a) Find the gradient of AB.

(b) The point A has coordinates $(4, -2)$ and a point C has coordinates $(-6, 4)$.
(i) Prove that AC is perpendicular to AB.
(ii) Find an equation for the line AC, expressing your answer in the form $px + qy = r$, where p, q and r are integers.

(c) The line with equation $x + 2y = 13$ also passes through the point B. Find the coordinates of B. [A]

7 The points A, B and C have coordinates $(1, 7)$, $(5, 5)$ and $(7, 9)$, respectively.

(a) Show that AB and BC are perpendicular.

(b) Find an equation for the line BC.

(c) The equation of the line AC is $3y = x + 20$ and M is the mid-point of AB.

(i) Find an equation of the line through M parallel to AC.

(ii) This line intersects BC at the point T. Find the coordinates of T. [A]

8 The point A has coordinates $(3, 5)$, B is the point $(-5, 1)$ and O is the origin.

(a) Find, in the form $y = mx + c$, the equation of the perpendicular bisector of the line segment AB.

(b) This perpendicular bisector cuts the y-axis at P and the x-axis at Q.

(i) Show that the line segment BP is parallel to the x-axis.

(ii) Find the area of triangle OPQ.

9 The points $A(-1, 2)$ and $C(5, 1)$ are opposite vertices of a parallelogram $ABCD$. The vertex B lies on the line $2x + y = 5$. The side AB is parallel to the line $3x + 4y = 8$. Find:

(a) the equation of the side AB,

(b) the coordinates of B,

(c) the equations of the sides AD and CD,

(d) the coordinates of D.

10 $ABCD$ is a rectangle in which the coordinates of A and C are $(0, 4)$ and $(11, 1)$, respectively, and the gradient of the side AB is -5.

(a) Find the equations of the sides AB and BC.

(b) Show that the coordinates of B is $(1, -1)$.

(c) Calculate the area of the rectangle.

(d) Find the coordinates of the point on the y-axis which is equidistant from points A and D. [A]

3

Key point summary

1 The distance between the points (x_1, y_1) and (x_2, y_2) is $\sqrt{(x_2 - x_1)^2 + (y_2 - y_1)^2}$. *p31*

2 The coordinates of the mid-point of the line segment joining (x_1, y_1) and (x_2, y_2) are $\left(\frac{x_1 + x_2}{2}, \frac{y_1 + y_2}{2}\right)$. *p33*

3 The gradient of a line joining the two points $A(x_1, y_1)$ and $B(x_2, y_2) = \frac{y_2 - y_1}{x_2 - x_1}$. *p35*

4 Lines with gradients m_1 and m_2: *p37*
- are parallel if $m_1 = m_2$,
- are perpendicular if $m_1 \times m_2 = -1$.

5 $y = mx + c$ is the equation of a straight line with gradient m and y-intercept c. *p38*

6 $ax + by + c = 0$ is the general equation of a line. *p38*
It has gradient $-\frac{a}{b}$ and y-intercept $-\frac{c}{b}$.

7 The equation of the straight line which passes through the point (x_1, y_1) and has gradient m is *p40*

$$y - y_1 = m(x - x_1).$$

8 The equation of the straight line which passes through the points (x_1, y_1) and (x_2, y_2) is *p42*

$$\frac{y - y_1}{y_2 - y_1} = \frac{x - x_1}{x_2 - x_1}.$$

9 A point lies on a line if the coordinates of the point satisfy the equation of the line. *p43*

10 Given accurately drawn graphs of the two intersecting straight lines with equations $ax + by + c = 0$ and $Ax + By + C = 0$, the coordinates of the point of intersection can be read off. These coordinates give the solution of the simultaneous equations $ax + by + c = 0$ and $Ax + By + C = 0$. *p43*

11 To find the coordinates of the point of intersection of the two lines with equations $ax + by + c = 0$ and $Ax + By + C = 0$, you solve the two equations simultaneously. *p43*

Test yourself	What to review
1 Calculate the distance between the points $(2, -3)$ and $(7, 9)$.	*Section 3.2*
2 State the coordinates of the mid-point of the line segment PQ where $P(3, -2)$ and $Q(7, 1)$.	*Section 3.3*
3 Find the gradient of the line joining the points A and B where A is the point $(-3, -2)$ and B is the point $(-5, 4)$.	*Section 3.4*
4 The lines CD and EF are perpendicular with points $C(1, 2)$, $D(3, -4)$, $E(-2, 5)$ and $F(k, 4)$. Find the value of the constant k.	*Section 3.5*
5 Find a Cartesian equation of the line which passes through the point $(-2, 1)$ and is perpendicular to the line $5y + 3x = 7$.	*Sections 3.5 and 3.7*
6 Find the point of intersection of the lines with equations $3x - 5y = 11$ and $y = 4x - 9$.	*Section 3.9*

3

Test yourself ANSWERS

1 13.

2 $(5, -\frac{1}{2})$.

3 -3.

4 $k = -5$.

5 $3y = 5x + 13$.

6 $(2, -1)$.

CHAPTER 4

C1: Quadratic functions and their graphs

Learning objectives

After studying this chapter, you should be able to:

- solve quadratic equations
- complete the square of a quadratic expression
- understand the effect of a translation on a quadratic graph
- find the equation of a parabola when translated by a given vector
- understand the term discriminant.

4.1 Parabolas and quadratic functions

When you throw a ball, its path is part of a **parabola**. The cross-section of a satellite dish also has a parabolic shape. Special features of a parabola are its **vertex** (the highest or lowest point of the curve) and its **axis of symmetry**.

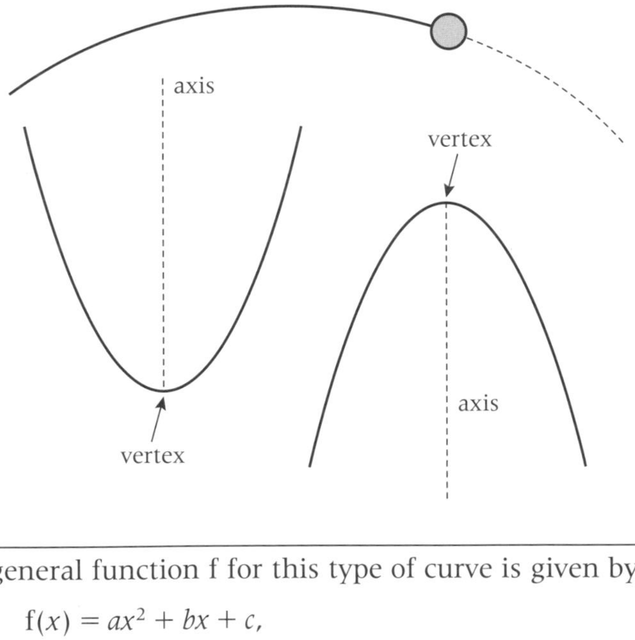

A general function f for this type of curve is given by

$$f(x) = ax^2 + bx + c,$$

where a is called the coefficient of x^2, b is the coefficient of x, and c is a constant. It is called a **quadratic function**. The graph of $y = ax^2 + bx + c$ is called a parabola.

In chapter 3, you learned how to identify the features of a line of the form $y = mx + c$, related to the values of m and c. In this chapter, you will learn how to sketch graphs of quadratic functions, dependent on the values of a, b and c.

4.2 Factorising quadratics and sketching graphs

The expression $(2x + 3)(x - 4)$ can be multiplied out to give $2x^2 - 5x - 12$ and the reverse process is called factorisation.

Although you might need a graphic calculator to draw the graph of $y = 2x^2 - 5x - 12$, you should be able to sketch the graph of $y = (2x + 3)(x - 4)$ without one.

Where does the graph cross the x-axis?

You need to solve $(2x + 3)(x - 4) = 0$.

The equation $p \times q = 0$ implies that either $p = 0$ or $q = 0$.

Either $(2x + 3) = 0 \Rightarrow x = -\frac{3}{2} = -1\frac{1}{2}$

or $(x - 4) = 0 \Rightarrow x = 4$.

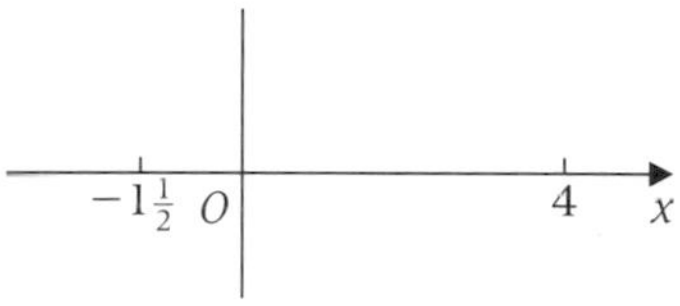

The values 4 and $-1\frac{1}{2}$ are called the **roots** of the equation and they are the values of x where the curve crosses the x-axis.

You know that the curve is either like ∧ or like ∨ .

By choosing another value of x, such as $x = 0$, $\Rightarrow y = (2 \times 0 + 3)(0 - 4) = -12$, you know that the parabola also passes through $(0, -12)$.

Since you know a parabola is symmetrical about its axis, you can now sketch the graph of $y = (2x + 3)(x - 4)$.

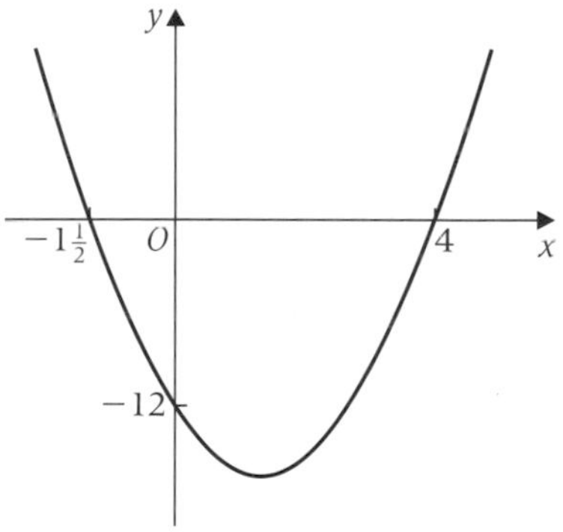

> A quadratic equation that can be written in the form $(x - p)(x - q) = 0$ has roots p and q.
> The graph of $(x - p)(x - q)$ crosses the x-axis at the points $(p, 0)$ and $(q, 0)$.

Worked example 4.1

Factorise $10 - 13x - 3x^2$ and hence sketch the graph of $y = 10 - 13x - 3x^2$.

Two numbers which multiply to give 10.

These two terms multiply to give $-3x^2$.

Solution

You need to form two brackets ( ) ( )

$$10 - 13x - 3x^2 = (5 + x)(2 - 3x)$$

Since $5 \times (-3x) + 2 \times x = -15x + 2x = -13x$

Now find where the graph crosses the x-axis.

$$(5 + x)(2 - 3x) = 0$$

$$\Rightarrow (5 + x) = 0 \text{ or } (2 - 3x) = 0$$

$$\Rightarrow x = -5 \text{ or } x = \frac{2}{3}$$

You can check the factorisation is correct by substituting these values into the original quadratic: $10 - 13 \times (-5) - 3 \times (-5)^2 = 10 + 65 - 75 = 0$ ✓

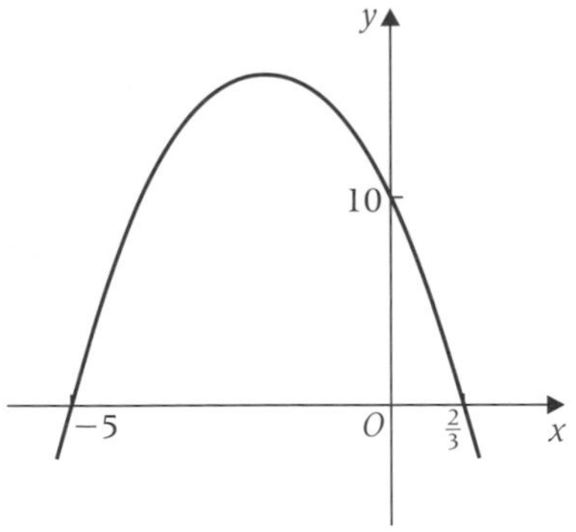

Check one other point such as when $x = 0$. $x = 0 \Rightarrow y = 10$.

Note that it is not always necessary to have equal axes on the scales in a sketch.

EXERCISE 4A

1 Factorise each of the following:

(a) $x^2 - 3x + 2$, **(b)** $x^2 - 7x - 8$,
(c) $x^2 + 7x + 12$, **(d)** $2x^2 - x - 3$,
(e) $3x^2 - 7x + 2$, **(f)** $4x^2 + x - 3$,
(g) $7 - 13x - 2x^2$, **(h)** $6x^2 + 5x - 6$,
(i) $6 + 5x - 4x^2$, **(j)** $5x - 2x^2$,
(k) $21 + 25x - 4x^2$, **(l)** $15 - 14x - 8x^2$,
(m) $12 + 16x - 3x^2$, **(n)** $3x + 4x^2$,
(o) $54 - 15x - 25x^2$.

2 Using your results from question **1**, find where the following parabolas cross the x-axis and sketch their graphs:

(a) $y = x^2 - 3x + 2$, **(b)** $y = x^2 - 7x - 8$,
(c) $y = x^2 + 7x + 12$, **(d)** $y = 2x^2 - x - 3$,
(e) $y = 3x^2 - 7x + 2$, **(f)** $y = 4x^2 + x - 3$,
(g) $y = 7 - 13x - 2x^2$, **(h)** $y = 6x^2 + 5x - 6$,
(i) $y = 6 + 5x - 4x^2$, **(j)** $y = 5x - 2x^2$,
(k) $y = 21 + 25x - 4x^2$, **(l)** $y = 15 - 14x - 8x^2$,
(m) $y = 12 + 16x - 3x^2$, **(n)** $y = 3x + 4x^2$,
(o) $y = 54 - 15x - 25x^2$.

3 By considering the graphs you have drawn in question **2**, comment on the shape of $y = ax^2 + bx + c$ in the cases:

(a) $a > 0$,
(b) $a < 0$.
(c) What is the shape of the graph when $a = 0$?

4 When a parabola crosses the x-axis at the points A and B, the axis of symmetry is the perpendicular bisector of the line joining the points A and B. In worked example 4.1, since A is the point $(-5, 0)$ and B is the point $(\frac{2}{3}, 0)$, the equation of the line of symmetry is $x = -\dfrac{13}{6}$.

Find the equation of the line of symmetry for each of the parabolas in question **2**.

5 A parabola passes through the three points given. Find their equations.

Hint. Consider an equation of the form:

$y = k(x - a)(x - b)$.

(a) $(3, 0)$, $(5, 0)$, $(0, 15)$,

(b) $(-2, 0)$, $(3, 0)$, $(0, -6)$,

(c) $(4, 0)$, $(6, 0)$, $(0, 48)$,

(d) $(1, 0)$, $(-1, 0)$, $(0, 1)$,

(e) $(2, 0)$, $(3, 0)$, $(0, 3)$,

(f) $(-2, 0)$, $(-5, 0)$, $(0, -30)$.

4.3 Completing the square

The expression $x^2 + 14x$ is not quite a perfect square. However, by adding 49 it can be written as $(x + 7)^2$.

Recall that

$(x + a)^2 = x^2 + 2ax + a^2$.

Hence, $x^2 + 14x = (x + 7)^2 - 49$

Writing a quadratic in this form is known as completing the square.

> $A(x + B)^2 + C$ is the completed square form.

Worked example 4.2

Express the quadratics:

(a) $x^2 + 10x + 31$, **(b)** $x^2 - 6x + 7$,

in the form $(x + p)^2 + q$, where p and q are constants.

Solution

(a) $(x + p)^2 + q = x^2 + 2px + p^2 + q$

Comparing this with $x^2 + 10x + 31$

Look at the coefficient of x in each case.

$2p = 10$

$\Rightarrow p = 5$

Comparing the constant term.

Also $p^2 + q = 31$

$\Rightarrow 5^2 + q = 31 \Rightarrow q = 6$

Therefore $x^2 + 10x + 31 \equiv (x + 5)^2 + 6$.

The sign $\equiv$ means 'is identical to' and is used sometimes rather than $=$ to indicate that the expressions are identical for all values of x.

A more direct approach to completing the square is to start with $x^2 + 10x + 31$ and to ask yourself what number p would make $(x + p)^2$ have the first two terms the same as those in $x^2 + 10x + 31$?

Since $(x + 5)^2 = x^2 + 10x + 25$, the value of p is 5.

$$x^2 + 10x = (x + 5)^2 - 25$$

$$\Rightarrow x^2 + 10x + 31 = (x + 5)^2 - 25 + 31 = (x + 5)^2 + 6$$

Notice that 5 is half of 10 and so in general you need to add half the coefficient of x.

(b) This time you need to consider $(x - 3)^2$.

$$(x - 3)^2 = x^2 - 6x + 9$$

$$x^2 - 6x = (x - 3)^2 - 9$$

$$\Rightarrow x^2 - 6x + 7 = (x - 3)^2 - 9 + 7$$

$$\Rightarrow x^2 - 6x + 7 \equiv (x - 3)^2 - 2$$

Since -3 is half of -6.

This is of the given form with $p = -3$ and $q = -2$.

Worked example 4.3

Express the quadratic $x^2 + 6x + 17$ in completed square form. Hence find the coordinates of the vertex and axis of symmetry of the parabola with equation $y = x^2 + 6x + 17$.

This technique can help you to find the equation of the line of symmetry and the coordinates of the vertex of any parabola.

Solution

Since 3 is half of 6.

$$x^2 + 6x = (x + 3)^2 - 9$$

$$\Rightarrow x^2 + 6x + 17 = (x + 3)^2 - 9 + 17 = (x + 3)^2 + 8$$

This form of the quadratic which involves a perfect square is called the completed square form.

The equation of the parabola $y = x^2 + 6x + 17$ can be written as $y = (x + 3)^2 + 8$

Since the coefficient of x^2 is positive, the graph is ⋃ shaped (see question 3 of exercise 4A)

Since $(x + 3)^2 \geqslant 0$, the least value of $x^2 + 6x + 17$ occurs when $x + 3 = 0 \Rightarrow x = -3$.

The corresponding y-coordinate is $y = 0 + 8 = 8$.

So the vertex of the parabola is at $(-3, 8)$.

The axis of symmetry has equation $x = -3$ and the parabola is sketched on the right.

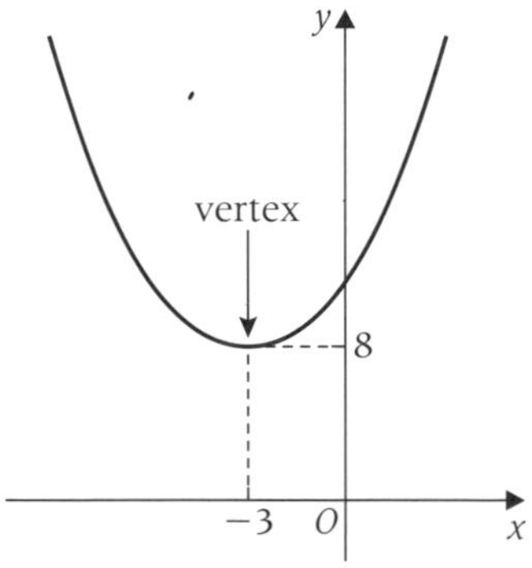

You will notice that this parabola does not cross the x-axis at all. A general condition for this to happen will be developed later.

Worked example 4.4

Complete the square for the quadratic $x^2 - 8x + 1$. Hence find the exact roots of the quadratic equation

$$x^2 - 8x + 1 = 0.$$

To complete the square, you need to find a perfect square with first two terms identical to those of $x^2 - 8x + 1$.

Solution

Consider $(x-4)^2 = x^2 - 8x + 16$

$$x^2 - 8x + 1 = (x-4)^2 - 16 + 1$$
$$= (x-4)^2 - 15$$

To solve $x^2 - 8x + 1 = 0$, this is equivalent to solving:

$(x-4)^2 - 15 = 0.$

$\Rightarrow (x-4)^2 = 15$

$\Rightarrow (x-4) = \pm\sqrt{15}$

$\Rightarrow x = 4 \pm \sqrt{15}$

You choose -4 because this is half of the coefficient of x, then consider $(x-4)^2$.

When you have written the expression in this form, you have completed the square.

4

EXERCISE 4B

1 Express each of the following quadratics in the form $(x+p)^2 + q$, where p and q are constants (some of the values may be negative).

(a) $x^2 + 8x + 19$, **(b)** $x^2 + 4x + 13$,

(c) $x^2 + 10x + 14$, **(d)** $x^2 - 10x + 30$,

(e) $x^2 - 8x + 3$, **(f)** $x^2 + 3x + 3$,

(g) $x^2 + x + 1$, **(h)** $x^2 - 5x + 7$,

(i) $x^2 - x + 2$, **(j)** $x^2 - 7x - 2$.

2 Express each of the following equations in the completed square form. Hence find the coordinates of the vertex and an equation for the line of symmetry for each parabola.

(a) $y = x^2 + 4x + 12$, **(b)** $y = x^2 + 12x + 40$,

(c) $y = x^2 - 6x + 2$, **(d)** $y = x^2 + 8x + 5$,

(e) $y = x^2 - 2x - 3$, **(f)** $y = x^2 - 14x + 32$,

(g) $y = x^2 + x + 3$, **(h)** $y = x^2 - 3x + 2$,

(i) $y = x^2 - 5x + 1$, **(j)** $y = x^2 - 9x + 15$.

3 By completing the square, solve the following quadratic equations, giving your answers in surd form.

(a) $x^2 + 4x - 3 = 0$, **(b)** $x^2 + 6x + 4 = 0$,

(c) $x^2 - 8x + 5 = 0$, **(d)** $x^2 - 2x - 4 = 0$,

(e) $x^2 - 10x - 3 = 0$, **(f)** $x^2 - 14x + 4 = 0$,

(g) $x^2 - x - 1 = 0$, **(h)** $x^2 + 3x - 5 = 0$,

(i) $x^2 - 3x + 1 = 0$, **(j)** $x^2 - 7x - 1 = 0$,

(k) $x^2 - 5x + 3 = 0$, **(l)** $x^2 - x - 3 = 0$.

Worked example 4.5

So far, all the examples have involved quadratics where the coefficient of x^2 has been 1. You can extend the procedure for more general quadratics.

(a) Complete the square for each of the following quadratics.

(i) $3x^2 + 6x + 17$, **(ii)** $12x - 2x^2 + 11$.

(b) Hence determine:

(i) the least value of $3x^2 + 6x + 17$ and the value of x at which this occurs,

(ii) the greatest value of $12x - 2x^2 + 11$ and the value of x at which this occurs.

Solution

(a) (i) $3x^2 + 6x + 17 = 3[x^2 + 2x] + 17$

Make the coefficient of x^2 equal to 1 by taking out the factor 3 from the first two terms.

Completing the square on $x^2 + 2x$

$= 3[(x + 1)^2 - 1^2] + 17$

$= 3(x + 1)^2 - 3 \times 1 + 17$

$= 3(x + 1)^2 + 14$

You can multiply this out and check it gives you the original expression.

(ii) $12x - 2x^2 + 11 = -2[x^2 - 6x] + 11$

Taking factor of −2 out of first two terms.

Completing the square on $x^2 - 6x$

$= -2[(x - 3)^2 - 9] + 11$

$= -2(x - 3)^2 + 18 + 11$

$= -2(x - 3)^2 + 29$

(b) (i) The least value of $(x + 1)^2$ is zero and occurs when $x = -1$.

This least value occurs when $x = -1$.

Hence the least value of $3x^2 + 6x + 17 = 3(x + 1)^2 + 14$ is 14.

(ii) The least value of $(x - 3)^2$ is zero.

Hence the **greatest** value of $12x - 2x^2 + 11 = 29 - 2(x - 3)^2$ is 29 and occurs when $x = 3$.

Notice this is the greatest value of the expression since the squared term is being subtracted.

Worked example 4.6

This technique can be used to find the shortest distance from a point to a straight line.

The distance from the point $A(-4, 3)$ to a general point $P(x, y)$ on the straight line with equation $y = 3x - 5$ is d.
Find an expression for d^2, in terms of x only.

By completing the square on the expression for d^2, find the closest point on the line $y = 3x - 5$ to the point $(-4, 3)$ and the shortest distance from the point to the line.

Solution

Since $y = 3x - 5$, the point P has coordinates $(x, 3x - 5)$

See chapter 3 section 3.2
$d = \sqrt{(x_2 - x_1)^2 + (y_2 - y_1)^2}$

$$\begin{aligned} d^2 = AP^2 &= (x + 4)^2 + ([3x - 5] - 3)^2 \\ &= (x + 4)^2 + (3x - 8)^2 \\ &= (x^2 + 8x + 16) + (9x^2 - 48x + 64) \\ &= 10x^2 - 40x + 80 \end{aligned}$$

Since A is the point $(-4, 3)$.

You need to complete the square on $10x^2 - 40x + 80$

$$\begin{aligned} 10x^2 - 40x + 80 &= 10(x^2 - 4x) + 80 \\ &= 10[(x - 2)^2 - 4] + 80 \\ &= 10(x - 2)^2 - 40 + 80 \\ &= 10(x - 2)^2 + 40 \end{aligned}$$

The least value of $(x - 2)^2$ is 0 and occurs when $x = 2$.

Least value of the expression for d^2 is 40.

Therefore the closest distance of the point to the line is $\sqrt{40} = 2\sqrt{10}$.

When $x = 2$, $y = 3x - 5 = 6 - 5 = 1$.
Closest point on the line to A is the point $(2,1)$.

4

EXERCISE 4C

1 Complete the square for each of the following

(a) $3x^2 + 6x - 2$, **(b)** $5x^2 + 40x + 7$,

(c) $2x^2 + 12x - 1$, **(d)** $4x^2 + 8x - 11$,

(e) $5x^2 + 5x + 11$, **(f)** $3x^2 + 9x - 8$,

(g) $2x^2 + 5x + 1$, **(h)** $3x^2 + 4x + 7$,

(i) $4x^2 + 7x - 2$.

2 (a) Express $3x^2 - 12x + 5$ in the form $A(x - B)^2 - C$, where A, B and C are constants whose values should be stated.

(b) Find the minimum value of each of the following expressions:

(i) $3x^2 - 12x + 5$,

(ii) $(3x^2 - 12x + 5)^2$.

(c) For each of the expressions in **(b)**, state a value of x which gives the minimum value. [A]

3 (a) Determine the value of each of the constants a and b such that $a - (3x + b)^2 = 1 - 30x - 9x^2$.

(b) Hence, or otherwise, determine the greatest value of $f(x) = 1 - 30x - 9x^2$, and state the value of x for which this greatest value occurs. [A]

4 A parabola has equation $y = ax^2 + bx + c$, where a, b and c are constants. Show that the axis of the parabola has equation $2ax + b = 0$, and find the coordinates of its vertex. Hence show that the vertex lies on the x-axis when $b^2 = 4ac$.

5 The function f is given by $f(x) = 3x^2 + 12x + 17$.

(a) Express $f(x)$ in the form $p(x+q)^2 + r$.

(b) Hence write down the least value of $f(x)$ and the value of x at which it occurs.

(c) Deduce the greatest value of $\dfrac{1}{f(x)}$.

6 Using the method of worked example 4.6, find the shortest distance from the point A to the given line in the cases:

(a) $A(5, 4)$ from $y = 2x + 9$,

(b) $A(-1, 2)$ from $y = x + 5$,

(c) $A(5, 4)$ from $x + y = 7$.

7 A piece of wire has length 20 cm and it is bent into a rectangle with one side equal to x cm. Show that the area of the rectangle, A cm^2, is given by $A = x(10 - x)$.

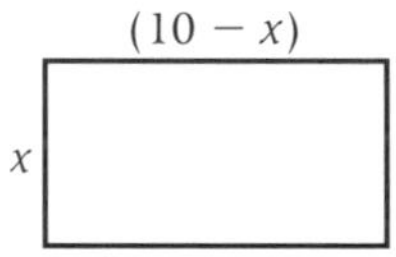

Complete the square on the expression for A and hence find the greatest value of A and the value of x for which this occurs.

(Was this value of x the value you expected?)

8 A farmer has 120 m of fencing. By using an existing wall, he decides to make a rectangular pen by adding three sides, with two sides equal to x m as shown in the diagram. Show that the area enclosed, A m^2, is given by $A = x(120 - 2x)$.

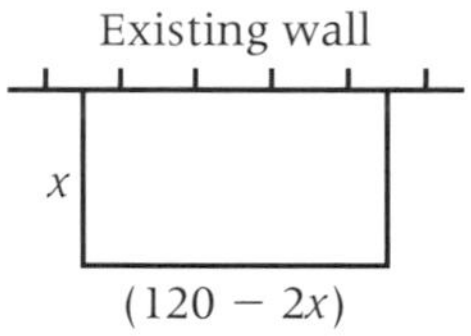

Complete the square on the expression for A and hence find the greatest value of A and the value of x for which this occurs.

(Was this value of x the value you expected?)

9 A pen is built onto an existing rectangular building (shaded) where AF is 10 units and FE is 5 units, as shown in the diagram. The perimeter $ABCDE$ is 65 units. Find a relationship between x and y, where $AB = x$ and $DE = y$.

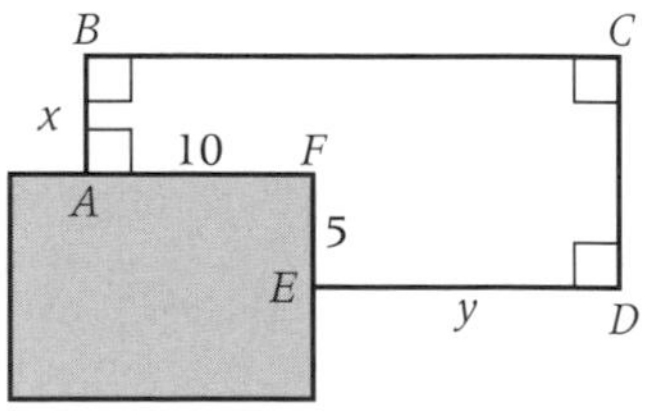

Show that the area enclosed by the pen is $125 + 30x - x^2$.

By completing the square, find the greatest possible area enclosed by the pen.

4.4 Translations parallel to the *x*-axis

The graph of $y = x^2$ is sketched in diagram **(a)** and the graph of $y = (x - 2)^2$ is shown in diagram **(b)**.

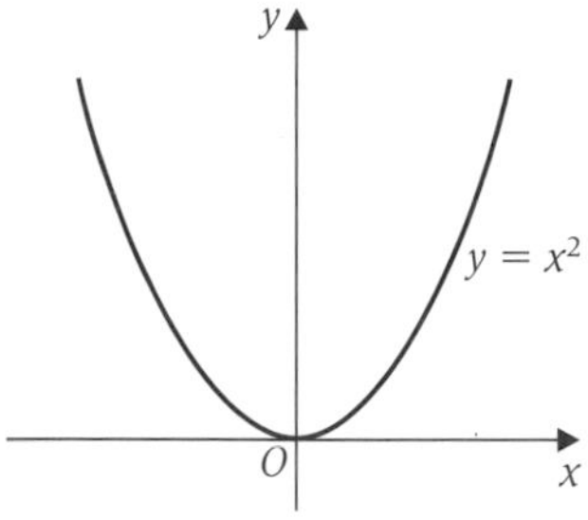

(a)

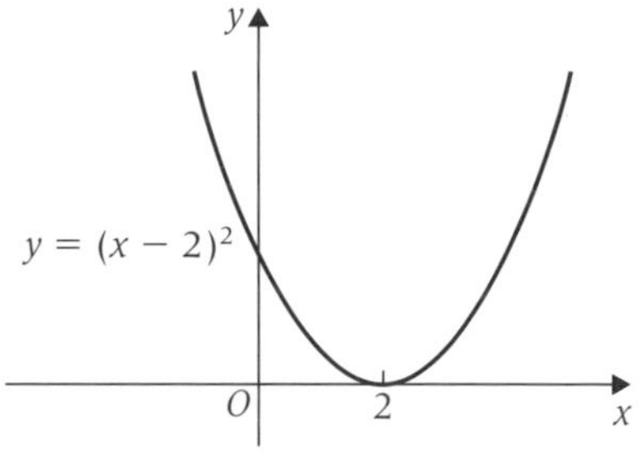

(b)

The vertex of $y = (x - 2)^2$ is at $(2, 0)$, whereas the vertex of $y = x^2$ is at the origin $(0, 0)$. A translation by the vector $\begin{bmatrix} 2 \\ 0 \end{bmatrix}$ moves the vertex of parabola **(a)** to the vertex of parabola **(b)**. In fact every point on $y = x^2$ is mapped onto the corresponding point on $y = (x - 2)^2$ by a translation with vector $\begin{bmatrix} 2 \\ 0 \end{bmatrix}$.

Similarly a translation with vector $\begin{bmatrix} -3 \\ 0 \end{bmatrix}$ maps the parabola $y = x^2$ onto the new parabola $y = (x + 3)^2$.

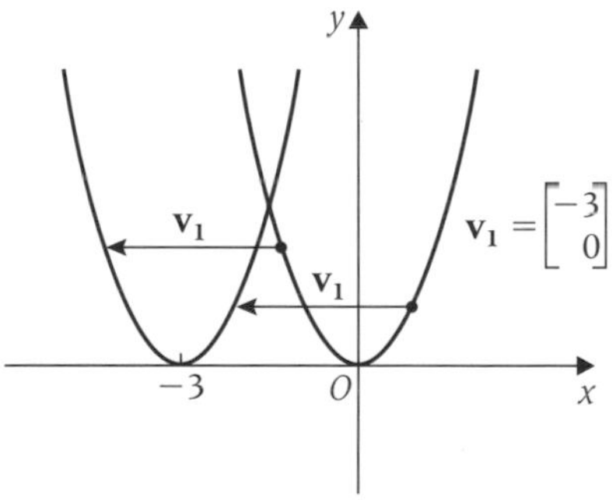

> In general, a translation of $\begin{bmatrix} a \\ 0 \end{bmatrix}$ transforms the graph of $y = f(x)$ into the graph of $y = f(x - a)$.

Worked example 4.7

The graph of $y = (x - 5)^2$ is translated by $\begin{bmatrix} -4 \\ 0 \end{bmatrix}$. Find the equation of the new graph.

Solution

Replace the variable x by $x - (-4) = x + 4$.

New graph has equation $y = (\{x + 4\} - 5)^2 = (x - 1)^2$.

4.5 General translations of quadratic graphs

Where is the vertex of the parabola with equation $y = (x - 4)^2 + 5$?

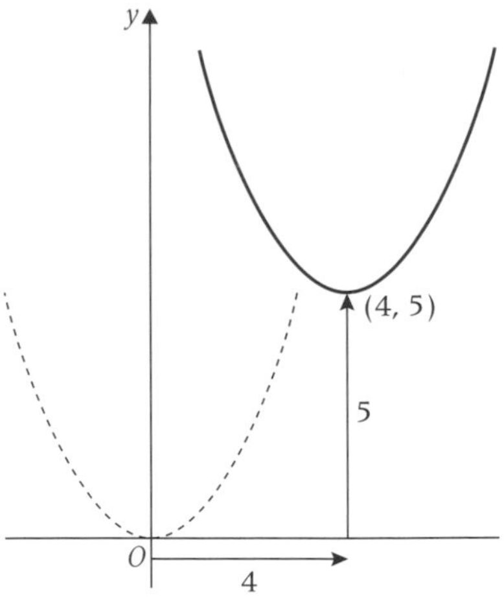

From what you discovered in section 4.3, the vertex is at (4, 5).

In order to draw its graph you could draw the graph of $y = x^2$ and translate it through the vector $\begin{bmatrix} 4 \\ 5 \end{bmatrix}$.

The reason for this is seen if you rearrange the equation in the form $y - 5 = (x - 4)^2$.

> In general, a translation of $\begin{bmatrix} a \\ b \end{bmatrix}$ transforms the graph of $y = \mathrm{f}(x)$ into the graph of $y - b = \mathrm{f}(x - a)$.

EXERCISE 4D

1 The following graphs have been transformed from the graph of $y = x^2$ by a translation. State the vector of the translation in each case.

A useful strategy is to find the vertex on each parabola that must correspond to the vertex of $y = x^2$, which is at the origin.

(a) $y = (x - 7)^2$, **(b)** $y = (x + 1)^2$,
(c) $y = (x - 6)^2$, **(d)** $y = (x + 2)^2$,
(e) $y - 5 = (x - 1)^2$, **(f)** $y = 3 + (x + 2)^2$,
(g) $y = (x - 1)^2 - 7$, **(h)** $y = 3 + (x + 8)^2$.

2 Find the equation of the graph $y = x^2$ after it has been translated by the vector given in each case.

(a) $\begin{bmatrix} 4 \\ 0 \end{bmatrix}$, **(b)** $\begin{bmatrix} 0 \\ 3 \end{bmatrix}$, **(c)** $\begin{bmatrix} 2 \\ 5 \end{bmatrix}$,
(d) $\begin{bmatrix} -1 \\ -1 \end{bmatrix}$, **(e)** $\begin{bmatrix} 3 \\ 2 \end{bmatrix}$, **(f)** $\begin{bmatrix} -1 \\ -3 \end{bmatrix}$,
(g) $\begin{bmatrix} -3 \\ -4 \end{bmatrix}$, **(h)** $\begin{bmatrix} 1 \\ -5 \end{bmatrix}$, **(i)** $\begin{bmatrix} 3 \\ 4 \end{bmatrix}$.

3 **(a)** Express $x^2 - 4x + 7$ in the form $(x - a)^2 + b$.

(b) Give the vector of the translation that transforms the first parabola into the second parabola for each of the cases below:

(i) $y = x^2$; $y = x^2 - 4x + 7$,
(ii) $y = x^2 - 4x + 7$; $y = x^2$,
(iii) $y = (x - 2)^2$; $y = x^2 - 4x + 7$,
(iv) $y = x^2 + 3$; $y = x^2 - 4x + 7$,
(v) $y = x^2 - 4x + 7$; $y = x^2 - 6x + 10$.

4.6 General quadratic equation formula

The general quadratic equation is $ax^2 + bx + c = 0$, $(a \neq 0)$.

$$\Rightarrow x^2 + \frac{b}{a}x + \frac{c}{a} = 0$$

Divide throughout by a since $a \neq 0$.

$$\Rightarrow \left(x + \frac{b}{2a}\right)^2 - \left(\frac{b}{2a}\right)^2 + \frac{c}{a} = 0$$

Completing the square.

$$\Rightarrow \left(x + \frac{b}{2a}\right)^2 = \frac{b^2}{4a^2} - \frac{c}{a}$$

$$\Rightarrow \left(x + \frac{b}{2a}\right)^2 = \frac{b^2 - 4ac}{4a^2}$$

$$\Rightarrow x + \frac{b}{2a} = \frac{\pm\sqrt{b^2 - 4ac}}{\sqrt{4a^2}}$$

Note the need for $\pm$ when you take square roots of both sides.

$$\Rightarrow x = \frac{-b}{2a} \pm \frac{\sqrt{b^2 - 4ac}}{2a}$$

$$\Rightarrow x = \frac{-b \pm \sqrt{b^2 - 4ac}}{2a}$$

This is a formula which needs to be learned off by heart so you can use it to solve quadratic equations which do not factorise.

Worked example 4.8

Find the exact solutions of the equation $3x^2 - 7x - 5 = 0$.

Solution

Comparison with $ax^2 + bx + c = 0$ gives $a = 3$, $b = -7$, $c = -5$.

Using $x = \dfrac{-b \pm \sqrt{b^2 - 4ac}}{2a}$

$$x = \frac{-(-7) \pm \sqrt{(-7)^2 - 4(3)(-5)}}{2(3)} = \frac{7 \pm \sqrt{49 + 60}}{6} = \frac{7 \pm \sqrt{109}}{6}$$

Answers should be left in this exact form since that was the request.

$$x = \frac{7 + \sqrt{109}}{6} \quad \text{or} \quad x = \frac{7 - \sqrt{109}}{6}$$

You can make a check on your calculator by taking one of the roots, e.g. $x = \dfrac{7 + \sqrt{109}}{6} \approx 2.9067$ and substituting into $3x^2 - 7x - 5 \approx 3 \times 2.9067^2 - 7 \times 2.9067 - 5 \approx -1.85 \times 10^{-4}$.

This is very close to zero and confirms that your answer is correct.

EXERCISE 4E

When the final values are in surd form, you will find that the two answers are **conjugate pairs**, such as $-3 \pm \sqrt{17}$ whenever a, b and c are rational numbers.

1 Find the exact values of the roots of the following equations, (in some cases there are no real roots and you should say so):

(a) $x^2 - 3x - 1 = 0$, **(b)** $2x^2 + 7x - 1 = 0$,

(c) $5x^2 - 7x + 3 = 0$, **(d)** $4x^2 - 5x - 2 = 0$,

(e) $7x^2 + 2x + 1 = 0$, **(f)** $5x^2 + 3x - 1 = 0$,

(g) $2x^2 - 5x + 1 = 0$, **(h)** $3x^2 + 2x + 1 = 0$,

(i) $4x^2 + 3x - 7 = 0$.

2 Find the exact solutions of the following equations:

(a) $x^2 - 4x - 1 = 0$, **(b)** $2x^2 + 5x - 6 = 0$,

(c) $4x^2 - 13x + 5 = 0$, **(d)** $5x^2 - 3x - 3 = 0$,

(e) $4x^2 + 7x + 2 = 0$, **(f)** $5x^2 + 7x - 1 = 0$,

(g) $2x^2 - 11x + 7 = 0$, **(h)** $3x^2 + 5x + 1 = 0$,

(i) $4x^2 + 7x - 1 = 0$.

4.7 The discriminant

The expression $b^2 - 4ac$ that appears in the quadratic equation formula is known as the **discriminant**, because it discriminates between the types of solutions the quadratic equation can have.

When $b^2 - 4ac$ is a perfect square, the solutions will be rational and it means that the original quadratic expression will factorise.

Worked example 4.9

Find the discriminant of each of the following equations and hence state whether the equation has rational or irrational roots:

(a) $3x^2 - 4x - 5 = 0$,

(b) $6x^2 - x - 12 = 0$.

Solution

(a) Comparing with $ax^2 + bx + c = 0$ gives $a = 3$, $b = -4$, $c = -5$.
Hence the discriminant $b^2 - 4ac = (-4)^2 - 4 \times 3 \times (-5)$ $= 16 + 60 = 76$.
Since 76 is **not** a perfect square the roots are irrational.

You could apply the formula $x = \dfrac{-b \pm \sqrt{b^2 - 4ac}}{2a}$ to find these irrational roots.

They are $\dfrac{4 \pm \sqrt{76}}{6}$ or $\dfrac{2 \pm \sqrt{19}}{3}$ (since $\sqrt{76} = 2\sqrt{19}$).

(b) In this case $a = 6$, $b = -1$, $c = -12$.
Hence the discriminant $b^2 - 4ac = (-1)^2 - 4 \times 6 \times (-12)$
$= 1 + 288 = 289$.
Since the discriminant is a perfect square, $17^2 = 289$, the roots of the quadratic are rational.

You could use the formulae to find the roots, but, since the discriminant is a perfect square, the quadratic will factorise.
$6x^2 - x - 12 = (3x + 4)(2x - 3) = 0$.
Hence, either $3x = -4$ or $2x = 3$.

The roots are $-\frac{4}{3}, \frac{3}{2}$ which are rational as predicted.

The number of real roots of the equation $ax^2 + bx + c = 0$ is determined by the discriminant.

$b^2 - 4ac > 0 \Rightarrow$ the equation has two real distinct roots
$b^2 - 4ac = 0 \Rightarrow$ the equation has one real (repeated) root
$b^2 - 4ac < 0 \Rightarrow$ the equation has no real roots

The graphical situation is illustrated below.

$b^2 - 4ac > 0$
Two points of intersection with x-axis.

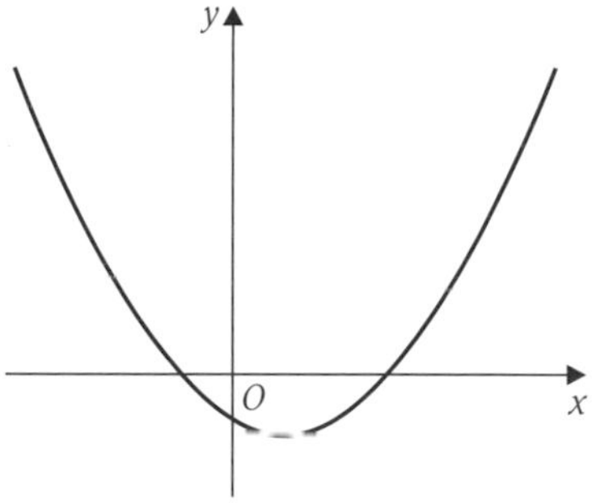

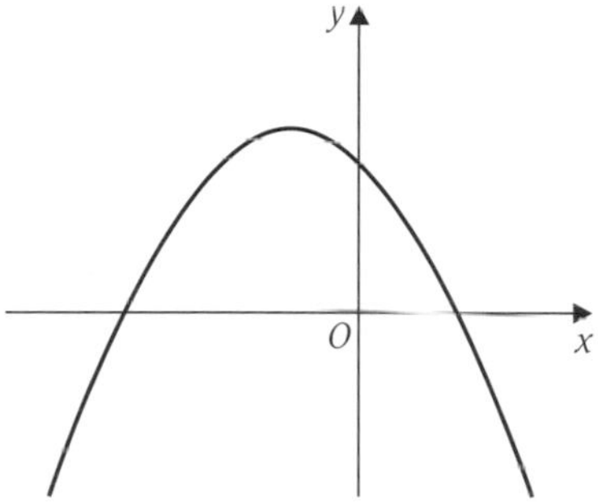

$b^2 - 4ac = 0$
Single point of intersection; curve touches x-axis.

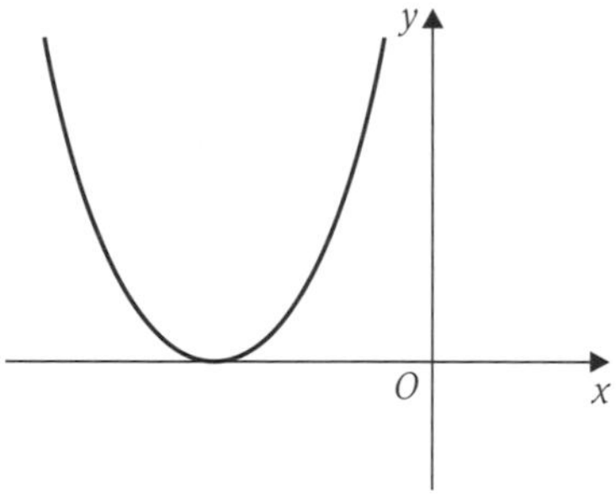

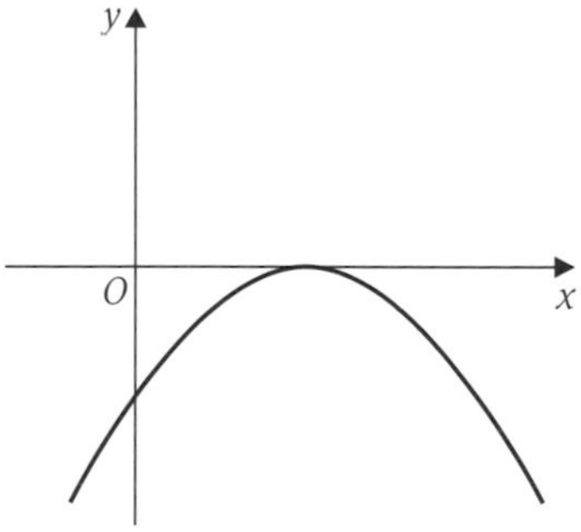

$b^2 - 4ac < 0$
No points of intersection with x-axis,

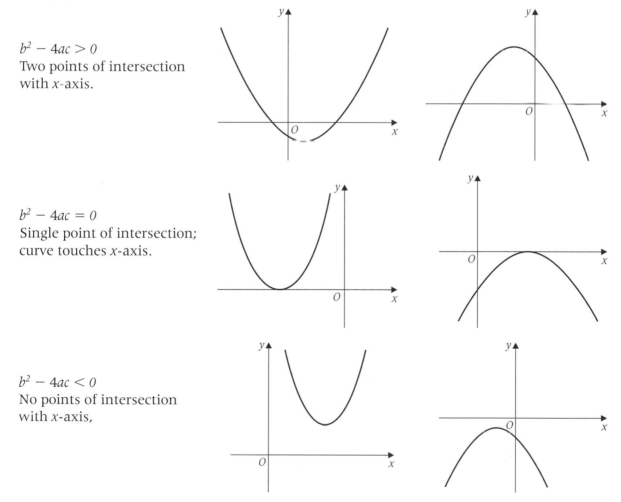

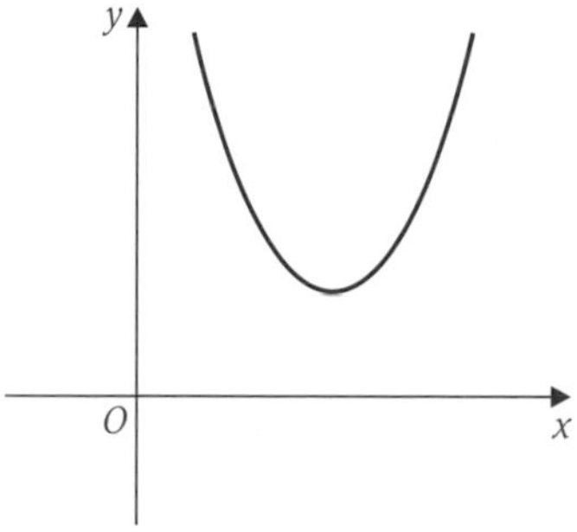

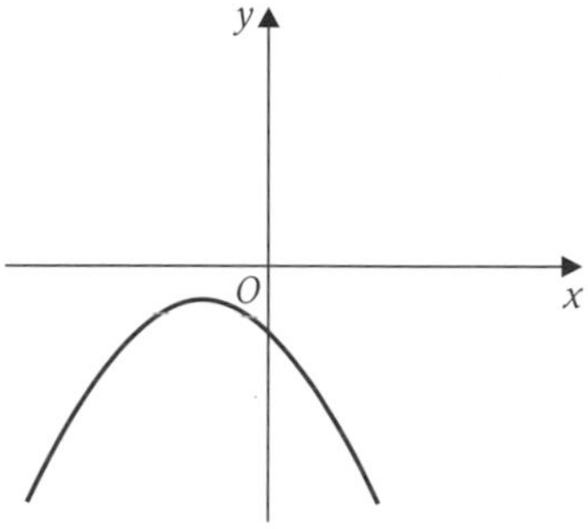

Worked example 4.10

Find the values of k for which the equation $kx^2 - 6x + (8 + k) = 0$ has equal roots.

This is another way of saying the equation has one (repeated) root.

Solution

For equal roots $(-6)^2 - 4k(8 + k) = 0$

Using $b^2 - 4ac = 0$.

$\Rightarrow 36 - 4k(8 + k) = 0 \Rightarrow 9 - k(8 + k) = 0$

$\Rightarrow 9 - 8k - k^2 = 0$

$\Rightarrow (9 + k)(1 - k) = 0$

$\Rightarrow k = -9$ or $k = 1$

Worked example 4.11

Find the condition on k for the equation $12x^2 - 8x + (5 + k) = 0$ to have two distinct real roots.

Solution

For two distinct real roots $(-8)^2 - 4(12)(5 + k) > 0$

Using $b^2 - 4ac > 0$.

$\Rightarrow 64 - 48(5 + k) > 0$

$\Rightarrow 64 > 48(5 + k) \Rightarrow 4 > 3(5 + k)$

$\Rightarrow 4 > 15 + 3k \Rightarrow -11 > 3k$

$\Rightarrow -\frac{11}{3} > k$ or $k < -\frac{11}{3}$.

EXERCISE 4F

You need to check whether the discriminant is a perfect square, in which case the roots are rational.

1 Find the discriminant of each of the following equations and hence state whether the equation has rational or irrational roots:

(a) $x^2 - 4x - 12 = 0$, **(b)** $x^2 - 5x - 6 = 0$,
(c) $x^2 - 3x - 7 = 0$, **(d)** $x^2 - 7x + 8 = 0$,
(e) $x^2 - 7x - 8 = 0$, **(f)** $2x^2 - 5x + 3 = 0$,
(g) $5x^2 - 7x + 2 = 0$, **(h)** $2x^2 - 3x - 4 = 0$,
(i) $24x^2 + 6x - 25 = 0$.

2 Obtain an inequality satisfied by k if the equations below have two distinct real roots:

(a) $x^2 - 2x - k = 0$, **(b)** $(3 + k)x^2 - 4x + 4 = 0$,
(c) $x^2 - 5x - 6k = 0$, **(d)** $(2 + k)x^2 - 7x - 8 = 0$,
(e) $5x^2 - 7x + (2k - 3) = 0$, **(f)** $(3k - 1)x^2 - 5x + 3 = 0$.

3 The following equations have repeated roots. Find the value(s) of the constant p, leaving your answer(s) in exact form.

(a) $px^2 - 5x + p = 0$, **(b)** $(3 - p)x^2 - 6x + 9 = 0$,
(c) $x^2 + 3px + 12 = 0$, **(d)** $(2p - 5)x^2 - 8x + 16 = 0$,
(e) $2x^2 - (p + 1)x + 8 = 0$, **(f)** $(p - 1)x^2 + 6x + (p + 7) = 0$.

4 Determine, by considering the discriminant, the number of points of intersection with the x-axis of the following parabolas.

(a) $y = x^2 - 7x + 3$, **(b)** $y = 3x^2 - 7x + 5$,
(c) $y = 3x^2 + 12x + 12$, **(d)** $y = 5x^2 - x - 5$,
(e) $y = 9x^2 - 6x + 1$, **(f)** $y = 4x^2 - 12x + 5$,
(g) $y = 5 - 7x - 2x^2$, **(h)** $y = 5 - 3x^2$,
(i) $y = 3x + 2x^2$, **(j)** $y = 3 + 2x^2$.

MIXED EXERCISE

1 (a) (i) Express $x^2 + 12x + 11$ in the form $(x + a)^2 + b$, finding the values of a and b.

(ii) State the minimum value of the expression $x^2 + 12x + 11$.

(b) Determine the values of k for which the quadratic equation

$$x^2 + 3(k - 2)x + (k + 5) = 0$$

has equal roots. [A]

2 (a) Express $2x^2 + 8x + 7$ in the form $A(x + B)^2 + C$, where A, B and C are constants.

(b) (i) State the minimum value of $2x^2 + 8x + 7$.

(ii) State the value of x which gives this minimum value. [A]

3 (a) Solve the equation $x^2 + 4x - 14 = 0$, giving your answers in the form $p + q\sqrt{2}$ where p and q are integers.

(b) Express $x^2 + 4x - 14$ in the form $(x + a)^2 - b$ and hence find the least value of the expression $x^2 + 4x - 14$.

4 The function f is defined by $f(x) = (1 - 2x)(1 + 2x)$.

(a) Find the coordinates of the points where the graph of $y = f(x)$ cuts the coordinate axes.

(b) Sketch the graph of $y = f(x)$.

(c) The graph of $y = f(x)$ is translated by 1 unit in the positive x-direction and 4 units in the positive y-direction to give the graph of $y = g(x)$. Find an expression for $g(x)$ in the form $ax^2 + bx + c$, where a, b and c are integers. [A]

5 The quadratic function f is defined by $f(x) = x^2 + 8x + 13$.

(a) (i) Express $f(x)$ in the form $(x + p)^2 + q$, where p and q are integers.

(ii) Write down the least value of $f(x)$ and the value of x for which this least value occurs.

(b) (i) State the geometrical transformation which maps the graph of $y = f(x)$ onto the graph of $y = f(x - 3)$.

(ii) Find an expression for $f(x - 3)$ in the form $ax^2 + bx + c$, where a, b and c are integers.

6 **(a)** Solve the equation $2x^2 + 32x + 119 = 0$. Write your answers in the form $p + q\sqrt{2}$, where p and q are rational numbers.

(b) **(i)** Express $2x^2 + 32x + 119$ in the form $2(x + m)^2 + n$, where m and n are integers.

(ii) Hence write down the minimum value of $2x^2 + 32x + 119$. [A]

Key point summary

1 An expression of the form $ax^2 + bx + c$ is a quadratic and the graph of $y = ax^2 + bx + c$ is called a parabola . *p52*

2 A quadratic equation that can be written in the form $(x - p)(x - q) = 0$ has roots p and q. *p53*

The graph of $y = (x - p)(x - q)$ crosses the x-axis at the points $(p, 0)$ and $(q, 0)$.

3 A quadratic can be written in the form $A(x + B)^2 + C$ and this is called the completed square form. *p55*

This form enables you to find the greatest or least values of the quadratic.

4 In general, a translation of $\begin{bmatrix} a \\ 0 \end{bmatrix}$ transforms the graph of $y = \mathrm{f}(x)$ into the graph of $y = \mathrm{f}(x - a)$. *p61*

5 The formula $x = \dfrac{-b \pm \sqrt{b^2 - 4ac}}{2a}$ can be used to find the solutions to any quadratic equation and this formula must be learned off by heart. *p63*

6 The expression $b^2 - 4ac$ is called the discriminant. *p65*

When $b^2 - 4ac$ is a perfect square, the roots of the quadratic equation are rational and the quadratic will factorise.

When $b^2 - 4ac > 0$, the quadratic equation has two distinct real roots.

When $b^2 - 4ac = 0$, the quadratic equation has one (repeated) root. This condition is sometimes called having equal roots.

When $b^2 - 4ac < 0$, the quadratic equation has no real roots.

Test yourself	What to review
1 Find the points where the curve with equation $y = 2x^2 - 3x + 1$ crosses the x-axis.	*Section 4.2*
2 A parabola has equation $y = x^2 - 8x + 11$. By completing the square, find the coordinates of its vertex.	*Section 4.3*
3 Express $f(x) = 3x^2 + 18x + 20$ in the form $a(x + b)^2 + c$ and hence find the least value of $f(x)$.	*Section 4.3*
4 Find the exact solutions of the equation $x^2 - 8x + 3 = 0$.	*Section 4.6*
5 Describe geometrically the transformation that maps the graph of $y = x^2$ onto the graph of $y = x^2 - 4x + 1$.	*Section 4.4*
6 Use the discriminant to state the number of real roots of each of the following equations. **(a)** $9x^2 - 6x + 1 = 0$, **(b)** $8x^2 - 6x + 1 = 0$, **(c)** $9x^2 - 5x + 1 = 0$.	*Section 4.7*
7 Find the condition on k for the equation $x^2 - 2x + (k - 3) = 0$ to have two real distinct roots.	*Section 4.7*

Test yourself ANSWERS

1 $(1, 0)$, $(\frac{1}{2}, 0)$.

2 $(4, -5)$.

3 $a = 3$, $b = 3$, $c = -7$; least value is -7.

4 $4 \pm \sqrt{13}$.

5 A translation with vector $\begin{bmatrix} 2 \\ -3 \end{bmatrix}$.

6 **(a)** 1; **(b)** 2; **(c)** 0.

7 $k < 4$.

CHAPTER 5

C1: Polynomials

Learning objectives

After studying this chapter, you should be able to:

- recognise polynomials
- add and subtract polynomials
- multiply polynomials
- find coefficients of polynomials
- compare or equate coefficients of polynomials.

5.1 Introduction and definitions

You should be familiar with linear and quadratic expressions such as these

- $3x - 4$ (linear)
- $x^2 + 5x - 10$ (quadratic)

You may also have met cubic expressions like

- $x^3 + 2x^2 - 6x - 2$

These types of expressions can be extended to include higher powers of x such as

- $x^5 + 3x^3 - 2x^2 + 3$
- $4x^6 + 2x^4 - x^3 + 5$

Polynomials do not have negative or fractional powers.

Expressions of this type are called **polynomials**.

An expression of the form

$$ax^n + bx^{n-1} + \ldots + px^2 + qx + r$$

(where $a, b, \ldots, p, q, r$ are constants)
is called a polynomial in x.

n is a non-negative integer.

The polynomial $2x^3 - 5x^2 + 3x + 1$ is written in *descending powers* of x. Sometimes it is more convenient to write the polynomial in *ascending powers* of x , namely in the form $1 + 3x - 5x^2 + 2x^3$. Another equivalent form is $3x - 5x^2 + 2x^3 + 1$, although it is more usual to write a polynomial in either ascending powers of x or descending powers of x.

> The **degree** of a polynomial is given by the highest power of the variable.

- $3x^5 + 3x^4 - 2x^2 - 1$ is a polynomial of degree 5.

So:

a **constant** expression is a polynomial of degree 0
a **linear** expression is a polynomial of degree 1
a **quadratic** expression is a polynomial of degree 2
a **cubic** expression is a polynomial of degree 3
etc.

5.2 Adding and subtracting polynomials

When adding or subtracting polynomials you will be primarily concerned with looking for 'like' terms

> Like terms are those which contain the same letter(s) and power.

- $4x^2$ and $7x^2$ are like terms because they both contain the same power of x.
- $5x^4$ and $-2x^4$ are like terms.
- $8x^3$ and $2x^5$ are not like terms since they have different powers of x.
- $3x^6$ and $2p^6$ are not like terms since they have different letters.

'like' terms can be collected together.

> Two or more polynomials can be added or subtracted by **'collecting like terms'**.

Worked example 5.1

Add together the two polynomials $5x^3 + 7x$ and $2x^3 - 3x$.

Solution

$(5x^3 + 7x) + (2x^3 - 3x)$
$= 5x^3 + 7x + 2x^3 - 3x$

Remove the brackets.

$- (5x^3 + 2x^3) + (7x - 3x)$
$= 7x^3 + 4x$

Collect like terms.

5 lots of x^3 plus 2 lots of x^3 makes 7 lots of x^3

7 lots of x minus 3 lots of x makes 4 lots of x

You must take care with the signs when subtracting polynomials. Remember: subtracting a negative number is the same as adding the corresponding positive number.

Worked example 5.2

Subtract $2x^4 - x^2$ from $6x^4 + 3x^2$.

Solution

$(6x^4 + 3x^2) - (2x^4 - x^2)$
$= 6x^4 + 3x^2 - 2x^4 + x^2$

Remove the brackets – take care with the signs.

$= (6x^4 - 2x^4) + (3x^2 + x^2)$
$= 4x^4 + 4x^2$

Collect like terms.

Worked example 5.3

Simplify:

(a) $(4x^3 + 5x^2 + 8x + 2) + (2x^3 - 2x^2 - 5x + 10)$,

(b) $(3t^5 + 2t^3 + 6t^2 - 3t) - (t^5 + 4t^4 - t^2 + 3)$

Solution

(a) $(4x^3 + 5x^2 + 8x + 2) + (2x^3 - 2x^2 - 5x + 10)$
$= 4x^3 + 5x^2 + 8x + 2 + 2x^3 - 2x^2 - 5x + 10$
$= (4x^3 + 2x^3) + (5x^2 - 2x^2) + (8x - 5x) + (2 + 10)$
$= 6x^3 + 3x^2 + 3x + 12$

(b) $(3t^5 + 2t^3 + 6t^2 - 3t) - (t^5 + 4t^4 - t^2 + 3)$
$= 3t^5 + 2t^3 + 6t^2 - 3t - t^5 - 4t^4 + t^2 - 3$
$= 2t^5 - 4t^4 + 2t^3 + 7t^2 - 3t - 3$

Notice that the variable does not have to be x. These are polynomials in t.

Notation

Polynomials are often represented in the same way as functions which you met in section 1.5.

- a polynomial can be denoted by $P(x)$ [or $Q(x)$, $f(x)$, etc.].

Worked example 5.4

Two polynomials are given by $P(x) = 4x^5 - 2x^2 + 8x - 4$ and $Q(x) = 6x^5 + 4x^2 - 2x + 6$.

Simplify:

(a) $P(x) + Q(x)$,

(b) $Q(x) - P(x)$,

(c) $3P(x) - 2Q(x)$.

Solution

(a) $P(x) + Q(x) = (4x^5 - 2x^2 + 8x - 4) + (6x^5 + 4x^2 - 2x + 6)$
$= 4x^5 - 2x^2 + 8x - 4 + 6x^5 + 4x^2 - 2x + 6$
$= 10x^5 + 2x^2 + 6x + 2$

(b) $Q(x) - P(x) = (6x^5 + 4x^2 - 2x + 6) - (4x^5 - 2x^2 + 8x - 4)$
$= 6x^5 + 4x^2 - 2x + 6 - 4x^5 + 2x^2 - 8x + 4$
$= 2x^5 + 6x^2 - 10x + 10$

(c) $3P(x) - 2Q(x) = 3(4x^5 - 2x^2 + 8x - 4) - 2(6x^5 + 4x^2 - 2x + 6)$
$= 12x^5 - 6x^2 + 24x - 12 - 12x^5 - 8x^2 + 4x - 12$
$= -14x^2 + 28x - 24$

EXERCISE 5A

1 Determine which of the following expressions are polynomials. For each polynomial, state its degree.

(a) $3 - 4x$, **(b)** $\sqrt{x} + 4x^2$, **(c)** $x^3 - 3x + 2x^{-2}$,

(d) $y^5 - 7$, **(e)** $2 - k - k^4$, **(f)** $x^3 + x - x^6 + 5x^2$,

(g) 5, **(h)** $\dfrac{4}{x^3 + 4x - 7}$.

2 Add together the two polynomials $4x^3 + 3x^2 - 2x + 10$ and $x^3 + 5x^2 + 3x - 7$.

3 Add $2x^4 - 7x^2 + x - 20$ to $5x^4 + 2x^3 - 3x^2 + 25$.

4 Subtract $3z^2 + 5z + 2$ from $7z^2 + 9z + 3$.

5 Subtract $5x^3 + 2x^2 - 3x - 10$ from $8x^3 + 6x^2 - 5x + 2$.

6 Simplify the following:

(a) $(x^3 + 4x + 5) + (3x^3 - 3x - 7)$,

(b) $(8h^4 - 5h^2 + 2) + (2h^4 + 3h^3 + 4)$,

(c) $(9x^6 + 4x^5 - 3x^3 + 2x - 1) + (4x^5 - 2x^3 - 5x)$,

(d) $3(5x^2 + 6x - 2) - (2x^2 + 7x + 3)$,

(e) $(4t^5 + 2t^4 - 9t^2 + 10) - 2(2t^5 - 3t^4 - t^2 + 12)$,

(f) $7(3x^3 + 9x^2 - 3x + 12) - 3(7x^3 + 7x^2 - 4x + 12)$.

7 Simplify the following:

(a) $(x^2 + 4x - 2) - (5x + 6) + (2x^2 + 3)$,

(b) $(5y^3 + 5y - 2) + 2(2y^2 + 3y - 4) - 5(y^3 + y^2)$,

(c) $(4x^3 + 3x^2 + 2x - 4) + 4(x^2 - 3x) - 3(x^3 - 2x^2 + 2)$,

(d) $(8q^2 + 2q - 5) - 4(2q^2 + 4q - 9) + 3(2q^3 - 3q^2 + 2q - 1)$.

5

8 Three polynomials are given by:
$P(x) = 3x^4 + 8x^3 - 2x^2 + 7x - 3$,
$Q(x) = 5x^3 - 4x^2 - 6x + 15$ and
$R(x) = 6x^2 - 7x - 2$

Simplify the following as far as possible:

(a) $P(x) + Q(x)$,

(b) $Q(x) - R(x)$,

(c) $2R(x) - P(x)$,

(d) $P(x) + 3Q(x) - 2R(x)$.

5.3 Multiplying polynomials

You will be familiar with the idea of multiplying two linear expressions together.

Suppose you want to expand and simplify

$$(x + 5)(x - 2)$$

The expression can be expanded as

$$\begin{aligned}(x + 5)(x - 2) &= x(x - 2) + 5(x - 2)\\ &= x^2 - 2x + 5x - 10\\ &= x^2 + 3x - 10\end{aligned}$$

The basic method consists of multiplying everything in the second bracket by each term in the first bracket in turn.

This basic method can be extended to multiply polynomials of any degree.

You must take careful note of the signs, remembering that

$$\begin{aligned}+ \times + &= +\\ - \times + &= -\\ + \times - &= -\\ - \times - &= +\end{aligned}$$

Worked example 5.5

Multiply the two polynomials $x + 4$ and $x^2 + 3x - 5$, simplifying your answer as far as possible.

Solution

$$\begin{aligned}(x + 4)(x^2 + 3x - 5) &= x(x^2 + 3x - 5) + 4(x^2 + 3x - 5)\\ &= x^3 + 3x^2 - 5x + 4x^2 + 12x - 20\\ &= x^3 + 7x^2 + 7x - 20\end{aligned}$$

This technique makes use of what is called the distributive law in algebra.

Worked example 5.6

Multiply the two polynomials $x^2 + 2x - 3$ and $2x^2 - 5x + 1$. Simplify your answer as far as possible.

Solution

$$\begin{aligned}&(x^2 + 2x - 3)(2x^2 - 5x + 1)\\&= x^2(2x^2 - 5x + 1) + 2x(2x^2 - 5x + 1) - 3(2x^2 - 5x + 1)\\&= 2x^4 - 5x^3 + x^2 + 4x^3 - 10x^2 + 2x - 6x^2 + 15x - 3\\&= 2x^4 - x^3 - 15x^2 + 17x - 3\end{aligned}$$

Worked example 5.7

Simplify $(t + 2)(t - 3)(t + 5)$

Solution

Firstly $(t + 2)(t - 3) = t^2 - 3t + 2t - 6 = t^2 - t - 6$

So that $(t + 2)(t - 3)(t + 5) = (t^2 - t - 6)(t + 5)$

$$\begin{aligned}&= t^2(t + 5) - t(t + 5) - 6(t + 5)\\&= t^3 + 5t^2 - t^2 - 5t - 6t - 30\\&= t^3 + 4t^2 - 11t - 30\end{aligned}$$

5.4 Finding coefficients

In the example above, the **coefficient** of t^3 is 1 , the coefficient of t^2 is 4 and the coefficient of t is -11. The number -30 is referred to as the **constant term**, but it is also the coefficient of t^0.

> For the polynomial $P(x) = ax^n + bx^{n-1} + \ldots + px^2 + qx + r$, the coefficient of x^n is a, …, the coefficient of x is q and the constant term is r.

Worked example 5.8

Find the coefficient of x^2 in the polynomial given by:

$$(x^3 + 2x^2 + 4x - 1)(x^3 - 3x^2 + 5x - 7)$$

> It is not necessary to find the full polynomial. Only the term involving x^2 is required.

Solution

Clearly if x^3 is multiplied by any term in the second bracket it is impossible to get a term in x^2.

The $2x^2$ term from the first bracket can multiply by the -7 in the second bracket to give $-14x^2$.

The term in x^2 is therefore given by:

$$\begin{aligned}&(2x^2) \times (-7) + (4x) \times (5x) + (-1) \times (-3x^2)\\&= -14x^2 + 20x^2 + 3x^2 = 9x^2\end{aligned}$$

> Remember that the coefficient is actually a constant – in this case the number in front of x^2.

The coefficient of x^2 is therefore 9.

Worked example 5.9

Two polynomials are multiplied together and the answer is $x^3 + 12x^2 + 34x - 12$. One of the polynomials is $x + 6$, find the other polynomial.

Solution

Firstly, call the unknown polynomial $P(x)$.

$$(x + 6) \times P(x) = x^3 + 12x^2 + 34x - 12$$

You should also notice that $P(x)$ must be a quadratic expression.

Since linear × quadratic = cubic.

So, let $P(x) = ax^2 + bx + c$ and your job is to find the values of a, b and c.

Now
$$\begin{aligned}(x + 6)(ax^2 + bx + c) &= x(ax^2 + bx + c) + 6(ax^2 + bx + c)\\ &= ax^3 + bx^2 + cx + 6ax^2 + 6bx + 6c\\ &= ax^3 + (b + 6a)x^2 + (c + 6b)x + 6c\end{aligned}$$

But this must be **identically equal to** the answer given by the question

$$\Rightarrow \quad ax^3 + (b + 6a)x^2 + (c + 6b)x + 6c \equiv x^3 + 12x^2 + 34x - 12$$

Since these two expressions must be the same, the number of x^3s, x^2s, xs and constants must be the same on both sides.

You can now compare the number of x^3s, etc. on each side. This process is called **comparing** or **equating coefficients.**

Equate coefficients of x^3: $a = 1$

Equate coefficients of x^2: $b + 6a = 12$

But you know that $a = 1$

$\Rightarrow$ $b + 6 = 12 \Rightarrow b = 6$

Equate coefficients of x: $c + 6b = 34$

But you know that $b = 6$

$\Rightarrow$ $c + 36 = 34 \Rightarrow c = -2$

Equate constant terms: $6c = -12 \Rightarrow c = -2$

(This confirms the earlier result.)

Therefore $a = 1$, $b = 6$, $c = -2$.

Hence,

$$P(x) = x^2 + 6x - 2.$$

This technique can be streamlined a little. For instance, you probably realised immediately that if

$$x^3 + 12x^2 + 34x - 12 \equiv (x + 6)(ax^2 + bx + c)$$

then since the coefficient of x^3 is 1, the value of a must be 1.

Similarly the constant term is -12 and this must come from $6 \times c$. So $c = -2$.

It is only really necessary to use the method above to find the value of b. This is illustrated in the next example.

Worked example 5.10

Find the polynomial Q(x) for which
$(2x - 3) \times Q(x) \equiv 6x^3 + x^2 - 23x + 12$.

Solution

The polynomial Q(x) must be a quadratic, so let

$$Q(x) = ax^2 + bx + c.$$

Hence $(2x - 3) \times (ax^2 + bx + c) \equiv 6x^3 + x^2 - 23x + 12$.

Comparing the coefficients of x^3: $2a = 6 \Rightarrow a = 3$.
Comparing the constant terms: $-3c = 12 \Rightarrow c = -4$.

Comparing the constant term is equivalent to putting $x = 0$ on both sides.

Consider $(2x - 3)(3x^2 + bx - 4) \equiv 6x^3 + x^2 - 23x + 12$.

The term in x^2 comes from $-3 \times 3x^2 + 2x \times bx = (-9 + 2b)x^2$.
Hence $-9 + 2b = 1 \Rightarrow b = 5$.

So $Q(x) = 3x^2 + 5x - 4$.

This can be checked by considering the coefficient of x in $(2x - 3)(3x^2 + 5x - 4)$, namely $2 \times (-4) + (-3) \times 5 = -23$ ✓

5

EXERCISE 5B

1 Multiply $(x + 4)$ by $(2x + 3)$.

2 Multiply $(x - 2)$ by $(3x - 7)$.

3 Multiply $(x^2 + 2x - 5)$ by $(x + 2)$.

4 Multiply $(y - 3)$ by $(3y^2 - y + 4)$.

5 Multiply $(k - 7)$ by $(-4k^2 + 2k - 5)$.

6 Multiply $(x^2 + 3x - 6)$ by $(3x^2 + 4x - 5)$.

7 Expand the following and simplify as far as possible:

(a) $(x - 4)(2x^2 + 4x - 12)$,
(b) $(8 - x)(3x^2 - 9x + 1)$,
(c) $(e^2 + 6e + 7)(e^2 - 5e - 3)$,
(d) $(4x^2 + 3x - 6)(x^3 + 2x^2 - 9x + 5)$,
(e) $(-6t^2 + 2t - 5)(3t^3 - 3t^2 + 8t + 11)$,
(f) $(x + 1)(x - 3)(x + 6)$,
(g) $(2 - x)(x + 4)(10 - x)$.

8 Find the coefficient of x^2 in each of the following:

(a) $(3x^2 + 5x - 7)(x^3 + 2x^2 - 4x + 2)$,
(b) $(2x^2 + 6x - 1)(x^3 - 3x^2 - 6x + 5)$,
(c) $(2x^3 + 3x - 1)(x^2 - 4x + 7)$,
(d) $(3 - x + 2x^2 + 5x^3)(2 - 3x - x^2)$.

9 Find the coefficient of x^3 in each of the polynomials in question **8**.

10 Find the constant term in each of the polynomials in question **8**.

11 In each of the following, find the polynomial $P(x)$:

(a) $(x^2 - 2x + 1) \times P(x) \equiv x^3 + 3x^2 - 9x + 5$,

(b) $(x + 2) \times P(x) \equiv x^3 + 5x^2 + 8x + 4$,

(c) $(x + 5) \times P(x) \equiv x^3 + 10x^2 + 22x - 15$,

(d) $(x - 3) \times P(x) \equiv 2x^3 - 19x + 3$,

(e) $(2x - 1) \times P(x) \equiv 8x^3 + 6x^2 - 19x + 7$.

12 (a) Given that $f(x) = (x - 1)(x + 3)(x + 7)(2x - 1)$, find all the possible values of a for which $f(a) = 0$.

(b) Given that $g(-5) = g(2) = 0$ and that

$$g(x) = (x - b)(x + c)(x^2 + 7x - 1),$$

where b and c are positive integers, find b and c.

Key point summary

1 An expression of the form $ax^n + bx^{n-1} + \dots + px^2 + qx + r$ (where $a, b, \dots, p, q, r$ are constants) is called a polynomial in x. *p70*

2 The **degree** of a polynomial is given by the highest power of the variable. *p71*

3 Polynomials can be added or subtracted by collecting **'like'** terms. *p71*

4 Two polynomials can be multiplied by taking the second bracket and multiplying it by each term in the first bracket. *p74*

5 For the polynomial $P(x) = ax^n + bx^{n-1} + \dots + px^2 + qx + r$, the coefficient of x^n is $a, \dots$, the coefficient of x is q and the constant term is r. *p75*

Test yourself	What to review
1 Simplify the following as far as possible: **(a)** $(3x^4 - 5x^3 + 7x^2 + 9x - 11) + (2x^4 - 3x^3 - 4x^2 - 2x + 1)$, **(b)** $(4x^3 + 3x^2 - 3x + 15) - (x^3 - 2x^2 + 4x - 2)$, **(c)** $(5x^5 + 2x^3 - 7x + 1) + (3x^4 + 6x^3) - (x^3 - 3x^2 + 1)$.	*Section 5.2*
2 Expand these brackets and simplify the answers as far as possible: **(a)** $(x - 2)(3x^2 + 2x - 6)$ **(b)** $(x^2 - 3x + 4)(2x^2 + 5x - 8)$.	*Section 5.3*
3 Find the coefficient of x^3 in the product $(4x^3 + x^2 - 3x + 1)(x^3 - 2x^2 + 3x - 2)$.	*Section 5.4*
4 Find the polynomial $P(x)$ if $(x + 2) \times P(x) = 2x^3 - 7x + 2$.	*Section 5.4*

Test yourself ANSWERS

1 **(a)** $5x^4 - 8x^3 + 3x^2 + 7x - 10$;
(b) $3x^3 + 5x^2 - 7x + 17$;
(c) $5x^5 + 3x^4 + 7x^3 + 3x^2 - 7x$.

2 **(a)** $3x^3 - 4x^2 - 10x + 12$;
(b) $2x^4 - x^3 - 15x^2 + 44x - 32$.

3 2.

4 $P(x) = 2x^2 - 4x + 1$.

CHAPTER 6

C1: Factors, remainders and cubic graphs

Learning objectives

After studying this chapter, you should be able to:

- recall and use the factor theorem
- factorise cubic polynomials having a factor of the form $(x - a)$
- sketch graphs of cubic functions
- divide a polynomial of degree less than 4 by a linear expression of the form $(x - a)$
- recall and use the remainder theorem for polynomials of degree less than 4 and divisor $(x - a)$.

6.1 Factorisation

Factorising is the reverse of expanding brackets.

Expanding $(x + 2)(x + 3)$ gives $x^2 + 5x + 6$

Reversing this process gives

$$x^2 + 5x + 6 = (x + 2)(x + 3)$$

$(x + 2)$ and $(x + 3)$ are said to be **factors** of $x^2 + 5x + 6$.

You will be very familiar with factorising certain types of polynomials already. For example,

- $x^3 + 2x^2 + 5x = x(x^2 + 2x + 5)$
- $x^2 - 9 = (x + 3)(x - 3)$
- $x^2 + x - 42 = (x + 7)(x - 6)$

$x^2 - 9$ is a **difference of two squares.**

Sometimes a mixture of the above techniques has to be used in order to factorise an expression completely.

$$x^3 + 8x^2 + 15x = x(x^2 + 8x + 15)$$

x is a common factor.

but the expression $x^2 + 8x + 15$ can itself be factorised

$$x^3 + 8x^2 + 15x = x(x + 3)(x + 5)$$

However, you need a method for factorising expressions such as $x^3 - 4x^2 - 7x + 10$. This is discussed in the next section.

6.2 The factor theorem

Consider $(x - 1)(x - 2)(x + 4)$

If you expand this you get the polynomial

$$P(x) = x^3 + x^2 - 10x + 8.$$

Given P(x), how can you work backwards and discover that the three linear factors are $(x - 1)$, $(x - 2)$ and $(x + 4)$?

The secret lies in noticing what happens when you substitute different values of x into the polynomial P(x).

Substituting $x = 1$ into P(x) gives:

$$P(1) = 1^3 + 1^2 - 10(1) + 8 = 1 + 1 - 10 + 8 = 0.$$

So substituting $x = 1$ into P(x) gives the answer 0. This is the key to factorising. The reason why the answer of 0 is so important becomes clearer when you consider P(x) written in factorised form.

$$P(x) = (x - 1)(x - 2)(x + 4)$$

When you substitute $x = 1$ into this form it is not difficult to see that the value of the first bracket is 0 and, hence, the overall value of P(x) will be 0

i.e. $P(1) = (1 - 1) \times (1 - 2) \times (1 + 4)$
$= 0 \times -1 \times 5$
$= 0$

You should also be able to see that substituting $x = 2$ and $x = -4$ gives an answer of 0 since the value of the second and third brackets, respectively, will be 0.

$P(2) = (2 - 1) \times (2 - 2) \times (2 + 4)$
$= 1 \times 0 \times 6$
$= 0$

and $P(-4) = (-4 - 1) \times (-4 - 2) \times (-4 + 4)$
$= -5 \times -6 \times 0$
$= 0$

This idea can be generalised and is known as the **factor theorem**.

> The **factor theorem** states that:
>
> $(x - a)$ is a factor of the polynomial $P(x) \Leftrightarrow P(a) = 0$.

In the exam for this module the factor theorem will not be applied directly to any polynomial of degree greater than 3.

The $\Leftrightarrow$ sign indicates that the result works both ways, i.e.

if $(x - a)$ is a factor of P(x) then P(a) = 0
or if P(a) = 0 then $(x - a)$ is a factor of P(x).

Since you will not be allowed to use a calculator in the exam for this module you should ensure that you know the values of x^3 and x^2 for integer values of x from -5 to $+5$.

Worked example 6.1

Use the factor theorem to show that $(x - 2)$ is a factor of $x^3 + x^2 - 7x + 2$.

Solution

Let $P(x) = x^3 + x^2 - 7x + 2$.
To show that $x - 2$ is a factor of $P(x)$ you must show that $P(2) = 0$.

$$\begin{aligned} P(2) &= 2^3 + 2^2 - 7(2) + 2 \\ &= 8 + 4 - 14 + 2 \\ &= 0 \end{aligned}$$

$P(2) = 0 \Rightarrow x - 2$ is a factor of $x^3 + x^2 - 7x + 2$.

Worked example 6.2

Substitute $x = 2$, $x = 3$, $x = 1$ and $x = -1$ into the polynomial $P(x) = x^3 - 4x^2 + x + 6$ and hence write down the three linear factors of $P(x)$.

Solution

$P(2) = 2^3 - 4(2)^2 + 2 + 6 = 8 - 16 + 2 + 6 = 0$
$P(2) = 0 \Rightarrow (x - 2)$ is a factor of $P(x)$

$P(3) = 3^3 - 4(3)^2 + 3 + 6 = 27 - 36 + 3 + 6 = 0$
$P(3) = 0 \Rightarrow (x - 3)$ is a factor of $P(x)$

$P(1) = (1)^3 - 4(1)^2 + 1 + 6 = 1 - 4 + 1 + 6 = 4$
$P(1) \neq 0 \Rightarrow (x - 1)$ is **not** a factor of $P(x)$

$P(-1) = (-1)^3 - 4(-1)^2 - 1 + 6 = -1 - 4 - 1 + 6 = 0$
$P(-1) = 0 \Rightarrow (x + 1)$ is a factor of $P(x)$

Therefore, the three linear factors of $P(x)$ are $(x - 2)$, $(x - 3)$ and $(x + 1)$.

Worked example 6.3

The polynomial $f(x) = x^3 + qx^2 + 5x - 12$ has $(x + 4)$ as a factor. Find the value of q.

Solution

If $(x + 4)$ is a factor then $f(-4) = 0$

$\Rightarrow \quad (-4)^3 + q(-4)^2 + 5(-4) - 12 = 0$

$\Rightarrow \quad -64 + 16q - 20 - 12 = 0$

$\Rightarrow \quad 16q = 96$

$\Rightarrow \quad q = 6$

Worked example 6.4

The polynomial $x^3 + ax^2 + bx - 20$ has factors $x + 2$ and $x - 5$. Find the value of a and the value of b.

It is a matter of personal preference as to whether the factor is contained within brackets such as $(x + 2)$ or without brackets such as $x + 2$. It is good to get used to seeing it used in either form in questions.

Solution

Let $P(x) = x^3 + ax^2 + bx - 20$.

Since $x + 2$ is a factor, $P(-2) = 0 \Rightarrow -8 + 4a - 2b - 20 = 0$
$\Rightarrow 4a - 2b = 28$

Since $x - 5$ is a factor, $P(5) = 0 \Rightarrow 125 + 25a + 5b - 20 = 0$
$\Rightarrow 25a + 5b = -105$

These two equations can be simplified to $2a - b = 14$
and $5a + b = -21$

Adding gives $7a = -7 \Rightarrow a = -1$. Hence $b = -16$.

EXERCISE 6A

1 Use the factor theorem to show that $(x - 2)$ is a factor of the polynomial $f(x) = x^3 - 5x + 2$.

2 Use the factor theorem to show that $(x + 1)$ is a factor of the polynomial $P(x) = x^3 - x^2 - 5x - 3$.

3 A polynomial is given by $P(t) = t^3 - 2t^2 - 17t + 10$. Use the factor theorem to show that $(t - 5)$ is a factor of $P(t)$.

4 The polynomial $P(x) = x^3 + 8x^2 + 16x + k$ has $(x + 3)$ as a factor. Find the value of k.

5 Given that $(x - 4)$ is a factor of the polynomial $f(x) = x^3 + px^2 - 3x - 4$, find the value of p.

6 The polynomial $P(x) = x^3 - 4x^2 + x + k$ has $(x - 2)$ as one of its factors.

(a) Find the value of k.

(b) Verify that $(x + 1)$ and $(x - 3)$ are also factors of $P(x)$.

7 Prove that $(x - 2)$ is a factor of $P(x) = x^3 + kx^2 - 2kx - 8$ for all values of the constant k.
Given that $(x + 1)$ is also a factor of $P(x)$, determine the value of k.

8 Given that $(x + 3)$ is a factor of $Q(x) = (x - 1)(x + 5)(x + 7) + c$, find the value of the constant c.

9 The polynomial $p(x) = x^3 - 3ax^2 + bx - 6$ has factors $(x - 1)$ and $(x - 3)$. Find the value of each of the constants a and b.

10 Find the value of each of the constants a and b for which $(x + 4)$ and $(x - 1)$ are each factors of the polynomial $x^3 + ax^2 + bx + 24$.

6.3 Further factorisation

Several of the techniques considered earlier can be used to factorise a polynomial completely.

Worked example 6.5

Use the factor theorem to show that $x + 3$ is a factor of $P(x) = x^3 - 6x + 9$. Hence factorise $P(x)$.

Solution

$P(-3) = (-3)^3 - 6(-3) + 9 = -27 + 18 + 9 = 0$
$P(-3) = 0 \Rightarrow x + 3$ is a factor of $x^3 - 6x + 9$.

To show that $x + 3$ is a factor of $P(x)$ you must show that $P(-3) = 0$.

You can write $x^3 - 6x + 9 \equiv (x + 3)(ax^2 + bx + c)$ and, by comparing the coefficients of x^3, you know that $a = 1$.

Similarly from the constant terms, $9 = 3c \Rightarrow c = 3$.

Hence $x^3 - 6x + 9 \equiv (x + 3)(x^2 + bx + 3)$.

The coefficient of x^2 on the left is zero.

$\Rightarrow 0 = b + 3 \Rightarrow b = -3$.

Alternatively, the coefficient of x on the right is $3 + 3b$. Hence $-6 = 3 + 3b$, giving the same answer $b = -3$.

$P(x) = (x + 3)(x^2 - 3x + 3)$.

Note. The discriminant of the quadratic $x^2 - 3x + 3$ is $(-3)^2 - 4 \times 1 \times 3 = -3$ and so the quadratic cannot be factorised.

Worked example 6.6

A polynomial is given by $P(x) = x^3 + 3x^2 - 6x - 8$.
Use the factor theorem to show that $x + 4$ is a factor of $P(x)$.
Given that $P(x)$ has two other linear factors with integer coefficients, factorise $P(x)$ completely.

Solution

$P(-4) = (-4)^3 + 3(-4)^2 - 6(-4) - 8 = -64 + 48 + 24 - 8 = 0$
$P(-4) = 0 \Rightarrow x + 4$ is a factor of $P(x)$.

To find the remaining factors, you need to consider the constant term. Suppose $P(x) = (x + 4)(x - a)(x - b)$.

Remember that the constant term is easily found by putting $x = 0$.

The constant term is given by $4 \times a \times b$.

Hence $4ab = -8 \Rightarrow ab = -2$.

Since a and b are whole numbers they must be ± 1 or ± 2.

Try $P(1) = 1 + 3 - 6 - 8 = -10$ so $x - 1$ is **not** a factor.

$P(-1) = -1 + 3 + 6 - 8 = 0 \Rightarrow x + 1$ is a factor.

$P(2) = 8 + 12 - 12 - 8 = 0 \Rightarrow x - 2$ is a factor.

You can write the factors in any order.

So that $P(x) = (x + 4)(x + 1)(x - 2)$.

Worked example 6.7

A polynomial is given by $P(x) = x^3 - 4x^2 - 5x + 24$.

(a) Use the factor theorem to show that $(x - 3)$ is a factor of $P(x)$ and express $P(x)$ in the form $(x - 3)Q(x)$, where $Q(x)$ is a quadratic.

(b) Find the exact solutions to the equation $P(x) = 0$.

Solution

(a) $P(3) = (3)^3 - 4(3)^2 - 5(3) + 24$
$= 27 - 36 - 15 + 24 = 0$

$P(3) = 0 \Rightarrow (x - 3)$ is a factor of $P(x)$.

Let $Q(x) = ax^2 + bx + c$.

Thus $(x - 3) \times Q(x) = x^3 - 4x^2 - 5x + 24$.

Now $(x - 3)(ax^2 + bx + c) = x(ax^2 + bx + c) - 3(ax^2 + bx + c)$
$= ax^3 + (b - 3a)x^2 + (c - 3b)x - 3c$

Thus $ax^3 + (b - 3a)x^2 + (c - 3b)x - 3c \equiv x^3 - 4x^2 - 5x + 24$.

Equate coefficients of x^3: $a = 1$

Equate coefficients of x^2: $b - 3a = -4$
$\Rightarrow b - 3 = -4 \Rightarrow b = -1$

Equate coefficients of x: $c - 3b = -5$
$\Rightarrow c + 3 = -5 \Rightarrow c = -8$

Equate constant terms: $-3c = 24 \quad \Rightarrow c = -8$

Thus $Q(x) = x^2 - x - 8$,

so that $P(x) = (x - 3)(x^2 - x - 8)$.

Thus P(x) has two factors, one linear $(x - 3)$ and one quadratic $(x^2 - x - 8)$.

(b) $P(x) = 0$

$\Rightarrow (x - 3)(x^2 - x - 8) = 0$

$\Rightarrow x - 3 = 0 \quad$ or $\quad x^2 - x - 8 = 0$

$\Rightarrow x = 3 \quad$ or $\quad x = \dfrac{1 \pm \sqrt{(-1)^2 - 4(1)(-8)}}{2}$

$x = \dfrac{1 \pm \sqrt{33}}{2}$.

The three solutions are $x = 3$, $x = \dfrac{1 + \sqrt{33}}{2}$ and $x = \dfrac{1 - \sqrt{33}}{2}$.

With practice, you will almost be able to write the quadratic factor down right away by realising that the coefficient of x^2 must be 1; the constant term must be -8; and with a little inspection of coefficients the quadratic must be $x^2 - x - 8$.

> In factorising a cubic polynomial, once a linear factor has been found, the remaining quadratic factor can be found by comparing coefficients.

Worked example 6.8

Factorise the polynomial $P(x) = x^3 + 5x^2 - 2x - 24$ completely.

Solution

You have no initial factor to work from here so you have to resort to searching for a first factor by substituting values into $P(x)$ and looking for 0 as the answer.

> The secret is to look at the constant term.
>
> Any linear factors must therefore be factors of -24. You need only try $\pm 1, \pm 2, \pm 3, \pm 4, \pm 6, \pm 8$, etc.
>
> In this case, it is possible to find all three factors by such a search, since $P(2) = 0$, $P(-3) = 0$ and $P(-4) = 0$. Hence $P(x) = (x - 2)(x + 3)(x + 4)$.

$P(1) = 1^3 + 5(1)^2 - 2(1) - 24 = 1 + 5 - 2 - 24 = -20$
$P(1) \neq 0 \Rightarrow (x - 1)$ is not a factor of $P(x)$

$P(-1) = (-1)^3 + 5(-1)^2 - 2(-1) - 24 = -1 + 5 + 2 - 24 = -18$
$P(-1) \neq 0 \Rightarrow (x + 1)$ is not a factor of $P(x)$

$P(2) = (2)^3 + 5(2)^2 - 2(2) - 24 = 8 + 20 - 4 - 24 = 0$
$P(2) = 0 \Rightarrow (x - 2)$ is a factor of $P(x)$

$(x - 2)$ is your first factor.

Thus $(x - 2) \times Q(x) = x^3 + 5x^2 - 2x - 24$
where $Q(x) = ax^2 + bx + c$

Now $(x - 2)(ax^2 + bx + c) = x^3 + 5x^2 - 2x - 24$

Clearly $a = 1$
and equating constant terms gives $-2c = -24 \Rightarrow c = 12$

$$(x - 2)(x^2 + bx + 12) = x^3 + 5x^2 - 2x - 24$$

Equating coefficients of x^2: $\Rightarrow b - 2 = 5 \Rightarrow b = 7$

> Checking the coefficient of x: $12 - 2b = -2 \Rightarrow b = 7$. ✓

Thus $Q(x) = x^2 + 7x + 12$.

This will factorise further $Q(x) = x^2 + 7x + 12 = (x + 3)(x + 4)$

so that $P(x) = (x - 2)(x + 3)(x + 4)$.

Sometimes the factorisation of a cubic leads to repeated linear factors. In such cases it is not possible to find three different values of a such that $P(a) = 0$. The next worked example considers such a case.

Worked example 6.9

(a) Factorise the polynomial $p(x) = x^3 + 6x^2 + 12x + 8$.

(b) Hence solve the equation $y^6 + 6y^4 + 12y^2 + 8 = 27$.

Solution

Factors of the constant 8 are ± 1, ± 2 and ± 4.
Clearly if x is positive, $p(x)$ will be greater than 8 and so
try $p(-1) = (-1)^3 + 6(-1)^2 + 12(-1) + 8 = -1 + 6 - 12 + 8 = 1$
$p(-1) \neq 0 \Rightarrow (x + 1)$ is not a factor of $p(x)$.

try $p(-2) = (-2)^3 + 6(-2)^2 + 12(-2) + 8 = -8 + 24 - 24 + 8 = 0$
$p(-2) = 0 \Rightarrow (x + 2)$ is a factor of $p(x)$.

Since for $p(x)$ the coefficient of x^3 is 1 and the constant term is 8 and $x + 2$ is a factor you can deduce that

$$(x + 2)(x^2 + kx + 4) = x^3 + 6x^2 + 12x + 8$$

Equating coefficients of x: $4 + 2k = 12 \Rightarrow k = 4$

$\Rightarrow$ $p(x) = (x + 2)(x^2 + 4x + 4) = (x + 2)[(x + 2)(x + 2)]$

$\Rightarrow$ $p(x) = (x + 2)^3$.

(b) Comparing $x^3 + 6x^2 + 12x + 8$

with $y^6 + 6y^4 + 12y^2 + 8$

you can see that the two expressions are identical if $x = y^2$.

Using **(a)**, the equation $y^6 + 6y^4 + 12y^2 + 8 = 27$

can be written as $(y^2 + 2)^3 = 27$

$\Rightarrow$ $y^2 + 2 = 3 \Rightarrow y^2 = 1 \Rightarrow y = \pm 1$

EXERCISE 6B

1 Use the factor theorem to show that $(x + 1)$ is a factor of the polynomial $f(x) = x^3 + 2x^2 - x - 2$ and hence factorise $f(x)$ completely.

2 A polynomial is given by $f(x) = x^3 + 3x^2 - 18x - 40$.

(a) Use the factor theorem to show that $(x + 5)$ is a factor of $f(x)$.

(b) Factorise $f(x)$ completely and hence find the solutions to the equation $x^3 + 3x^2 - 18x - 40 = 0$.

3 A polynomial is given by $P(x) = x^3 - 8x^2 + 9x + 18$.

(a) Use the factor theorem to show that $(x - 3)$ is a factor of $P(x)$.

(b) Factorise $P(x)$ completely and hence solve the equation $P(x) = 0$.

4 Use the factor theorem to verify that $(x - 1)$ is a factor of the polynomial $P(x) = x^3 + x^2 - 6x + 4$. Factorise $P(x)$ completely and, hence, find the exact solutions of the equation $x^3 + x^2 - 6x + 4 = 0$.

5 A polynomial is given by $P(x) = x^3 + kx^2 - 10x - 24$, where k is a constant.

(a) Given that $x + 2$ is a factor of $P(x)$, find the value of k.

(b) Factorise $P(x)$ completely.

6 A polynomial is given by $f(x) = x^3 - 3x^2 - 6x + 8$. Factorise $f(x)$ completely.

7 **(a)** Find a linear and a quadratic factor of the polynomial $P(x) = x^3 - 27$.

(b) Factorise $x^3 - 8$.

(c) Factorise $x^3 - a^3$.

8 (a) Find a linear and a quadratic factor of the polynomial $Q(x) = x^3 + 64$.

(b) Factorise $x^3 + 125$.

(c) Factorise $x^3 + b^3$.

9 The polynomial $P(x)$ has factors $x - 1$ and $x + 2$, where $P(x) = x^3 + cx^2 + dx - 8$, and c and d are constants.

(a) Find the value of c and the value of d.

(b) Express $P(x)$ as the product of linear factors.

10 A polynomial is given by $p(x) = x^3 - 5x^2 - 33x - 27$.

(a) Use the factor theorem to show that $x + 3$ is a factor of $P(x)$.

(b) (i) Express $P(x)$ as a product of three linear factors.

(ii) Hence factorise $y^6 - 5y^4 - 33y^2 - 27$ completely.

11 A polynomial is given by $p(x) = x^3 - 6x^2 + 9x - 4$.

(a) Factorise $P(x)$ completely.

(b) Hence solve the equation $y^6 - 6y^4 + 9y^2 - 4 = 0$.

6.4 Graphs of cubic functions

In chapter 4 you sketched the graphs of quadratic functions. In this section you will sketch the graphs of cubic functions.

The phrase 'cubic functions' may be replaced by 'cubic polynomials'.

The simplest cubic function is x^3.

The graph of $y = x^3$ has rotational symmetry of order 2 about the origin.

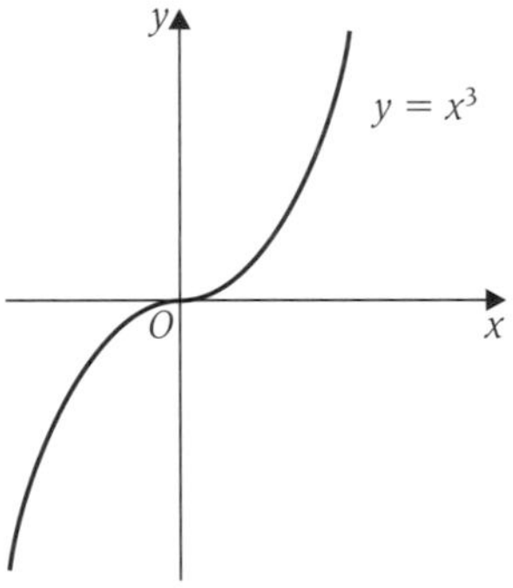

The graph of any cubic function, where the coefficient of x^3 is positive, has the shape of the graph of $y = x^3$ at its extremities (where x is negatively large and where x is positively large).

Reflecting the graph of $y = x^3$ in the x-axis gives the graph of $y = -x^3$.

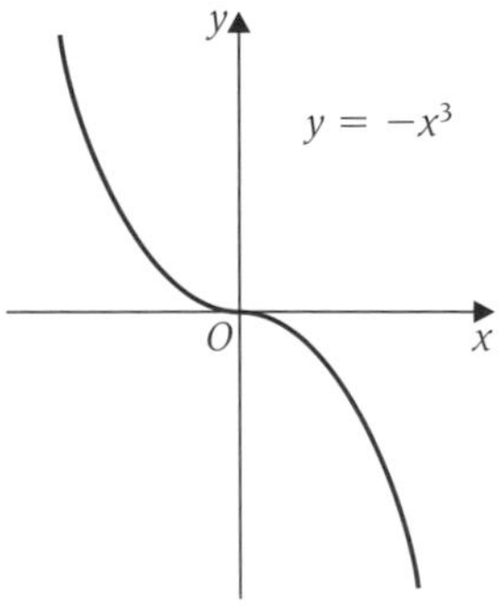

The graph of any cubic function, where the coefficient of x^3 is negative, has the shape of the graph of $y = -x^3$ at its extremities.

Since the graphs of cubic functions are continuous and the extremities are on opposite sides of the x-axis, it is clear that all cubic graphs must cross the x-axis at least once. Similarly, since the extremities are on opposite sides of the y-axis, all cubic graphs cross the y-axis, but only once.

To sketch the graph of a cubic function:

Step 1: Find the sign of the coefficient of x^3. This gives the shape of the graph at the extremities.

Step 2: Find the point where the graph crosses the y-axis by finding the value of y when $x = 0$.

Step 3: Find the point (or points) where the graph crosses the x-axis by finding the value of x when $y = 0$. (If there is a repeated root the graph will touch the x-axis.)

Step 4: Calculate values of y for some values of x. This is particularly useful to determine the quadrant in which the graph might turn close to the y-axis.

Step 5: Complete the sketch of the graph by joining the sections. (Your sketch should show the main features of the graph and also, where possible, values where the graph intersects the coordinate axes.)

6

Worked example 6.10

Sketch the graph of $y = (2x + 3)(x + 4)(5 - 2x)$.

Solution

$y = (2x + 3)(x + 4)(5 - 2x)$

For large values of x, $y \approx (2x)(x)(-2x)$
The coefficient of x^3 is -4 so the shape of the graph is the same as $y = -x^3$ when x is numerically very large.

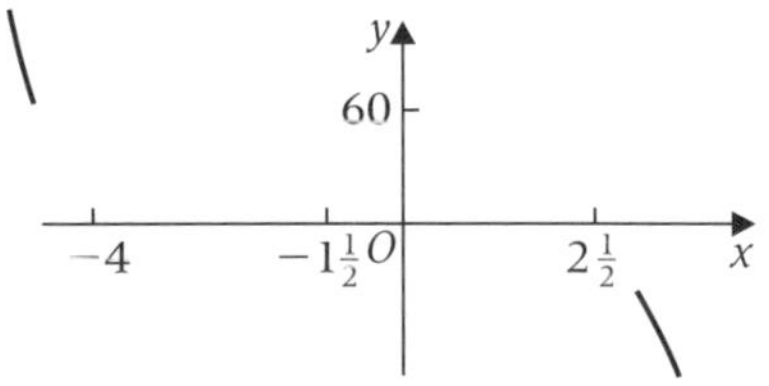

When $x = 0$, $y = (3)(4)(5) = 60$ so the graph crosses the y-axis at 60.

When $y = 0$, $(2x + 3)(x + 4)(5 - 2x) = 0 \Rightarrow x = -\frac{3}{2}, -4$ or $\frac{5}{2}$.

When $x = 1$, $y = (5)(5)(3) = 75 > 60$ so the graph must turn in the first quadrant.

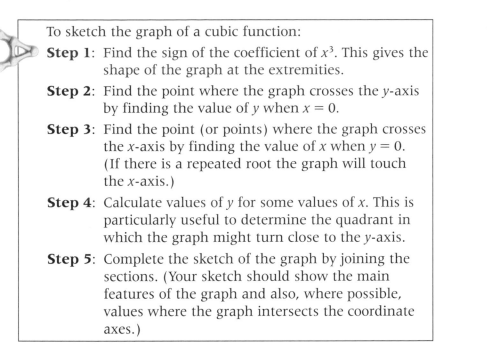

To decide whether this graph turns in the first or second quadrant you can find the value of y when $x = 1$.

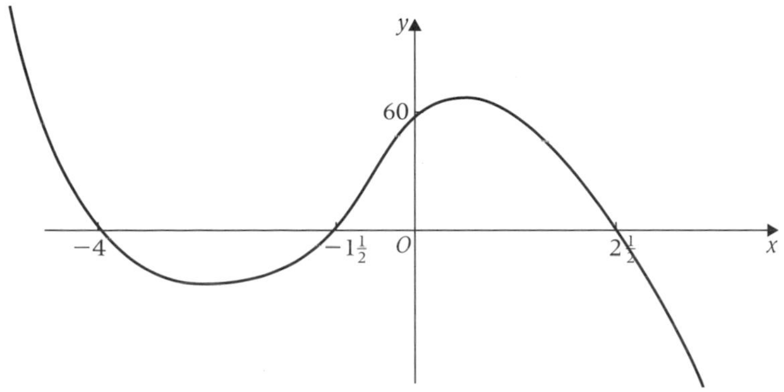

Worked example 6.11

Sketch the graph of $y = (x + 1)(x^2 - x + 1)$.

Solution

The coefficient of x^3 is $+1$ so the shape of the graph is the same as $y = x^3$ when x is numerically very large.
When $x = 0$, $y = (1)(1) = 1$ so the graph crosses the y-axis at 1.
When $y = 0$, $(x + 1)(x^2 - x + 1) = 0 \Rightarrow x = -1$ or $x^2 - x + 1 = 0$.
Now $x^2 - x + 1 = 0$ has no real roots (since $b^2 - 4ac = 1 - 4 = -3 < 0$) so the graph crosses the x-axis only at -1.

See section 4.7.

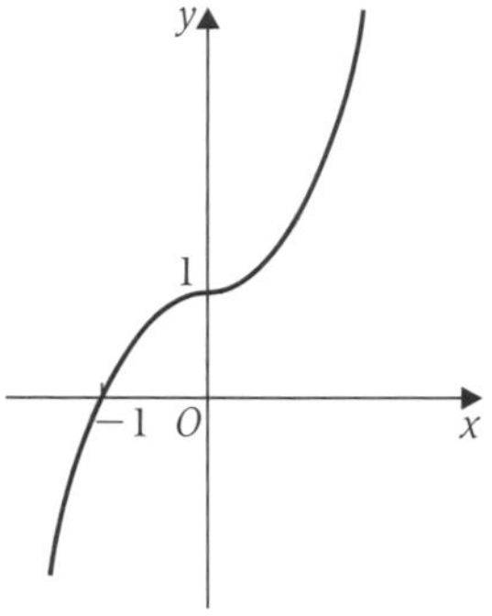

Multiplying out the brackets gives the equation of the graph as $y = x^3 + 1$. The graph of $y = x^3 + 1$ is obtained by moving (translation) the graph of $y = x^3$ by 1 unit upwards parallel to the y-axis.

Worked example 6.12

Sketch the graph of $y = (2x + 3)(x - 1)^2$.

Solution

The coefficient of x^3 is $+2$ so the shape of the graph is the same as $y = x^3$ when x is numerically very large.
When $x = 0$, $y = (3)(-1)^2 = 3$ so the graph crosses the y-axis at 3.

When $y = 0$, $(2x + 3)(x - 1)^2 = 0 \Rightarrow x = -\frac{3}{2}$ and $x = 1$ (repeated root) so the graph crosses the x-axis at $x = -\frac{3}{2}$ and touches the x-axis at $x = 1$.

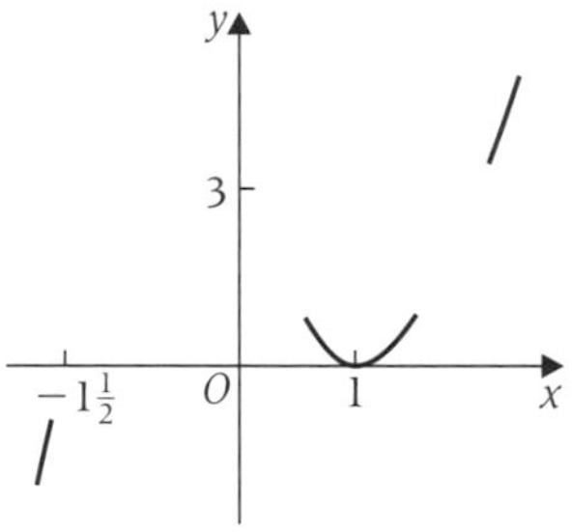

When $x = -1$, $y = (1)(-2)^2 = 4 > 3$ so the graph must turn in the second quadrant.

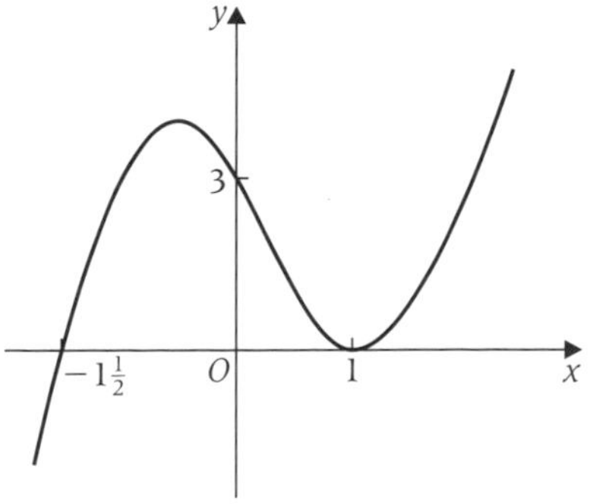

EXERCISE 6C

1 Sketch the graphs of the following:

(a) $y = 2x^3$, **(b)** $y = x^3 - 1$, **(c)** $y = (x - 1)^3$,

(d) $y = -(x + 2)^3$, **(e)** $y = 8 - x^3$, **(f)** $y = (3 - x)^3$.

2 Sketch the graphs of the following:

(a) $y = (x - 2)(x^2 + 2x + 4)$,

(b) $y = (1 + x)(2 + x)(3 + x)$,

(c) $y = (x - 4)(2 - x)(x - 5)$,

(d) $y = (x + 1)(x - 4)^2$,

(e) $y = (2x - 3)(x + 2)(x + 7)$,

(f) $y = (x + 2)(2 - x)^2$.

3 Use the factor theorem to show that $(x - 2)$ is a factor of $P(x) = x^3 - 3x^2 - 4x + 12$.
Write $P(x)$ as a product of three linear factors.
Hence sketch the graph of the curve $y = x^3 - 3x^2 - 4x + 12$.

6

6.5 Dividing a polynomial by a linear expression

In this module the division of a polynomial by a linear expression will be restricted to the cases:

$\frac{x+b}{x-a}$, $\frac{x^2+bx+c}{x-a}$, $\frac{x^3+bx^2+cx+d}{x-a}$, where a, b, c and d are integers.

The next worked example shows you how to divide a linear expression by another linear expression.

Worked example 6.13

(a) Divide $x + 4$ by x.

(b) Divide x by $x - 3$.

(c) Divide $x + 2$ by $x + 5$.

Solution

(a) $\frac{x+4}{x} = \frac{x}{x} + \frac{4}{x}$

$\Rightarrow \frac{x+4}{x} = 1 + \frac{4}{x}$

Compare $\frac{9}{5} = \frac{5+4}{5} = \frac{5}{5} + \frac{4}{5}$
$= 1 + \frac{4}{5}$

(b) $\frac{x}{x-3} = \frac{x-3+3}{x-3} = \frac{x-3}{x-3} + \frac{3}{x-3}$

$\Rightarrow \frac{x}{x-3} = 1 + \frac{3}{x-3}$

Write the first part of the numerator to match the denominator.

(c) $\dfrac{x+2}{x+5} = \dfrac{x+5-3}{x+5} = \dfrac{x+5}{x+5} - \dfrac{3}{x+5}$

$\Rightarrow \quad \dfrac{x+2}{x+5} = 1 - \dfrac{3}{x+5}$

$x + 2 = x + 5 - 3$

In the remainder of this section we extend this process to polynomials of degree 2 (quadratic) and polynomials of degree 3 (cubic).

Worked example 6.14

Divide $x^2 + 3x + 1$ by $x + 2$.

divisor

Solution

The first step is to write the polynomial $x^2 + 3x + 1$ in the form $(x + 2)(x + q) + r$.

$$x^2 + 3x + 1 \equiv (x + 2)(x + q) + r$$
$$x^2 + 3x + 1 \equiv x^2 + (2 + q)x + (2q + r)$$

Comparing the coefficients of x: $3 = 2 + q \Rightarrow q = 1$
Comparing the constant terms: $1 = 2q + r \Rightarrow 1 = 2 + r \Rightarrow r = -1$

So $x^2 + 3x + 1 \equiv (x + 2)(x + 1) - 1$
polynomial = divisor × quotient + remainder

Quotient is $(x + 1)$, remainder is -1.

Dividing both sides of the identity by $(x + 2)$ leads to

$$\frac{x^2 + 3x + 1}{x + 2} \equiv x + 1 - \frac{1}{x + 2}$$

Worked example 6.15

Divide $x^3 + 2x^2 - x + 3$ by $x - 2$.

Solution

As a first step, write the polynomial $x^3 + 2x^2 - x + 3$ in the form $(x - 2)(x^2 + px + q) + r$

$$x^3 + 2x^2 - x + 3 \equiv (x - 2)(x^2 + px + q) + r$$
$$x^3 + 2x^2 - x + 3 \equiv x^3 + (-2 + p)x^2 + (q - 2p)x + (-2q + r)$$

Comparing the coefficients of x^2: $2 = -2 + p \Rightarrow p = 4$
Comparing the coefficients of x: $-1 = q - 2p \Rightarrow -1 = q - 8$
$\Rightarrow q = 7$
Comparing the constant terms: $3 = -2q + r \Rightarrow 3 = -14 + r$
$\Rightarrow r = 17.$

So $x^3 + 2x - x + 3 \equiv (x - 2)(x^2 + 4x + 7) + 17$
polynomial = divisor × quotient + remainder

Quotient $x^2 + 4x + 7$; remainder 17.

Dividing both sides of the identity by $(x - 2)$ leads to

$$\frac{x^3 + 2x^2 - x + 3}{x - 2} \equiv x^2 + 4x + 7 + \frac{17}{x - 2}$$

Worked example 6.16

Divide $x^3 + 2x^2 - 1$ by $x - 1$.

divisor

Solution

Write $x^3 + 2x^2 - 1$ in the form $(x - 1)(x^2 + px + q) + r$

$$x^3 + 2x^2 + 0x - 1 \equiv (x - 1)(x^2 + px + q) + r$$

$$x^3 + 2x^2 + 0x - 1 \equiv x^3 + (-1 + p)x^2 + (q - p)x + (-q + r)$$

Write missing terms as 0 coefficients.

Comparing the coefficients of x^2: $2 = -1 + p \Rightarrow p = 3$

Comparing the coefficients of x: $0 = q - p \Rightarrow 0 = q - 3 \Rightarrow q = 3$

Comparing the constant terms: $-1 = -q + r \Rightarrow -1 = -3 + r \Rightarrow r = 2.$

So $x^3 + 2x^2 - 1 \equiv (x - 1)(x^2 + 3x + 3) + 2$

polynomial = divisor × quotient + remainder

Quotient $x^2 + 3x + 3$, remainder 2.

Dividing both sides of the identity by $(x - 1)$ leads to

$$\frac{x^3 + 2x^2 - 1}{x - 1} \equiv x^2 + 3x + 3 + \frac{2}{x - 1}$$

EXERCISE 6D

1 Divide:

(a) $x + 6$ by x,
(b) x by $x + 1$,
(c) $x + 3$ by $x + 2$,
(d) $x - 4$ by $x + 2$,
(e) $x^2 + x + 1$ by $x + 1$,
(f) $x^2 + 2x - 4$ by $x + 1$,
(g) $x^2 + 3x - 5$ by $x - 1$,
(h) $x^2 + 2x + 2$ by $x - 3$,
(i) $x^2 - 6x + 4$ by $x - 4$,
(j) $x^2 - 6$ by $x - 2$.

2 Divide:

(a) $x^3 + x^2 + x - 1$ by $x - 1$,
(b) $x^3 + x^2 + x - 1$ by $x + 1$,
(c) $x^3 + 2x^2 - x - 1$ by $x + 2$,
(d) $x^3 + x^2 + 3x - 5$ by $x - 2$,
(e) $x^3 + 3x^2 + x - 1$ by $x + 3$,
(f) $x^3 + 4x^2 - x + 3$ by $x - 3$,
(g) $x^3 - 4x^2 + x + 5$ by $x + 4$,
(h) $x^3 - 5x^2 + 2x - 7$ by $x - 4$,
(i) $x^3 - 24x + 6$ by $x + 5$,
(j) $x^3 - 4x^2 + 4x$ by $x - 5$.

3 Find **(i)** the quotient and **(ii)** the remainder when the polynomial $P(x)$ is divided by $(x - 1)$. Compare your answer to **(ii)** with the value of $P(1)$ and state your conclusion.

(a) $P(x) = x^2 + x - 3$,
(b) $P(x) = x^3 + 3x^2 + x - 3$,
(c) $P(x) = x^3 - 2x^2 + x - 3$.

4 Find **(i)** the quotient and **(ii)** the remainder when the polynomial $\mathrm{P}(x)$ is divided by $(x - 2)$. Compare your answer to **(ii)** with the value of $\mathrm{P}(2)$ and state your conclusion.

(a) $\mathrm{P}(x) = x^2 + x - 3$,

(b) $\mathrm{P}(x) = x^3 + 3x^2 - 10x - 3$,

(c) $\mathrm{P}(x) = x^3 - 2x^2 + x - 3$.

5 When the polynomial $x^3 + 2x^2 + x + k$, where k is a constant, is divided by $x + 2$ the remainder is 1. Find **(a)** the value of k and **(b)** the quotient.

6.6 The remainder theorem

In the previous section we noted that dividing a polynomial by a linear expression led to the identity

Polynomial = Divisor × Quotient + Remainder

When the divisor is a linear expression (degree 1), the remainder is a constant (degree 0) so

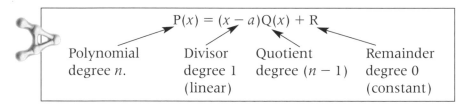

$$\mathrm{P}(x) = (x - a)\mathrm{Q}(x) + \mathrm{R}$$

Polynomial	Divisor	Quotient	Remainder
degree n.	degree 1 (linear)	degree $(n - 1)$	degree 0 (constant)

Since this identity is true for all values of x, consider the case when $x = a$,

so $\quad \mathrm{P}(a) = (a - a)\mathrm{Q}(a) + \mathrm{R}$

$\Rightarrow \quad \mathrm{P}(a) = 0 + \mathrm{R}$

$\Rightarrow \quad \mathrm{R} = \mathrm{P}(a)$

This is called the **remainder theorem**.

If a polynomial $\mathrm{P}(x)$ is divided by $(x - a)$, the remainder is $\mathrm{P}(a)$.

This will be extended in MPC4 to consider divisors of the form $(ax - b)$.

If the remainder is 0 then $(x - a)$ is a factor of $\mathrm{P}(x)$ so $\mathrm{P}(a) = 0 \Leftrightarrow (x - a)$ is a factor of $\mathrm{P}(x)$. This is the factor theorem which was considered at the start of this chapter.

Worked example 6.17

When $x^3 - 4x^2 + 6x + k$ is divided by $x - 2$ the remainder is 5. Find the value of the constant k.

Solution

Let $\quad P(x) = x^3 - 4x^2 + 6x + k$

$P(x) = (x - 2)Q(x) + R$

$\Rightarrow \quad P(2) = 0 + R$

$R = P(2) = (2)^3 - 4(2)^2 + 6(2) + k$

$\Rightarrow \quad 5 = 8 - 16 + 12 + k$

$\Rightarrow \quad k = 1$

Worked example 6.18

The polynomial $P(x) = x^3 - ax^2 + bx + 3$ when divided by $x + 2$ leaves a remainder of 9. Given also that $(x - 1)$ is a factor of $P(x)$ find the values of the constants a and b. Hence find the roots of the equation $P(x) = 0$.

Solution

$P(x) = (x + 2)Q(x) + R$

$\Rightarrow \quad 9 = P(-2) = (-2)^3 - a(-2)^2 + b(-2) + 3$

$\Rightarrow \quad 9 = -8 - 4a - 2b + 3$

$\Rightarrow \quad 4a + 2b = -14 \quad \Rightarrow \quad 2a + b = -7 \qquad$ [A]

$(x - 1)$ is a factor of $P(x) \Rightarrow P(1) = 0$

$\Rightarrow \quad 1 - a + b + 3 = 0 \quad \Rightarrow \quad a - b = 4 \qquad$ [B]

Solving the simultaneous equations [A] and [B] leads to $a = -1$ and $b = -5$.

$\Rightarrow \quad P(x) = x^3 + x^2 - 5x + 3 = (x - 1)(x^2 + px + q)$

$x^3 + x^2 - 5x + 3 = x^3 + (p - 1)x^2 + (q - p)x - q$

Comparing the coefficients of x^2: $\quad 1 = p - 1 \Rightarrow p = 2$

Comparing the constant terms: $\quad 3 = -q \Rightarrow q = -3$

$\Rightarrow \quad P(x) = (x - 1)(x^2 + 2x - 3) = (x - 1)(x - 1)(x + 3)$

$P(x) = 0 \Rightarrow \quad (x - 1)^2(x + 3) = 0$

$\Rightarrow \quad x = 1, -3$

EXERCISE 6E

1 Find the remainder when the polynomial $x^3 - 3x^2 + 4x - 6$ is divided by:

(a) $x - 1$, **(b)** $x - 2$, **(c)** $x + 1$,

(d) $x - 3$, **(e)** $x + 2$, **(f)** x.

2 Find the remainder when the polynomial $P(x)$ is divided by the linear expression $f(x)$:

(a) $P(x) = x^2 - 3x + 2 \qquad f(x) = x - 3$,

(b) $P(x) = x^3 + 2 \qquad f(x) = x + 1$,

(c) $P(x) = x^3 + x^2 - 3x + 2 \qquad f(x) = x - 2$,

(d) $P(x) = x^3 + 3x^2 - 7 \qquad f(x) = 3 + x$.

3 The polynomial $P(x)$ is defined by $P(x) = x^3 - 3x^2 + 10x + k$, where k is a constant. When $P(x)$ is divided by $x - 1$ the remainder is -9. Show that $k = -17$.

4 When $x^3 + kx^2 + 6x - 2$ is divided by $x + 1$ the remainder is 2. Find the value of the constant k.

5 The polynomial $P(x) = x^3 + ax^2 - x + 12$ leaves a remainder of 9 when divided by $x - 3$. Find the remainder when $P(x)$ is divided by $x - 2$.

6 The polynomial $P(x) = x^3 + 3x^2 - 2x + k$ has a factor of $x + 1$. Find the remainder when $P(x)$ is divided by $x - 3$.

7 Given that $P(x)$, where $P(x) = x^3 + 3x^2 + kx + 4$ and k is a constant, is such that the remainder on dividing $P(x)$ by $(x - 1)$ is three times the remainder on dividing $P(x)$ by $(x + 1)$, find the value of k.

8 The polynomial $P(x) = x^3 - 4x^2 + kx - 4$ leaves a remainder of -2 when divided by $x - 1$.

(a) Find the value of the constant k.

(b) Show that $x - 2$ is **not** a factor of $P(x)$.

9 When divided by $(x - 1)$ the polynomial $P(x) = x^3 + ax^2 + bx + 12$ leaves a remainder of 6. When divided by $(x + 3)$, $P(x)$ leaves a remainder of -30.

(a) Find the values of the constants a and b.

(b) Show that $(x - 3)$ is a factor of $P(x)$.

(c) Hence find the roots of the equation $P(x) = 0$.

10 The polynomials $f(x)$ and $g(x)$ are defined by $f(x) = x^3 + px^2 - x + 5$, $g(x) = x^3 - x^2 + px + 1$, where p is a constant. When $f(x)$ and $g(x)$ are divided by $x - 2$, the remainder is R in each case. Find the values of p and R. [A]

11 The polynomial $P(x) = x^3 - 4x^2 + 2x + k$ has $(x - 2)$ as a factor.

(a) Find the value of the constant k.

(b) Calculate the remainder when $P(x)$ is divided by $(x + 3)$.

(c) Find the exact values of the roots of the equation $P(x) = 0$. [A]

Key point summary

1 The **factor theorem** states that $(x - a)$ is a factor of the polynomial $P(x) \Leftrightarrow P(a) = 0$. *p81*

2 In factorising a cubic polynomial, once a linear factor has been found, the remaining quadratic factor can be found by equating coefficients. *p85*

3 To sketch the graph of a cubic function: *p89*

Step 1: Find the sign of the coefficient of x^3. This gives the shape of the graph at the extremities.

Step 2: Find the point where the graph crosses the y-axis by finding the value of y when $x = 0$.

Step 3: Find the point (or points) where the graph crosses the x-axis by finding the value of x when $y = 0$. (If there is a repeated root the graph will touch the x-axis.)

Step 4: Calculate values of y for some values of x. This is particularly useful to determine the quadrant in which the graph might turn close to the y-axis.

Step 5: Complete the sketch of the graph by joining the sections. (Your sketch should show the main features of the graph and also, where possible, values where the graph intersects the coordinate axes.)

4 Polynomial = divisor × quotient + remainder *p94*

5 When a polynomial is divided by a linear expression the remainder will always be a constant and the quotient will always be one degree less than the polynomial. *p94*

6 **The remainder theorem:** *p94*
If a polynomial $P(x)$ is divided by $(x - a)$, the remainder is $P(a)$.

Test yourself	What to review
1 Given that $x + 1$ is a factor of the polynomial $x^3 + kx^2 - 2x + 7$, find the value of the constant k.	*Section 6.2*
2 A polynomial is given by $P(x) = x^3 - 9x^2 + 2x + 48$. **(a)** Use the factor theorem to show that $(x - 3)$ is a factor of $P(x)$. **(b)** Factorise $P(x)$ completely and solve the equation $P(x) = 0$.	*Section 6.3*
3 Factorise the polynomial $f(x) = x^3 + 9x^2 + 6x - 56$ completely.	*Section 6.3*
4 Sketch the graph $y = (x - 4)(x - 1)(x + 4)$.	*Section 6.4*
5 Divide $x^3 + x^2 - 2x - 7$ by $x - 2$.	*Section 6.5*
6 The polynomial $x^3 + ax^2 + bx + c$ leaves the remainders 2, 2 and 6 when divided by $(x + 1)$, x and $(x - 1)$, respectively. Find the values of the constants a, b and c.	*Section 6.6*

Test yourself ANSWERS

1 $k = -8$.

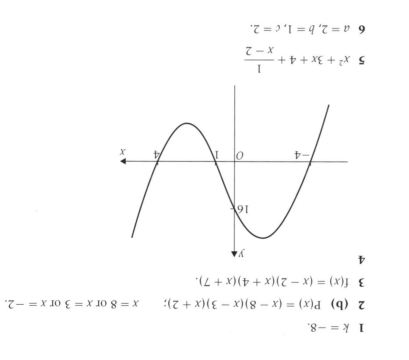

2 **(b)** $P(x) = (x - 8)(x - 3)(x + 2)$; $x = 8$ or $x = 3$ or $x = -2$.

3 $f(x) = (x - 2)(x + 4)(x + 7)$.

4

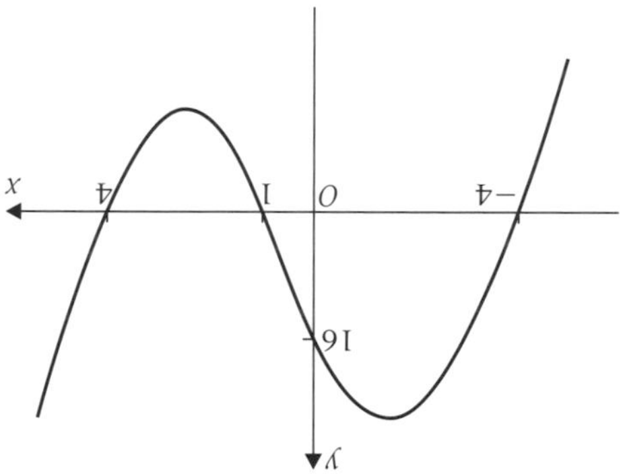

5 $x^2 + 3x + 4 + \frac{1}{x - 2}$

6 $a = 2$, $b = 1$, $c = 2$.

CHAPTER 7

C1: Simultaneous equations and quadratic inequalities

Learning objectives

After studying this chapter, you should be able to:

- solve simultaneous equations with one linear equation and one equation having degree two
- find points of intersection of lines and curves when the resulting equation is quadratic
- interpret geometrically the implication of equal roots, distinct real roots or no real roots of the resulting quadratic equation when solving simultaneous equations
- solve quadratic inequalities.

7.1 Simultaneous equations

You have already learnt how to solve a pair of linear equations. In this section you will tackle more difficult examples of simultaneous equations. One of the equations will be linear but the other equation will have degree two. That means it may involve x^2, y^2 or xy, for example.

Worked example 7.1

Solve the simultaneous equations: $\begin{array}{c} y = x + 1 \\ x^2 + y^2 = 5 \end{array}$

Solution

Substituting $y = x + 1$ into the second equation requires you to work out

$$y^2 = (x + 1)^2 = x^2 + 2x + 1$$

Hence $x^2 + y^2 = 5 \;\Rightarrow\; x^2 + (x^2 + 2x + 1) = 5$

$\Rightarrow\; 2x^2 + 2x - 4 = 0$

$\Rightarrow\; x^2 + x - 2 = 0$

$(x + 2)(x - 1) = 0 \;\Rightarrow\; x = -2, \quad \text{or} \quad x = 1$

You should expect the resulting quadratic to factorise.

Since $y = x + 1$, when $x = -2$, $y = -1$ and when $x = 1$, $y = 2$.

The final solution is then usually written grouping the pairs of values

$$\left.\begin{array}{l} x = -2 \\ y = -1 \end{array}\right\} \quad \left.\begin{array}{l} x = 1 \\ y = 2 \end{array}\right\}$$

> Simultaneous equations can usually be solved by eliminating one of the variables, such as y, and solving for the other variables.

Worked example 7.2

Solve the simultaneous equations $\begin{aligned} 3x - y &= 7 \\ 7y^2 - 6xy + 8 &= 0 \end{aligned}$

Solution

You need to rearrange the first equation to make x or y the subject. Since the second equation involves y^2 and xy, it is easier to eliminate x from the first equation.

Rearranging the first equation gives $x = \dfrac{y+7}{3}$.

Substituting $x = \dfrac{y+7}{3}$ into the second equation gives

$$7y^2 - \frac{6(y+7)y}{3} + 8 = 0$$

Hence $7y^2 - 2y(y+7) + 8 = 0$

$\Rightarrow \quad 7y^2 - 2y^2 - 14y + 8 = 0$

$\Rightarrow \quad 5y^2 - 14y + 8 = 0$

$(y-2)(5y-4) = 0 \quad \Rightarrow \quad y = 2, \quad \text{or} \quad y = \dfrac{4}{5}$

Since $x = \dfrac{y+7}{3}$, when $y = 2$, $x = 3$ and when $y = \frac{4}{5}$, $x = 2\frac{3}{5}$.

$7\frac{4}{5} \div 3 = \frac{39}{5} \times \frac{1}{3} = \frac{13}{5} = 2\frac{3}{5}$

Hence the solution is $\left.\begin{aligned} x &= 3 \\ y &= 2 \end{aligned}\right\} \quad \left.\begin{aligned} x &= 2\tfrac{3}{5} \\ y &= \tfrac{4}{5} \end{aligned}\right\}$

You can expect at least one of the pairs to involve whole numbers and these can easily be checked in the original equations.

Worked example 7.3

Solve the simultaneous equations $\begin{aligned} x + 3y &= 7 \\ x^2 - y &= 15 \end{aligned}$

Solution

Since the second equation involves x^2 but only y, it is easier to eliminate y from the second equation.

Rearrange $x + 3y = 7$ to give $3y = 7 - x$ and hence $y = \dfrac{1}{3}(7 - x)$.

Substituting into the second equation: $x^2 - \dfrac{1}{3}(7 - x) = 15$.

Notice the need to use brackets.

Multiplying throughout by 3 gives $3x^2 - (7 - x) = 45$.

Hence $3x^2 - 7 + x = 45 \quad \Rightarrow \quad 3x^2 + x - 52 = 0$.

You could use the formula to solve the quadratic if you do not spot the factors.

$(3x + 13)(x - 4) = 0 \quad \Rightarrow \quad x = -\dfrac{13}{3}, \quad \text{or} \quad x = 4$

Taking each of these x-values in turn and substituting into $y = \frac{1}{3}(7 - x)$;

When $x = -\frac{13}{3}$, $y = \frac{1}{3}\left(7 + \frac{13}{3}\right) = \frac{34}{9}$ and when $x = 4$, $y = \frac{1}{3}(7 - 4) = 1$.

The solution is $\left.\begin{matrix} x = 4 \\ y = 1 \end{matrix}\right\}$ $\left.\begin{matrix} x = -4\frac{1}{3} \\ y = 3\frac{7}{9} \end{matrix}\right\}$

The answers to the second pair could have been left as $x = -\frac{13}{3}$, $y = \frac{34}{9}$.

EXERCISE 7A

1 Solve the following sets of simultaneous equations:

(a) $y = x - 8$, $x^2 + y = 4$

(b) $x = 5 - y$, $y^2 + 2x = 18$

(c) $y = 3 - x$, $x^2 + y^2 = 5$

(d) $x = 4 - y$, $x^2 + y^2 = 10$

2 The numbers x and y satisfy the simultaneous equations

$y = 2x + 1$
$xy = 3$.

(a) Show that $2x^2 + x - 3 = 0$.

(b) Hence solve the simultaneous equations. [A]

3 Solve the simultaneous equations

$y = 2 - x$
$x^2 + 2xy = 3$. [A]

4 Solve the simultaneous equations

$y = 6 - 2x$
$xy + x = 3$. [A]

5 Solve the following sets of simultaneous equations:

(a) $x + 2y = 5$, $x^2 - xy = 6$

(b) $3x + 2y = 4$, $y^2 + 2xy = 0$

(c) $4x = y + 7$, $3x^2 - 4y = 7$

(d) $5x + 3y = 4$, $5x^2 - 3y^2 = 8$

7.2 The intersection of a line and curve

> When a line intersects a curve, the corresponding equations can be solved simultaneously in order to find the point of intersection.

Worked example 7.4

Calculate the points of intersection of the line with equation $y = 3x + 5$ and the parabola with equation $y = 3x^2 + 2x + 1$.

At a point of intersection, the x-coordinates of the line and curve are equal and so are their y-coordinates.

Solution

Substitute $y = 3x + 5$ into $y = 3x^2 + 2x + 1$ to eliminate y.

$$\Rightarrow \quad 3x + 5 = 3x^2 + 2x + 1$$
$$\Rightarrow \quad 0 = 3x^2 - x - 4$$
$$\Rightarrow \quad (3x - 4)(x + 1) = 0$$
$$\Rightarrow \quad x = \frac{4}{3} \text{ or } x = -1$$

Because $b^2 - 4ac = 49$ is a perfect square, the quadratic factorises.

Substitute values of x into $y = 3x + 5$.

When $x = \frac{4}{3}$, $y = 3 \times \frac{4}{3} + 5 = 9$ giving the point $\left(\frac{4}{3}, 9\right)$.

When $x = -1$, $y = (3 \times -1) + 5 = 2$ giving the point $(-1, 2)$.

The two points of intersection are $\left(\frac{4}{3}, 9\right)$ and $(-1, 2)$.

You could check that these points also satisfy the quadratic: e.g. $3 \times (-1)^2 + 2 \times (-1) + 1 = 3 - 2 + 1 = 2$ ✓

Worked example 7.5

Determine whether the line $3x - 2y = 9$ and the curve $y = x^2 - 5x + 7$ intersect.

Solution

Eliminating y,

$$\Rightarrow \quad 3x - 2(x^2 - 5x + 7) = 9$$
$$\Rightarrow \quad 3x - 2x^2 + 10x - 14 = 9$$
$$\Rightarrow \quad 0 = 2x^2 - 13x + 23$$

Substitute the expression $y = x^2 - 5x + 7$ into the equation of the line.

Consider the discriminant with $a = 2$, $b = -13$, $c = 23$:

$$b^2 - 4ac = (-13)^2 - 4 \times 2 \times 23 = 169 - 184 = -15$$

Because the discriminant is negative, there are no real solutions to the quadratic equation.

Hence there are no real points of intersection of the line and curve.

The line and curve do not intersect.

In this section, while solving for points of intersection of a straight line and a parabola it was necessary to solve a quadratic equation of the form $ax^2 + bx + c = 0$.

The last two worked examples demonstrated the following situations:

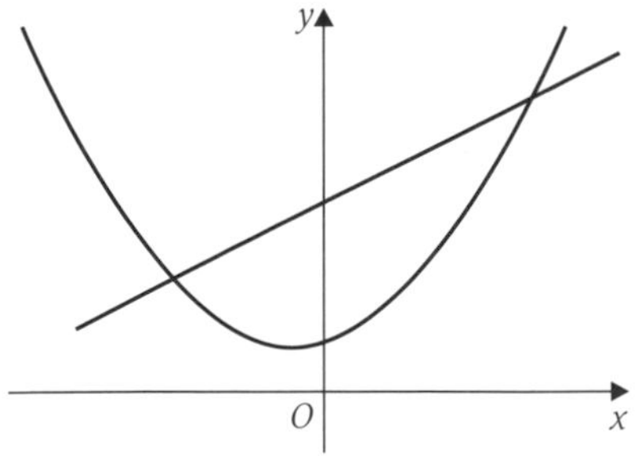

The curve and line intersect in two distinct points. This occurs when $b^2 - 4ac > 0$.

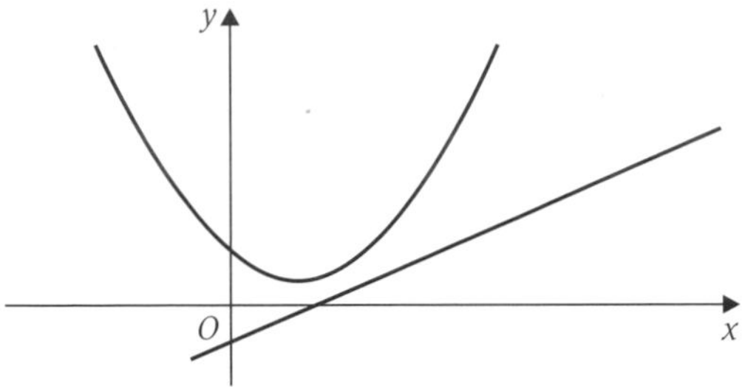

The curve and line do not intersect.
This occurs when $b^2 - 4ac < 0$.

7.3 The condition for a line to be a tangent to the graph of a quadratic function

Worked example 7.6

Show that the line $y = 3x - 5$ intersects the curve with equation $y = x^2 - 3x + 4$ at a single point. Interpret your solution geometrically.

Solution

Eliminating y gives $x^2 - 6x + 9 = 0$.
Hence $(x - 3)^2 = 0$ which has the repeated root $x = 3$.
Since the point lies on the line $y = 3x - 5$, when $x = 3$, $y = 4$.

There is a single point of intersection at (3, 4).

The line actually touches the curve at the point (3, 4). When this situation occurs we say that the line is a **tangent** to the curve at this point.

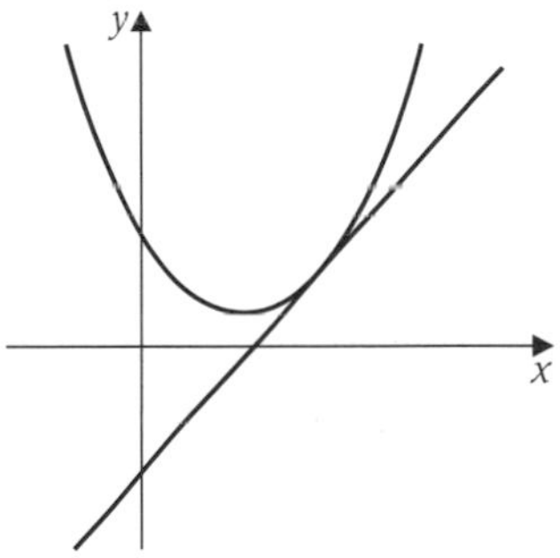

> When solving for points of intersection of a straight line and a parabola, it is necessary to solve a quadratic equation of the form $ax^2 + bx + c = 0$. If $b^2 - 4ac = 0$, the line is a tangent to the parabola.

7

Worked example 7.7

Given that the line $y = mx + 5$ is a tangent to the parabola with equation $y = 2x^2 - 5x + 7$, find the possible values of the constant m.

Solution

Solving for intersection points gives $mx + 5 = 2x^2 - 5x + 7$.
Hence $2x^2 - 5x - mx + 2 = 0 \quad \Rightarrow \quad 2x^2 - (5 + m)x + 2 = 0$.
Comparing with $ax^2 + bx + c = 0$ gives $a = 2$, $b = -(5 + m)$, $c = 2$.

The condition for the line to be a tangent is that $b^2 - 4ac = 0$.

So $(5 + m)^2 - 4 \times 2 \times 2 = 0 \quad \Rightarrow \quad (5 + m)^2 = 16$
Therefore $5 + m = 4$ or $5 + m = -4$.
Hence the possible values for m are -1 and -9.

EXERCISE 7B

1 Find the points of intersection of:

(a) the line $y = x + 2$ and the curve $y = x^2 - 4$,

(b) the line $x + y = 4$ and the curve $y = 2x^2 - 7x + 8$,

(c) the line $3x - 2y = 7$ and the curve $y = x^2 - 5x - 7$,

(d) the line $5x + y + 6 = 0$ and the curve $y = 3x^2 + 5x - 6$,

(e) the line $2x + 3y = 4$ and the curve $y = 5x^2 - 3x - 14$.

2 By considering the discriminant of the resulting quadratic, determine the number of points of intersection of the following lines and curves:

(a) the line $y = x - 1$ and the curve $y = x^2 - x$,

(b) the line $y = x + 4$ and the curve $y = x^2 - 7x + 3$,

(c) the line $2x + y = 2$ and the curve $y = 3x^2 - 5x + 3$,

(d) the line $2x + y + 6 = 0$ and the curve $y = 4x^2 + 7x - 3$,

(e) the line $3x - 2y = 5$ and the curve $y = 3x^2 - 4x - 2$.

If there are no points of intersection you must show why that is the case.

3 Find the coordinates of the points where the line $y = 2x + 5$ intersects the curve $y = x^2 + 2x + 2$, giving your answers in surd form. [A]

4 Find the exact value of the x-coordinate at each of the points of intersection of the following:

(a) the line $y = 3x + 1$ and the curve $y = 2x^2 - 5x - 7$,

(b) the line $2x + y = 7$ and the curve $y = 4x^2 - 8x - 5$,

(c) the line $3x + 4y = 9$ and the curve $y = 3x^2 - 7x + 2$.

5 Show that the quadratic equation resulting from the solution of the simultaneous equations $y = 3x - 5$ and $y = 2x^2 - 5x + 3$ has a single repeated root.
State the geometrical relationship between the curve with equation $y = 2x^2 - 5x + 3$ and the line $y = 3x - 5$.

6 Find the value of the constant c for which the line $y = 3x + c$ is a tangent to the curve with equation $y = x^2 - 4x + 7$.

7 (a) Show that the curve with equation $y = 3(x + 1)^2$ and the line with equation $y = kx - 9$ intersect when $3x^2 + (6 - k)x + 12 = 0$.

(b) (i) Find the values of k for which the quadratic equation $3x^2 + (6 - k)x + 12 = 0$ has equal roots.

(ii) State the geometrical relationship between the line $y = kx - 9$ and the curve $y = 3(x + 1)^2$ for these values of k.

8 A straight line has equation $y = 4x + k$, where k is a constant, and a parabola has equation $y = 3x^2 + 12x + 7$. Show that the x-coordinate of any points of intersection of the line and the parabola satisfy $3x^2 + 8x + 7 - k = 0$. Hence find the range of values of k for which the line and parabola do not intersect.

7.4 Quadratic inequalities

You learned how to solve linear inequalities in the first chapter. In this section, you will extend the idea to quadratic inequalities. These can be solved algebraically or by consideration of a graph.

> Quadratic inequalities can be solved by drawing a graph and considering when the parabola is above or below the x-axis.

Worked example 7.8

Solve the inequality $(2x - 3)(1 - 5x) > 0$.

Solution

Let $f(x) = (2x - 3)(1 - 5x)$ and consider the graph of $y = f(x)$.

The graph is a parabola and you need to know when the graph is above the x-axis, or when $y > 0$.

The graph of $y = (2x - 3)(1 - 5x)$ cuts the x-axis when $2x - 3 = 0$ and when $1 - 5x = 0$, namely at $(\frac{3}{2}, 0)$ and at $(\frac{1}{5}, 0)$.

Also when $x = 0$, $y = -3$ so the parabola passes through $(0, -3)$. You can sketch its graph.

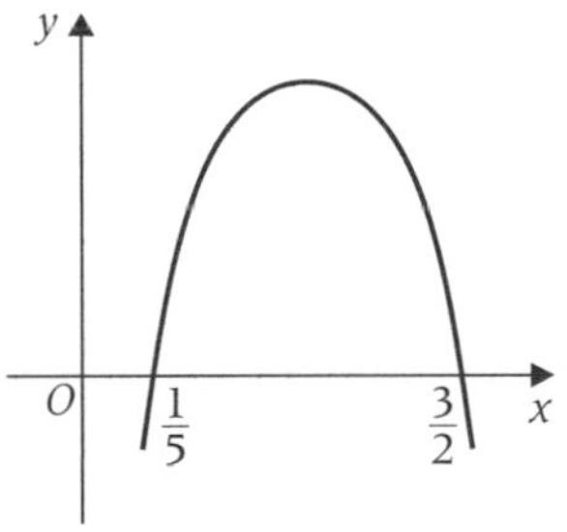

From the graph, you can see when $y > 0$. This is when x lies between $\frac{1}{5}$ and $\frac{3}{2}$. The original inequality is strict. It does not include an equals sign.

Solution is $\frac{1}{5} < x < \frac{3}{2}$.

You can check your answer by choosing values in this interval. For instance, $x = 1 \Rightarrow (2x - 3)(1 - 5x) = (-1) \times (-4) = 4 > 0$. ✓

Worked example 7.9

Solve the inequality $3x^2 + 5x - 2 \geqslant 0$.

You can simplify matters so that you do not actually need to draw a graph.

Instead, you consider the sign of the expression over the different intervals.

Solution

Factorising the quadratic gives $(3x - 1)(x + 2) \geqslant 0$.

Let $f(x) = (3x - 1)(x + 2)$.

$f(x) = 0$ when $x = \frac{1}{3}$ or when $x = -2$.

These are called **critical values**.

Consider a number line with these critical values marked on.

There are now three separate zones to consider.

Choose a value in each of these zones and work out $f(x)$.

For $x < -2$, choose $x = -3$, say. $f(-3) = (-10)(-1) = 10 > 0$.

For x between -2 and $\frac{1}{3}$, choose $x = 0$ so $f(0) = -2 < 0$.

For $x > \frac{1}{3}$, choose $x = 1$, say. $f(1) = 2 \times 3 = 6 > 0$.

These are the only values of x when the expression is zero.

Therefore the expression must always be positive or always negative throughout the separate intervals shown.

You can now draw a diagram indicating when $f(x)$ is positive or negative.

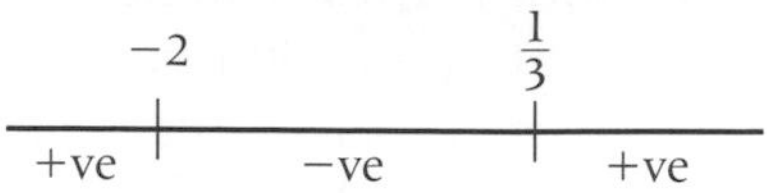

The question asks when the expression is greater than or equal to zero.

There are two separate intervals forming the solution.

$x \leqslant -2$ or $x \geqslant \frac{1}{3}$

When you get your final answer you can always check you have put the signs the right way round.

Solution predicts $x = 2$ lies in the solution interval.

$3 \times 2^2 + 5 \times 2 - 2 = 20$
(> 0 so ✓)

Also $x = -4$ should satisfy the inequality

$3 \times (-4)^2 + 5 \times (-4) - 2 = 26$
(> 0 so ✓)

Worked example 7.10

Solve the inequality $2x^2 < 15 - x$.

Solution

Rearranging the inequality gives $2x^2 + x - 15 < 0$.

Factorising $(x + 3)(2x - 5) < 0$.

Critical values are $x = -3$ and $x = 2\frac{1}{2}$.

Critical values are given by $x + 3 = 0$ and $2x - 5 = 0$.

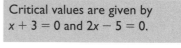

Indicating on a diagram the sign of $(x + 3)(2x - 5)$.

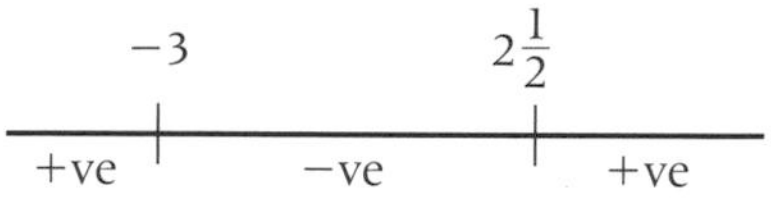

Solution is $-3 < x < 2\frac{1}{2}$.

EXERCISE 7C

Solve each of the following inequalities. The quadratics will all factorise.

1 $(x - 3)(2x + 7) < 0$
2 $(1 - x)(2x - 3) \leqslant 0$
3 $x^2 - x - 6 > 0$
4 $x^2 - 5x - 6 < 0$
5 $x^2 + 3x - 10 \geqslant 0$
6 $x^2 + 7x + 12 < 0$
7 $2x^2 - x - 1 > 0$
8 $x^2 - 16 > 0$
9 $3x^2 - 4x - 4 \leqslant 0$
10 $4x^2 - 5x + 1 > 0$
11 $2x^2 > 3x + 9$
12 $x^2 \leqslant 25$
13 $2x + 3 \leqslant x^2$
14 $5x^2 + 3 > 8x$
15 $3x + 7 < 4x^2$
16 $11x - 3 > 8x^2$
17 $5x + 2x^2 > 3$
18 $12x^2 - 13x < 22$
19 $20x^2 + 3 > 16x$
20 $5 + 9x^2 \geqslant 18x$

You may find a sketch helpful in the first instance, but examiners will regard a diagram showing the sign of the expression as sufficient working, provided you state the critical values.

7.5 Quadratics with irrational roots

In the previous exercise, all the quadratics factorised, and you could use a simple sketch or a sign diagram to help you solve them.

> The values of x for which the quadratic is equal to zero are the critical values. You can easily solve a quadratic inequality by means of a sign diagram, indicating when the expression is positive or negative.

If the quadratic has two irrational roots then you can proceed in a similar way.

Suppose the two critical values are p and q where $p < q$.

Mark the critical values on a diagram.

Consider the value of the expression for each of the three zones and solve the inequality.

Worked example 7.11

Solve the inequality $2x^2 < 3x + 7$.

Solution

Rearrange to get $2x^2 - 3x - 7 < 0$.

Solving the quadratic equation $2x^2 - 3x - 7 = 0$,

$$\Rightarrow x = \frac{-b \pm \sqrt{b^2 - 4ac}}{2a} \text{ where } a = 2,\ b = -3,\ c = -7$$

$$\Rightarrow x = \frac{3 \pm \sqrt{9 + 56}}{4}$$

$$\Rightarrow x = \frac{3 + \sqrt{65}}{4} \approx \frac{11}{4} \text{ or } x = \frac{3 - \sqrt{65}}{4} \approx -\frac{5}{4}$$

If you cannot use a calculator, you may wish to approximate $\sqrt{65}$ as 8 to get an idea of the approximate values of the roots.

The critical values are $\frac{1}{4}(3 + \sqrt{65})$ and $\frac{1}{4}(3 - \sqrt{65})$ but the approximations are easier to deal with on a sign diagram.

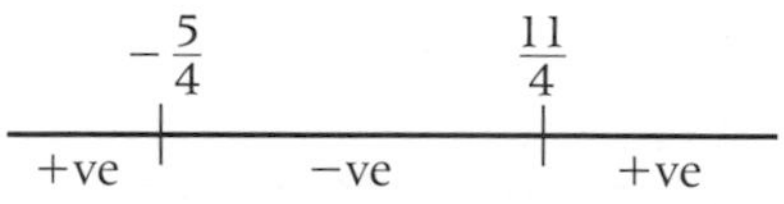

Let

$$\begin{aligned} f(x) &= 2x^2 - 3x - 7 \\ f(-3) &= 18 + 9 - 7 = 20 > 0 \\ f(0) &= 0 + 0 - 7 = -7 < 0 \\ f(10) &= 200 - 30 - 7 = 163 > 0 \end{aligned}$$

You have probably realised that after finding a **single** value of the function, you can complete the sign diagram. A second value is useful as a check.

The inequality $2x^2 - 3x - 7 < 0$ has solution

$$\tfrac{1}{4}(3 - \sqrt{65}) < x < \tfrac{1}{4}(3 + \sqrt{65}).$$

7.6 Completing the square

All quadratic inequalities can be solved by completing the square. You need to make use of one of the following standard results.

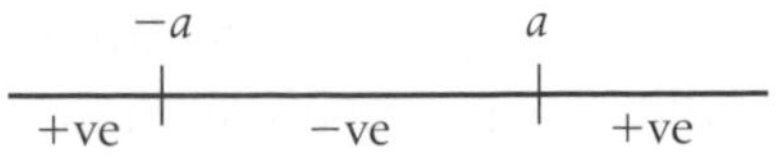

Consider the inequality $y^2 > a^2$.

$$y^2 - a^2 > 0 \Rightarrow (y - a)(y + a) > 0$$

> $y^2 > a^2 \Rightarrow y > a$ or $y < -a$

Similarly you can use the sign diagram to solve the inequality $y^2 < a^2$.

> $y^2 < a^2 \Rightarrow -a < y < a$

Worked example 7.12

By completing the square, solve the inequalities:

(a) $x^2 + 4x - 1 \geqslant 0$, **(b)** $2x^2 - 12x - 5 < 0$.

Solution

(a) $x^2 + 4x - 1 = (x + 2)^2 - 4 - 1$

$\Rightarrow x^2 + 4x - 1 \geqslant 0 \Rightarrow (x + 2)^2 \geqslant 5$

Using the result in the first box above

$$x + 2 \geqslant \sqrt{5} \text{ or } x + 2 \leqslant -\sqrt{5}$$

$$x \geqslant \sqrt{5} - 2 \text{ or } x \leqslant -\sqrt{5} - 2$$

Subtracting 2 from both sides of each linear inequality.

(b) $2x^2 - 12x - 5 = 2(x^2 - 6x) - 5 = 2[(x - 3)^2 - 9] - 5$

$= 2(x - 3)^2 - 23$

$2x^2 - 12x - 5 < 0 \Rightarrow 2(x - 3)^2 - 23 < 0$

$$\Rightarrow (x - 3)^2 < \frac{23}{2} \Rightarrow (x - 3)^2 < \left(\sqrt{\frac{23}{2}}\right)^2$$

Using the result in the second box above,

$$\Rightarrow -\sqrt{\frac{23}{2}} < x - 3 < \sqrt{\frac{23}{2}}$$

$$\Rightarrow 3 - \sqrt{\frac{23}{2}} < x < 3 + \sqrt{\frac{23}{2}}$$

Adding 3 to both sides of each linear inequality.

EXERCISE 7D

Solve each of the following inequalities.

1 $x^2 - 6x - 4 < 0$

2 $x^2 + 8x + 3 \leqslant 0$

3 $x^2 - 4x - 1 > 0$

4 $x^2 - 2x - 5 < 0$

5 $x^2 - 10x + 3 \geqslant 0$

6 $x^2 + 3x + 1 < 0$

7 $2x^2 - 4x - 1 > 0$

8 $3x^2 - 6x - 7 > 0$

9 $3x^2 - 12x - 5 \leqslant 0$

10 $4x^2 - 12x + 3 > 0$

7.7 The discriminant revisited

At times, it is necessary to solve quadratic inequalities when you are considering the conditions for a quadratic equation to have real roots, for example.

Recall from Chapter 4 that the quadratic $ax^2 + bx + c = 0$ has

- two distinct real roots when $b^2 - 4ac > 0$;
- two equal roots when $b^2 - 4ac = 0$.

Therefore the condition for real roots is $b^2 - 4ac \geqslant 0$.

Worked example 7.13

Find the possible values of k for the quadratic equation

$$(2k - 1)x^2 + 8kx + 2(5k + 3) = 0$$

to have real roots.

Solution

$(8k)^2 - 4(2k - 1) \times 2(5k + 3) \geqslant 0$

Using $b^2 - 4ac \geqslant 0$.

$\Rightarrow 64k^2 - 8(10k^2 + k - 3) \geqslant 0$

$\Rightarrow -8k^2 + (10k^2 + k - 3) \leqslant 0$

Divide both sides by -8 and reverse the inequality sign.

$\Rightarrow 2k^2 + k - 3 \leqslant 0$

$\Rightarrow (2k + 3)(k - 1) \leqslant 0$

The critical points are $k = -\frac{3}{2}$ and $k = 1$. The sign diagram is shown in the margin.

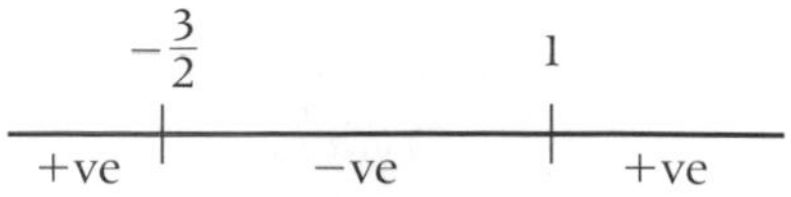

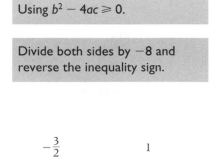

Solution is $-1\frac{1}{2} \leqslant k \leqslant 1$.

Worked example 7.14

Find the condition on k for the quadratic equation

$$3kx^2 - (k+3)x + (k-2) = 0$$

to have **no** real roots.

Solution

Using $b^2 - 4ac < 0$.

$\Rightarrow (k+3)^2 - 4 \times 3k \times (k-2) < 0$

$\Rightarrow (k+3)^2 < 12k(k-2)$

$\Rightarrow k^2 + 6k + 9 < 12k^2 - 24k$

$\Rightarrow 0 < 11k^2 - 30k - 9$

$\Rightarrow 0 < (11k+3)(k-3)$

A different approach from multiplying by a negative number is to take terms to the opposite side of the inequality.

The critical points are $k = 3$ and $k = -\frac{3}{11}$ with the sign diagram for the expression $(11k+3)(k-3)$ shown in the margin.

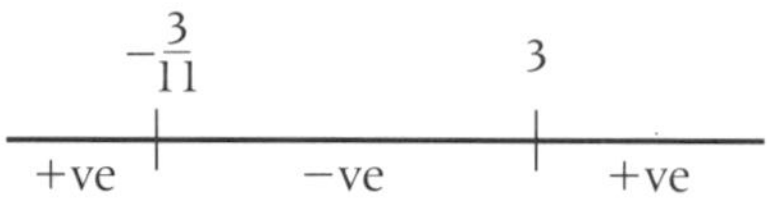

Solution is $k > 3$, $k < -\frac{3}{11}$.

EXERCISE 7E

1 Find the values of k for which the quadratic equation

$$(k-3)x^2 + (k+2)x + (2k+1) = 0$$

has real roots.

2 Determine the condition on k for the quadratic equation

$$(k+1)x^2 + (3k+2)x - (k+6) = 0$$

to have real roots.

3 Find the possible values of k for the quadratic equation

$$2kx^2 - (k+2)x + (3-k) = 0$$

to have two distinct real roots.

4 Find the values of k for which the quadratic equation

$$(k+3)x^2 - 4kx + (5-4k) = 0$$

has no real roots.

5 Determine the condition on k for the quadratic equation

$$(2k-5)x^2 - (k-1)x + (k-2) = 0$$

to have two distinct real roots.

6 Determine the condition on k for the quadratic equation

$$(3k+4)x^2 - (3+k)x + (2k+3) = 0$$

to have real roots.

MIXED EXERCISE

1 The quadratic function f is defined by $f(x) = x^2 + 6x + 11$.

(a) (i) Express $f(x)$ in the form $(x + p)^2 + q$, where p and q are integers.

(ii) State the value of x for which $f(x)$ is least.

(b) Solve the inequality $f(x) > 3$.

(c) State the geometrical transformation which maps the curve $y = f(x)$ onto $y = f(x + 2)$. [A]

2 (a) Express $x^2 - 6x + 7$ in the form $(x + a)^2 + b$, finding the values of a and b.

(b) Hence, or otherwise, find the range of values of x for which $x^2 - 6x + 7 < 0$. [A]

3 (a) Express $x^2 + 4x - 5$ in the form $(x + a)^2 + b$, finding the values of the constants a and b.

(b) Find the values of x for which $x^2 + 4x - 5 > 0$. [A]

4 Find the range of values of the constant c for which the line $y = 3x + c$ does not intersect the parabola with equation $y = 2x^2 + 5x - 7$.

5 The parabola with equation $y = 2x^2 + 4x - 8$ intersects the line $y + kx + 4k = 0$, where k is a constant.

(a) Show that any points of intersection have x-coordinates which satisfy $2x^2 + (k + 4)x + 4(k - 2) = 0$.

(b) Hence find the condition on k such that:

(i) the line is a tangent to the parabola,

(ii) the line cuts the parabola in two distinct points.

6 (a) Solve $2x^2 + 8x + 7 = 0$, giving your answers in surd form.

(b) Hence solve $2x^2 + 8x + 7 > 0$. [A]

7 Determine the condition on k for which the equation

$$(k - 2)x^2 + 2(2k - 3)x + (5k - 6) = 0$$

has real roots.

8 Express $kx^2 + 2kx + 2k - 3$ in the form $k(x + a)^2 + f(k)$. Hence find the range of values of k for which $kx^2 + 2kx + 2k - 3$ is always positive.

Key point summary

1 Simultaneous equations can usually be solved by eliminating one of the variables, such as y, and solving for the other variable *p100*

2 When a line intersects a curve, the corresponding equations can be solved simultaneously in order to find the point of intersection. *p101*

3 When solving for points of intersection of a straight line and a parabola, it is necessary to solve a quadratic equation of the form $ax^2 + bx + c = 0$. *p103*
If $b^2 - 4ac = 0$, the line is a tangent to the parabola.
If $b^2 - 4ac > 0$, the line and parabola intersect at two distinct points.
If $b^2 - 4ac < 0$, the line and parabola do not intersect.

4 Quadratic inequalities can be solved by drawing a graph and considering when the parabola is above or below the x-axis. *p105*

5 $y^2 > a^2 \Rightarrow y > a$ or $y < -a$ *p108*

6 $y^2 < a^2 \Rightarrow -a < y < a$ *p108*

7 The values of x for which the quadratic is equal to zero are the critical values. You can easily solve a quadratic inequality by means of a sign diagram, indicating when the expression is positive or negative. *p107*

Test yourself	What to review
1 Solve the simultaneous equations $x + y = 3$, $x^2 + 3y = 7$.	*Section 7.1*
2 Find the points of intersection of the line $y = 3x - 2$ and the curve with equation $xy = 8$.	*Section 7.2*
3 Prove that the line with equation $x + 2y = 2$ and the curve with equation $y = 3x^2 - 2x + 2$ have no points of intersection.	*Section 7.2*
4 Find the value of the constant k so that the line $y = 2x + k$ is a tangent to the parabola with equation $y = x^2 - 6x + 2$.	*Section 7.3*
5 Solve the inequality $3x^2 < 4x + 4$.	*Section 7.4*
6 Solve the inequality $x^2 - 12x > 4$ by completing the square.	*Section 7.6*
7 Find the values of k for which the quadratic equation $2x^2 - (k + 1)x + (3k - 7) = 0$ has real roots.	*Section 7.7*

Test yourself ANSWERS

1 $x = 2, y = 1$ or $x = 1, y = 2$.

2 $(2, 4)$, $(-\frac{4}{3}, -6)$.

3 Discriminant $= -39$, $\therefore$ no real roots.

4 $k = -14$.

5 $-\frac{2}{3} < x < 2$.

6 $x > 6 + 2\sqrt{10}$, $x < 6 - 2\sqrt{10}$.

7 $k \geq 19$, $k \leq 3$.

CHAPTER 8

C1: Coordinate geometry of circles

Learning objectives

After studying this chapter, you should be able to:

- recognise a Cartesian equation of a circle and be able to write down its equation given its centre and radius
- use the method of completing the square to find the radius and centre of a given circle and also to sketch the circle
- recognise that the circle $(x - a)^2 + (y - b)^2 = r^2$ is a translation of the circle $x^2 + y^2 = r^2$ and be able to write down the equation of the circle formed by applying a given translation on a given circle
- use the circle properties **(i)** that the angle in a semicircle is a right angle and **(ii)** that the perpendicular from the centre to the chord bisects the chord, to find the equation of the circle passing through given points
- determine if a line meets a circle
- find the coordinates of any points of intersection of a line and a circle whose equations are given
- find the equation of the tangent and normal at a given point to a circle by using the circle property 'the tangent to a circle is perpendicular to the radius at its point of contact'.

8.1 The Cartesian equation of a circle

In your GCSE course a circle was defined to be the locus of points which are at a constant distance from a fixed point. The constant distance is the radius and the fixed point is the centre of the circle. In this section you will derive, recognise and use the equation of a circle in Cartesian coordinates.

Consider a circle of radius r and centre the origin $O(0, 0)$.

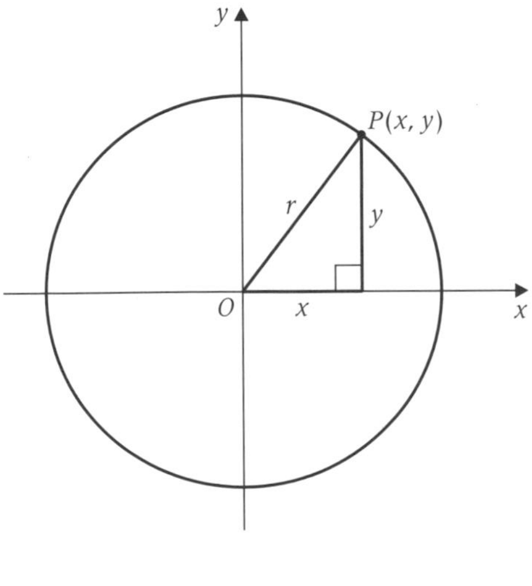

Let $P(x, y)$ be any point on the circle then $OP = r$.

$\Rightarrow \quad OP^2 = r^2$.

But $OP^2 = (x - 0)^2 + (y - 0)^2$,

$\Rightarrow \quad x^2 + y^2 = r^2$.

> The equation of a circle with centre $(0, 0)$ and radius r is $x^2 + y^2 = r^2$.

This equation is **not** given in the examination formulae booklet.

Next, consider a circle of radius r and centre $C(a, b)$.

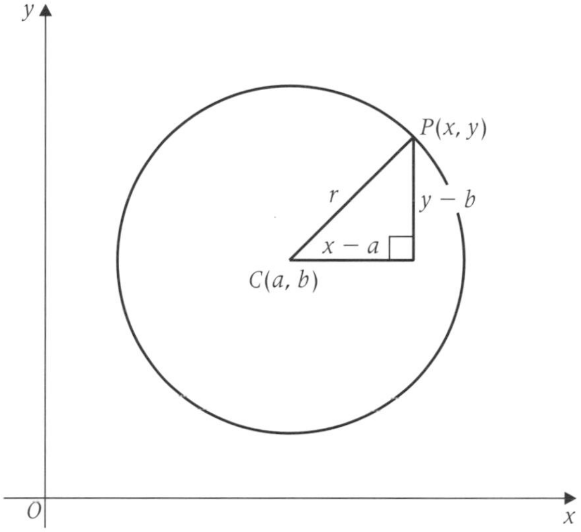

Let $P(x, y)$ be any point on the circle then $CP = r$

$\Rightarrow \quad CP^2 = r^2$.

But $CP^2 = (x - a)^2 + (y - b)^2$

$\Rightarrow \quad (x - a)^2 + (y - b)^2 = r^2$.

> The equation of a circle with centre (a, b) and radius r is $(x - a)^2 + (y - b)^2 = r^2$.

This equation is **not** given in the examination formulae booklet.

The equation $(x - a)^2 + (y - b)^2 = r^2$ can be expanded and written as $x^2 + y^2 - 2ax - 2by + a^2 + b^2 - r^2 = 0$.

An equation of this form represents a circle.

Note that the equation is of degree 2 and the coefficients of x^2 and y^2 are equal. Also there are no xy terms.

Worked example 8.1

(a) Find an equation of the circle with centre the origin and radius $\sqrt{2}$.

(b) Verify that the point (1, 1) lies on the circle.

8

Solution

(a) The centre is (0, 0) $\Rightarrow$ $a = 0, b = 0$.
The radius is $r = \sqrt{2}$.
The equation of the circle is $(x - 0)^2 + (y - 0)^2 = (\sqrt{2})^2$, which can be written as $x^2 + y^2 = 2$.

(b) The point (1, 1) lies on the circle if $x = 1, y = 1$ satisfies the equation of the circle. Since $1^2 + 1^2 = 2$, you can deduce that the point (1, 1) lies on the circle.

Worked example 8.2

(a) Write down an equation of the circle with centre (−1, 3) and radius 5.

(b) Find the coordinates of the points where the circle intersects the x-axis.

Solution

(a) The circle with centre (a, b) and radius r has equation $(x - a)^2 + (y - b)^2 = r^2$.

Centre (−1, 3) $\Rightarrow a = -1, b = 3$.

Radius, $r = 5$.

An equation of the circle with centre (−1, 3) and radius 5 is $[x - (-1)]^2 + (y - 3)^2 = (5)^2$,

$\Rightarrow \quad (x + 1)^2 + (y - 3)^2 = 25$.

(b) On the x-axis, $y = 0$.
The circle intersects the x-axis when $(x + 1)^2 + (0 - 3)^2 = 25$

$\Rightarrow \quad (x + 1)^2 = 16 \quad \Rightarrow \quad (x + 1) = \pm 4$
$\Rightarrow \quad x = 3, x = -5$

The circle intersects the x-axis at the points (3, 0) and (−5, 0).

For the circle whose equation is given in the form $(x - a)^2 + (y - b)^2 = r^2$, you can write down its centre as (a, b) and its radius as r. But how can you find the centre and radius when the equation of the circle is given in the form $x^2 + y^2 + px + qy + c = 0$? The next worked example shows you how to use the method of 'completing of the square', which you studied in section 4.3, to write $x^2 + y^2 + px + qy + c = 0$ in the form $(x - a)^2 + (y - b)^2 = r^2$ and hence to find the centre and radius of the circle.

Worked example 8.3

A circle has equation $x^2 + y^2 - 8x + 10y = 0$. Find the radius and the coordinates of the centre of the circle.

Solution

Writing the equation of the circle as $[x^2 - 8x] + [y^2 + 10y] = 0$ and completing the square for both parts gives

$$[x^2 - 8x + (-4)^2] + [y^2 + 10y + (5)^2] = (-4)^2 + (5)^2$$

$\Rightarrow \quad (x - 4)^2 + (y + 5)^2 = 16 + 25$

$\Rightarrow \quad (x - 4)^2 + (y + 5)^2 = (\sqrt{41})^2$

Compare with $(x - a)^2 + (y - b)^2 = r^2$ leads to $a = 4$, $b = -5$ and $r = \sqrt{41}$.

The circle $x^2 + y^2 - 8x + 10y = 0$ has radius $\sqrt{41}$ and centre $(4, -5)$.

EXERCISE 8A

1 Find an equation of the circle with centre C and radius r:

(a) $C(0, 0)$, $r = 7$,

(b) $C(2, 0)$, $r = 2$,

(c) $C(0, 2)$, $r = 5$,

(d) $C(2, 3)$, $r = 4$,

(e) $C(-3, -2)$, $r = 3$,

(f) $C(2, -1)$, $r = \sqrt{3}$.

2 Determine the coordinates of the centre and the radius of the circle with equation:

(a) $x^2 + y^2 - 4x = 0$,

(b) $x^2 + y^2 + 2y = 0$,

(c) $x^2 + y^2 - 2x + 4y = 0$,

(d) $x^2 + y^2 - 6x + 4y = 3$,

(e) $x^2 + y^2 - 12x + 4y = 10$,

(f) $x^2 + y^2 + 8x - 10y + 5 = 0$,

(g) $x^2 + y^2 + 4x + 2y - 6 = 0$,

(h) $2x^2 + 2y^2 - 16x + 4y = 1$.

3 **(a)** Write down an equation of the circle with centre $(-3, 4)$ and radius 5.

(b) Find the coordinates of the points where the circle intersects the coordinate axes.

4 Find, as a multiple of π, the area of the circle with equation:

(a) $x^2 + y^2 = 9$;

(b) $(x - 1)^2 + y^2 - 8 = 0$.

5 A circle, C_1, has equation $x^2 + y^2 - 6y = 3$. Another circle, C_2 has equation $x^2 + 2x + y^2 = 23$. Show that the area of C_2 is twice the area of C_1.

8

8.2 Sketching and applying translations on circles

In section 8.1 you considered the graphs of the circles
C_1: $x^2 + y^2 = r^2$
C_2: $(x - a)^2 + (y - b)^2 = r^2$

C_1 and C_2 are circles of the same size (radius) but different centres.

C_2 was 'formed' by moving C_1 from centre (0, 0) to centre (a, b); a units in the positive x-direction then b units in the positive y-direction.

> Moving a curve without altering its shape is called a **translation**. The translation vector $\begin{bmatrix} a \\ b \end{bmatrix}$ represents the move, a units in the positive x-direction then b units in the positive y-direction.

You have also used other geometrical transformations in your GCSE course: reflections, rotations, and perhaps stretches.

A translation is a geometric transformation.

> The circle $(x - a)^2 + (y - b)^2 = r^2$ can be obtained from the circle $x^2 + y^2 = r^2$ by applying the translation $\begin{bmatrix} a \\ b \end{bmatrix}$.

The two circles must have the same radius.

Worked example 8.4

Describe the geometrical transformation by which the circle C_2, with equation $(x - 1)^2 + (y + 2)^2 = 9$, can be obtained from the circle C_1 with equation $x^2 + y^2 = 9$.

Solution

C_1 has centre (0, 0) and radius 3.
C_2 has centre (1, −2) and radius 3.

Since C_1 and C_2 have the same radius, C_2 is obtained by moving C_1 from centre (0, 0) to centre (1, −2), that is 1 unit in the positive x-direction then 2 units in the negative y-direction.

The geometrical transformation is a translation of $\begin{bmatrix} 1 \\ -2 \end{bmatrix}$.

Worked example 8.5

A translation of $\begin{bmatrix} 2 \\ -1 \end{bmatrix}$ transforms the graph of the circle C_1, with equation $(x - 3)^2 + (y - 4)^2 = 16$, into the graph of the circle C_2. Find the equation of C_2.

Solution

The centre of C_1 is (3, 4). The radius of C_1 is 4.
A translation does not change the size of the circle so the radius of C_2 is also 4.

The translation $\begin{bmatrix} 2 \\ -1 \end{bmatrix}$ moves the centre of C_1 2 units in the positive x-direction then 1 unit in the negative y-direction, so the centre of C_2 is (5, 3).

An equation for C_2 is $(x - 5)^2 + (y - 3)^2 = 4^2$
or $x^2 + y^2 - 10x - 6y + 18 = 0$.

> In general, a translation of $\begin{bmatrix} a \\ b \end{bmatrix}$ transforms the graph of the circle $(x - p)^2 + (y - q)^2 = r^2$ into the graph of $(x - p - a)^2 + (y - q - b)^2 = r^2$.

Worked example 8.6

The circle C_2, with equation $x^2 + y^2 = 6y + 7$, can be obtained from the circle C_1 whose centre is (0, 0) by applying a translation. Find the equation of C_1 and describe the translation.

Solution

C_2: $x^2 + y^2 - 6y = 7$

$x^2 + [y^2 - 6y + (-3)^2] = 7 + (-3)^2$

$x^2 + (y - 3)^2 = 16$

Comparing with $(x - a)^2 + (y - b)^2 = r^2$, gives $a = 0$, $b = 3$ and $r = 4$.
C_2 has radius 4 and centre (0, 3).
Since a translation does not change the radius of the circle, C_1 must also have radius 4. Since its centre is (0, 0), the equation of C_1 is $x^2 + y^2 = 16$.
The translation moves the centre from (0, 0) to (0, 3), that is a move of 3 units in the positive y-direction, so the translation is $\begin{bmatrix} 0 \\ 3 \end{bmatrix}$.

We now consider how to sketch a circle:

> To sketch a circle:
> **1** find the radius and coordinates of the centre of the circle;
> **2** indicate the centre;
> **3** mark the four points which show the ends of the horizontal and vertical diameters;
> **4** draw the circle to pass through these four points;
> **5** if any intercepts with the coordinate axes are integers, normally they should also be indicated.

8

Worked example 8.7

Sketch the circle whose equation is $x^2 + y^2 - 2x - 4y - 4 = 0$.

Solution

Writing the equation as $[x^2 - 2x] + [y^2 - 4y] = 4$.
Completing the squares gives
$[x^2 - 2x + (-1)^2] + [y^2 - 4y + (-2)^2] = 4 + (-1)^2 + (-2)^2$
$\Rightarrow \quad (x - 1)^2 + (y - 2)^2 = 9$

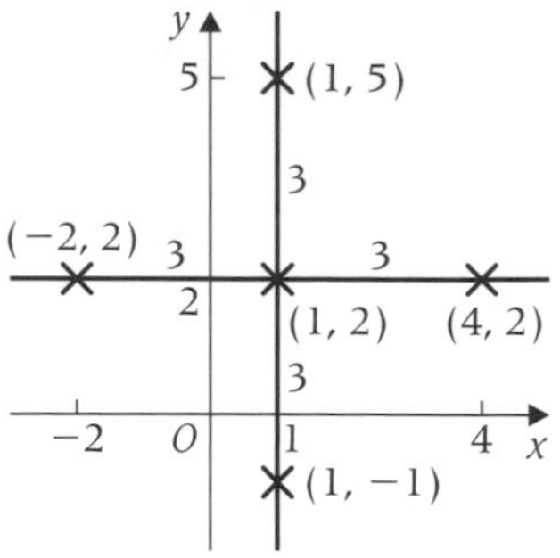

The circle has centre (1, 2) and radius 3.
The ends of the horizontal diameter are (1 − 3, 2) and (1 + 3, 2).
The ends of the vertical diameter are (1, 2 − 3) and (1, 2 + 3).
Mark the points (−2, 2) and (4, 2) and also (1, −1) and (1, 5).

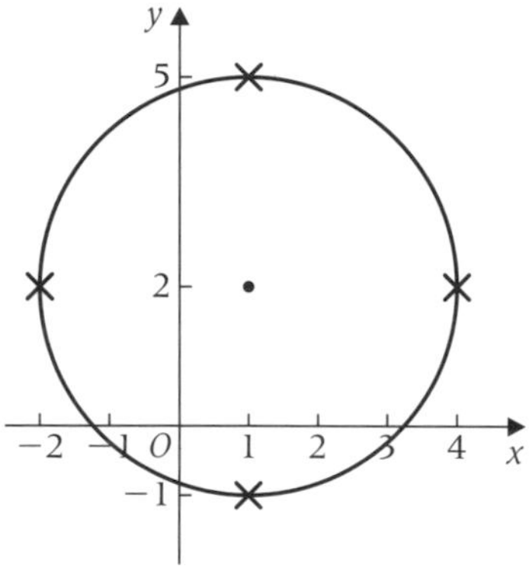

You should note that the circle $x^2 + y^2 - 2x - 4y - 4 = 0$ can be obtained by applying the translation $\begin{bmatrix} 1 \\ 2 \end{bmatrix}$ to the circle $x^2 + y^2 = 9$.

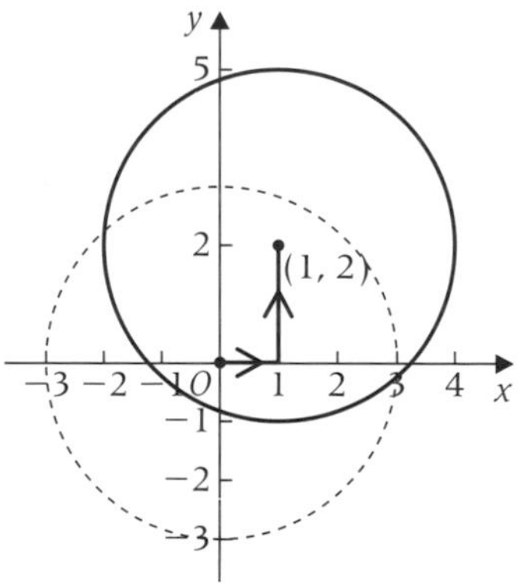

EXERCISE 8B

1 Describe the geometrical transformation which has been applied to the circle $x^2 + y^2 = 4$ to obtain the circle:

(a) $(x - 2)^2 + y^2 = 4$,

(b) $x^2 + (y + 3)^2 = 4$,

(c) $(x - 2)^2 + (y + 3)^2 = 4$,

(d) $(x + 1)^2 + (y - 4)^2 = 4$,

(e) $x^2 + y^2 - 4x + 6y + 9 = 0$,

(f) $x^2 + y^2 + 10x - 4y + 25 = 0$.

2 Write down the equation of the circle which is obtained by applying the given translation to the given circle:

(a) $\begin{bmatrix} 0 \\ 3 \end{bmatrix}$, $x^2 + y^2 = 49$,

(b) $\begin{bmatrix} -2 \\ 0 \end{bmatrix}$, $x^2 + y^2 = 1$,

(c) $\begin{bmatrix} 2 \\ 5 \end{bmatrix}$, $x^2 + y^2 = 16$,

(d) $\begin{bmatrix} -4 \\ 1 \end{bmatrix}$, $x^2 + y^2 - 81 = 0$,

(e) $\begin{bmatrix} 0 \\ 2 \end{bmatrix}$, $(x - 1)^2 + (y + 2)^2 = 9$,

(f) $\begin{bmatrix} -4 \\ 0 \end{bmatrix}$, $(x + 4)^2 + (y - 2)^2 = 25$,

(g) $\begin{bmatrix} -2 \\ 1 \end{bmatrix}$, $(x - 2)^2 + y^2 = 10$,

(h) $\begin{bmatrix} 3 \\ -4 \end{bmatrix}$, $(x + 2)^2 + (y + 3)^2 = 8$,

(i) $\begin{bmatrix} 1 \\ 0 \end{bmatrix}$, $x^2 + y^2 + 4x - 2y = 2$,

(j) $\begin{bmatrix} -1 \\ 2 \end{bmatrix}$, $x^2 + y^2 - 4x + 2y = 4$.

3 The circle C_2, with equation $x^2 + y^2 - 4x = 5$, can be obtained from the circle C_1, whose centre is (0, 0), by applying a translation. Find the equation of C_1 and describe the translation.

4 The circle C_2, with equation $x^2 + y^2 + 6x - 4y = 12$, can be obtained from the circle C_1 whose centre is (0, 0), by applying a translation. Find the equation of C_1 and describe the translation.

5 Sketch the graph of:

(a) $x^2 + y^2 = 16$,
(b) $(x - 1)^2 + y^2 = 1$,
(c) $x^2 + (y + 1)^2 = 1$,
(d) $(x - 2)^2 + (y + 1)^2 = 4$,
(e) $x^2 + y^2 - 4y = 5$,
(f) $x^2 + y^2 + 2x + 2y + 1 = 0$,
(g) $x^2 + y^2 - 4x + 6y = 12$,
(h) $x^2 + y^2 - 6x + 8y = 0$.

6 Describe the geometrical transformation which has been applied to the circle $x^2 + y^2 = 2x - 4y + 11$ to obtain the circle $x^2 + y^2 + 6x = 7$.

8.3 Finding the equation of a circle using circle properties

In GCSE you used some circle properties to solve geometrical problems. In this section you will learn how to apply these properties to find the equation of circles.

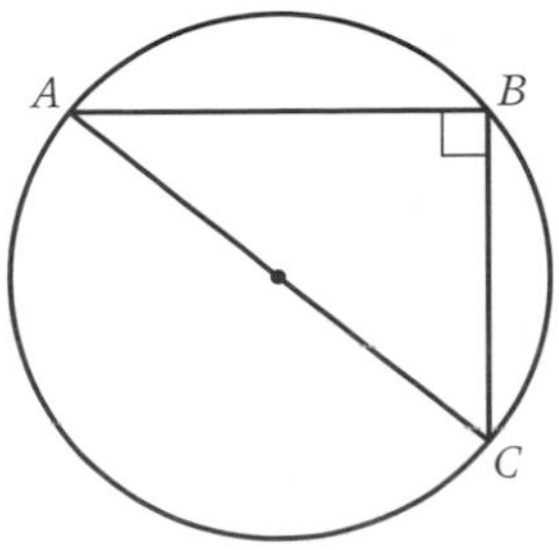

The angle in a semicircle is a right angle:
AC is a diameter $\Rightarrow$ angle $ABC = 90°$.
also
Angle $ABC = 90° \Rightarrow AC$ is a diameter.

Worked example 8.8

A right-angled triangle PQR has vertices $P(2, 14)$, $Q(-6, 2)$ and $R(12, -10)$. Find the equation of the circle which passes through the three points P, Q and R.

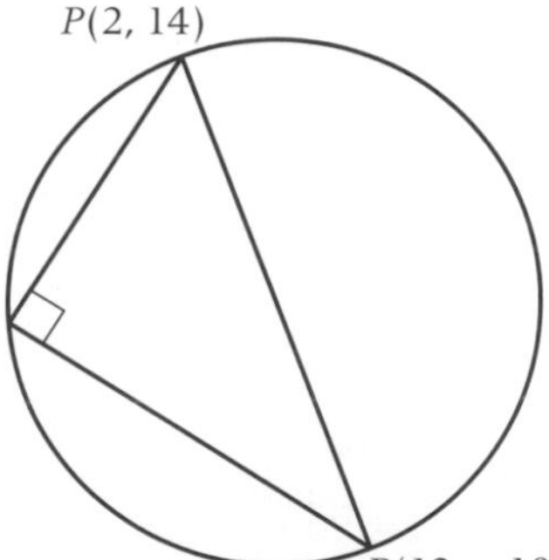

Solution

Angle $PQR = 90° \Rightarrow PQR$ is a semicircle $\Rightarrow PR$ is a diameter.

So the centre, C, of the circle is the mid-point of PR.

C is $\left(\dfrac{2+12}{2}, \dfrac{14+(-10)}{2}\right) = C(7, 2)$.

The radius of the circle $= QC = \sqrt{(7-(-6))^2 + (2-2)^2} = 13$.

You could also use radius $= PC$ or radius $= RC$.

The equation of the circle with centre $(7, 2)$ and radius 13 is $(x-7)^2 + (y-2)^2 = 13^2$,

Used $(x-a)^2 + (y-b)^2 = r^2$.

so the circle through the points P, Q and R has equation $x^2 + y^2 - 14x - 4y - 116 = 0$.

Unless told otherwise, both forms of the equation will be acceptable in the examination.

If you know the coordinates of the end points of a diameter you could use the same method as in worked example 8.8 to find the equation of the circle.

To find the equation of a circle given the coordinates of A and B, the end points of the diameter AB:

(i) find the coordinates of the mid-point of AB (see section 3.3) – this gives the centre $C(a, b)$ of the circle;

(ii) find the distance CA (or CB) (see section 3.2) – this gives the radius r of the circle;

(iii) use the equation of the circle is $(x-a)^2 + (y-b)^2 = r^2$.

But how could you find the equation of a circle given the coordinates of three points on the circle? To do this you will need to recall another circle property.

You found the equation of a perpendicular bisector in exercise 3F.

The perpendicular from the centre of the circle to a chord bisects the chord.

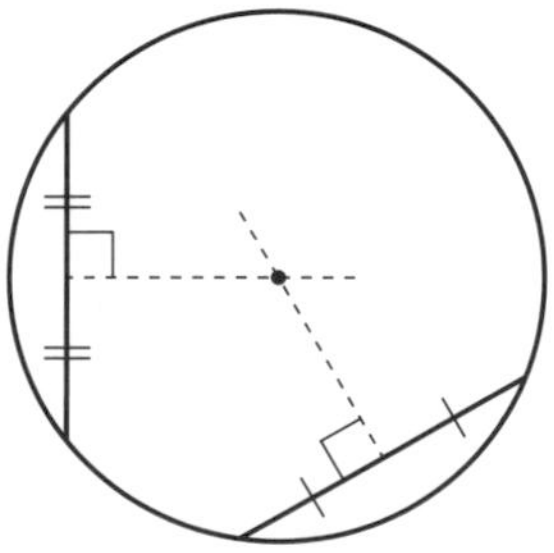

The perpendicular bisectors of two chords intersect at the centre of the circle.

By finding the equations of the perpendicular bisectors of two chords and solving the two equations simultaneously you can find the centre of the circle and hence the equation of the circle.

The next worked example shows you how to find the equation of a circle given the coordinates of three points on the circle.

Worked example 8.9

A circle passes through the three points $O(0, 0)$, $A(0, 8)$ and $B(-2, 4)$.

(a) Find the equations of the perpendicular bisectors of the chords OA and OB.

(b) Hence find the coordinates of the centre of the circle.

(c) Find the equation of the circle.

Solution

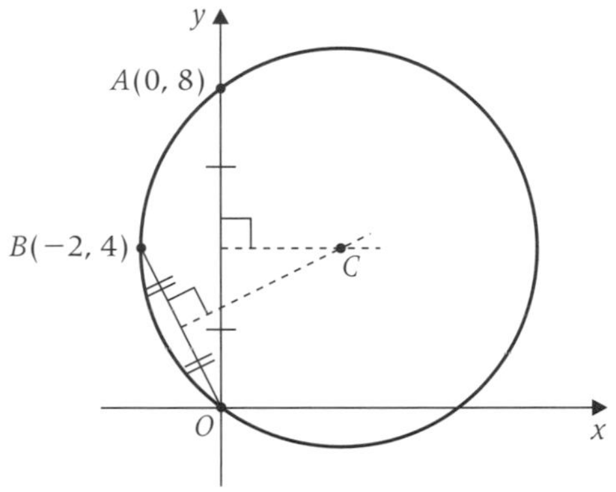

(a) Mid-point of OA is $(0, 4)$.
OA is vertical so the perpendicular is horizontal (gradient $= 0$).
Equation of the perpendicular bisector of OA is $y = 4$.

$y - 4 = 0(x - 0)$

Mid-point of $OB = \left(\frac{-2+0}{2}, \frac{4+0}{2}\right) = (-1, 2)$.

Gradient of $OB = \frac{4-0}{-2-0} = -2$.

Gradient of perpendicular bisector of OB is $\frac{1}{2}$.

Used $m_1 \times m_2 = -1$.

Equation of the perpendicular bisector of OB is
$y - 2 = \frac{1}{2}(x - (-1))$
or $2y = x + 5$.

(b) The centre of the circle is the point of intersection of the perpendicular bisectors of the chords OA and OB.
Solving $y = 4$ and $2y = x + 5$ simultaneously gives $x = 3$, $y = 4$. The centre of the circle is $C(3, 4)$.

(c) Radius of the circle $= OC = \sqrt{(3-0)^2 + (4-0)^2} = \sqrt{25} = 5$.
The equation of the circle is $(x-3)^2 + (y-4)^2 = 5^2$
or $x^2 + y^2 - 6x - 8y = 0$.

Check $r = CB = 3 - (-2) = 5$.

Worked example 8.10

The origin O, and the point P, lie on the circle $x^2 + y^2 - 6x + 8y = 0$. Given that OP is a diameter of the circle, find the coordinates of P.

Solution

$x^2 - 6x + y^2 + 8y = 0$

Completing the square gives $(x^2 - 6x + 9) + (y^2 + 8y + 16) = 9 + 16$
or $(x - 3)^2 + (y + 4)^2 = 5^2$.

The centre of the circle is $C(3, -4)$.
Point C is the mid-point of the diameter OP, where O is $(0, 0)$.

Let P be the point (p, q) then $\dfrac{0 + p}{2} = 3$ and $\dfrac{0 + q}{2} = -4 \Rightarrow p = 6$ and $q = -8$.

P is the point $(6, -8)$.

EXERCISE 8C

1 The points $A(4, 0)$ and $B(0, 4)$ lie on the circle $x^2 + y^2 = 16$. The perpendicular from the centre of the circle meets the chord AB at the point N.

(a) Write down the coordinates of N.

(b) Find the distance from the centre of the circle to the point N.

2 Triangle PQR has vertices $P(-2, -1)$, $Q(-7, 4)$ and $R(-1, 1)$.

(a) Show that PR is perpendicular to QR.

(b) Find the equation of the circle which passes through the three points P, Q and R.

3 PQ is a diameter of a circle. Find an equation of the circle when:

(a) P is $(2, 0)$, Q is $(6, 0)$,

(b) P is $(0, 5)$, Q is $(0, -1)$,

(c) P is $(2, 1)$, Q is $(6, 4)$,

(d) P is $(3, 5)$, Q is $(7, -3)$,

(e) P is $(-2, -3)$, Q is $(6, 2)$,

(f) P is $(1, 3)$, Q is $(0, -4)$.

4 Points A and B lie on the circle $(x - 3)^2 + (y - 1)^2 = 25$ such that AB is a diameter of the circle. Given that A is the point $(7, -2)$ find the coordinates of B.

5 The origin, O, and the point A lie on the circle $x^2 + y^2 - 4x + 2y = 0$. Given that OA is a diameter of the circle find the coordinates of A.

6 A circle passes through the three points $O(0, 0)$, $A(4, 0)$ and $B(2, 4)$.

(a) Find the equations of the perpendicular bisectors of the chords OA and OB.

(b) Hence find the coordinates of the centre of the circle.

(c) Find the equation of the circle.

7 Find the equation of the circle which passes through the points $(2, -1)$, $(4, -1)$ and $(0, 1)$.

8.4 Conditions for a line to meet a circle

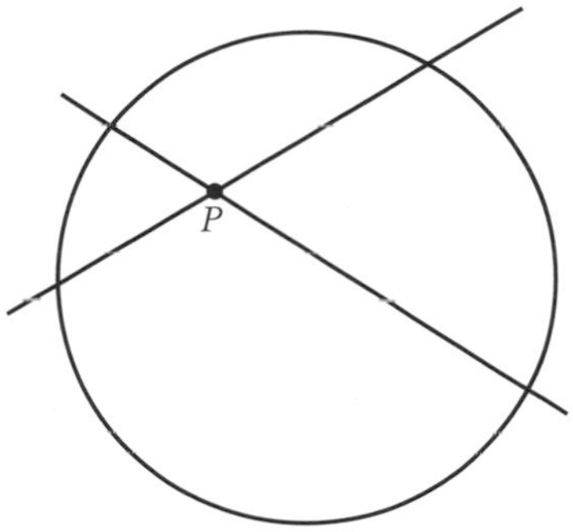

Clearly any line, which passes through a given point that lies inside a circle, will intersect the circle in two points no matter what the gradient of the line is. If the given point lies outside the circle the problem is less trivial. In this section you will be shown the conditions **(i)** for the line to intersect the circle, **(ii)** for the line to be a tangent to the circle, and **(iii)** for the line not to meet the circle.

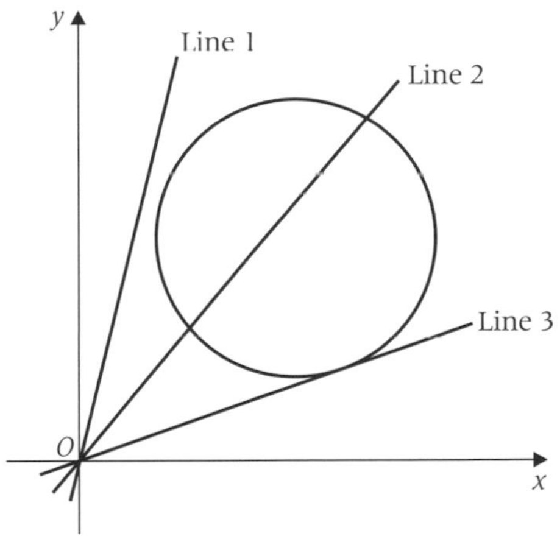

The diagram shows three lines drawn through the origin.

Line 1 does not intersect the circle;

Line 2 intersects the circle in two distinct points;

Line 3 touches the circle at a single point and is called a tangent.

In section 7.2 you were shown how to find the points of intersection of lines and quadratic graphs. When applied to a line and a circle you have the following general method:

- from the equation of the line make x (or y) the subject;
- substitute into the equation of the circle to get a quadratic equation in y (or x) of the form $ay^2 + by + c = 0$ (or $ax^2 + bx + c = 0$);
- **(i)** if the discriminant $b^2 - 4ac < 0$, there are no real roots and the line does not intersect the circle (**Line 1** type),
  (ii) if the discriminant $b^2 - 4ac > 0$, there are two real distinct roots and the line intersects the circle at two points (**Line 2** type),
  (iii) if the discriminant $b^2 - 4ac = 0$, there is one real (repeated) root and the line touches the circle at one point, the line is a tangent to the circle (**Line 3** type);
- if asked to find the coordinates of the points of intersection, solve the quadratic equation and substitute the found value(s) into the equation of the line to find the other coordinate(s).

Worked example 8.11

Show that the line $x + y + 3 = 0$ and the circle $x^2 + y^2 = 4$ do not intersect.

Solution

Rearranging the equation of the line gives $y = -x - 3$.

Substituting into $x^2 + y^2 = 4$ gives

$$x^2 + (-x - 3)^2 = 4$$
$$\Rightarrow \quad x^2 + x^2 + 6x + 9 = 4$$
$$\Rightarrow \quad 2x^2 + 6x + 5 = 0$$
$$\Rightarrow \quad \text{discriminant } b^2 - 4ac = 6^2 - 4(2)(5) = -4 < 0$$

so the line does not intersect the circle.

Worked example 8.12

Show that the line $x + y + 3 = 0$ and the circle $x^2 + y^2 = 5$ intersect, and find the coordinates of the points of intersection.

Solution

Rearranging the equation of the line gives $y = -x - 3$.
Substituting into $x^2 + y^2 = 5$ gives

$$x^2 + (-x - 3)^2 = 5$$
$$\Rightarrow \quad x^2 + x^2 + 6x + 9 = 5$$
$$\Rightarrow \quad 2x^2 + 6x + 4 = 0$$
$$\Rightarrow \quad \text{discriminant } b^2 - 4ac = 6^2 - 4(2)(4) = 4 > 0$$

so the line intersects the circle in two points.

Factorising the quadratic gives $2(x + 2)(x + 1) = 0$
$\Rightarrow \quad x = -2, x = -1.$

When $x = -2$, $y = -(-2) - 3 = -1$.

When $x = -1$, $y = -(-1) - 3 = -2$.

Line $x + y + 3 = 0$ intersects the circle $x^2 + y^2 = 5$ at $(-1, -2)$ and $(-2, -1)$.

Worked example 8.13

The line $y = mx - 3$ is a tangent to the circle $x^2 + y^2 = 5$. Find the possible values of m.

Solution

Substituting $y = mx - 3$ into $x^2 + y^2 = 5$ gives

$$x^2 + (mx - 3)^2 = 5$$
$$\Rightarrow \quad x^2 + m^2x^2 - 6mx + 9 = 5$$
$$\Rightarrow \quad (1 + m^2)x^2 - 6mx + 4 = 0$$

For the line to be a tangent to the circle the roots of this quadratic must be real and equal,

$$\Rightarrow \quad b^2 - 4ac = 0$$
$$\Rightarrow \quad 36m^2 - 4(1 + m^2)(4) = 0$$
$$\Rightarrow \quad 20m^2 = 16$$
$$\Rightarrow \quad m^2 = \frac{4}{5} \Rightarrow m = \pm\frac{2}{\sqrt{5}}.$$

EXERCISE 8D

1 Show that the line $2y + x = 5$ intersects the circle $x^2 + y^2 = 25$ and find the coordinates of the points of intersection.

2 Show that the line $2y + x = 7$ does not meet the circle $x^2 + y^2 = 9$.

3 Show that the line $y - 2x = 5$ is a tangent to the circle $x^2 + y^2 = 5$ and find the coordinates of the point of contact.

4 Show that the line $8x + y + 10 = 0$ is a tangent to the circle $x^2 + y^2 - 6x + 3y - 5 = 0$ and find the coordinates of the point of contact.

5 Show that the line $2y + x = 9$ does not meet the circle $x^2 + y^2 + 8y + 1 = 0$.

6 Show that the line $y + x = 3$ intersects the circle $x^2 + y^2 - 6x + 8y - 1 = 0$ and find the coordinates of the points of intersection.

7 The line $y = mx$ is a tangent to the circle $x^2 + y^2 - 10y + 16 = 0$.

(a) Find the two possible values of m.

(b) The tangents meet the circle at points A and B. Find the length of AB.

8.5 The length of the tangents from a point to a circle

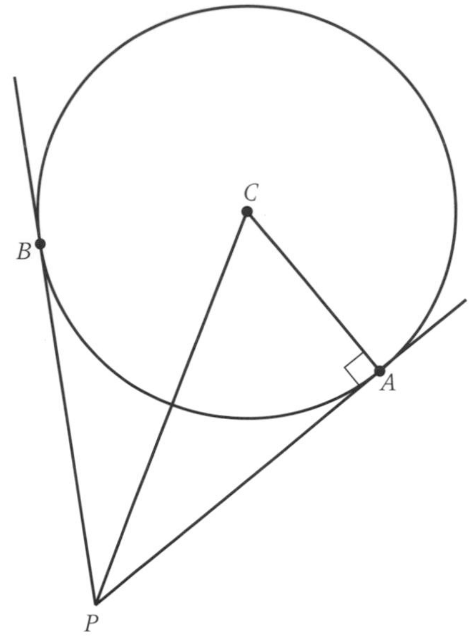

Tangents PA and PB are drawn from point P to the circle with centre C. The radius CA is perpendicular to the tangent PA.

The radius is always perpendicular to the tangent.

To find the length of the tangent PA use Pythagoras' theorem $PA^2 = CP^2 - CA^2$.

Tangents from a point to a circle have equal lengths.

Worked example 8.14

Find the length of the tangents from the point $P(5, 6)$ to the circle $(x - 1)^2 + (y - 2)^2 = 9$.

Solution

The circle $(x - 1)^2 + (y - 2)^2 = 9$ has centre $C(1, 2)$ and radius 3.

P is the point $(5, 6)$ so
$CP^2 = [(5 - 1)^2 + (6 - 2)^2] = 32$.

The distance between (x_1, y_1) and (x_2, y_2) is $\sqrt{(x_2 - x_1)^2 + (y_2 - y_1)^2}$

If A is the point of contact of one of the tangents then PA is the length of the tangent and CA is the radius.

$CAP = 90°$ so $PA^2 = CP^2 - CA^2$.

$\Rightarrow \quad PA^2 = 32 - 3^2 = 23$

$\Rightarrow \quad PA = \sqrt{23}$.

Radius and tangent are perpendicular. Pythagoras in triangle CAP.

$\Rightarrow$ Lengths of the tangents are $\sqrt{23}$.

Tangents from a point to circle are equal.

EXERCISE 8E

1 Find the lengths of the tangents from the point (3, 4) to the circle $x^2 + y^2 = 4$.

2 Find the lengths of the tangents from the point (5, 7) to the circle $x^2 + y^2 - 2x - 4y - 4 = 0$.

3 The lengths of the tangents from the point $(4, k)$ to the circle $x^2 + y^2 = 9$ are $4\sqrt{2}$. Find the possible values of the constant k.

4 Show that the tangents from the point $(3, -4)$ to the circles $x^2 + y^2 + 2x - 4y + 4 = 0$ and $x^2 + y^2 + 4x - 4y - 2 = 0$ are equal in length.

5 Show that the lengths of the tangents from the point (h, k) to the circle $x^2 + y^2 + 2fx + 2gy + c = 0$ are $\sqrt{h^2 + k^2 + 2fh + 2gk + c}$.

8.6 The equation of the tangent and the equation of the normal at a point on a circle

A **tangent** at a point on a curve is a line which touches the curve at that point.
A **normal** at a point on a curve is the line which is perpendicular to the tangent at that point.

In the diagram, AT is the tangent at P and BN is the normal at P. Usually you will need to use differentiation (see chapter 10) to find the equations of normals and tangents but in this section you will learn how to find the equations of tangents and normals to circles without using differentiation.
The method makes use of the circle property
'the tangent to a circle is perpendicular to the radius at its point of contact'.
Since the radius is perpendicular to the tangent it follows that

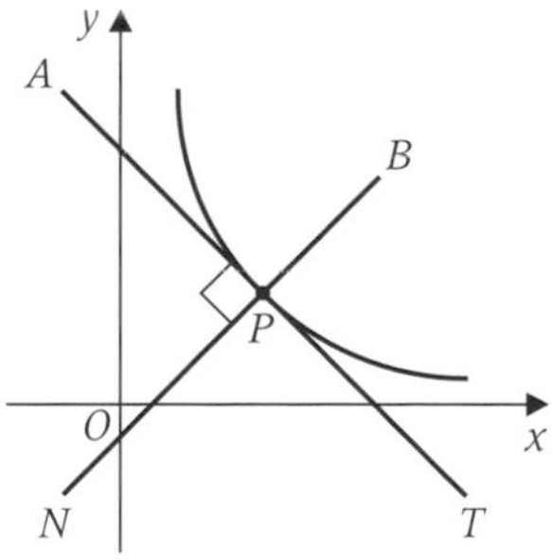

> The normal at a given point P on a circle with centre C is the same as the radius CP.

Worked example 8.15

The point $P(1, 3)$ lies on the circle $x^2 + y^2 = 10$.

(a) Find the equation of the normal at P.

(b) Find the equation of the tangent at P.

Solution

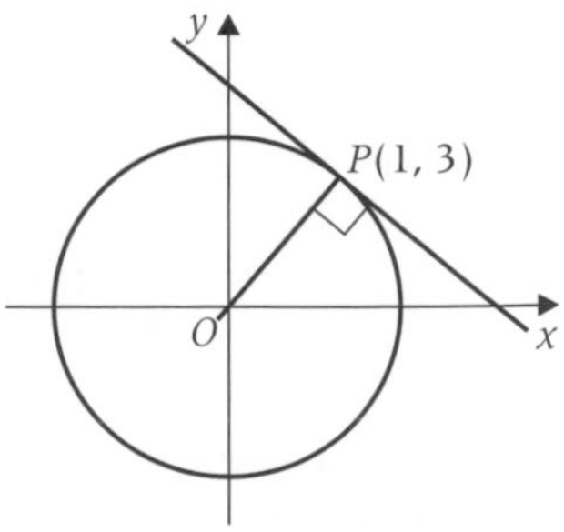

(a) The centre of the circle $x^2 + y^2 = 10$ is $O(0, 0)$.

The normal to the circle at P is the radius OP.

The gradient of $OP = \dfrac{3 - 0}{1 - 0} = 3$.

The equation of the normal OP is $y - 0 = 3(x - 0)$ or $y = 3x$.

(b) The tangent at P is perpendicular to OP so

the gradient of the tangent at P is $-\dfrac{1}{3}$.

Used $m_1 \times m_2 = -1$.

The equation of the tangent at $P(1, 3)$ is $y - 3 = -\dfrac{1}{3}(x - 1)$

or $3y + x = 10$.

> To find the equations of tangents and normals to circles at the point $P(x_1, y_1)$:
>
> **(i)** find the centre C of the circle;
> **(ii)** find the gradient, m_1, of CP;
> **(iii)** the normal is CP, so its equation can be found using $y - y_1 = m_1(x - x_1)$;
> **(iv)** find the gradient, m_2, of the tangent using $m_1 \times m_2 = -1$;
> **(v)** the equation of the tangent is $y - y_1 = m_2(x - x_1)$.

Worked example 8.16

Find the equation of the tangent at the point $P(4, -2)$ on the circle $x^2 + y^2 - 4x + 10y + 16 = 0$.

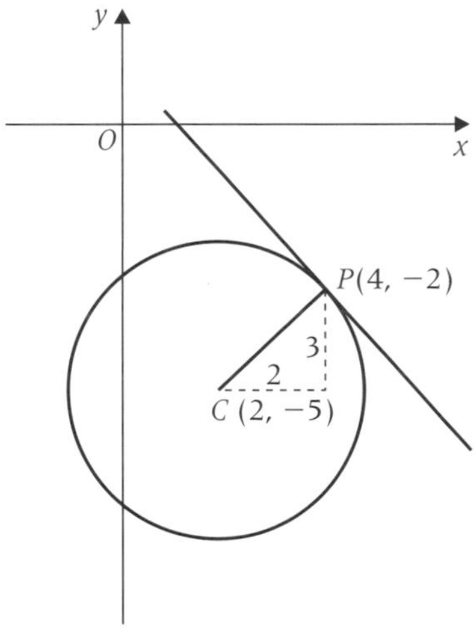

Solution

$x^2 + y^2 - 4x + 10y + 16 = 0 \Rightarrow (x - 2)^2 + (y + 5)^2 = 13$.
The centre of the circle is $C(2, -5)$.

The gradient of the radius $CP = \dfrac{-2 - (-5)}{4 - 2} = \dfrac{3}{2}$,

so the gradient of the tangent at $P(4, -2)$ is $-\dfrac{2}{3}$.

Used $m_1 \times m_2 = -1$, since the radius and tangent are perpendicular.

The equation of the tangent at $P(4, -2)$ is

$y - (-2) = -\dfrac{2}{3}(x - 4)$

Used $y - y_1 = m(x - x_1)$.

or $3y + 2x = 2$.

The next worked example shows you how to find the equation of the normal when given the equation of the tangent. The method can also be applied to find the equation of the tangent when the normal is given.

Worked example 8.17

The line $y = x + 3$ is a tangent to a circle at the point $P(-1, 2)$. Find the equation of the normal to the circle at P.

Solution

The gradient of the tangent $y = x + 3$ is 1.

Compared with $y = mx + c$.

$\Rightarrow$ the gradient of the normal at P is -1.

Used $m_1 \times m_2 = -1$.

The equation of the normal at $P(-1, 2)$ is $y - 2 = -1[x - (-1)]$
or $y + x = 1$.

EXERCISE 8F

1 The point $P(3, 0)$ lies on the circle $(x - 1)^2 + (y - 2)^2 = 8$.

(a) Write down the coordinates of the centre of the circle.

(b) Find the equation of the normal to the circle at P.

(c) Find the equation of the tangent to the circle at P.

2 The point $P(4, -3)$ lies on the circle with equation $x^2 + y^2 = 25$.

(a) By sketching the circle write down the equations of the two vertical tangents to the circle.

(b) Find the equation of the tangent to the circle at P.

3 The equation of the normal to a circle at the point $P(3, 2)$ is $2y - x = 1$. Find the equation of the tangent to the circle at the point P.

4 Find the equation of the tangent at the point $P(6, 2)$ on the circle $x^2 + y^2 - 6x - 2y = 0$.

5 The circle with equation $(x - 5)^2 + (y - 7)^2 = 25$ has centre C.
The point $P(2, 3)$ lies on the circle.

(a) Determine the gradient of PC.

(b) Hence, find the equation of the tangent to the circle at P.

6 The point $P(3, 3)$ lies on the circle $x^2 + y^2 = 18$.

(a) Find the equation of the tangent to the circle at P.

(b) Find the equation of the normal to the circle at P.

(c) This normal intersects the circle again at the point Q. Find the coordinate of Q.

7 The equation of the tangent to a circle at the point P(5, 3) is $2y + x = 11$. Find the equation of the normal to the circle at the point P.

8 **(a)** Determine the coordinates of the centre and the radius of the circle with equation $x^2 + y^2 - 6x + 8y = 0$.

(b) Find the coordinates of the points P and Q, where the line $x = 7$ intersects this circle.

(c) Find the equations of the tangents to the circle at P and Q.

9 A circle C has its centre at the point $Q(-1, 4)$. The point $P(3, 0)$ lies on the circle C.

(a) Find the gradient of the radius PQ.

(b) Find the equation of the tangent to C at the point P.

(c) Find the equation of the circle C.

10 The line $4y + 3x = 75$ is a tangent to a circle at the point $P(9, 12)$.

(a) Show that the gradient of the normal to the circle at P is $\frac{4}{3}$.

(b) Find the equation of the normal to the circle at the point P.

(c) The centre of the circle is $C(6, k)$. Show that $k = 8$.

(d) Find the equation of the circle.

MIXED EXERCISE

1 A circle with centre C has equation $x^2 + y^2 - 6x + 4y = 7$.

(a) By using the method of completing the square write this equation in the form $(x - a)^2 + (y - b)^2 = r^2$.

(b) Write down the coordinates of C and the radius of the circle.

(c) Find the x-coordinates of the points A and B where the circle cuts the x-axis.

(d) Find the coordinates of the mid-point of AB and hence find length of the perpendicular from C to the chord AB.

2 The points A and P have coordinates (7, 4) and (2, 5), respectively, and O is the origin.

(a) Find the equations of the two circles passing through P, one with centre O, the other with centre A.

(b) Find the equations of the tangents at P to both circles.

3 A circle C has equation $(x - 1)^2 + (y + 3)^2 = 16$.

(a) Describe a geometrical transformation by which C can be obtained from the circle with equation $x^2 + y^2 = 16$.

(b) By sketching the circle C, write down the equations of the two horizontal tangents to the circle C.

4 **(a)** Find the equation of the circle, centre $(-3, 2)$ and radius 5.

(b) Calculate the coordinates of the points P and Q, where this circle cuts the y-axis.

(c) Find the gradients of the tangents to the circle at P and Q.

(d) Calculate the coordinates of the point of intersection, R, of the tangents at P and Q. [A part]

5 A circle with centre C has equation $x^2 + y^2 - 12x + 6y - 20 = 0$.

(a) By using the method of completing the square find the coordinates of C and the radius of the circle.

(b) The origin O is the mid-point of a chord PQ of this circle.
(i) Show that the gradient of the chord PQ is 2.
(ii) Find the distance of O from the centre of the circle.
(iii) Find the length of the chord PQ.

6 **(a)** Solve the simultaneous equations

$$2x + y = 9,$$
$$(x - 2)^2 + y^2 = 5.$$

(b) Hence state the nature of the geometrical relationship between the line with equation $2x + y = 9$, and the circle with equation $(x - 2)^2 + y^2 = 5$ at the point $(4, 1)$. [A]

7 The circle C has equation $(x - 4)^2 + (y - 3)^2 = r^2$, where r is a positive constant, and line L has equation $x = 10$.
Given that C passes through the origin:

(a) find the value of r,

(b) show that L and C have no points of intersection,
[A part]

8 A circle C has equation $(x - 3)^2 + (y + 2)^2 = 9$ and a straight line L has equation $3x + 4y = 1$.

(a) Show that the centre of the circle C lies on the line L.

(b) Find, in surd form, the x-coordinates of the points where the circle C crosses the x-axis. [A part]

9 The points $A(2, 5)$ and $B(-4, 13)$ lie at opposite ends of a diameter of a circle with centre C.

(a) Find the radius of the circle.

(b) Find the equation of the circle in the form $x^2 + y^2 + px + qy + c = 0$.

(c) Find the equation of the normal to the circle at the point A.

10 A circle C has equation $x^2 + y^2 - 2x - 17 = 0$.

(a) By completing the square find the radius and coordinates of the centre of the circle C.

(b) The circle C can be obtained by applying a translation to a circle whose centre is the origin. Describe the translation.

(c) The point $P(-2, 3)$ lies on the circle C.

(i) Find the equation of the normal to the circle at P.

(ii) Show that the equation of the tangent to the circle at P is $y = x + 5$.

(iii) This tangent intersects the x-axis at A and the y-axis at B.

Show that the length of $AB = \frac{5}{3} \times$ the radius of C.

Key point summary

1 The equation of a circle with centre (0, 0) and radius r is $x^2 + y^2 = r^2$. p115

2 The equation of a circle with centre (a, b) and radius r is $(x - a)^2 + (y - b)^2 = r^2$. p115

3 Moving a curve without altering its shape is called a **translation**. The translation vector $\begin{bmatrix} a \\ b \end{bmatrix}$ represents the move, a units in the positive x-direction then b units in the positive y-direction. p118

4 The circle $(x - a)^2 + (y - b)^2 = r^2$ can be obtained from the circle $x^2 + y^2 = r^2$ by applying the translation $\begin{bmatrix} a \\ b \end{bmatrix}$. p118

5 In general, a translation of $\begin{bmatrix} a \\ b \end{bmatrix}$ transforms the graph of the circle $(x - p)^2 + (y - q)^2 = r^2$ into the graph of $(x - p - a)^2 + (y - q - b)^2 = r^2$. p119

6 To sketch a circle: p119

(i) find the radius and coordinates of the centre of the circle;

(ii) indicate the centre;

(iii) mark the four points which show the ends of the horizontal and vertical diameters;

(iv) draw the circle to pass through these four points;

(v) if any intercepts with the coordinate axes are integers, normally they should also be indicated.

7 To find the equation of a circle given the coordinates of A and B, the end points of the diameter AB: p122
(i) find the coordinates of the mid-point of AB (see section 3.3) – this gives the centre $C(a, b)$ of the circle;
(ii) find the distance CA (or CB) (see section 3.2) – this gives the radius r of the circle;
(iii) use the equation of the circle as $(x - a)^2 + (y - b)^2 = r^2$.

8 The perpendicular bisectors of two chords intersect at the centre of the circle. p122

9 To determine the conditions for a line to intersect a circle: p126
- from the equation of the line make x (or y) the subject;
- substitute into the equation of the circle to get a quadratic equation in y (or x) of the form $ay^2 + by + c = 0$ (or $ax^2 + bx + c = 0$);
- (i) if the discriminant $b^2 - 4ac < 0$, there are no real roots and the line does not intersect the circle,
  (ii) if the discriminant $b^2 - 4ac > 0$, there are two real distinct roots and the line intersects the circle at two points,
  (iii) if the discriminant $b^2 - 4ac = 0$, there is one real (repeated) root and the line touches the circle at one point, the line is a tangent to the circle;
- if asked to find the coordinates of the points of intersection, solve the quadratic equation and substitute found value(s) into the equation of the line to find the other coordinate(s).

10 The normal at a given point P, on a circle with centre C is the same as the radius CP. p129

11 To find the equations of tangents and normals to circles at the point $P(x_1, y_1)$: p130
(i) find the centre C of the circle;
(ii) find the gradient, m_1, of CP;
(iii) the normal is CP, so its equation can be found using $y - y_1 = m_1(x - x_1)$;
(iv) find the gradient, m_2, of the tangent using $m_1 \times m_2 = -1$;
(v) the equation of the tangent is $y - y_1 = m_2(x - x_1)$.

8

Test yourself	What to review
1 A circle C has centre (2, 1) and radius 10. Write down an equation for C and verify that the point (8, −7) lies on the circle.	*Section 8.1*
2 A circle with centre C has equation $x^2 + y^2 + 6x - 2y + 9 = 0$. **(a)** By using the method of completing the square write this equation in the form $(x - a)^2 + (y - b)^2 = r^2$. **(b)** Write down the coordinates of C and the radius of the circle. **(c)** By sketching the circle show that the x-axis is a tangent to the circle and state the coordinates of the point of contact.	*Sections 8.1 and 8.2*
3 A translation of $\begin{bmatrix} -2 \\ 1 \end{bmatrix}$ transforms the graph of the circle C_1 with equation $(x - 2)^2 + (y - 1)^2 = 16$ into the graph of the circle C_2. Find an equation of C_2.	*Section 8.2*
4 A right-angled triangle PQR has vertices $P(2, 0)$, $Q(4, 1)$ and $R(2, 5)$. Find the equation of the circle which passes through the three points P, Q and R.	*Section 8.3*
5 Find the possible values for the constant c so that the line $y = 2x + c$ is a tangent to the circle $x^2 + y^2 = 4$.	*Section 8.4*
6 **(a)** By solving the equations $y = 5 - 2x$ and $x^2 + y^2 = 5$ simultaneously, show that the line $y = 5 - 2x$ is a tangent to the circle $x^2 + y^2 = 5$ and state the coordinates of the point P where the line meets the circle. **(b)** Find the equation of the normal to the circle at the point P.	*Section 8.4 and 8.6*
7 Find the equation of the tangent at the point $P(2, 4)$ on the circle $x^2 + y^2 - 10x = 0$.	*Section 8.6*
8 Point C is the centre of the circle $(x - 1)^2 + (y + 2)^2 = 16$. Tangents from the point $P(7, 6)$ meet the circle at points A and B. Show that the perimeter of the kite $PACB$ is $4(2 + \sqrt{21})$.	*Section 8.5*

Test yourself ANSWERS

1 $(x - 2)^2 + (y - 1)^2 = 100$.

2 **(a)** $(x + 3)^2 + (y - 1)^2 = 1$; **(b)** $C(-3, 1)$ radius = 1;
(c) $(-3, 0)$.

3 $x^2 + (y - 2)^2 = 16$.

4 $x^2 + y^2 - 4x - 5y + 4 = 0$.

5 $c = \pm\sqrt{20}$.

6 **(a)** (2, 1); **(b)** $y = \frac{1}{2}x$.

7 $4y - 3x = 10$.

CHAPTER 9

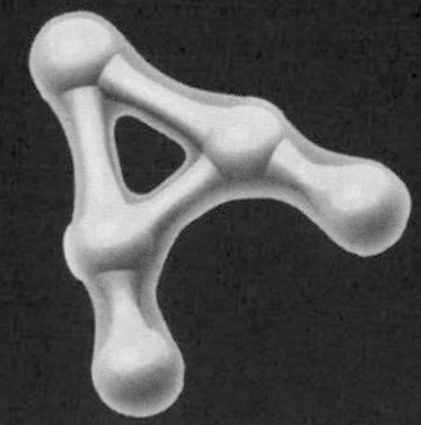

C1: Introduction to differentiation: gradient of curves

Learning objectives

After studying this chapter, you should be able to:

- understand the terms 'chord' and 'tangent' to a curve and how they are related
- understand the term 'derivative' and how you can find gradients of curves
- use differentiation to find the gradient of a curve at any point on a curve
- simplify expressions before differentiating.

9.1 Introduction

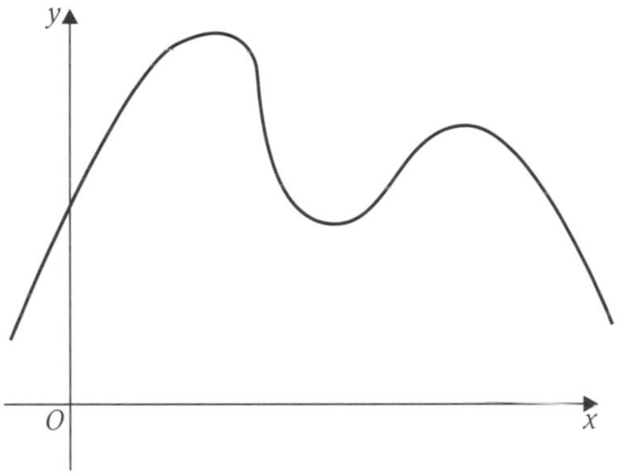

Imagine that the curve above represents part of a roller coaster track and that a pole extends out in front of the car which indicates the direction of motion. Clearly, as you are riding the roller coaster, the direction of the pole will change: at times you may be climbing and the pole will be pointing upwards, it may be momentarily horizontal at the top of a curve or the pole will be pointing downwards as you plunge towards the ground.

All of this tells you that the direction of the curve is constantly changing which, in turn, means that the gradient of the curve is always changing.

How do you find the gradient of a curve if it is different at each point on the curve ?

The aim of this chapter is to develop a systematic way of finding the gradient of a curve at any point you choose.

9.2 Gradients of chords

A straight line that joins two points on a curve is called a chord.

Since the chord is a straight line you can use:

$$\text{gradient of chord} = \frac{\text{difference in } ys}{\text{difference in } xs} = \frac{y_2 - y_1}{x_2 - x_1}$$

See Section 3.4 of chapter 3

Using the definitions :

δx is the 'increase in x' δy is the 'increase in y'

> Gradient of chord $= \dfrac{\delta y}{\delta x}$.

The Greek letter δ (delta) [or the capital letter Δ] is used as an abbreviation for difference.

The word difference can be a little confusing since it does not take into account the sign of the result.

Worked example 9.1

The diagram shows part of the curve $y = x^3$.
Find the gradient of the chord joining the two points A, where $x = 1$, and B, where $x = 3$.

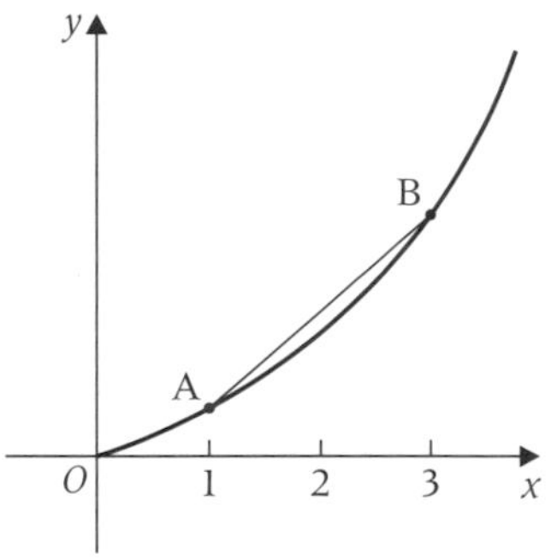

Solution

First, you need to find the y-coordinates of the points A and B.

at A, $x = 1 \Rightarrow y = 1^3 = 1$

at B, $x = 3 \Rightarrow y = 3^3 = 27$

Since the curve has the equation $y = x^3$.

So A is the point $(1, 1)$ and B is the point $(3, 27)$.

For the chord joining $A(1, 1)$ to $B(3, 27)$ we have

$\delta x = 3 - 1 = 2$ and $\delta y = 27 - 1 = 26$.

Note that to find the 'increase' you take the second value and subtract the first value.

The gradient of the chord $AB = \dfrac{\delta y}{\delta x} = \dfrac{26}{2} = 13$.

Worked example 9.2

A curve has equation $y = 13 - x^2$. Find the gradient of the chord joining the points on the curve with x-coordinates 2 and 3.

Solution

When $x = 2$, $y = 13 - 2^2 = 9$.
When $x = 3$, $y = 13 - 3^2 = 4$.

The coordinates of the points are $(2, 9)$ and $(3, 4)$.

Hence $\delta x = 3 - 2 = 1$ and $\delta y = 4 - 9 = -5$

Notice that this is a negative answer since the y-value has actually decreased.

The gradient of the chord $= \dfrac{\delta y}{\delta x} = \dfrac{-5}{1} = -5$.

EXERCISE 9A

1 A curve has equation $y = x^2$. Find the gradient of the chord joining the points on the curve with x-coordinates 3 and 4.

2 A curve has equation $y = x^3$. Find the gradient of the chord joining the points with x-coordinates 2 and 3.

3 Find the gradient of the chord through the points with x-coordinates -1 and 2 on the curve $y = 3x^2 + 2$.

Hint. Watch out for double negatives.

4 A curve has equation $y = 7 - x^3$. Find the gradient of the chord through the points with x-coordinates 1 and 1.5.

9.3 The tangent to a curve at a point

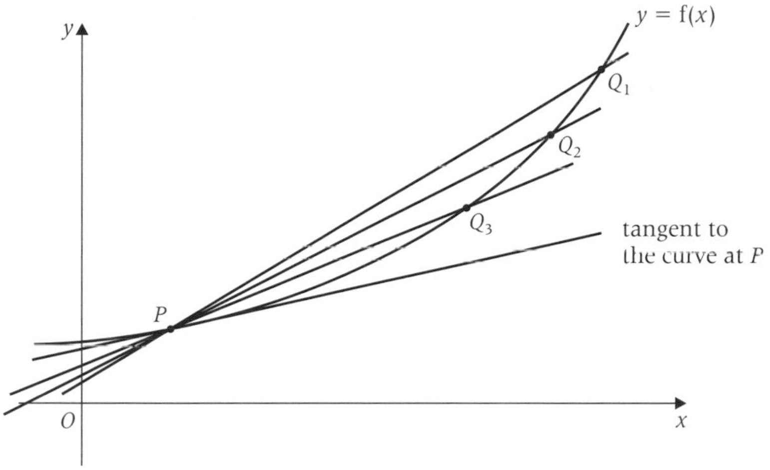

Consider a point P on the curve with equation $y = f(x)$.

The position of P is fixed but a second point Q moves along the curve so that it gets closer and closer to P.
Assume that Q starts at position Q_1 then moves to position Q_2 then to Q_3, etc. Clearly, the directions of the chords PQ_1, PQ_2, PQ_3, … are different.
You can continue this process of moving Q closer to P until they become infinitely close together.
When the two points are infinitely close together, the resulting chord is called the **tangent to the curve at P**.

> The tangent to a curve at any point P is the straight line which just touches the curve at the point P.

This idea of making the two points on the curve get closer and closer to each other will be essential in developing a systematic method of finding the gradient of a curve at any point.

9

In fact, the gradient of the tangent to the curve is used in order to define the gradient of the curve itself.

> The gradient of a curve at any point P is equal to the gradient of the tangent to the curve at P.

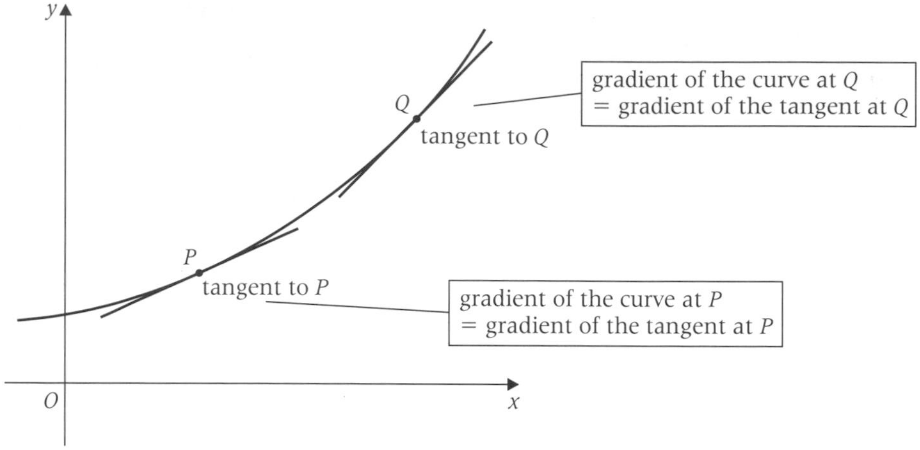

This definition forms the basis of a method for finding the gradient of a curve at any point.

Worked example 9.3

A curve has equation $y = 4x^2$.

Find the gradient of the chord joining the following points:

(a) $x = 1$ and $x = 1.1$,

(b) $x = 1$ and $x = 1.01$,

(c) $x = 1$ and $x = 1.001$.

Use your answers to **(a)**, **(b)** and **(c)** to suggest what the gradient of the curve is when $x = 1$.

Solution

> You need the y-coordinates of each of the points:
> when $x = 1$, $y = 4(1)^2 = 4$;
> when $x = 1.1$, $y = 4(1.1)^2 = 4.84$;
> when $x = 1.01$, $y = 4(1.01)^2 = 4.0804$;
> when $x = 1.001$, $y = 4(1.001)^2 = 4.008004$.

(a) $(1, 4)$ and $(1.1, 4.84)$

Hence $\delta x = 1.1 - 1 = 0.1$ and $\delta y = 4.84 - 4 = 0.84$

the gradient of the chord $= \dfrac{\delta y}{\delta x} = \dfrac{0.84}{0.1} = 8.4$

(b) $(1, 4)$ and $(1.01, 4.0804)$

$\delta x = 1.01 - 1 = 0.01$ and $\delta y = 4.0804 - 4 = 0.0804$

The gradient of the chord $= \dfrac{\delta y}{\delta x} = \dfrac{0.0804}{0.01} = 8.04.$

(c) (1, 4) and (1.001, 4.008004)

$\delta x = 1.001 - 1 = 0.001$ and $\delta y = 4.008004 - 4 = 0.008004$

The gradient of the chord $= \dfrac{\delta y}{\delta x} = \dfrac{0.008004}{0.001} = 8.004.$

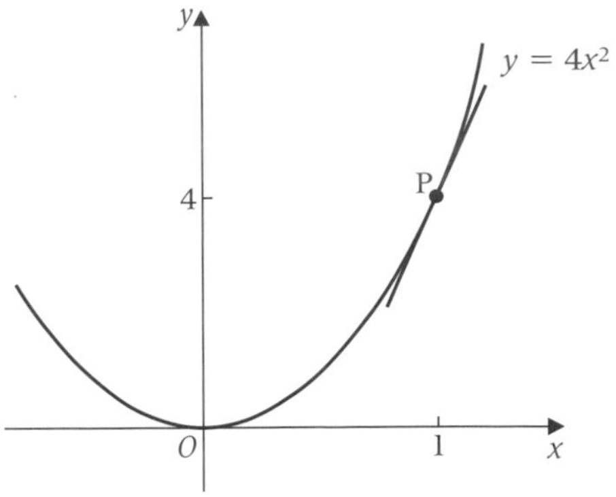

You are taking two points on the curve and making the second point get closer and closer to (1, 4).

This process eventually gives the gradient of the tangent at (1, 4) and, therefore, the gradient of the curve at (1, 4). Consider the sequence formed by the gradients of the chords:

8.4, 8.04, 8.004, …

We say that the **limit** of the sequence is 8.

It seems sensible to suggest that the sequence is converging to 8.

In other words, this process suggests that the gradient of the curve $y = 4x^2$ at the point (1, 4) is 8.

While this offers strong evidence that the gradient is 8, it cannot be considered as a proof.

A more general method which can prove this result is considered later.

EXERCISE 9B

1 Find the gradient of the chord joining the points with x-coordinates 2 and 2.1 on the curve $y = x^2 - 5$.

2 Find the gradient of the chord joining the points with x-coordinates 1 and 1.1 on the curve $y = x^4$.

3 A curve has equation $y = x^5$.

Find the gradients of the chords joining the following pairs of points on the curve:

(a) $x = 1$ and $x = 1.1$,

(b) $x = 1$ and $x - 1.01$,

(c) $x = 1$ and $x = 1.001$.

Using your answers to **(a)**, **(b)** and **(c)**, make a guess at what you think the gradient of the curve is at the point where $x = 1$?

You need to consider two points on the curve: P and another point close to it and let the second point get closer to P.

4 Use a similar approach to question **3** to find the gradient of the curve $y = x^3$ at the point P where $x = 2$.

9.4 The gradient of a curve as the limit of the gradient of the chord

Section 9.4 will not be tested in the C1 examination. The section shows how you can find the gradient of a curve at a point from first principles.

There are two problems with the method above using a sequence of numerical values:

(i) it only offers strong evidence that the gradient has a particular value (it does not prove the result),

(ii) it is time consuming.

The method can be generalized by using h to represent a small increase in x.

h is used for the 'increase in x' instead of δx. This is because the notation becomes easier later on:
writing h^2 instead of $(\delta x)^2$, etc. looks a little less cumbersome.

Worked example 9.4

A curve has equation $y = x^2$ and the points P and Q have coordinates (2, 4) and $(2 + h, (2 + h)^2)$, respectively. Show that the gradient of PQ is $4 + h$ and deduce the gradient of the curve at P.

Solution

The diagram illustrates the situation.

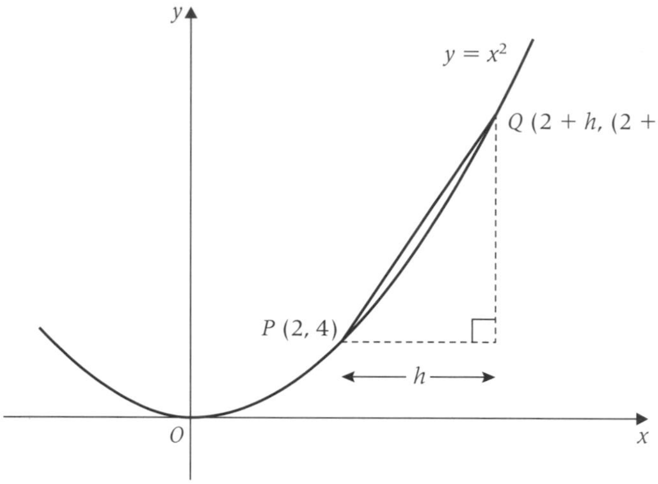

The coordinates of P are (2, 4) and the coordinates of Q are $(2 + h, (2 + h)^2)$ since both points lie on the curve $y = x^2$.

$$\Rightarrow \text{ Gradient of chord } PQ = \frac{(2+h)^2 - 4}{(2+h) - 2}$$

$$= \frac{(4 + 4h + h^2) - 4}{(2+h) - 2}$$

Note how the bracket in the numerator has been expanded.

$$= \frac{4h + h^2}{h}$$

$$= \frac{h(4+h)}{h}$$

h is a common factor in the numerator.

$$= 4 + h$$

h has been cancelled in the top and bottom.

$$\Rightarrow \quad \text{Gradient of } PQ = 4 + h$$

To find the **gradient of the curve at *P***, you must find the **gradient of the tangent at *P***.

To do this we use the basic idea of moving point Q closer and closer to P.
In doing this, it is important to realise that the value of h gets smaller and smaller.

It is not difficult to see that as ***h* tends to zero**, the gradient gets closer and closer to 4

gradient of the curve at P = gradient of PQ as $h \to 0$ ('tends to')

gradient of the curve at $P = \lim_{h \to 0} (4 + h)$ (limit as h tends to 0)

Therefore, the gradient of the curve at P = 4 + 0 = 4.

The very same method can be used to generate the gradient of a curve at **any point** where $x = a$.

Worked example 9.5

The points P and Q lie on the curve $y = x^3$ and have x-coordinates $x = a$ and $x = a + h$, respectively.

Show that the gradient of the chord PQ is $3a^2 + 3ah + h^2$. Hence find the gradient of the curve at the point P.

Solution

The coordinates of P are (a, a^3).

The coordinates of Q are $(a + h, (a + h)^3)$.

$$\text{Gradient of } PQ = \frac{(a+h)^3 - a^3}{(a+h) - a}$$

$$= \frac{a^3 + 3a^2h + 3ah^2 + h^3 - a^3}{h}$$

$$= \frac{3a^2h + 3ah^2 + h^3}{h} = 3a^2 + 3ah + h^2$$

To find the gradient of the curve, let $h \to 0$

$\Rightarrow$ gradient $= 3a^2 + 3(0) + 0 = 3a^2$

i.e., let Q get closer to P.

EXERCISE 9C

Examination questions will **not** be set on this C1 material. It forms an introduction to what will be tested in FP1.

1 A curve has equation $y = 5x^2$. The points P and Q have coordinates $(1, 5)$ and $(1+h, 5(1+h)^2)$, respectively. Find the gradient of the chord PQ in as simplified a form as possible. Hence find the gradient of the curve at the point P.

2 A curve has equation $y = x^4$. Find the gradient of the chord PQ and deduce the gradient of the curve at the point P in the following cases.

(a) $P(1, 1)$ and $Q(1 + h, (1 + h)^4)$,

(b) $P(2, 16)$ and $Q(2 + h, (2 + h)^4)$,

(c) $P(a, a^4)$ and $Q(a + h, (a + h)^4)$.

9.5 The derivative or derived function

Using the result from worked example 9.5, the gradient of the curve $y = x^3$ at the point where $x = a$ is $3a^2$.

It follows that the gradient of the curve at **any point** $P(x, y)$ on the curve $y = x^3$ is actually $3x^2$. This is a very interesting result since the gradient of the curve turns out to be a function of the variable x. It is commonly known as the **derivative** or **derived function**.

This function allows us to substitute different values of x and obtain the gradient of the curve at that point.
Since P was any point on the curve, this gives a method for finding the gradient of the curve $y = x^3$ at any point, e.g.,

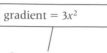

gradient $= 3x^2$

at the point where $x = 2$, the gradient $= 3(4) = 12$

at the point where $x = -5$, the gradient $= 3(25) = 75$

From question **2(c)** of Exercise 9C, you should have found that the gradient of the curve $y = x^4$ at the point where $x = a$ is $4a^3$.

Hence, the gradient of the curve at **any point** $P(x, y)$ on the curve $y = x^4$ is in fact $4x^3$.

Can you predict what the derived function will be for the curve with equation $y = x^5$?

9.6 Differentiation notation

The process of finding derivatives is called **differentiation**. It was discovered by two mathematicians: *Sir Isaac Newton* and *Gottfried Wilhelm von Leibniz*. While they used different notation and were using the idea to solve very different problems they had in fact developed **differential calculus** at around the same time. It was Leibniz who published his calculus first in 1684 but it is generally thought that Newton had been using the idea earlier. A bitter argument erupted between the supporters of each mathematician as to who should be given credit for the technique. It raged for many years but over time both men have been credited equally for their development of one of the most powerful and significant advances in the history of mathematics.

Recall that if a curve has equation $y = x^4$, then the derived function $= 4x^3$.

We say:

x^4 is **differentiated with respect to x** to give $4x^3$
or $4x^3$ is the derivative of x^4.

Since the variable is x.

Rather than having to write the word 'derivative' every time, various ways of denoting the derivative have been developed:

This was the notation used by Leibniz and is the value of $\frac{\delta y}{\delta x}$ as $\delta x \to 0$.

(i) $\frac{dy}{dx}$ — This is read as 'dy by dx'

For example

$$y = x^3 \Rightarrow \frac{dy}{dx} = 3x^2.$$

Sometimes this may be seen as

$$\frac{d}{dx}(x^3) = 3x^2$$

(ii) $f'(x)$ — This is read as 'f dashed x'

So that

$$f(x) = x^4 \Rightarrow f'(x) = 4x^3.$$

Newton's 'fluxion' notation.

You need to be familiar with both types of notation and you should be able to use them appropriately.

9.7 General rules for differentiation

Having to use differentiation from first principles every time you need to find the gradient of a curve can become very time-consuming. Fortunately, shortcuts have been developed in the form of a simple strategy which allows you to differentiate functions very quickly and easily.

> If $y = x^n$, then $\dfrac{dy}{dx} = nx^{n-1}$.

Suppose you need to differentiate $y = x^5$

$$\frac{dy}{dx} = 5x^4$$

Index has been reduced by 1

The 5 brought down in front.

Differentiate the function and multiply by the scaling factor, the constant c.

> If $y = cx^n$, then $\dfrac{dy}{dx} - cnx^{n-1}$ (where c is a constant).

For example:

The constant here is 7.

differentiate $y = 7x^3$

$$\frac{dy}{dx} = 7 \times 3x^2 = 21x^2$$

Bring the power down in front and reduce the power of x by 1, then multiply the answer by 7.

If the function is made up of the sum or difference of two (or more) simpler functions, then differentiate term by term.

> If $y = f(x) \pm g(x)$, then $\dfrac{dy}{dx} = f'(x) \pm g'(x)$.

Differentiate $y = x^7 - x^4 + 5x^3$

$5(3x^2) = 15x^2$

$$\frac{dy}{dx} = 7x^6 - 4x^3 + 15x^2$$

Differentiate each term separately.

Worked example 9.6

Differentiate the following with respect to x:

(a) $y = 6x^3$, **(b)** $y = -3x^5$, **(c)** $y = 9x$,
(d) $y = 6$, **(e)** $y = 3x^5 - 4x^2 + 8x + 15$.

Solution

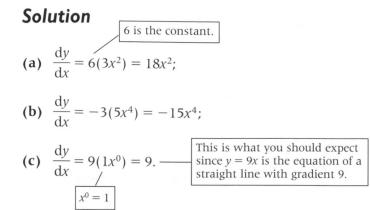

6 is the constant.

(a) $\frac{dy}{dx} = 6(3x^2) = 18x^2;$

(b) $\frac{dy}{dx} = -3(5x^4) = -15x^4;$

(c) $\frac{dy}{dx} = 9(1x^0) = 9.$

This is what you should expect since $y = 9x$ is the equation of a straight line with gradient 9.

$x^0 = 1$

In $y = 9x$ the power of x is 1.

(d) There does not seem to be a power of x here and so it looks like you cannot use the quick method of differentiating.

You must have $y = x^{\text{something}}$ in order to use this method.

This is not the case since you can rewrite y as

$$y = 6x^0 \quad (\text{since } x^0 = 1)$$

$$\Rightarrow \quad \frac{dy}{dx} = 6(0x^{-1}) = 0$$

You now have a power of x.

This is what you should expect since $y = 6$ is a horizontal line and therefore has a gradient of 0.

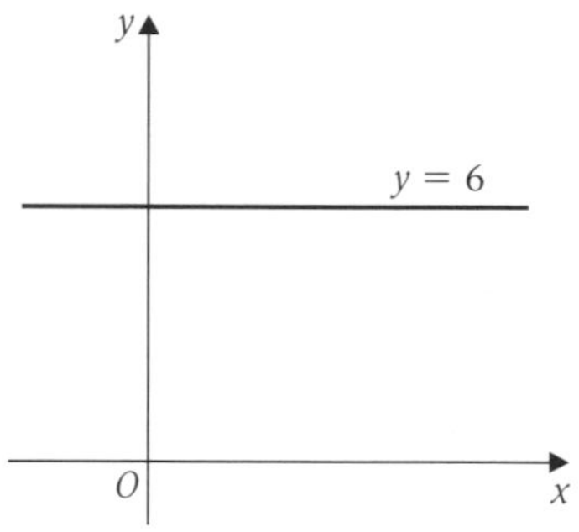

Any line with the equation $y = k$, where k is a constant, has a gradient of 0.

(e) $\frac{dy}{dx} = 3(5x^4) - 4(2x^1) + 8(1x^0) + 0$
$= 15x^4 - 8x + 8$

You are not restricted to differentiating with respect to x. For instance, the variable A may be given in terms of t.

Worked example 9.7

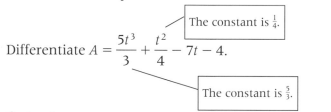

The constant is $\frac{1}{4}$.

Differentiate $A = \frac{5t^3}{3} + \frac{t^2}{4} - 7t - 4.$

The constant is $\frac{5}{3}$.

Solution

Differentiate A with respect to t.

Since the variable is t.

$$\frac{dA}{dt} = \frac{5}{3}(3t^2) + \frac{1}{4}(2t^1) - 7(1t^0) - 0$$

$$= \frac{15t^2}{3} + \frac{2t}{4} - 7$$

$$= 5t^2 + \frac{t}{2} - 7$$

EXERCISE 9D

1 Find $\dfrac{dy}{dx}$ for each of the following:

(a) $y = 5$
(b) $y = 3x^3 - 7x^2$
(c) $y = x^6 - 4x^2 + 5x$
(d) $y = x^6 + x^3$
(e) $y = 3 - 2x + 5x^2$
(f) $y = 7x^3 - 9$
(g) $y = 1 + 4x^2 - 9x^3$
(h) $y = 8x^{10} - 7x^6 + 42$
(i) $y = 8x^2 + 4x - 3$
(j) $y = 6x^2 - 4x + 11$
(k) $y = x^7 - 3x^2 + 2x - 1$
(l) $y = x^4 - 3x^3 - x - 5$

2 Find $f'(x)$ for each of these functions:

(a) $f(x) = \dfrac{x^3}{3}$
(b) $f(x) = 6x^5 - 8x^2 + \dfrac{x^2}{6}$
(c) $f(x) = 7x^2 - \dfrac{x}{5} + 2$
(d) $f(x) = \dfrac{2}{3}x^6 - \dfrac{3}{4}x^8 + 9$
(e) $f(x) = \dfrac{3x^5}{4} - \dfrac{5x^2}{9} + \dfrac{1}{2}$
(f) $f(x) = \dfrac{2x^{12}}{3} - \dfrac{7x^3}{6} + \dfrac{5x}{4}$

3 Differentiate each of the following with respect to the appropriate variable:

(a) $p = 6q^3 - 7q$
(b) $y = 3t^9 - 0.4t + 7$
(c) $m = \dfrac{4n^2}{3} - \dfrac{4n}{7}$
(d) $r = 7s^3 - \dfrac{s}{4} + 3$
(e) $t = \dfrac{3w^2}{4} - \dfrac{4w^7}{7}$
(f) $z = 5p^3 - \dfrac{p^8}{4} + 2p$

9.8 Finding gradients at specific points on a curve

In order to find the gradient at a particular point on the curve:
(i) differentiate the equation of the curve;
(ii) substitute into the derivative the x-coordinate of the point at which we need the gradient.

Worked example 9.8

(a) Find the gradient of the curve $y = 3x - 8x^2$ at the point $P(3, -63)$.

(b) Find the gradient of the curve $y = 4x^3 + 20x^2$ at the point Q where $x = -2$.

Solution

Differentiate to find the derivative.

(a) $\frac{dy}{dx} = 3 - 16x$

$\Rightarrow$ gradient at $P = 3 - 16(3) = -45$

Substitute the x-coordinate of P which is 3.

(b) $\frac{dy}{dx} = 12x^2 + 40x$

$\Rightarrow$ gradient at $Q = 12(-2)^2 + 40(-2)$

$= 48 - 80$

$= -32$

Substitute $x = -2$.

EXERCISE 9E

1 Find the gradient of the curve $y = x^4 + 1$ at the point where $x = 2$.

2 Calculate the gradient of the tangent to the curve with equation $y = x^2 + 3$ at the point (4, 19).

3 Find the gradient of the curve $y = 3x^3 - 5x^2 - 10$ at the point where $x = 3$.

4 Calculate the gradient of the tangent to the curve with equation $y = 6x - x^3$ at the point (1, 5).

5 Find the gradient of the curve $y = 2x^3 - 5x^2 + 7x - 10$ at the point where $x = -1$.

6 Find the gradient of:

(a) $y = x^2 - 3x + 1$ at the point (3, 1),

(b) $y = 5 - x - 4x^2$ at the point (1, 0),

(c) $y = x^4 - 3x^2$ at the point (1, −2),

(d) $y = 2x^5 + 8x^2$ at the point (−2, −32),

(e) $y = 39 - x^2 - 7x$ at the point (3, 9).

7 Find the gradient of the following curves at the given point:

(a) $y = x^3 + 5x^2 - 7$, where $x = 2$,

(b) $y = 4 - 3x^2 - x^3$ where $x = -1$,

(c) $y = x^5 - 12x^3 + 7x - 3$ where $x = 1$,

(d) $y = 3x^4 - 5x^3 + 4x - 2$ where $x = 2$.

8 The equation of a curve is $y = x^2 - 2x - 24$.
Find the gradient of the curve:

(a) at the point where the curve crosses the y-axis,

(b) at each of the points where the curve crosses the x-axis.

9 A curve has equation $y = x^3 - 3x + 7$. Find the gradient of the curve at the point where $x = 1$. What can you say about the tangent to the curve at this point and what might the curve look like around this point?

9.9 Finding points on a curve with a given gradient

Sometimes you may need to work the other way round. You may need to find the points where a curve has a particular gradient.

This will be particularly important in chapter 11 where you will be looking for points where the tangent is parallel to the x-axis in order to find points on a curve with zero gradient.

Worked example 9.9

(a) Determine the points on the curve with equation $y = x^3 + 5x + 4$ where the gradient is equal to 17.

(b) Find the point on the curve $y = 2x^2 - 3x + 4$ at which its tangent is parallel to the line $y = 7x - 15$.

Solution

(a) Since $y = x^3 + 5x + 4$

$$\Rightarrow \quad \frac{dy}{dx} = 3x^2 + 5$$

You need the points where the gradient = 17.

$$3x^2 + 5 = 17$$
$$\Rightarrow \quad 3x^2 = 12$$
$$x^2 = 4$$
$$\Rightarrow \quad x = \pm 2$$

To find the corresponding y-coordinates, we substitute each x-coordinate into the original equation.

So there are two points where the gradient equals 17:

when $x = 2, \quad y = 2^3 + 5 \times 2 + 4 = 22$
when $x = -2, \quad y = (-2)^3 + 5 \times (-2) + 4 = -14$

The points where the gradient is 17 are (2, 22) and (−2, −14).

(b) You can find the gradient of the tangent to the curve by differentiation.

$$y = 2x^2 - 3x + 4 \quad \Rightarrow \quad \frac{dy}{dx} = 4x - 3$$

Also, for $y = 7x - 15$, the gradient is $\frac{dy}{dx} = 7$.

Or recognize this as a straight line with gradient 7.

The gradients must be equal

$$\Rightarrow \quad 4x - 3 = 7$$

Parallel lines have equal gradients.

$$\Rightarrow \quad x = 2.5$$

when $x = 2.5, \quad y = 2(2.5)^2 - 3(2.5) + 4 = 9$

Substitute $x = 2.5$ in the equation of the **curve** to find the y-coordinate.

The tangent to the curve $y = 2x^2 - 3x + 4$ is parallel to the line $y = 7x - 15$ at the point (2.5, 9).

EXERCISE 9F

1 Find the point on the curve $y = x^2 - 6x + 12$ where the gradient is equal to 2.

2 Find the point on the curve $y = 3x^2 - 5x + 3$ where the gradient is equal to 7.

3 The tangent at the point P to the curve $y = 4x^2 - 4x - 15$ is parallel to the line $y = 8 - 20x$. Find the coordinates of the point P.

4 Determine the x-coordinates of the two points on the curve with equation $y = x^3 - 5x^2 + 7x - 2$ where the gradient is zero.

5 Determine the two points on the curve with equation $y = \frac{x^3}{3} - \frac{x^2}{2} + 6x - 1$ where the gradient is equal to 12.

6 Find the coordinates of the point on the curve with equation $y = 3x^2 - 12x + 3$ at which:

(a) the gradient equals 9,

(b) the gradient equals 0.

7 Find the coordinates of the points on the curve $y = x^3 - 9x^2 + 24x - 3$ at which the tangent is parallel to the x-axis.

9.10 Simplifying expressions before differentiating

Sometimes it is necessary to simplify expressions before you can differentiate them.

You may have to expand brackets to obtain a line of expressions of the form kx^n.

Worked example 9.10

Differentiate the following with respect to x:

(a) $y = (x - 6)(x + 2)$,

(b) $y = (3x + 2)^2$,

(c) $y = x(x - 2)(2x + 3)$.

Solution

(a) $y = (x - 6)(x + 2)$ — Multiply out the brackets.

$y = x^2 - 4x - 12$ — Remember, the expression must be written with terms that are powers of x.

$\Rightarrow \frac{dy}{dx} = 2x - 4$ — Now you differentiate.

(b) $y = (3x + 2)(3x + 2) = 9x^2 + 12x + 4$

$\Rightarrow \dfrac{dy}{dx} = 18x + 12$

Expand the brackets first.

(c) $y = (x^2 - 2x)(2x + 3) = 2x^3 - x^2 - 6x$

$\Rightarrow \dfrac{dy}{dx} = 6x^2 - 2x - 6$

$x(x - 2) = x^2 - 2x$

EXERCISE 9G

1 Find the gradient of the curve $y = 2x(x - 5)$ at the point where $x = 2$.

2 Find the gradient of the curve $y = (x - 5)(x^2 - 10)$ at the point where $x = 3$.

3 Calculate the gradient of the tangent to the curve with equation $y = x(2 - x)(1 + 3x)$ at the point (1, 4).

4 The equation of a curve is $y = (x - 8)(x - 2)$.
Find the gradient of the curve:

(a) at the point where the curve crosses the y-axis,

(b) at each of the points where the curve crosses the x-axis.

5 Find the point on the curve $y = (x - 3)(x + 5)$ where the gradient is equal to 8.

6 The tangent at the point P to the curve $y = (3x - 5)(x + 2)$ is parallel to the line $y = 7x + 1$. Find the coordinates of the point P.

7 Determine the two points on the curve with equation $y = (x^2 + 3)(x - 5)$ where the gradient is equal to -5.

8 Find the coordinates of the point on the curve with equation $y = (4 - 3x)(2 - x)$ at which:

(a) the gradient equals 8,

(b) the gradient equals 0.

Key point summary

1 The gradient of a chord $= \dfrac{\delta y}{\delta x}$, where δx is the 'increase in x' and δy is the 'increase in y'. p138

2 The tangent to a curve at any point P is the straight line which just touches the curve at the point P. p139

3 The gradient of a curve $y = f(x)$ at a point P is defined to be equal to the gradient of the tangent to the curve at P. p140

4 The process of finding the gradient of the tangent to a curve or the derived function is called differentiation. *p143*

5 The derivative is commonly denoted by $\frac{dy}{dx}$ or $f'(x)$. *p144*
$\frac{dy}{dx}$ is called the derivative of y with respect to x.
It is the rate of change of y with respect to x.

6 When differentiating, you can use the following strategy: *p145*

(i) if $y = x^n$, then $\frac{dy}{dx} = nx^{n-1}$,

(ii) if $y = cx^n$ (c is a constant), then $\frac{dy}{dx} = cnx^{n-1}$,

(iii) if $y = f(x) \pm g(x)$, then $\frac{dy}{dx} = f'(x) \pm g'(x)$.

7 To find the gradient of a curve at a specific point, substitute the x-coordinate of the point into the expression for the derivative. *p147*

Test yourself	What to review
1 Find the gradient of the chord joining the points with x-coordinates 2 and 3 on the curve with equation $y = 3x^2 - 7x - 2$.	*Section 9.2*
2 Differentiate the following functions with respect to x: **(a)** $y = 8x^3 - 2x + 8$, **(b)** $y = 7x - x^8$, **(c)** $y = \frac{1}{3}x^3 + \frac{3}{4}x^8$.	*Section 9.7*
3 Find the gradient of the curve $y = 6x^2 - 5x - 2$ at the point (3, 37).	*Section 9.8*
4 Find the points on the curve $y = x^3 - 3x^2 + 9x - 5$, where the gradient is equal to 9.	*Section 9.9*
5 Differentiate the following functions with respect to x: **(a)** $y = x(2x - 3)$, **(b)** $y = x(2x - 3)(x + 2)$, **(c)** $y = (x^2 + 1)(x - 4)$.	*Section 9.10*

Test yourself ANSWERS

1 8.

2 **(a)** $\frac{dy}{dx} = 24x^2 - 2$; **(b)** $\frac{dy}{dx} = 7 - 8x^7$; **(c)** $\frac{dy}{dx} = x^2 + 6x^7$.

3 31. **4** (0, −5), (2, 9).

5 **(a)** $\frac{dy}{dx} = 4x - 3$; **(b)** $\frac{dy}{dx} = 6x^2 + 2x - 6$; **(c)** $\frac{dy}{dx} = 3x^2 - 8x + 1$.

CHAPTER 10

C1: Applications of differentiation: tangents, normals and rates of change

Learning objectives

After studying this chapter, you should be able to:

- find the equation of the tangent to a curve at a point on the curve
- find the equation of the normal to a curve at any given point
- appreciate that derivatives represent rates of change
- understand how derivatives can tell you whether a function is increasing or decreasing.

10.1 The equation of a tangent to a curve

In the previous chapter you will have discovered how to find the **gradient** of the tangent to a curve. In this section you will learn how to find an **equation** for a tangent at a given point on the curve.

> The equation of a tangent to a curve at the point $P(x_1, y_1)$ is given by
>
> $$y - y_1 = m(x - x_1),$$
>
> where m is the gradient of the curve at P.

Worked example 10.1

A curve has equation $y = x^3 + 3x^2 - 7$. Find an equation of the tangent to the curve at the point where $x = -1$.

Solution

Since $y = x^3 + 3x^2 - 7$, $\dfrac{dy}{dx} = 3x^2 + 6x$.

When $x = -1$, $\dfrac{dy}{dx} = 3 - 6 = -3$.
The tangent must also have gradient -3.

All points on the curve satisfy $y = x^3 + 3x^2 - 7$; so when $x = -1$, $y = -1 + 3 - 7 = -5$.

The tangent must also pass through $(-1, -5)$.

Hence an equation for the tangent is

$$y - (-5) = -3(x - (-1))$$
$$\text{or } y + 5 = -3(x + 1)$$
$$\text{or } 3x + y + 8 = 0$$

You learnt in chapter 3 that a line with gradient m, passing through the point (x_1, y_1) has equation $y - y_1 = m(x - x_1)$.

There are several ways of writing this equation and that is why the question asks for 'an' equation of the tangent to the curve.

EXERCISE 10A

1 A curve has equation $y = x^2 + 3x + 2$.

(a) Show that the point $(1, 6)$ lies on the curve.

(b) Find the gradient of the curve at the point $(1, 6)$.

(c) Find an equation of the tangent to the curve at the point $(1, 6)$.

2 Find an equation of the tangent to each of the following curves at the point indicated:

(a) $y = x^3 - 4x + 5$ at $(2, 5)$,

(b) $y = 4 - 3x^2 - x^3$ at $(1, 0)$,

(c) $y = x^5 - 3x^4 + 2x$ at $(-1, -6)$.

3 Find an equation of the tangent to each of the following curves at the point where x has the given value:

(a) $y = x^3 + 3x - 1$, $x = 3$,

(b) $y = 5 - x^3$, $x = 2$,

(c) $y = 3x^4 + 5x - 2$, $x = 0$,

(d) $y = 2x^3 - 6x^2 - 5$, $x = 1$,

(e) $y = x^6 - 2x^4 - 3x$, $x = -1$.

4 Find an equation of the tangent to each of the following curves at the point where x has the given value:

(a) $y = x(x + 3)(x - 1)$, $x = 1$,

(b) $y = (5 - x)(x + 4)$, $x = 2$,

(c) $y = 3x(x^2 + 5)$, $x = -1$,

(d) $y = (2x - 3)(6x^2 - 5)$, $x = 1$.

5 The curve with equation $y = (2x + 1)(x - 3)$ cuts the x-axis at the point $A(-\frac{1}{2}, 0)$ and at the point B.

(a) Show that the tangent to the curve at the point A has equation $14x + 2y + 7 = 0$.

(b) Find an equation of the tangent to the curve at the point B.

6 **(a)** Find the gradient of the curve $y = 12 - 3x^2$ at the point $P(2, 0)$.

(b) Find the coordinates of the point Q where the tangent at P crosses the y-axis. [A]

7 A curve has equation $y = x^2(4x + 5)$ and the point P has coordinates $(-1, 1)$.

(a) Prove that the tangent to the curve at P has equation $y = 2x + 3$.

(b) Show that this tangent intersects the curve again at the point with x-coordinate equal to $\frac{3}{4}$.

8 Find the equations of the tangents to the curve with equation $y = (x + 1)(5 - x)$ at the points $P(-1, 0)$ and $Q(5, 0)$. Prove that these two tangents intersect at the point $(2, 18)$.

9 Find the equations of the tangents to the curve with equation $y = x^3 + 2x^2 - 3x$ at the points $A(-2, 6)$ and $B(1, 0)$. The tangents at A and B intersect at the point C. Find the coordinates of the point C.

10 The curve with equation $y = (x - 1)(x - 2)(x - 3)$ cuts the y-axis at the point P.

(a) Find the equation of the tangent to the curve at the point P.

(b) This tangent intersects the x-axis at the point Q. Find the area of triangle OPQ, where O is the origin.

11 The point $A(2, 1)$ lies on the curve with equation $y = x^3 - 4x^2 + 3x + 3$.

(a) Find an equation of the tangent to the curve at the point A.

(b) Find an equation of the straight line passing through A which is perpendicular to this tangent.

10.2 The equation of a normal to a curve

The normal to a curve at a point P is a straight line that is perpendicular to the tangent at P.

When the gradient of the tangent is m, the gradient of the normal is $-\frac{1}{m}$.

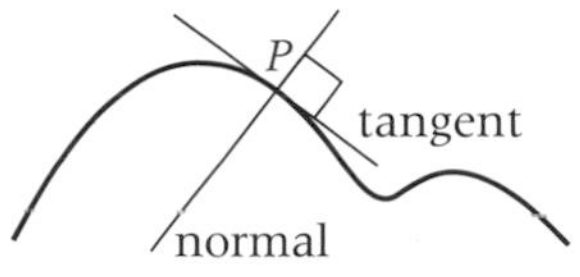

Worked example 10.2

A curve has equation $y = x^3 - 4x^2 - 7x - 1$. Find an equation of the normal to the curve at the point $P(-1, 1)$.

Solution

Since $y = x^3 - 4x^2 - 7x - 1$, $\dfrac{dy}{dx} = 3x^2 - 8x - 7$.

When $x = -1$, $\dfrac{dy}{dx} = 3 + 8 - 7 = 4$.

The tangent to the curve at P has gradient 4. Therefore the gradient of the normal at P is $-\frac{1}{4}$.

Recall that the gradients m_1 and m_2 of two perpendicular lines are such that $m_1 \times m_2 = -1$.

This normal passes through the point $P(-1, 1)$.

Therefore the normal has equation

$$y - 1 = -\frac{1}{4}(x + 1)$$

It is perfectly correct to leave the equation in this form but often the equation is rewritten with integer coefficients.

Multiplying both sides by 4 gives $4y - 4 = -x - 1$
or $x + 4y = 3$.

EXERCISE 10B

1 A curve has equation $y = 3x^4 - 5x^3 - 2$ and the point $A(2, 6)$ lies on the curve.

(a) (i) Find the value of $\dfrac{dy}{dx}$ when $x = 2$.

(ii) Hence find the gradient of the normal to the curve at A.

(b) Find an equation of the normal to the curve at the point A.

2 Find an equation of the normal to each of the following curves at the point indicated:

(a) $y = x^2 - 5x + 1$ at $(4, -3)$,

(b) $y = x^3 + 5x^2 - 7x + 3$ at $(1, 2)$,

(c) $y = 6x - x^4$ at $(2, -4)$,

(d) $y = x^5 - 3x^2 + 1$ at $(-1, -3)$.

3 Find an equation for the normal to each of the following curves at the point where x has the given value:

(a) $y = x^3 + 3x - 11$, $x = 2$,

(b) $y = x^3 - 5x + 2$, $x = 1$,

(c) $y = x^5 - 6x + 3$, $x = 0$,

(d) $y = 2x(x - 5)(x - 3)$, $x = 2$,

(e) $y = (x^3 - 6)(2 - 7x)$, $x = -1$.

4 (a) Find an equation of the normal to the curve $y = x^3 - 4x^2 + 5$ at the point $(1, 2)$.

(b) This normal cuts the x-axis at the point X and the y-axis at the point Y. Find the area of the triangle OXY, where O is the origin.

5 The points $P(-1, k)$ and $Q(4, 6)$ lie on the curve with equation $y = x^2 - 4x + 6$.

(a) Find the value of the constant k.

(b) Show that the length of PQ is $5\sqrt{2}$.

(c) Find an equation for the normal to the curve at the point Q.

6 Find the equations of the normals to the curve with equation $y = x^3 + 2x^2 - 3x$ at the points $A(-2, 6)$ and $B(1, 0)$. The normals at A and B intersect at the point N. Find the coordinates of the point N.

7 A curve has equation $y = x^3 - 4x^2 + 7x - 2$. The point $P(1, 2)$ lies on the curve.

(a) Find an equation of the normal to the curve at the point P.

(b) The point Q is where the curve crosses the y-axis. Find the equation of the normal at the point Q.

(c) Determine the coordinates of the point where the normals to the curve at the points P and Q intersect.

10.3 Rates of change

You have seen how the gradient of a curve at any point is a measure of the steepness of the curve at that point. The gradient of the curve can also be thought of as a measure of how quickly the value of y is changing with respect to x.

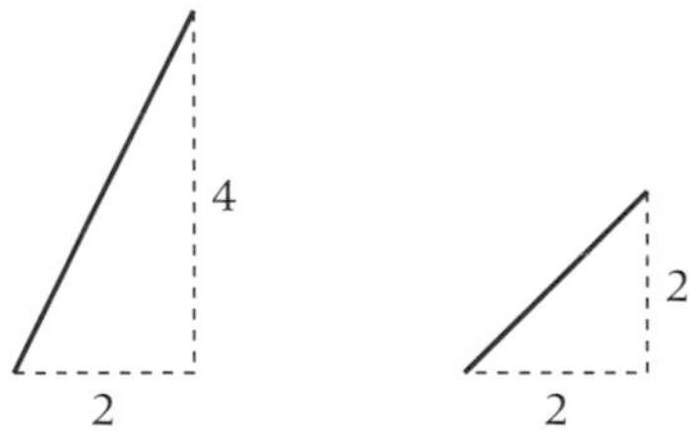

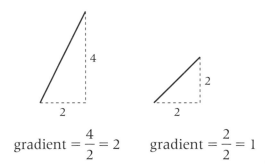

gradient $= \frac{4}{2} = 2$ gradient $= \frac{2}{2} = 1$

The line on the left is clearly steeper than the line on the right since the value of y is increasing at a quicker rate.
In general, the gradient of a curve is found by differentiating.
Thus, the gradient of a curve, $\frac{dy}{dx}$, is a measure of the rate that y is changing with respect to x.

> $\frac{dy}{dx}$ is the rate of change of y with respect to x.

It should also be clear that if the gradient is positive then y is increasing with respect to x and if the gradient is negative then y is decreasing with respect to x.

$$\frac{dy}{dx} > 0 \Rightarrow y \text{ is increasing with respect to } x.$$
$$\frac{dy}{dx} < 0 \Rightarrow y \text{ is decreasing with respect to } x$$

Many 'real-world' problems are concerned with the rate at which quantities are changing: for example, the rate at which the area of rain forest is decreasing, the rate at which we are using up the natural resources of the planet, the rate at which our cars consume petrol, etc.. Many such problems can be modelled mathematically and the various **rates of change can be expressed as derivatives**.

In general,

$\frac{dP}{dQ}$ represents the rate of change of P with respect to Q.
A positive gradient represents a rate of increase whereas a negative gradient represents a rate of decrease.

In many problems you will be concerned with rates of change with respect to time.

Worked example 10.3

The volume, $V\,\text{m}^3$, of water in a tank is given by $V = t^3 - 5t^2 + 6t + 2$, where t is the time in hours measured from a particular instant and $0 \leqslant t \leqslant 4$.

Find the rate of change of volume with respect to time after:
(a) 0.5 hours, **(b)** 2 hours, **(c)** 3 hours,
and comment on whether the volume is increasing or decreasing at that moment.

Solution

$$\frac{dV}{dt} = 3t^2 - 10t + 6$$

(a) When $t = 0.5$, $\frac{dV}{dt} = 3(0.5)^2 - 10(0.5) + 6 = 1.75$

$\Rightarrow$ after 0.5 hours the volume is **increasing** at a rate of $1.75\,\text{m}^3\,\text{h}^{-1}$.

(b) When $t = 2$, $\frac{dV}{dt} = 3(2)^2 - 10(2) + 6 = -2$

$\Rightarrow$ after 2 hours the volume is **decreasing** at a rate of $2\,\text{m}^3\,\text{h}^{-1}$.

(c) When $t = 3$, $\frac{dV}{dt} = 3(3)^2 - 10(3) + 6 = 3$

$\Rightarrow$ after 3 hours the volume is **increasing** at a rate of $3\,\text{m}^3\,\text{h}^{-1}$.

Worked example 10.4

The radius, r cm, of a circular ink spot is given by $r = \frac{t^2}{8} + 1$, where t is the number of seconds after it first appears.

(a) Find the radius of the ink spot when it first appears.

(b) Find the time taken for the radius to become 9 cm.

(c) Find the rate of increase of the radius after 4 s.

Solution

(a) The initial radius will be given when $t = 0$.

$$r = \frac{0^2}{8} + 1 = 1$$

The initial radius is 1 cm.

(b) You need to know which value of t gives $r = 9$.

$$r = \frac{t^2}{8} + 1$$
$$\Rightarrow \quad 9 = \frac{t^2}{8} + 1$$
$$\Rightarrow \quad t^2 = 64$$
$$\Rightarrow \quad t = 8 \text{ s}$$

$t \neq -8$ since t is positive.

(c) To find the rate of increase you need to differentiate.

$$\frac{\mathrm{d}r}{\mathrm{d}t} = \frac{2t}{8} = \frac{t}{4}$$

After 4 s the rate of increase is $\frac{4}{4} = 1 \text{ cm s}^{-1}$.

Worked example 10.5

A particle is travelling in a straight line. Its distance s metres from a fixed point at time t seconds is given by the expression

$$s = t^3 - 3t^2 + 5t$$

(a) Find $\frac{\mathrm{d}s}{\mathrm{d}t}$ when $t = 3$ and interpret this result.

(b) Prove that s increases as t increases for all values of t.

Solution

(a) $\frac{\mathrm{d}s}{\mathrm{d}t} = 3t^2 - 6t + 5$

When $t = 3$, $\frac{\mathrm{d}s}{\mathrm{d}t} = 3(3)^2 - 6(3) + 5 = 14$.

Speed is the rate of change of distance with respect to time. Hence after 3 seconds the speed is 14 m s^{-1}.

(b) You can complete the square for the expression $\frac{ds}{dt}$.

$$\frac{ds}{dt} = 3t^2 - 6t + 5 = 3[(t-1)^2 - 1] + 5$$

Hence $\frac{ds}{dt} = 3(t-1)^2 + 2$.

Since this expression will always be greater than or equal to 2, it is clear that $\frac{ds}{dt} > 0$.

Hence s increases as t increases for all values of t.

EXERCISE 10C

1 Find the rate of change of y with respect to x in each of the following cases:

(a) $y = 10x^3(3 - x)$, **(b)** $y = (4 - x)(3 - x^2)$.

2 Given that $E = 7q - q^4$, find the rate of change of E with respect to q at the instant when $q = 2$.

3 The radius, r m, of a circular oil slick t minutes after it first forms is given by $r = t^3 + 6t + 4$.

(a) Find the initial radius of the slick.

(b) Find the rate of increase of the radius after 4 minutes.

4 A balloon is being blown up. The volume, V cm^3, of air in the balloon after t seconds is given by $V = \frac{4}{3}\pi t^3$. Find the rate at which the volume of air is increasing after 2 seconds.

5 The volume, V m^3, of water in a tank on a certain day is given by $V = t^3 - 37t^2 + 342t + 20$, where t is the number of hours since midnight.

Find $\frac{dV}{dt}$ at:

(a) 4.00 a.m., **(b)** 9.00 a.m., **(c)** 3.30 p.m., **(d)** 10.00 p.m.

Comment on the significance of the sign of each of your answers.

6 A ball is propelled vertically at a speed of 30 m s^{-1} After t seconds its height, s metres, is given by

$$s = 30t - 5t^2$$

Find how quickly the ball is travelling after 2 seconds.

7 The current in an electrical circuit is defined as the rate of change of the charge, Q, flowing in the circuit. In a particular circuit the charge flowing at any time in the first 15 seconds is given by $Q = \frac{t^3}{3} - 5t^2 + 10t + 100$.

Find the current in the circuit after:

(a) 4 seconds, **(b)** 8 seconds, **(c)** 12 seconds.

Comment on the significance of the sign of each of your answers.

10.4 Increasing and decreasing functions

Consider the function f defined by $f(x) = x^4$. Its graph is shown below.

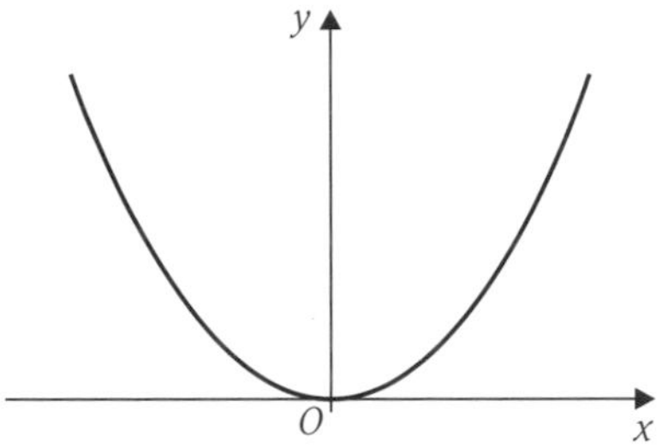

You can see from the graph that when $x > 0$, as x increases then so does the value of x^4.

Therefore, in the interval $x > 0$, the function is said to be **increasing**.

It should also be clear that for $x < 0$, as x increases then the value of x^4 decreases.

The function is said to be **decreasing** in the interval $x < 0$.

When $x = 0$ the function is neither increasing nor decreasing. It is at a **stationary point** and such points will be discussed in detail in the next chapter.

You should also be able to see that in the interval $x > 0$, where the function is increasing, the gradient is always positive, whereas in the interval $x < 0$, where the function is decreasing, the gradient is always negative.

When $f'(x) > 0$ for all values of x in a given interval, the function f is said to be an **increasing function** over this interval.

When $f'(x) < 0$ for all values of x in a given interval, the function f is said to be a **decreasing function** over this interval.

Worked example 10.6

A function f is defined by $f(x) = 2x^3 - 7$.

(a) Differentiate the function to obtain $f'(x)$.

(b) Use your answer from **(a)** to decide if the function is increasing or decreasing when $x < 0$.

Solution

(a) $f'(x) = 6x^2$

(b) When $x < 0$, the expression $6x^2$ is always positive since x^2 is positive.
Therefore, $f'(x) > 0$ and hence f is an increasing function when $x < 0$.

Worked example 10.7

The function f is defined for all values of x by $f(x) = x^2 + 4x + 1$. Find the set of values of x for which f is decreasing.

Solution

$$f'(x) = 2x + 4$$

For the function to be decreasing, $f'(x) < 0$:

$\Rightarrow \quad 2x + 4 < 0$
$\Rightarrow \quad x < -2$ is the set of values of x for which f is decreasing.

Worked example 10.8

A curve with equation $y = 3x^5 + 7x - 4$ is defined for all values of x. Show that y increases as x increases.

Solution

$$y = 3x^5 + 7x - 4$$

$$\Rightarrow \quad \frac{dy}{dx} = 15x^4 + 7$$

The expression $15x^4 + 7$ is always positive since x^4 is always positive.

$$\Rightarrow \quad \frac{dy}{dx} > 0 \quad \Rightarrow \quad y \text{ increases as } x \text{ increases.}$$

Worked example 10.9

The function f is defined for all real values of x by $f(x) = x^3 + 4x^2 - 3x - 4$.
Find the set of values of x for which f is decreasing.

Solution

$$f'(x) = 3x^2 + 8x - 3$$

For the function to be decreasing you need $f'(x) < 0$.

$\Rightarrow \quad 3x^2 + 8x - 3 < 0$
$\Rightarrow \quad (3x - 1)(x + 3) < 0$

This is a quadratic inequality.

The critical values are $x = \frac{1}{3}$ and $x = -3$.

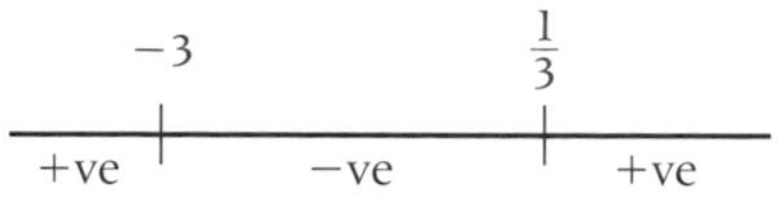

$\Rightarrow$ the function is decreasing when $-3 < x < \frac{1}{3}$.

Worked example 10.10

A function f is defined for all real values of x by $f(x) = x^3 + 9x^2 + 32x - 18$.

(a) Differentiate $f(x)$ with respect to x to obtain $f'(x)$.

(b) Write $f'(x)$ in the form $a(x + b)^2 + c$, where a, b and c are constants and, hence, decide whether f is an increasing or decreasing function giving a reason.

Solution

(a) $f'(x) = 3x^2 + 18x + 32$

(b) The form $a(x + b)^2 + c$ suggests that you need to complete the square.

$$3x^2 + 18x + 32 = 3[x^2 + 6x] + 32$$

Complete the square:

$$3[(x + 3)^2 - 9] + 32$$
$$= 3(x + 3)^2 - 27 + 32$$
$$= 3(x + 3)^2 + 5$$

The minimum value of this expression is 5 and, hence, the expression is always positive.

$\Rightarrow$ $f'(x) > 0$ $\Rightarrow$ f is an increasing function.

EXERCISE 10D

1 Show that the function f defined for all real values of x by $f(x) = -2x^3 - 7x$ is a decreasing function.

2 A function g is defined by $g(x) = (x - 2)^2$.

(a) Find the derivative $g'(x)$.

(b) Explain why g is an increasing function for $x > 2$.

Remember to multiply out the brackets before differentiating.

3 A function f is defined by $f(x) = 4 - 3x - x^3$.

(a) Find the derivative $f'(x)$.

(b) Use your answer from **(a)** to deduce whether the function is an increasing or a decreasing function.

4 Find the set of values of x for which the function h, where $h(x) = 2 - (x + 3)^2$ is increasing.

5 Find the values of x for which the function f defined by $f(x) = x^3 - x^2 - 8x + 6$ is decreasing.

6 The function f is defined for all real values of x by $f(x) = x^3 - 6x^2 + 13x - 5$.

(a) Differentiate $f(x)$ with respect to x to obtain $f'(x)$.

(b) Express $f'(x)$ in the form $a(x - b)^2 + c$, where a, b and c are constants whose values should be stated. Hence show that f is an increasing function. [A]

Key point summary

1 The equation of the tangent to a curve at the point $P(x_1, y_1)$ is given by $y - y_1 = m(x - x_1)$, where m is the gradient of the curve at P. *p153*

2 The normal to a curve at the point P is a straight line that is perpendicular to the tangent at P. When the gradient of the tangent is m, the gradient of the normal is $-\frac{1}{m}$. *p155*

3 $\frac{dP}{dQ}$ represents the rate of change of P with respect to Q. A positive gradient represents a rate of increase whereas a negative gradient represents a rate of decrease. *p158*

4 $\frac{dy}{dx} > 0 \Rightarrow$ y is increasing with respect to x *p158*

$\frac{dy}{dx} < 0 \Rightarrow$ y is decreasing with respect to x.

5 When $f'(x) > 0$ in a given interval, the function f is said to be an **increasing function** over this interval. *p161*

When $f'(x) < 0$ in a given interval, the function f is said to be a decreasing function over this interval.

Test yourself	What to review
1 Find an equation of the tangent to the curve $y = x^3 - 3x^2 - 4$ at the point $(3, -4)$.	*Section 10.1*
2 Find an equation of the normal to the curve $y = x^4 - 5x^2 + 6$ at the point where $x = 1$.	*Section 10.2*
3 The volume, V cm^3, of air in a balloon is given by $V = \frac{4}{9}\pi t^3$, where t is the time elapsed in seconds. Find the rate at which the volume is changing after 3 seconds.	*Section 10.3*
4 A function f is defined by $f(x) = x^3 - 6x^2 + 14x - 7$. **(a)** Differentiate the function to find $f'(x)$. **(b)** Write $f'(x)$ in the form $a(x - b)^2 + c$, where a, b and c are constants, and hence show that $f(x)$ is an increasing function.	*Section 10.4*

Test yourself ANSWERS

1 $y = 9x - 31$.

2 $6y - x = 11$.

3 12π cm^3 s^{-1}.

4 **(a)** $f'(x) = 3x^2 - 12x + 14$; **(b)** $f'(x) = 3(x - 2)^2 + 2$.

$f'(x) \geqslant 2$ so the derivative is always positive. Therefore f is an increasing function.

CHAPTER 11

C1: Maximum and minimum points and optimisation problems

Learning objectives

After studying this chapter, you should be able to:

- understand what is meant by stationary point
- find maximum and minimum points
- find second derivatives
- use the second derivative to determine whether a stationary point is a maximum or a minimum point
- solve practical optimisation problems using differentiation.

11.1 The least value of a quadratic function

You have already learned how to find the least value of a quadratic function by completing the square.

For example, $x^2 - 2x - 3$ can be written as $(x - 1)^2 - 4$. Hence the equation $y = x^2 - 2x - 3$ can be written as $y = (x - 1)^2 - 4$.

The graph of $y = (x - 1)^2 - 4$ is obtained by translating the graph of $y = x^2$ by the vector $\begin{bmatrix} 1 \\ -4 \end{bmatrix}$.

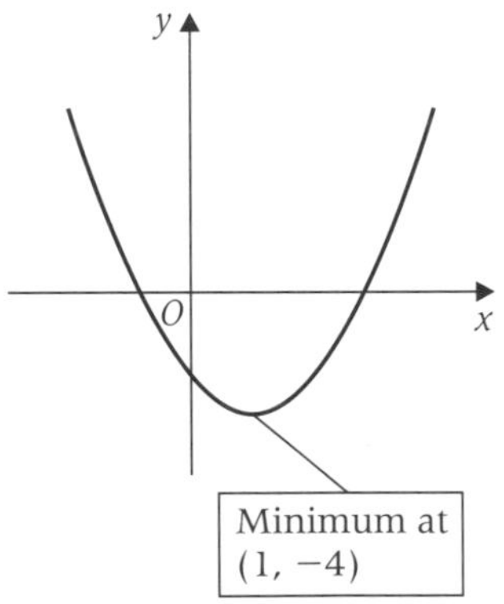

The minimum point on the graph is at (1, −4).

Alternatively, you could differentiate $y = x^2 - 2x - 3$ and locate the point where the gradient is zero.

$\frac{dy}{dx} = 2x - 2$, therefore $\frac{dy}{dx} = 0$ when $x = 1$.

When $x = 1$, $y = 1 - 2 - 3 = -4$.

The minimum point is at $(1, -4)$.

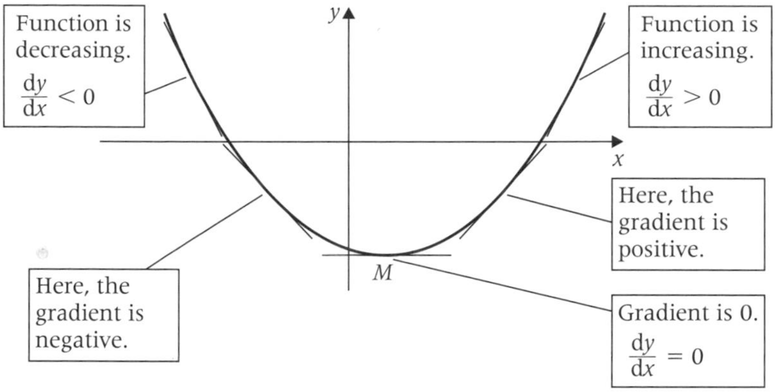

To the right of the minimum point M, the function is increasing and to the left of M the function is decreasing.

The point M is called a stationary point since the gradient is zero at M. For this particular curve, the stationary point is a minimum point.

Stationary points are points on a curve where **the tangent is horizontal**.

> Stationary points occur when $\frac{dy}{dx} = 0$.

Worked example 11.1

Find the coordinates of the stationary point on the curve with equation $y = 7 - 12x - 3x^2$.

Solution

$$\frac{dy}{dx} = -12 - 6x$$

at stationary points, gradient = 0

$$\Rightarrow \quad -12 - 6x = 0$$
$$\Rightarrow \quad x = -2$$

When $x = -2$, $y = 7 - 12(-2) - 3(-2)^2 = 7 + 24 - 12 = 19$

To find the y-coordinate, you substitute into the original equation.

$\Rightarrow$ there is a stationary point at $(-2, 19)$

11

11.2 Maximum and minimum points

These points, sometimes known as turning points, can be described as the 'hills' and 'valleys' of the curve. At each of these points the gradient is zero.

In this section you will consider two types of stationary point.

Maximum points

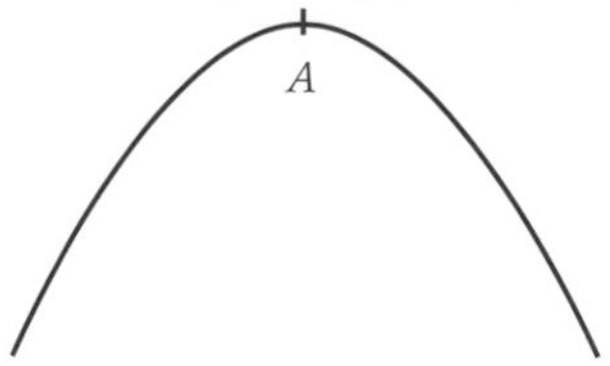

Notice that the following applies:

At a maximum point, A

Just to the left of A	$\frac{dy}{dx} > 0$	(positive gradient)
at A	$\frac{dy}{dx} = 0$	(zero gradient)
Just to the right of A	$\frac{dy}{dx} < 0$	(negative gradient)

At a maximum point, the gradient goes from positive to zero to negative.

Minimum points

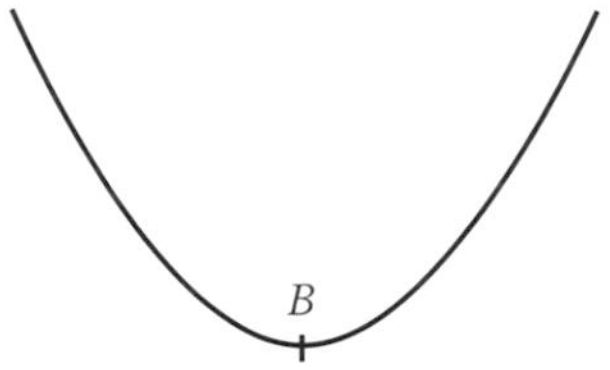

Notice that the following applies:

At a minimum point, B

Just to the left of B	$\frac{dy}{dx} < 0$	(negative gradient)
at B	$\frac{dy}{dx} = 0$	(zero gradient)
Just to the right of B	$\frac{dy}{dx} > 0$	(positive gradient)

At a minimum point, the gradient goes from negative to zero to positive.

Worked example 11.2

Find the stationary point of the curve with equation

$$y = x^2 - 8x + 14$$

and determine whether it is a maximum or a minimum point.

Solution

$$\frac{dy}{dx} = 2x - 8$$

At stationary points, gradient = 0

$\Rightarrow$ $2x - 8 = 0$

$\Rightarrow$ $x = 4$

When $x = 4$, $y = 4^2 - 8(4) + 14 = -2$

$\Rightarrow$ there is a stationary point at $(4, -2)$

To find the y-coordinate, you substitute into the original equation.

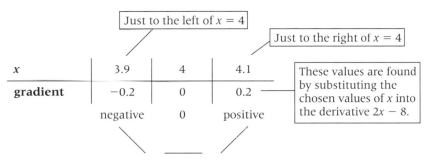

To decide the nature of the stationary point, you can find the value of the gradient just to the left and just to the right of the stationary point.

Near $x = 4$,

Just to the left of $x = 4$

Just to the right of $x = 4$

x	3.9	4	4.1
gradient	-0.2	0	0.2
	negative	0	positive

These values are found by substituting the chosen values of x into the derivative $2x - 8$.

A diagram like this showing the gradient at three points will help you to visualise whether you have a maximum or minimum point.

$\Rightarrow$ $(4, -2)$ is a minimum point.

$\Rightarrow$ The curve $y = x^2 - 8x + 14$ has one stationary point, a minimum point at $(4, -2)$.

EXERCISE 11A

1 Find the coordinates of the stationary points of each of the following curves:

(a) $y = x^2 - 4x + 5$, **(b)** $y = 3x^2 - 6x + 7$,

(c) $y = 8 - 10x - x^2$, **(d)** $y = x^4 - 32x + 2$.

2 Find the stationary point of the curve with equation

$$y = 6 + 10x - 2x^2.$$

By considering gradients of points close to the stationary point, determine whether it is a maximum or minimum point.

3 Find the maximum and minimum points of these curves.

(a) $y = 3x^2 - 30x + 4$, **(b)** $y = 9 + 40x - 4x^2$,

(c) $y = x^3 - 3x + 5$, **(d)** $y = x^3 - 48x$.

4 For each of the following curves find any stationary points and determine whether they are maximum or minimum points:

(a) $y = 20 + 90x - 15x^2$,

(b) $y = 5x^2(x - 6)$,

(c) $y = 4x(x^2 - 3)$.

5 Find the points on the curve $y = 4x^3 - 3x^2 - 19$ at which the tangents are parallel to the x-axis.

6 Show that the curve with equation $y = x^3 + 3x - 7$ has no real stationary points.

7 The function f is defined for all real values of x by $f(x) = (x^2 + 4)(2x - 1)$.

(a) Prove that the curve with equation $y = f(x)$ crosses the x-axis at only one point and state the x-coordinate of this point.

(b) **(i)** Differentiate $f(x)$ with respect to x to obtain $f'(x)$.

(ii) Hence show that the gradient of the curve $y = f(x)$ is 12 at the point where $x = 1$.

(iii) Prove that the curve $y = f(x)$ has no stationary points. [A]

8 The curve with equation $y = x^3 - 3x^2 - 9x + 4$ is shown below.

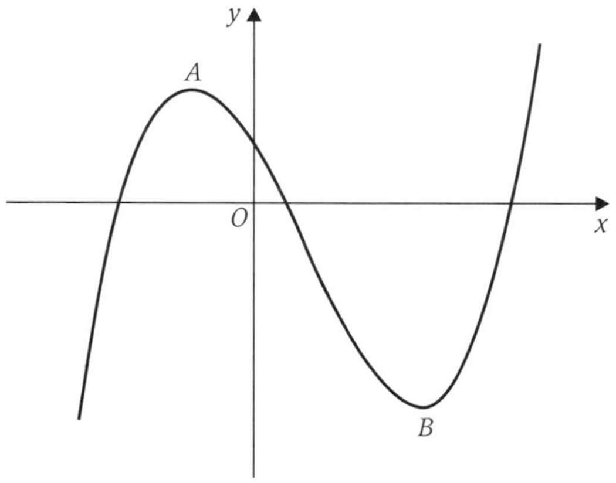

Find the coordinates of the stationary points A and B.

9 A stone is thrown and its height y metres while in the air is given by the equation $y = \dfrac{15x}{2} - x^2$.

This is shown in the diagram opposite.

Find the maximum height that the stone reaches.
Verify that this point is a maximum point on the curve.

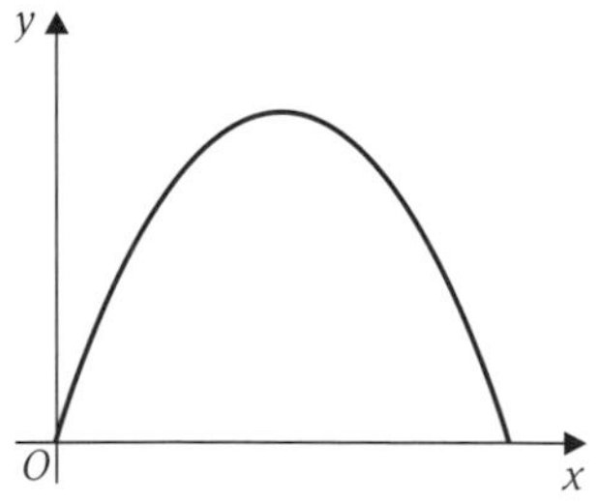

11.3 Second derivatives

You are now very familiar with the idea of differentiating an expression to obtain the derivative or derived function.

$$y = 4x^3 - 6x^2 + 7x - 15$$

$$\Rightarrow \quad \frac{dy}{dx} = 12x^2 - 12x + 7$$

There is nothing to prevent you from differentiating once more to obtain $24x - 12$.

> The expression obtained by differentiating $\frac{dy}{dx}$ is known as the **second derivative** and is denoted by $\frac{d^2y}{dx^2}$.

Read as d two y by d x squared.

> If function notation is used then the second derivative of f is denoted by $f''(x)$.

f double dashed.

Worked example 11.3

Given that $y = 2x^4 + 3x^3 - 6x^2 + 8x - 14$, find $\frac{dy}{dx}$ and $\frac{d^2y}{dx^2}$.

Solution

$$y = 2x^4 + 3x^3 - 6x^2 + 8x - 14$$

$$\Rightarrow \quad \frac{dy}{dx} = 8x^3 + 9x^2 - 12x + 8$$

$$\Rightarrow \quad \frac{d^2y}{dx^2} = 24x^2 + 18x - 12$$

11.4 The second derivative test

You will recall from the earlier sections that the stationary points of a curve occur when the gradient is zero.

For example

$$y = x^3 - 9x^2 + 24x - 10$$

$$\Rightarrow \quad \frac{dy}{dx} = 3x^2 - 18x + 24$$

At stationary points, $\frac{dy}{dx} = 0$

$3x^2 - 18x + 24 = 0$

$\Rightarrow \quad x^2 - 6x + 8 = 0$

$\Rightarrow \quad (x - 2)(x - 4) = 0$

$\Rightarrow$ there are two stationary points: $x = 2$ and $x = 4$.

11

Previously, to determine the nature of the stationary points you considered the sign of the gradient just to the left and right of each point.

x	1	2	3	
gradient	+ve	0	−ve	$\Rightarrow x = 2$ is a maximum point

x	3	4	5	
gradient	−ve	0	+ve	$\Rightarrow x = 4$ is a minimum point

This task of finding the gradient on either side of a stationary point is rather tedious and can be very difficult without a calculator.

There is, however, a more elegant and efficient method for deciding whether a stationary point is a maximum or a minimum which involves the use of the **second derivative**.

Consider the graph of $y = x^3 - 9x^2 + 24x - 10$ and the graph of its derivative.

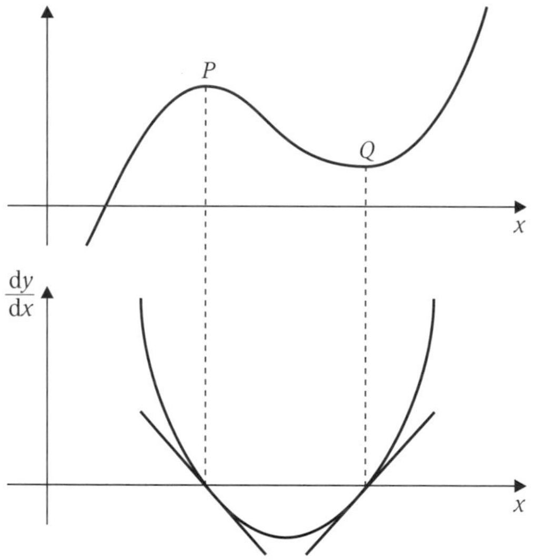

The second derivative can be thought of as the gradient of the second graph where $\frac{dy}{dx}$ has been plotted.

The second derivative is therefore negative at the maximum point P. The second derivative is positive at the minimum point Q.

> If P is a point where $\frac{dy}{dx} = 0$ and $\frac{d^2y}{dx^2}$ is negative then P is a maximum point.
> If Q is a point where $\frac{dy}{dx} = 0$ and $\frac{d^2y}{dx^2}$ is positive then Q is a minimum point.

Worked example 11.4

Find the stationary point on the curve $y = 32x - x^4 - 45$ and determine whether it is a maximum or a minimum point.

Solution

$$y = 32x - x^4 - 45$$

$$\Rightarrow \quad \frac{dy}{dx} = 32 - 4x^3$$

At stationary points, gradient = 0

$$\Rightarrow \quad 32 - 4x^3 = 0$$
$$\Rightarrow \quad x^3 = 8$$
$$\Rightarrow \quad x = 2$$

To determine the nature of the stationary point at $x = 2$ you can use the second derivative test.

$$\frac{d^2y}{dx^2} = -12x^2$$

When $x = 2$, $\frac{d^2y}{dx^2} = -12 \times 4 = -48$ which is negative.

Also when $x = 2$, $y = 64 - 16 - 45 = 3$

$\Rightarrow$ (2, 3) is a maximum point

Note that the test is inconclusive when $\frac{d^2y}{dx^2} = 0$.

For instance, the graph of $y = x^4 + 2$ has a single minimum point at the point (0, 2).

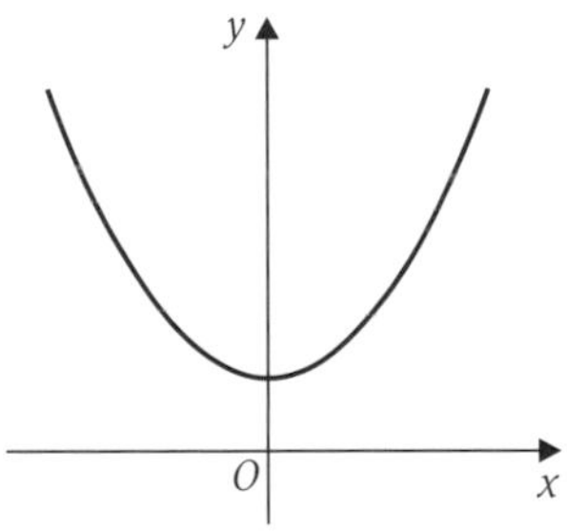

However $\frac{dy}{dx} = 4x^3$ and hence stationary points occur when $4x^3 = 0$ or $x = 0$.

But $\frac{d^2y}{dx^2} = 12x^2 = 0$ when $x = 0$.

This is neither negative or positive and so the second derivative test cannot be used.
Inspection of the gradient on either side of the origin shows that (0, 2) is a minimum point.

Worked example 11.5

A curve has equation $y = x^3 + 3x^2 - 9x - 21$.

(a) Find the coordinates of the two stationary points of the curve.

(b) Find the second derivative and, hence, determine the nature of the stationary points.

(c) Hence sketch the curve. (*There is no need to indicate the values of* x *where the curve crosses the* x*-axis.*)

Solution

(a) $y = x^3 + 3x^2 - 9x - 21$

$\Rightarrow \quad \dfrac{dy}{dx} = 3x^2 + 6x - 9$

At stationary points:

$3x^2 + 6x - 9 = 0$

$\Rightarrow \quad x^2 + 2x - 3 = 0$

$\Rightarrow \quad (x + 3)(x - 1) = 0$

$\Rightarrow \quad x = -3$ or $x = 1$

The coordinates of the two stationary points are $(-3, 6)$ and $(1, -26)$.

To find the y-coordinates, substitute $x = -3$ and $x = 1$ into the equation of the curve.

(b) $\dfrac{d^2y}{dx^2} = 6x + 6$

At $x = -3$, $\dfrac{d^2y}{dx^2} = 6(-3) + 6 = -12 < 0$

$\Rightarrow \quad (-3, 6)$ is a maximum point.

At $x = 1$, $\dfrac{d^2y}{dx^2} = 6(1) + 6 = 12 > 0$

$\Rightarrow \quad (1, -26)$ is a minimum point.

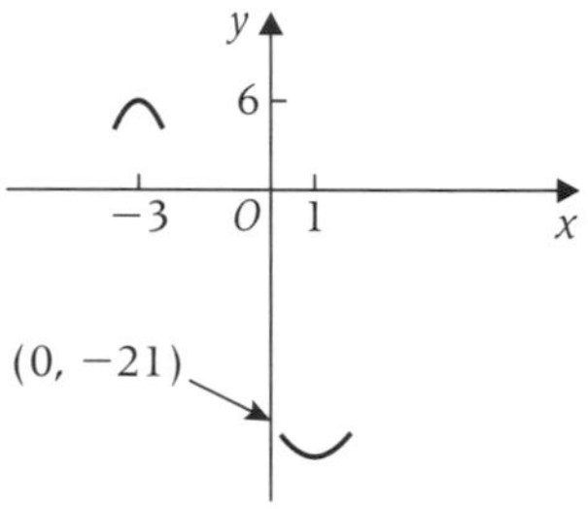

(c) Since you know the curve has a maximum point at $(-3, 6)$ and a minimum point at $(1, -26)$, you can indicate these first of all on your sketch.
You also know where the curve crosses the y-axis. Putting $x = 0$ gives $y = -21$.

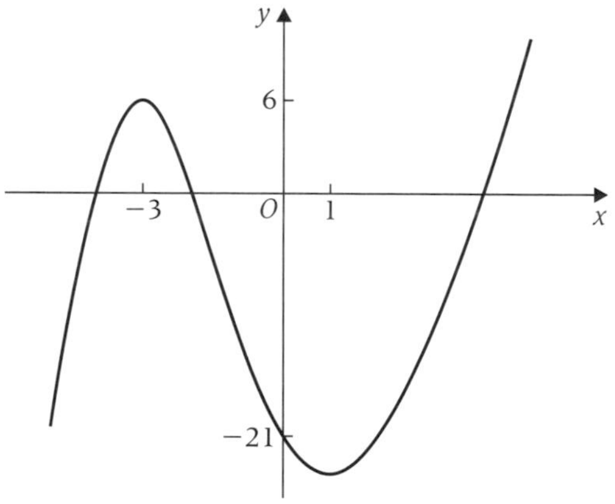

EXERCISE 11B

1 For each of the following expressions, find:

(i) $\dfrac{dy}{dx}$, **(ii)** $\dfrac{d^2y}{dx^2}$.

(a) $y = 2x^3 + 6x^2 - 9x + 10$

(b) $y = 10x^6 - 9x^3 + 2x^2 - 7x + 15$

(c) $y = x^2(4x^3 - 9)$

2 Find $f''(x)$ for each of the following functions:

(a) $f(x) = 8x + 2x^5$

(b) $f(x) = (x^2 + 6)^2$

(c) $f(x) = (x + 2)(x - 1)(x - 2)$

3 A curve has equation $y = x^2 - 8x + 5$.

(a) Find $\frac{dy}{dx}$ and, hence, find the coordinates of the stationary point of the curve.

(b) Use the second derivative test to determine whether the stationary point is a maximum or a minimum.

4 Find the two stationary points on the curve with the equation $y = x^3 - 3x^2 - 24x + 10$.

Find the value of $\frac{d^2y}{dx^2}$ at each of the stationary points and hence determine whether they are maximum or minimum points. Hence sketch the curve.

5 Find the stationary points on the curve with the equation $y = 2x^3 - 9x^2 + 12x - 5$.

Find the value of $\frac{d^2y}{dx^2}$ at each of the stationary points and, hence, determine whether they are maximum or minimum points. Hence sketch the curve.

6 For each of the following curves:

(i) find all the stationary points of the curve,

(ii) find the value of the second derivative at each stationary point and hence determine their nature,

(iii) sketch the curve.

(a) $y = x^2 - 5x$

(b) $y = (5 - 2x)^2$

(c) $y = x^4 - 108x + 240$

(d) $y = x(x - 3)^2$

7 Find the two stationary points of the curve $y = 3x^2(x - 3)$ and determine which is a maximum point and which is a minimum point. Sketch the curve.

8 An office worker can leave home at any time between 6.00 a.m. and 10.00 a.m. each morning. When he leaves home x **hours** after 6.00 a.m., $(0 \leqslant x \leqslant 4)$, his journey time to the office is y **minutes**, where $y = x^4 - 8x^3 + 16x^2 + 8$.

(a) Find $\frac{dy}{dx}$.

(b) Find the **three** values of x for which $\frac{dy}{dx} = 0$.

(c) Show that y has a maximum value when $x = 2$.

(d) Find the time at which the office worker arrives at the office on a day when his journey time is a maximum. [A]

9 The size of a population, P, of birds on an island is modelled by $P = 59 + 117t + 57t^2 - t^3$, where t is the time in years after 1970.

(a) Find $\frac{dP}{dt}$.

(b) (i) Find the positive value of t for which P has a stationary value.

(ii) Determine whether this stationary value is a maximum or a minimum.

(c) State the year when the model predicts that the population will reach its maximum value. [A]

Note

Consider the curve opposite.

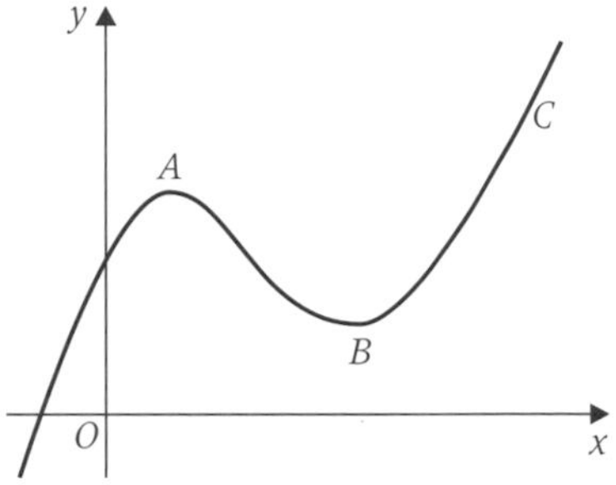

This curve has two stationary points: a maximum point at A and a minimum point at B. Notice that whilst there is a maximum point at A, this is not the maximum value that the curve reaches.

For example, as x increases beyond point C, the curve will continue to grow without limit.

For this reason A is sometimes called a **local maximum point**. Similarly, B is called a **local minimum point**.

11.5 Stationary points of inflection

There is a third type of point where the gradient is equal to zero. These are called **stationary points of inflection**.

Stationary points of inflection can appear as two types:

This section is included for completeness. No questions will be set in the examination involving stationary points of inflection.

(a)

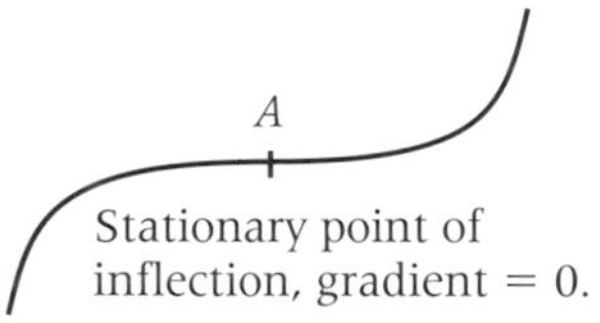

Notice that the following applies:

At a stationary point of inflection, A –

Just to the left of A, $\frac{dy}{dx} > 0$;

At A, $\frac{dy}{dx} = 0$;

Just to the right of A, $\frac{dy}{dx} > 0$.

(b)

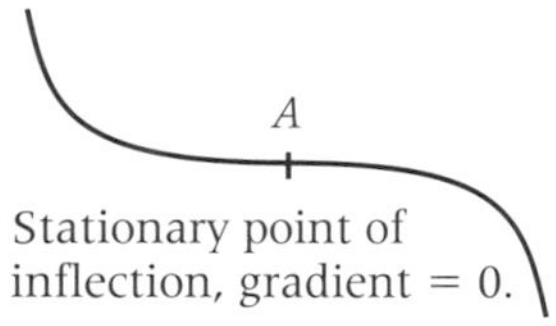

Notice that the following applies:

At a stationary point of inflection, A –

Just to the left of A, $\frac{dy}{dx} < 0$;

At A, $\frac{dy}{dx} = 0$;

Just to the right of A, $\frac{dy}{dx} < 0$.

So a stationary point of inflection can be recognised since the sign of the gradient is the same on each side of the point.

Worked example 11.6

Find the stationary points on the curve with equation $y = x^3(4 - x)$ and determine their nature.

Solution

Since $y = x^3(4 - x) = 4x^3 - x^4$,

$$\frac{dy}{dx} = 12x^2 - 4x^3.$$

Stationary values occur when $\frac{dy}{dx} = 0 \quad \Rightarrow \quad 12x^2 - 4x^3 = 0.$

$4x^2(3 - x) = 0$. Therefore $x = 0$ or $x = 3$.

$$\frac{d^2y}{dx^2} = 24x - 12x^2.$$

When $x = 3$, $\frac{d^2y}{dx^2} = 72 - 108 = -36 < 0$. Hence the curve has a maximum point at (3, 27).

However, when $x = 0$, $\frac{d^2y}{dx^2} = 0$. Therefore the second derivative test cannot be used.

Examining the gradient for values near to $x = 0$:

x	-0.5	0	0.5
gradient	3.5	0	2.5

Hence the point (0, 0) is a stationary point of inflection.

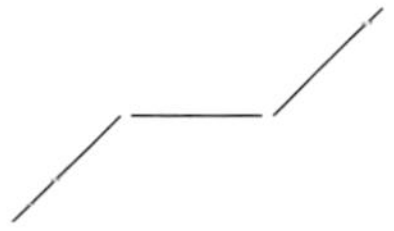

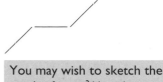

You may wish to sketch the graph of $y = x^3(4 - x)$ using computer software or a graphics calculator to verify these results.

11.6 Optimisation

Many practical problems can be solved using differentiation.

Worked example 11.7

A farmer wants to build a rectangular pen for some animals and has 100 m of fencing. One side of the pen is to be an existing wall. Find the dimensions of the pen enclosing the maximum area.

Solution

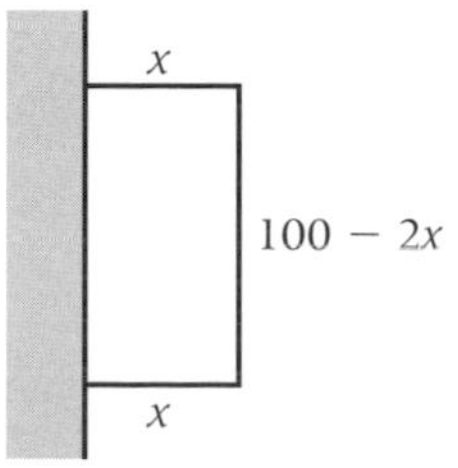

Let the width of the pen be x metres.
The length can then be written as $(100 - 2x)$ metres.

This is better than introducing a new variable for the length as it is advisable to have as few variables as possible.

Area of pen is A m^2, where

$$A = x(100 - 2x) = 100x - 2x^2$$

In order to find the maximum area you have to differentiate.

$$\frac{dA}{dx} = 100 - 4x$$

At a stationary point, $\frac{dA}{dx} = 0$

$\Rightarrow \quad 100 - 4x = 0$

$\Rightarrow \quad x = \frac{100}{4} = 25$

$$\frac{d^2A}{dx^2} = -4 \text{ for all values of } x.$$

Since this is negative, the area must take its maximum value when $x = 25$.

When the width = 25 m, length = $100 - 2(25) = 50$ m

In order to maximise the area inside the pen, its dimensions should be 25 m by 50 m.

Worked example 11.8

The sum of two positive integers is 12. Find the two numbers such that the product of one number and the cube of the other number is as large as possible.

Solution

Let the two numbers be x and y.
You are asked to maximise the product of one number and the cube of the other:

Product $P = xy^3$

You could write $P = x^3y$ and arrive at the same solution.

This contains two variables and in order to differentiate you must have an expression that contains just one variable.
You also know that $x + y = 12$

$\Rightarrow \quad x = 12 - y$

Substituting this into the product expression gives

$$P = (12 - y)y^3 = 12y^3 - y^4$$

$\Rightarrow \quad \frac{dP}{dy} = 36y^2 - 4y^3$

At a stationary point, $\frac{dP}{dy} = 0$

$\Rightarrow \quad 36y^2 - 4y^3 = 0$

$\Rightarrow \quad 4y^2(9 - y) = 0$

$\Rightarrow \quad y = 0, y = 9$

But the two integers must be positive so $y = 0$ is not a solution

$$\Rightarrow \qquad y = 9$$

$\dfrac{d^2P}{dy^2} = 72y - 12y^2$ and when $y = 9$, $\dfrac{d^2P}{dy^2} = 72 \times 9 - 12 \times 81 = -324$.

This negative value proves that you are finding the maximum value of P.

When $y = 9$, $x = 12 - 9 = 3$

$x + y = 12$

The two integers that satisfy the given conditions are 3 and 9 (9 is the number that is cubed).

Worked example 11.9

The diagram below shows a rectangular sheet of metal 10 cm by 16 cm.

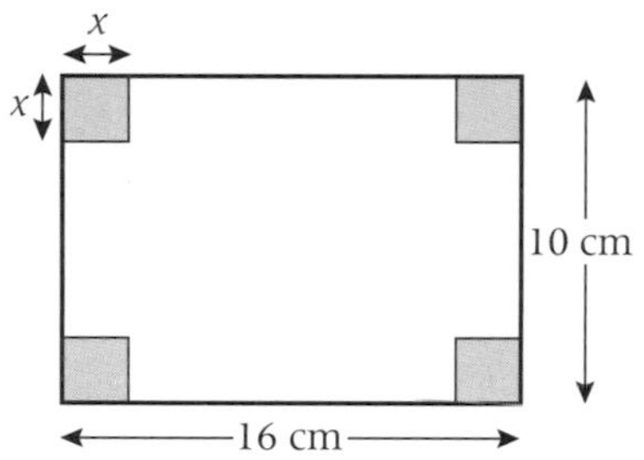

A square of side x cm is cut from each corner and the metal is then folded to make an open box (i.e., the box does not have a lid).

(a) Show that the volume, V cm³, the box can hold is given by

$$V = 4x^3 - 52x^2 + 160x.$$

(b) Differentiate V with respect to x and hence find the values of x for which $\dfrac{dV}{dx} = 0$.

(c) Show that there is a single value of x for which V is stationary. Verify that this value of x gives a maximum value of V.

(d) Calculate the maximum possible volume of the box.

Solution

(a) You should be able to see that the dimensions of the box when folded are

width: $(10 - 2x)$ length: $(16 - 2x)$ height: x

$$\begin{aligned}\Rightarrow \quad \text{Volume } V &= x(10 - 2x)(16 - 2x)\\ &= x(160 - 20x - 32x + 4x^2)\\ &= x(4x^2 - 52x + 160)\\ &= 4x^3 - 52x^2 + 160x \quad \text{as required.}\end{aligned}$$

(b) $$\frac{dV}{dx} = 12x^2 - 104x + 160$$

At a stationary point, $\frac{dV}{dx} = 0$

$\Rightarrow \quad 12x^2 - 104x + 160 = 0$

$\Rightarrow \quad 3x^2 - 26x + 40 = 0$

$\Rightarrow \quad (3x - 20)(x - 2) = 0$

$\Rightarrow \quad x = \frac{20}{3} = 6\frac{2}{3} \quad \text{or} \quad x = 2$

You could also use the quadratic formula.

(c) If you cut squares of side $6\frac{2}{3}$ cm from each corner there will be no metal left so $x = 6\frac{2}{3}$ cannot be a solution.

$\Rightarrow$ The single stationary value occurs when $x = 2$.

To verify this is a maximum value you can use the second derivative test.

$$\frac{d^2V}{dx^2} = 24x - 104$$

When $x = 2$, $\frac{d^2V}{dx^2} = 24(2) - 104 = -56$

Since $\frac{d^2V}{dx^2}$ is negative, $x = 2$ gives a maximum value of V.

(d) The maximum volume is given by substituting $x = 2$ into the volume expression.

Maximum volume $V = 4(2)^3 - 52(2)^2 + 160(2) = 144 \text{ cm}^3$

EXERCISE 11C

1 A rectangle has one side x cm and its perimeter is 44 cm. Show that the area of the rectangle, A cm^2, is given by $A = 22x - x^2$.
Find the maximum value of A and give the dimensions of the rectangle for which the maximum area occurs.

2 The sum of two positive integers is 21. Find the two integers such that the product of one number and the square of the other number is a maximum.

3 The sum of two positive integers is 40. Find the two integers such that the product of the square of one number and the cube of the other number is a maximum.

4 Sam wants to make a rectangular enclosure bordering a stream. He has 60 m of fencing and he will use three straight sections of fencing, two of which are of length x m, as shown in the diagram.

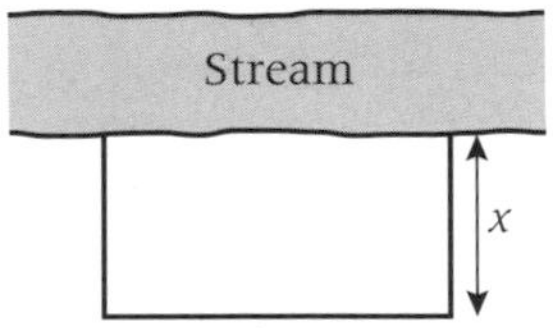

(a) Show that the area, A m^2, of the rectangular enclosure is given by $A = 60x - 2x^2$.

(b) Find the value of x for which A has a stationary value. Show that this value of x gives a maximum value and not a minimum value.

(c) Hence state the maximum area of the rectangular enclosure.

5 Pat has £120 to spend on a decorative fence to enclose a rectangular garden. The fencing for three sides of the garden costs £2 per metre and the fencing for the fourth side, which is of length x metres, costs £3 per metre.

(a) Given that the rectangle has dimensions x metres by y metres, show that $5x + 4y = 120$.

(b) Hence show that the area of the garden is A m², where

$$A = \frac{x}{4}(120 - 5x).$$

(c) Find the value of x for which A is maximum and state the maximum area of the garden.

6 An open box is made from a rectangular piece of cardboard 6 cm by 16 cm by cutting a square of side x cm from each corner and then folding the resulting shape to form the box.

(a) Show that the volume, V cm³, of the box is given by

$$V = 4x^3 - 44x^2 + 96x.$$

(b) Given that x can vary, find the values of x for which $\frac{dV}{dx} = 0$. Show that only one of these values is acceptable.

(c) Verify that this value of x gives a maximum volume.

(d) Find the maximum possible volume of the box.

7 Small trays are to be made from rectangular pieces of card. Each piece of card is 8 cm by 5 cm and the tray is formed by removing squares of side x cm from each corner and folding the remaining card along the dashed lines, as shown in the diagram, to form an open-topped box.

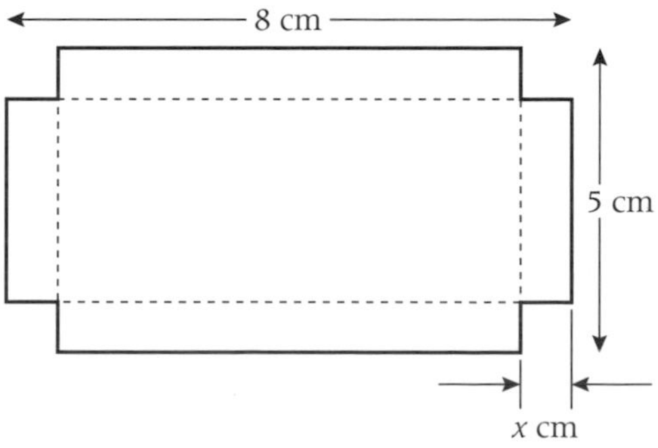

(a) Explain why $0 < x < 2.5$.

(b) Show that the volume, V cm^3, of a tray is given by

$$V = 4x^3 - 26x^2 + 40x.$$

(c) Find the value of x for which $\frac{dV}{dx} = 0$.

(d) Calculate the greatest possible volume of a tray. [A]

8 An open-topped box has height h cm and a square base of side x cm.

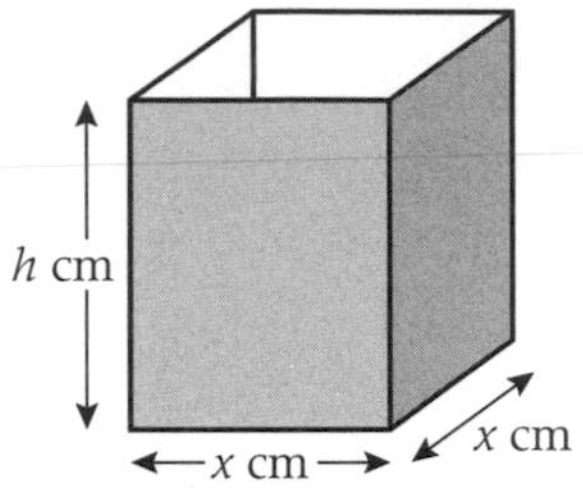

The box has capacity V cm^3. The area of its **external** surface, consisting of its horizontal base and four vertical faces, is A cm^2.

(a) Find expressions for V and A in terms of x and h.

(b) It is given that $A = 3000$.

(i) Show that $V = 750x - \frac{1}{4}x^3$.

(ii) Find the positive value of x for which $\frac{dV}{dx} = 0$, giving your answer in surd form.

(iii) Hence find the maximum possible value of V, giving your answer in the form $p\sqrt{10}$, where p is an integer. (*You do not need to show your answer is a maximum.*) [A]

Key point summary

1 Stationary points occur when $\frac{dy}{dx} = 0$. *p167*

2 There are three types of stationary point: *pp168, 176*

(i) maximum point;

(ii) minimum point;

(iii) stationary point of inflection.

3 The nature of a stationary point can be determined by considering the value of the gradient just to the left and right of the point. *pp168, 176*

Gradient changes from negative to zero to positive – minimum point

Gradient changes from positive to zero to negative – maximum point

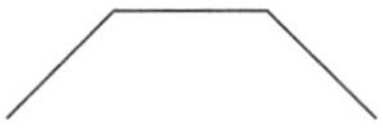

Gradient changes from positive to zero to positive – stationary point of inflection

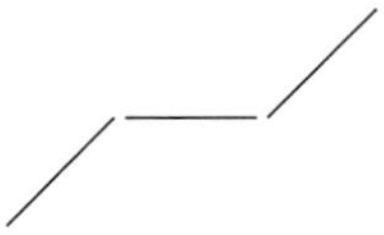

Gradient changes from negative to zero to negative – stationary point of inflection

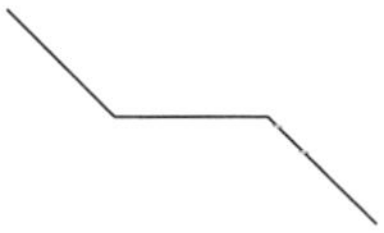

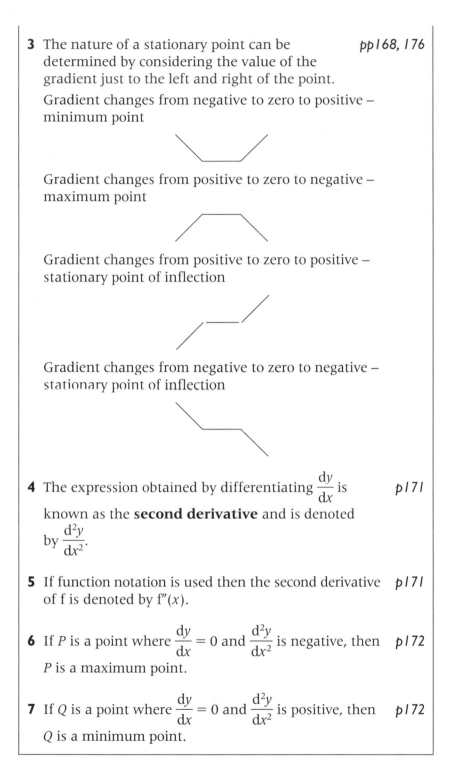

4 The expression obtained by differentiating $\frac{\mathrm{d}y}{\mathrm{d}x}$ is known as the **second derivative** and is denoted by $\frac{\mathrm{d}^2y}{\mathrm{d}x^2}$. *p171*

5 If function notation is used then the second derivative of f is denoted by $\mathrm{f}''(x)$. *p171*

6 If P is a point where $\frac{\mathrm{d}y}{\mathrm{d}x} = 0$ and $\frac{\mathrm{d}^2y}{\mathrm{d}x^2}$ is negative, then P is a maximum point. *p172*

7 If Q is a point where $\frac{\mathrm{d}y}{\mathrm{d}x} = 0$ and $\frac{\mathrm{d}^2y}{\mathrm{d}x^2}$ is positive, then Q is a minimum point. *p172*

Test yourself	What to review
1 Find the stationary points on each of the following curves: **(a)** $y = 2x^2 + 12x - 7$, **(b)** $y = 6x - 2x^3$.	*Section 11.1*
2 Find the coordinates of the stationary point on the curve with equation $y = (x - 8)(x - 6)$ and determine whether it is a maximum or minimum point.	*Section 11.2*
3 A curve has equation $y = x^3 + 6x^2 - 36x + 25$. Find the x-coordinates of the stationary points of the curve and determine their nature.	*Section 11.4*
4 Two numbers have a sum of 12. Find the maximum value of the square of one number multiplied by the other.	*Section 11.6*

Test yourself ANSWERS

1 **(a)** $(-3, -25)$; **(b)** $(1, 4)$ and $(-1, -4)$.

2 $(7, -1)$ minimum.

3 minimum point when $x = 2$ and maximum point when $x = -6$.

4 256.

CHAPTER 12

C1: Integration

Learning objectives

After studying this chapter, you should be able to:

- understand that integration is the reverse process of differentiation
- be able to integrate expressions of the form kx^n, where n is a non-negative integer
- understand the term 'indefinite integral' and realise the need to add an arbitrary constant
- be able to find the equation of a curve given its derivative and the coordinates of a point on the curve
- understand how to evaluate a 'definite integral'
- use definite integrals to find areas of regions bounded by curves and lines.

12.1 Finding a function from its derivative

You saw earlier how to find a derivative by differentiation, so that

$$\begin{aligned} &\mathrm{f}(x) = 2x^3 - 5x^2 + 3x - 7 \\ \Rightarrow\ &\mathrm{f}'(x) = 6x^2 - 10x + 3 \end{aligned}$$

You learned to reduce the index by one and to multiply at the front by the index. See section 9.7 of chapter 9.

Can the process be reversed?

In general, $y = x^n \Rightarrow \dfrac{\mathrm{d}y}{\mathrm{d}x} = nx^{n-1}$.

Reversing the process, it would seem that

Increase the index by 1 and **divide** at the front by the new index.

$$\frac{\mathrm{d}y}{\mathrm{d}x} = x^m \ \Rightarrow\ y = \frac{1}{m+1}x^{m+1}$$

The general process of finding a function from its derivative is known as **integration**.

As a general guide, so that you do not confuse the two techniques,

when you **In**tegrate, you **In**crease the index by 1, whereas

when you **D**ifferentiate, you **D**ecrease the index by 1.

Worked example 12.1

Given that $\frac{dy}{dx} = 12x^2 + 4x - 5$, find an expression for y.

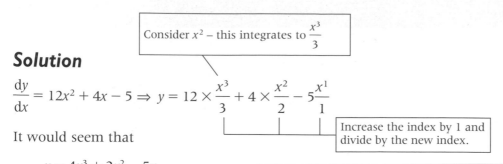

Consider x^2 – this integrates to $\frac{x^3}{3}$

Solution

$$\frac{dy}{dx} = 12x^2 + 4x - 5 \Rightarrow y = 12 \times \frac{x^3}{3} + 4 \times \frac{x^2}{2} - 5\frac{x^1}{1}$$

Increase the index by 1 and divide by the new index.

It would seem that

$$y = 4x^3 + 2x^2 - 5x$$

but that is not quite the complete answer.

Whenever you differentiate a constant you get zero,

$$\text{e.g. } y = 7 \Rightarrow \frac{dy}{dx} = 0, \text{ or } y = -19 \Rightarrow \frac{dy}{dx} = 0$$

and so the expression for y above could have any constant on the end and still satisfy $\frac{dy}{dx} = 12x^2 + 4x - 5$.

The answer to this example is therefore

$$y = 4x^3 + 2x^2 - 5x + c, \text{ where } c \text{ is a constant.}$$

The constant c is called an **arbitrary constant.**

12.2 Use of an arbitrary constant

You can use any letter (other than those being used for the variables) to represent an arbitrary constant, but it is customary to use either c or k.

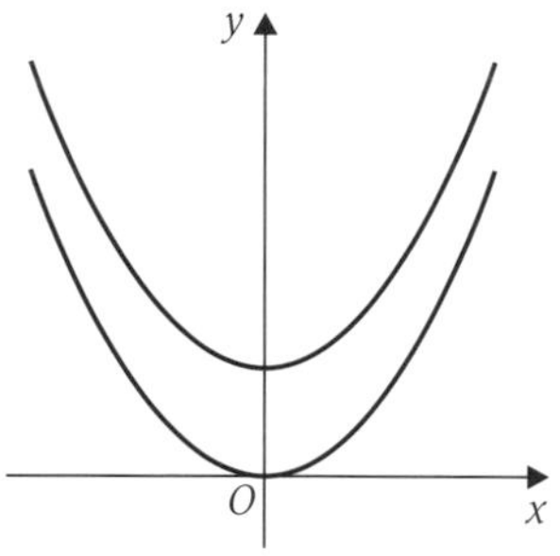

The graphs of $y = x^4$ and $y = x^4 + 5$, for example, are parallel. They both have the same gradient function, namely $4x^3$.

In fact, you could translate $y = x^4$ in the y-direction by k units to obtain $y = x^4 + k$ and they would still have the same derivative, $4x^3$.

Hence, if you are given that $\frac{dy}{dx} = 4x^3$, you can only conclude that $y = x^4 + k$, where k is an arbitrary constant.

Worked example 12.2

Given that $f'(x) = x^4 - 12x^5$, find an expression for $f(x)$.

Solution

$f'(x) = x^4 - 12x^5$

$$\Rightarrow \quad f(x) = \frac{1}{5}x^5 - \frac{12}{6}x^6 + k = \frac{x^5}{5} - 2x^6 + k$$

Integrating: Increase the index by 1 and divide by the new index.

Don't forget the arbitrary constant.

12.3 Finding the equation of a curve

If you are given the coordinate of one point on the curve, you can usually find the actual equation of the curve.

Worked example 12.3

A curve passes through the point (1, 5) and $\frac{dy}{dx} = 16x^7 - 6x$.

Find its equation

Solution

Since $\frac{dy}{dx} = 16x^7 - 6x$, integration gives

$$y = 16 \times \frac{x^8}{8} - 6 \times \frac{x^2}{2} + c, \quad \text{where } c \text{ is a constant}$$

$$\Rightarrow \quad y = 2x^8 - 3x^2 + c$$

Since the curve passes through (1, 5) you can find the value of c.

$x = 1$ and $y = 5$ simultaneously, $\Rightarrow 5 = 2 - 3 + c \quad \Rightarrow \quad 5 = -1 + c$

$$\Rightarrow \quad c = 6$$

The equation of the curve is $y = 2x^8 - 3x^2 + 6$.

EXERCISE 12A

1 Find y in terms of x for each of the following:

(a) $\frac{dy}{dx} = 12x^5$, **(b)** $\frac{dy}{dx} = 2x^3$,

(c) $\frac{dy}{dx} = 4x^7$, **(d)** $\frac{dy}{dx} = 12$,

(e) $\frac{dy}{dx} = 3x^2 - 2x + 5$, **(f)** $\frac{dy}{dx} = 6x^5 - 18x^2 - x$,

(g) $\frac{dy}{dx} = 4x^7 - 12x^5 - 4$, **(h)** $\frac{dy}{dx} = 6x^{11}$,

(i) $\frac{dy}{dx} = 5x^3$, **(j)** $\frac{dy}{dx} = 3x^8 - 2$,

(k) $\frac{dy}{dx} = 2 - 6x^7$, **(l)** $\frac{dy}{dx} = 3x^5 - 4x$.

Don't forget the arbitrary constant.

You can always check your final answer by differentiating and making sure you get the original expression for $\frac{dy}{dx}$ or $f'(x)$.

2 Find an expression for $f(x)$ in each of the following cases:

(a) $f'(x) = 6x^2 - 4x + 5$, **(b)** $f'(x) = 3x^2 + 10x + 4$,

(c) $f'(x) = 10x^4 - 3x^5 + 7$, **(d)** $f'(x) = 10x^9 - 8x^3 + 1$,

(e) $f'(x) = 2x - 3x^5$, **(f)** $f'(x) = 10x^4 - 12x^3 + 7x$,

(g) $f'(x) = x^4 + 3x^2 - 5$, **(h)** $f'(x) = 2x^3 - 12x^7$,

(i) $f'(x) = 14x^6 - 5x^4 + 3x$, **(j)** $f'(x) = 8x^3 + 9x^5 - 4$.

3 The graph of $y = f(x)$ passes through the point $(1, -3)$ and $f'(x) = 12x^2 + 8x - 3$. Find $f(x)$.

4 A curve passes through the point $(-1, 2)$ and is such that $\frac{dy}{dx} = 8x^3 - 12x^2 + 3$. Find the equation of the curve.

5 A curve passes through the point $(1, -7)$ and is such that $\frac{dy}{dx} = 3 - 8x + 5x^4$. Find the equation of the curve.

6 Given that $\frac{dy}{dx} = x^n$, for almost all values of n, $y = \frac{1}{n+1}x^{n+1} + c$. There is, however, one value of n for which this cannot be true. What is this value of n?

12.4 Simplifying expressions before integrating

Sometimes it is necessary to obtain an expression involving terms of the form kx^n before integrating.

Worked example 12.4

The graph of $y = f(x)$ passes through the point (2, 6) and $f'(x) = (3x - 4)(3 - 2x)$. Find $f(x)$.

Solution

$$f'(x) = (3x - 4)(3 - 2x) = 9x - 6x^2 - 12 + 8x = 17x - 6x^2 - 12$$

Multiplying out the brackets.

$$\Rightarrow f(x) = 17 \times \frac{x^2}{2} - 6 \times \frac{x^3}{3} - 12 \times \frac{x^1}{1} + c$$

$$= \frac{17}{2}x^2 - 2x^3 - 12x + c$$

c is an arbitrary constant.

When $x = 2$, $y = f(x) = 6$, $\Rightarrow 6 = \left(\frac{17}{2} \times 4\right) - (2 \times 8) - (12 \times 2) + c$

$$\Rightarrow 6 = 34 - 16 - 24 + c$$

$$\Rightarrow 6 = -6 + c \Rightarrow c = 12$$

$$\Rightarrow f(x) = \frac{17}{2}x^2 - 2x^3 - 12x + 12$$

EXERCISE 12B

1 Find an expression for y in terms of x and a constant in each of the following cases:

(a) $\dfrac{dy}{dx} = x(3x - 2)$, **(b)** $\dfrac{dy}{dx} = x^3(10x + 6)$,

(c) $\dfrac{dy}{dx} = 4x^5(2x^2 + 3)$, **(d)** $\dfrac{dy}{dx} = x^4(3x + 5)$,

(e) $\dfrac{dy}{dx} = (1 - x)(3x - 2)$, **(f)** $\dfrac{dy}{dx} = x(x - 2)(x + 3)$,

(g) $\dfrac{dy}{dx} = x^2(x + 2)(x + 1)$, **(h)** $\dfrac{dy}{dx} = x(x + 3)(x + 5)$.

2 Find $f(x)$ in terms of x and an arbitrary constant for each of the following where:

(a) $f'(x) = (2x - 5)(3 - 2x)$, **(b)** $f'(x) = x(6x - 4)$,

(c) $f'(x) = 8x(x + 1)(x - 1)$, **(d)** $f'(x) = 4x(3x^2 - 1)$,

(e) $f'(x) = x^2(x^2 - 3)$, **(f)** $f'(x) = 12x^3(x + 1)(x - 1)$

(g) $f'(x) = x^4(x - 1)(x + 2)$, **(h)** $f'(x) = 28x^3(x^2 + 4)(x - 1)$.

3 A curve passes through the point $(-1, 3)$ and is such that $\dfrac{dy}{dx} = x^3(25x - 16)$. Find the equation of the curve.

4 A curve passes through the point $(2, -5)$ and is such that $\dfrac{dy}{dx} = (2x - 3)(3x + 5)$. Find the equation of the curve.

5 A curve passes through the point $(1, 2)$ and is such that $\dfrac{dy}{dx} = 12x(x - 3)(3x + 5)$. Find the equation of the curve.

12.5 Indefinite integrals

The special symbol $\int$ is used to denote integration.

It is rather like an elongated letter S and, as you will see in the next section, it stands for summation (actually the first letter of the Latin word *summa*). The symbol dates back to the 1600s when a letter S actually looked like this.

When you need to integrate x^3, for example, you write

$$\int x^3 \, dx = \frac{1}{4}x^4 + c$$

and read as

the integral of x^3 with respect to x

It is a common mistake to forget the dx symbol. Try to remind yourself to include it.

> Because the answer contains an arbitrary constant and so does not have a unique value, it is called an **indefinite integral**.

There is nothing new in this section apart from this new notation.

Worked example 12.5

Find **(a)** $\int x(x-3)\,dx$, **(b)** $\int (x+2)(3x-1)\,dx$.

Solution

(a) $\int x(x-3)\,dx = \int x^2 - 3x\,dx$

$$= \frac{x^3}{3} - \frac{3}{2}x^2 + c$$

(b) $\int (x+2)(3x-1)\,dx = \int (3x^2 + 5x - 2)\,dx$

$$= 3\frac{x^3}{3} + 5\frac{x^2}{2} - 2x + k$$

$$= x^3 + \frac{5}{2}x^2 - 2x + k,$$

where k is a constant.

Notice that you keep the integral sign in your working while you simplify the expression.

EXERCISE 12C

Find each of the following:

1 $\int 12x^5\,dx$.

2 $\int 20x^4\,dx$.

3 $\int 2x\,dx$.

4 $\int x^2(4x+6)\,dx$.

5 $\int 88x^{10}\,dx$.

6 $\int 20x^3\,dx$.

7 $\int 3x^2\,dx$.

8 $\int x^3(5+6x^2)\,dx$.

9 $\int (5x^4 - 3x)\,dx$.

10 $\int (3x^4 - 2x + 1)\,dx$.

11 $\int (x^2-1)(x+1)\,dx$.

12 $\int (x^3 - 3x + 1)\,dx$.

13 $\int (x+1)(3x+5)\,dx$.

14 $\int (x-1)(6x-5)\,dx$.

15 $\int (3-2x)(x+3)\,dx$.

16 $\int x(3-2x)(x+3)\,dx$.

17 $\int (3+2x)^2\,dx$.

18 $\int (2-3x)^2\,dx$.

19 $\int (x^3+2)(5x-1)\,dx$.

20 $\int (x^3-1)^2\,dx$.

21 $\int (5-6x)^2\,dx$.

22 $\int x(x^3-1)^2\,dx$.

12.6 Integration as an area

Consider the curve with equation $y = f(x)$. How can you find the area of the region bounded by the curve, the x-axis and the lines with equations $x = a$ and $x = b$?

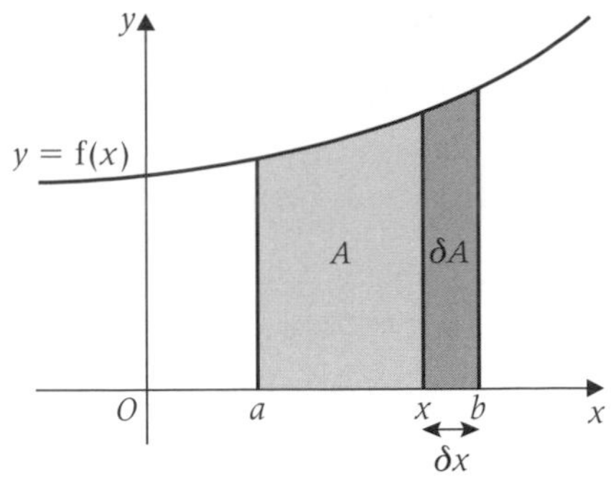

Let the region under the curve from $x = a$ to a general position shown be A.

Now as x increases by δx, y becomes $y + \delta y$ and A becomes $A + \delta A$.

By considering the two rectangles, it is clear that

$$y\delta x < \delta A < (y + \delta y)\delta x$$

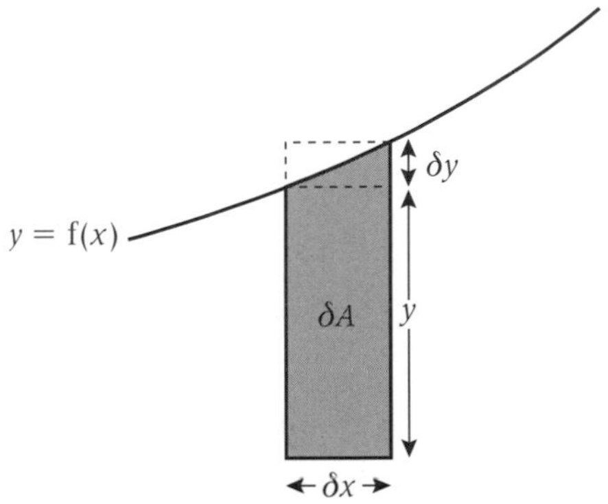

or, dividing throughout by δx.

$$y < \frac{\delta A}{\delta x} < y + \delta y$$

As $\delta x \to 0$, $\delta y \to 0$ and $\dfrac{\delta A}{\delta x} \to \dfrac{dA}{dx}$.

Hence $\dfrac{dA}{dx} = y$

Suppose $\int y \, dx = \int f(x) \, dx = F(x) + c$

then $\quad A = F(x) + c$

Since A is obviously zero when $x = a$,

$$0 = F(a) + c \Rightarrow c = -F(a)$$

$$\Rightarrow A = F(x) - F(a)$$

Hence the area of the region up to the point where $x = b$ is given by substituting $x = b$ into the formula for A.

Area, $A = F(b) - F(a)$.

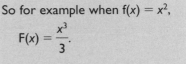

So for example when $f(x) = x^2$,

$F(x) = \dfrac{x^3}{3}$.

The function $F(x)$ is called a primitive of $f(x)$.

Worked example 12.6

Find the area bounded by the curve with equation $y = 8x^3 + 6x + 2$, the x-axis and the lines $x = 1$ and $x = 2$, shaded in the diagram.

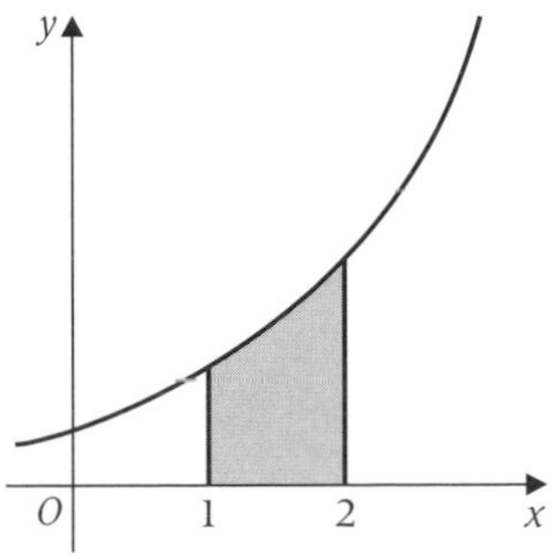

Solution

Using the notation from above,

$$f(x) = 8x^3 + 6x + 2$$

$$\Rightarrow F(x) = 8 \times \frac{x^4}{4} + 6 \times \frac{x^2}{2} + 2x = 2x^4 + 3x^2 + 2x$$

Area is given by $F(2) - F(1)$

$$F(2) = 32 + 12 + 4 = 48; \qquad F(1) = 2 + 3 + 2 = 7.$$

Area of region is $48 - 7 = 41$.

Note, there is no need to add $+c$ when finding $F(x)$.
If you do, it will cancel out when you find $F(b) - F(a)$.

12.7 Definite integrals

Finding the area under a curve is made easier by using the notation of a **definite integral**.

The values of the endpoints $x = a$ and $x = b$ are placed at the bottom and top of the integral sign and these are called **limits**.

A definite integral is of the form $\int_a^b f(x)\,dx$.

The function $f(x)$ is called the integrand.

The area calculated in worked example 12.6 is given by the definite integral $\int_1^2 8x^3 + 6x + 2\,dx$.

This example can be reworked using a square bracket notation.

$$\int_1^2 8x^3 + 6x + 2\,dx = [2x^4 + 3x^2 + 2x]_1^2$$

$$= (32 + 12 + 4) - (2 + 3 + 2) = 48 - 7 = 41$$

Equivalent to F(2)   Equivalent to F(1)

The procedure can be streamlined without introducing the function $F(x)$. Instead square brackets are used after integration with the limits placed at the top and bottom of the final bracket.

Most graphic calculators find approximate values of definite integrals and this might be a helpful check on your answer.

Worked example 12.7

Find the value of the definite integral $\int_2^3 8x^3 + 2x\,dx$.

Solution

(a) $\int_2^3 8x^3 + 2x\,dx = [2x^4 + x^2]_2^3$

$= (2 \times 3^4 + 3^2) - (2 \times 2^4 + 2^2)$

$= 171 - 36 = 135$

Integrate the expression and put the limits at the end of the square brackets.

Evaluate at top limit then subtract the value of the primitive at the bottom limit.

Worked example 12.8

Find the area of the region bounded by the curve with equation $y = 4x^2(x + 1)$, the x-axis from $x = 1$ to $x = 2$ and the lines $x = 1$ and $x = 2$.

Solution

The area of the region can be written as a definite integral:

$$\int_1^2 4x^2(x + 1)\,dx = \int_1^2 (4x^3 + 4x^2)\,dx$$

Multiply out the brackets.

$$= \left[x^4 + \frac{4}{3}x^3\right]_1^2$$

Integrating each term.

$$= \left[2^4 + \frac{4}{3} \times 2^3\right] - \left[1^4 + \frac{4}{3} \times 1^3\right]$$

'top limit value' minus 'bottom limit value'.

$$= \left(16 + \frac{32}{3}\right) - \left(1 + \frac{4}{3}\right) = 15 + \frac{28}{3} = 24\tfrac{1}{3}$$

The area of the region is therefore $24\frac{1}{3}$.

EXERCISE 12D

Evaluate each of the definite integrals from (1) to (10).

Don't worry if some of the final answers are negative.

1 $\int_0^1 6x\,dx$.

2 $\int_1^3 10x^4\,dx$.

3 $\int_0^1 (5x^4 - 6x)\,dx$.

4 $\int_1^4 2x^3\,dx$.

5 $\int_1^3 (10x - 6x^2)\,dx$.

6 $\int_1^2 (x - 1)(x + 1)\,dx$.

7 $\int_0^1 (3x - 4)(x + 1)\,dx$.

8 $\int_2^4 (3x - x^3)\,dx$.

9 $\int_0^1 (x - 2)(x + 3)\,dx$.

10 $\int_0^1 x(x - 2)(x + 3)\,dx$.

11 Calculate the area of the region bounded by the curve $y = x^3$, the x-axis from the origin to $x = 4$ and the line $x = 4$.

12 Calculate the area of the region bounded by the curve with equation $y = 2x + 12x^2$, the x-axis from $x = 2$ to $x = 3$ and the lines $x = 2$ and $x = 3$.

You may wish to sketch the curves and regions using a graphic calculator.

13 Calculate the area of the region bounded by the curve with equation $y = x^3 + 3x + 1$, the x-axis from $x = 1$ to $x = 2$ and the lines $x = 1$ and $x = 2$.

14 Calculate the area of the region bounded by the curve with equation $y = x^3(5x + 8)$, the x-axis from $x = 1$ to $x = 2$ and the lines $x = 1$ and $x = 2$.

15 Calculate the area of the region bounded by the curve with equation $y = x^2(4x + 6)$, the x-axis from $x = -1$ to $x = 2$ and the lines $x = -1$ and $x = 2$.

16 Calculate the area of the region bounded by the curve with equation $y = (x + 3)(x + 5)$, the x-axis from $x = -3$ to $x = 0$ and the lines $y = 0$ and $x = 0$.

12.8 Regions bounded by lines and curves

Sometimes a region is bounded by more than one curve or a curve and a line as the following example illustrates.

Worked example 12.9

Find the points of intersection of the curve $y = x^2 - 3x + 4$ and the line $y = x + 1$.
Calculate the area of the finite region bounded by this curve and this line, shaded in the diagram.

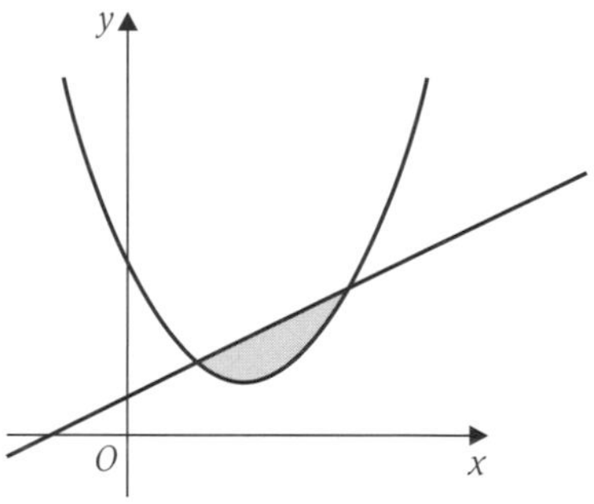

Solution

Eliminating y to find points of intersection,

$$\Rightarrow x + 1 = x^2 - 3x + 4$$
$$\Rightarrow 0 = x^2 - 4x + 3$$
$$\Rightarrow 0 = (x - 1)(x - 3)$$
$$\Rightarrow x = 1 \text{ or } x = 3.$$

When $x = 1$, $y = 2$, and when $x = 3$, $y = 4$.

Points of intersection are $(1, 2)$ and $(3, 4)$.

The area of the shaded region shown in the diagram can be thought of as the difference of two areas.

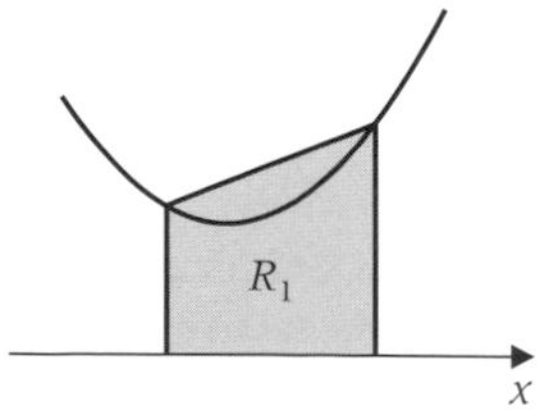

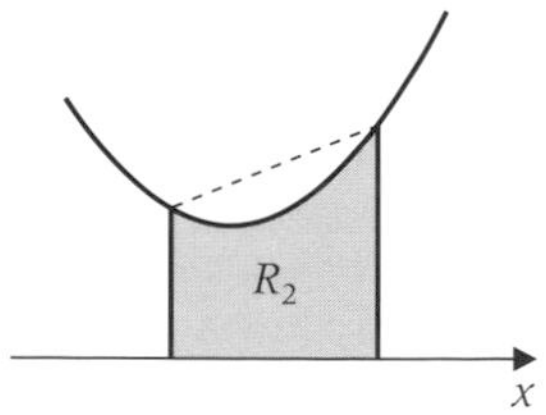

Area of $R_1 = \int_1^3 x + 1\,dx = \left[\frac{x^2}{2} + x\right]_1^3 = \left(\frac{9}{2} + 3\right) - \left(\frac{1}{2} + 1\right) = 6$

Area of $R_2 = \int_1^3 x^2 - 3x + 4\,dx = \left[\frac{x^3}{3} - 3\frac{x^2}{2} + 4x\right]_1^3$

$= \left(9 - \frac{27}{2} + 12\right) - \left(\frac{1}{3} - \frac{3}{2} + 4\right) = \frac{14}{3}$

Area of shaded region $= 6 - \frac{14}{3} = \frac{4}{3}$.

Since R_1 is a trapezium, its area could have been found more easily by using

$\frac{1}{2}(2 + 4) \times (3 - 1) = 6$

12.9 Regions below the x-axis

At times the region being considered lies partly below the x-axis. For example, suppose you are finding the area of the shaded region bounded by the curve $y = x(x - 3)$ and the x-axis.

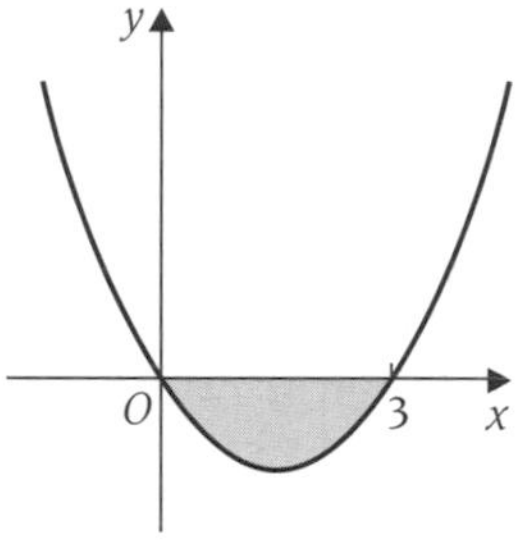

$$\int_0^3 x(x - 3)\,dx = \int_0^3 (x^2 - 3x)\,dx = \left[\frac{x^3}{3} - 3\frac{x^2}{2}\right]_0^3 = \left(9 - \frac{27}{2}\right) - 0 = -\frac{9}{2}$$

The negative sign indicates that the region was below the x-axis. The area is therefore $4\frac{1}{2}$.

Because the summation involves y-coordinates that are negative, the integral has a negative value. Care must be taken to consider separately any regions where the y-value is negative.

Worked example 12.10

Sketch the graph of $y = (x - 2)(x - 3)$. Find the area of the region bounded by the curve and the x-axis from $x = 2$ to $x = 3$.

Solution

The curve is a parabola and cuts the x-axis at $x = 2$ and $x = 3$.

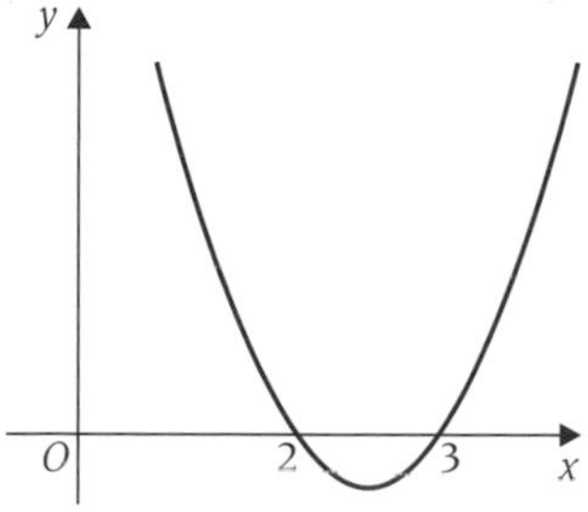

Multiplying out the brackets gives

$y = x^2 - 5x + 6$

Consider

$$\int_2^3 (x^2 - 5x + 6)\mathrm{d}x = \left[\frac{x^3}{3} - \frac{5x^2}{2} + 6x\right]_2^3$$

$$= \left[\frac{27}{3} - \frac{45}{2} + 18\right] - \left[\frac{8}{3} - \frac{20}{2} + 12\right] = \frac{19}{3} - \frac{25}{2} + 6 = -\frac{1}{6}$$

The negative answer occurs because the region is below the x-axis.

The area of the region is therefore $\frac{1}{6}$.

EXERCISE 12E

1 Sketch the graph of the parabola with equation $y = x^2 + 1$ and the line with equation $y = 3x - 1$. Find the two points of intersection of the line and the curve. Calculate the area of the finite region bounded by the parabola and the line.

2 Show that the curve with equation $y = 18 + 3x - x^2$ and the line with equation $y = 18 - 3x$ intersect on the x-axis and on the y-axis, and state the points of intersection.
Calculate the area of the finite region bounded by the line and the curve.

3 The line $x + y = 9$ and the curve with equation $y = x^2 - 2x + 3$ intersect at the points P and Q. Find the coordinates of P and Q and hence determine the area of the finite region bounded by the curve and the line.

4 Two curves have equations $y = x^2 + 8$ and $y = x(10 - x)$. Sketch their graphs and find their points of intersection. Calculate the area of the finite region bounded by the two curves.

5 The two curves $y = x^2 - 2x + 3$ and $y = 3 - 4x - x^2$ intersect at A and B. Sketch their graphs and find the points A and B. Calculate the area of the finite region bounded by the two curves.

6 Find the points where the curve $y = (x - 1)(x - 2)$ crosses the x-axis. Sketch the curve and determine the area of the finite region below the x-axis bounded by the curve and the x-axis.

7 Find the points where the curve $y = x^2 + x - 6$ crosses the x-axis. Sketch the curve and determine the area of the finite region bounded by the curve and the x-axis.

8 The two curves $y = x(x - 5)$ and $y = 2x^2 - 10x$ intersect at the origin and the point F. Sketch the curves and state the coordinates of F.
Calculate the area of the finite region bounded by the two curves.

9 Find the area of the finite region bounded by the parabola with equation $y = x(x - 4)$ and the line $x + y = 0$.

10 Sketch the graphs of the curves with equations $y = 16 - x^2$ and $y = x^2 - 5x + 13$ on the same axes. Find the coordinates of the two points of intersection.
Calculate the area of the finite region bounded by the two curves.

11 The sketch shows the graph of the curve with equation $y = (x + 2)(x + 1)(x - 1)$.

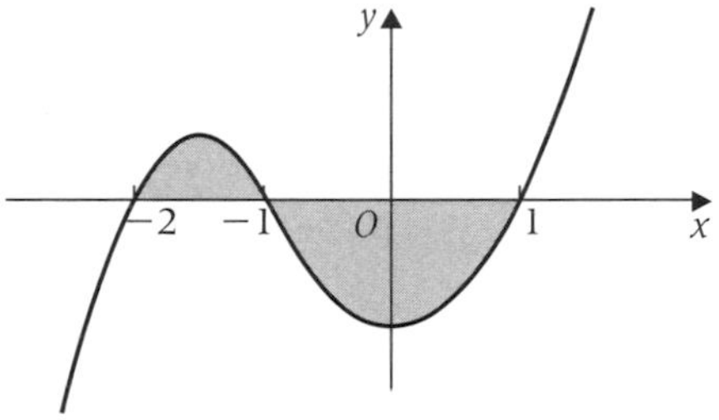

Find the total area of the two regions, shaded on the diagram, bounded by the curve and the x-axis.

Why would you not obtain the correct answer by simply finding

$$\int_{-2}^{1} (x + 2)(x + 1)(x - 1)\,dx?$$

12 The diagram shows a sketch of the curve $y = x^2 - 4$ and the line $y = x - 2$. The curve intersects the x-axis at the points A and B and the curve and line intersect at a further point C.

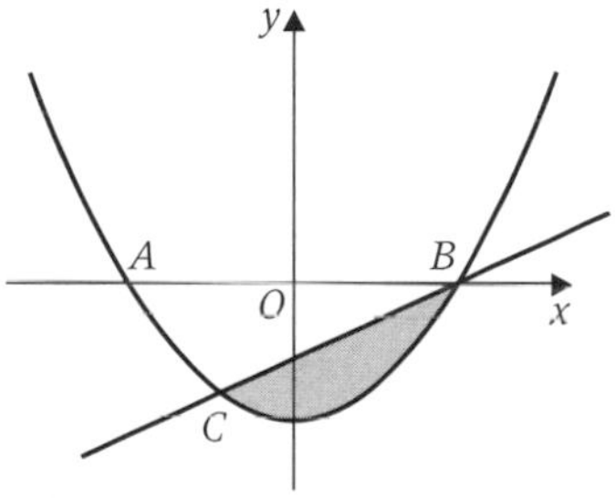

(a) Find the coordinates of the points A, B and C.

(b) Calculate the area of the shaded region bounded by the line and curve.

The following exercise contains some examination style questions involving differentiation and integration which should serve as a useful revision exercise.

CALCULUS REVISION EXERCISE

1 The diagram shows a part of the graph of $y = x - 2x^4$.

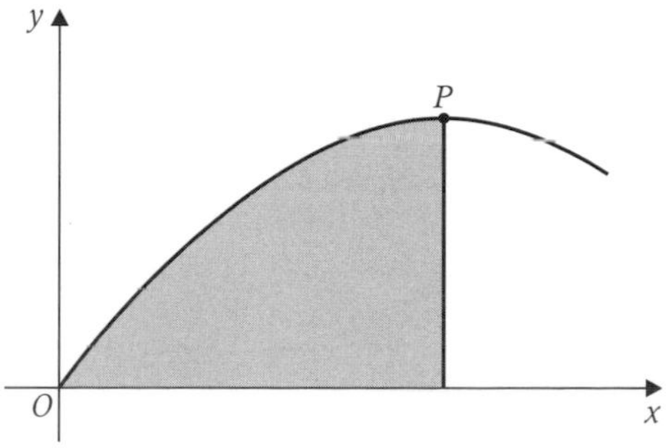

(a) **(i)** Find $\frac{dy}{dx}$.

(ii) Show that the x-coordinate of the stationary point P is $\frac{1}{2}$.

(iii) Find the y-coordinate of P.

(b) **(i)** Find $\int (x - 2x^4)\,dx$.

(ii) Hence find the area of the shaded region. [A]

2 The curve with equation $y = x^3 - 6x^2 + 9x + 16$ is sketched below.
The curve crosses the x-axis at the point $A(-1, 0)$.

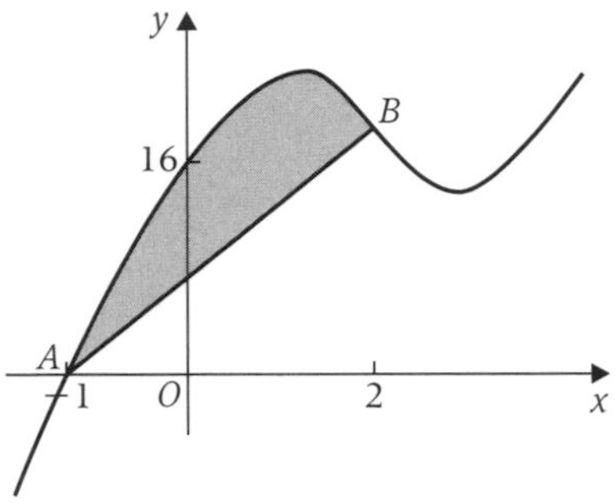

(a) **(i)** Find $\frac{dy}{dx}$.

(ii) Hence find the x-coordinates of the stationary points of the curve.

(b) **(i)** Find $\int_{-1}^{2} (x^3 - 6x^2 + 9x + 16)\,dx$.

(ii) The point $B(2, 18)$ lies on the curve. Find the area of the shaded region bounded by the curve and the line AB. [A]

3 The points $P(1, 10)$ and $Q(4, 4)$ lie on the curve $y = x^3 - 6x^2 + 7x + 8$ as shown in the diagram.

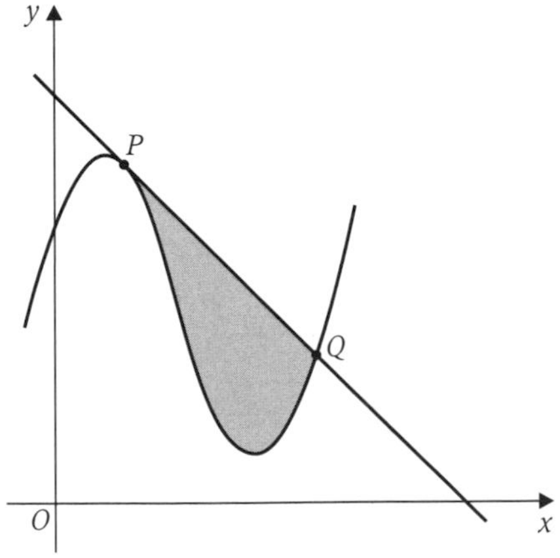

Find the area of the shaded region between the curve and the line PQ. [A]

4 The diagram shows the graphs of $y = 2x$ and $y = -2x^2 + x + 6$ intersecting at two points P and Q.

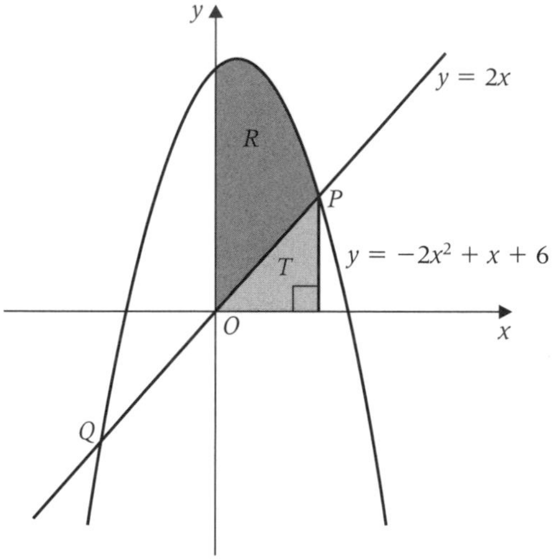

(a) Show that P has x-coordinate $\frac{3}{2}$ and find the x-coordinate of Q.

(b) Calculate the area of the shaded triangle T.

(c) (i) Find $\int (-2x^2 + x + 6)\,dx$.

(ii) Hence find the area of the shaded region R. [A]

5 The function f is defined for all values of x by $f(x) = x^3 - 7x^2 + 14x - 8$.
It is given that $f(1) = 0$ and $f(2) = 0$.

(a) Find the values of $f(3)$ and $f(4)$.

(b) Write $f(x)$ as a product of three **linear** factors.

(c) The diagram shows the graph of $y = x^3 - 7x^2 + 14x - 8$.

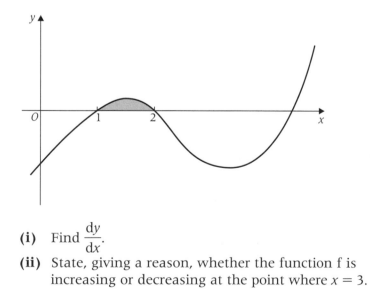

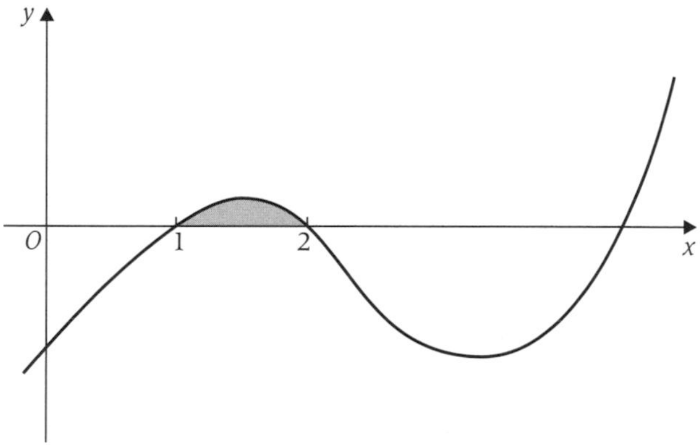

(i) Find $\frac{dy}{dx}$.

(ii) State, giving a reason, whether the function f is increasing or decreasing at the point where $x = 3$.

(iii) Find $\int (x^3 - 7x^2 + 14x - 8)\,dx$.

(iv) Hence find the area of the shaded region enclosed by the graph of $y = f(x)$, for $1 \leqslant x \leqslant 2$, and the x-axis. [A]

6 The diagram shows the curve $y = x^2 - 4x + 6$, the points $P(-1, 11)$ and $Q(4, 6)$ and the line PQ.

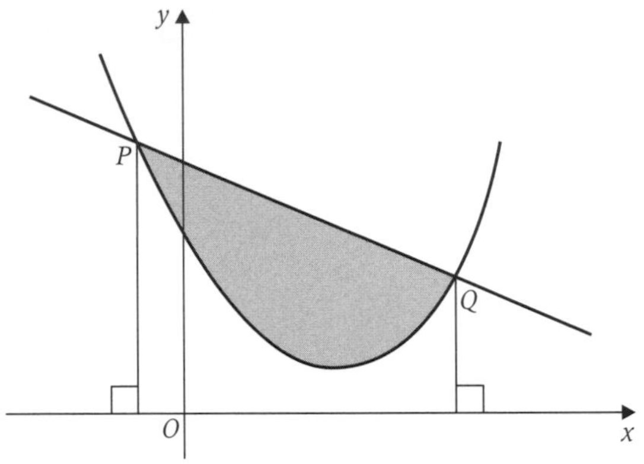

(a) Show that the length of PQ is $5\sqrt{2}$.

(b) Find the equation of the tangent to the curve at Q in the form $y = mx + c$.

(c) Find the area of the shaded region in the diagram. [A]

7 The line $y = 2x + 5$ intersects the curve $y = x^2 + 2x + 2$ at the points P and Q, as shown in the diagram.

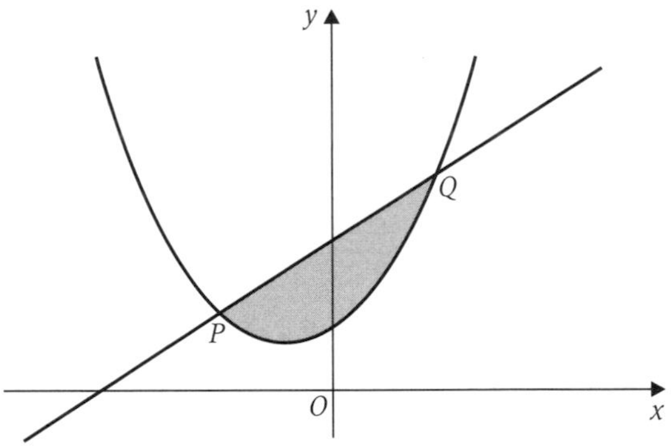

(a) Find the coordinates of P and Q, giving your answers in surd form.

(b) Find the area of the shaded region, giving your answer in surd form. [A]

8 The diagram shows the graph of $y = 12 - 3x^2$ and the tangent to the curve at the point $P(2, 0)$. The region enclosed by the tangent, the curve and the y-axis is shaded.

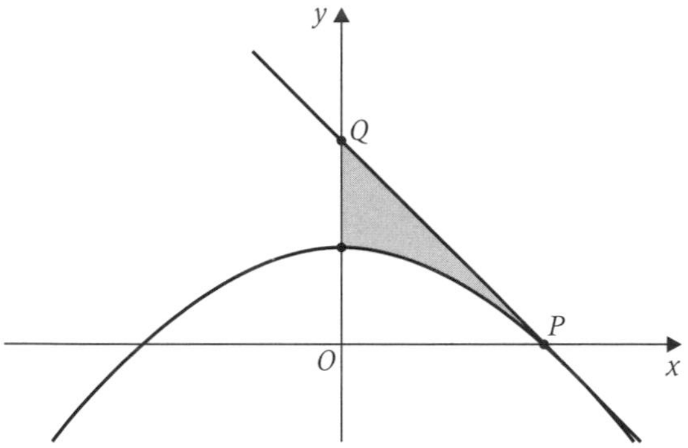

(a) Find $\int_0^2 (12 - 3x^2)\, dx$.

(b) **(i)** Find the gradient of the curve $y = 12 - 3x^2$ at the point P.

(ii) Find the y-coordinate of the point Q where the tangent at P crosses the y-axis.

(c) Find the area of the shaded region. [A]

9 The diagram shows the graph of $y = x^3 - x, \quad x \geq 0$. The points on the graph for which $x = 1$ and $x = 2$ are labelled A and B, respectively.

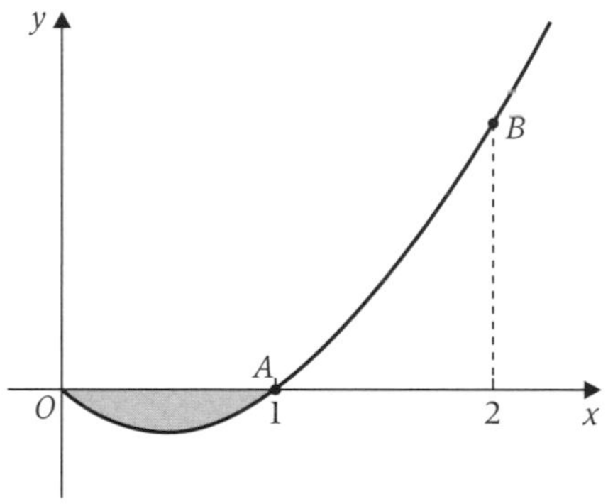

(a) Find the y-coordinate of B and hence find the equation of the straight line AB, giving your answer in the form $ax + by + c = 0$.

(b) Find, by integration, the area of the shaded region.

10

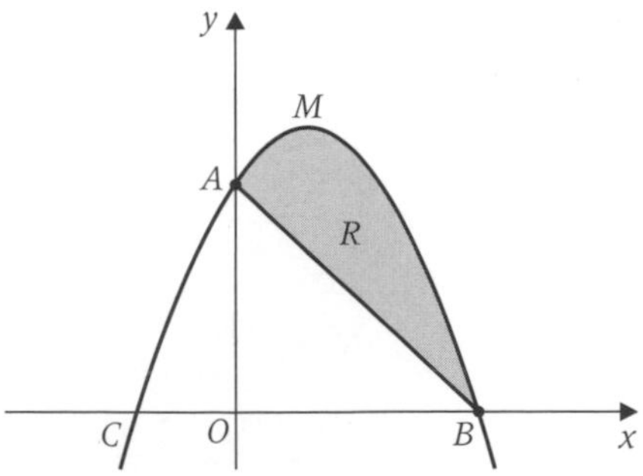

The curve with equation $y = 12 + 4x - x^2$ cuts the y-axis at A, the positive x-axis at B and the negative x-axis at C, as shown in the diagram. The point O is the origin and the maximum point of the curve is M. The shaded region R is bounded by the line AB and the curve.

The point B has coordinate $(6, 0)$.

(a) Show that $x = 2$ at the point M.

(b) Find the coordinates of C.

(c) Show that triangle OAB and the region R have equal areas.

[A]

Key point summary

1 Integration is the reverse process of differentiation. *p185*

2 $\dfrac{dy}{dx} = x^n \ (n \neq -1) \Rightarrow y = \dfrac{1}{n+1}x^{n+1} + c$, where c is an arbitrary constant. *p185*

3 An indefinite integral has no limits and is of the form $\int f(x)\,dx$. You must remember to include an arbitrary constant in your answer. *p189*

4 A definite integral is of the form $\int_a^b f(x)\,dx$, where a and b are the limits. *p192*

5 After integrating a definite integral, you obtain $[F(x)]_a^b$ and this is evaluated as $F(b) - F(a)$. *p192*

6 The area under a curve $y = f(x)$ from $x = a$ to $x = b$ is given by *p191*

$$\int_a^b f(x)\,dx$$

7 You can find the area of a region bounded by a curve and a line by finding the areas of separate regions and subtracting one from the other. *p194*

8 When a region lies entirely below the x-axis, the value of the definite integral is negative. If the integral has value $-A$, then the area of the region is A. *p195*

Test yourself	What to review
1 Find $f(x)$ when $f'(x) = 12x^2(x + 1)$.	*Section 12.2*
2 Obtain the equation of the curve passing through $(-1, 5)$ which has gradient $\frac{dy}{dx} = 6x^2 - 2x + 1$.	*Section 12.3*
3 Find $\int (x - 3)(3x + 5)\,dx$.	*Section 12.5*
4 Evaluate $\int_0^1 (5x^4 - 6x)\,dx$.	*Section 12.7*
5 Find the area bounded by the curve with equation $y = x^3 + 2$, the x-axis and the lines $x = 1$ and $x = 3$.	*Section 12.8*
6 Find the area of the finite region bounded by the x-axis and the curve with equation $y = 2(x + 1)(x - 2)$.	*Section 12.9*

Test yourself ANSWERS

1 $f(x) = 3x^4 + 4x^3 + c$.

2 $y = 2x^3 - x^2 + x + 9$.

3 $x^3 - 2x^2 - 15x + c$.

4 -2.

5 24.

6 9.

Exam style practice paper for C1

Time allowed 1 hour 30 minutes

Answer **all** questions

1 **(a)** Solve the inequality $6(y - 2) < 3 - 4(y - 5)$ (3 marks)

(b) Solve the inequality $(2x + 5)(x - 2) \geqslant 0$ (3 marks)

2 Express each of the following in the form $p + q\sqrt{3}$, where p and q are integers:

(a) $(3 - \sqrt{3})(5 + 3\sqrt{3})$, (3 marks)

(b) $\dfrac{44}{5 - \sqrt{3}}$. (3 marks)

3 A curve C has equation $y = 2x^3 - 9x^2 + 12x$.

(a) Find $\dfrac{dy}{dx}$ and $\dfrac{d^2y}{dx^2}$. (5 marks)

(b) The curve C has two stationary points. By finding the values of x for which $\dfrac{dy}{dx} = 0$, find the x-coordinates of these two stationary points. (3 marks)

(c) Find the value of $\dfrac{d^2y}{dx^2}$ at each stationary point and hence write down the coordinate of the minimum point of the curve C. (3 marks)

4 The points A and B have coordinates $(0, 2)$ and $(3, 8)$, respectively.

(a) **(i)** Find the gradient of AB. (1 mark)

(ii) Find an equation of the line AB. (1 mark)

(iii) Show that AB has length $p\sqrt{5}$, where p is an integer. (3 marks)

(b) The mid-point of AB is M. Find the coordinate of M. (2 marks)

(c) The two lines with equations $2x + 3y = 7$ and $y = \dfrac{1}{2}x + 7$ intersect at the point R. Show that RM is parallel to the x-axis. (4 marks)

5 The polynomial p(x) is given by $p(x) = x^3 + 4x^2 + x - 6$.

(a) Write down the value of the remainder when p(x) is divided by x. (1 mark)

(b) Use the factor theorem to show that $(x - 1)$ is a factor of p(x). (2 marks)

(c) Divide p(x) by $(x - 1)$ to find a quadratic factor of p(x). (2 marks)

(d) Write p(x) as a product of three linear factors. (2 marks)

(e) Sketch the curve $y = x^3 + 4x^2 + x - 6$, indicating the coordinates of the points where the curve intersects the coordinate axes. (3 marks)

6 A circle C has equation $x^2 + 2x + y^2 - 6y = 15$.

(a) By completing the square, express this equation in the form $(x - a)^2 + (y - b)^2 = k$, where a, b and k are integers. (3 marks)

(b) Show that the radius of the circle is 5 and write down the coordinates of its centre. (2 marks)

(c) Describe a geometrical transformation by which C can be obtained from the circle with equation $x^2 + (y - 1)^2 = 25$. (3 marks)

(d) The point $P(2, -1)$ lies on the circle C.

(i) Show that the equation of the line from the centre of C to the point P has equation $3y + 4x = 5$. (3 marks)

(ii) Find the equation of the tangent to the circle C at the point P. Give your answer in the form $ax = by + c$, where a, b and c are positive integers. (3 marks)

7

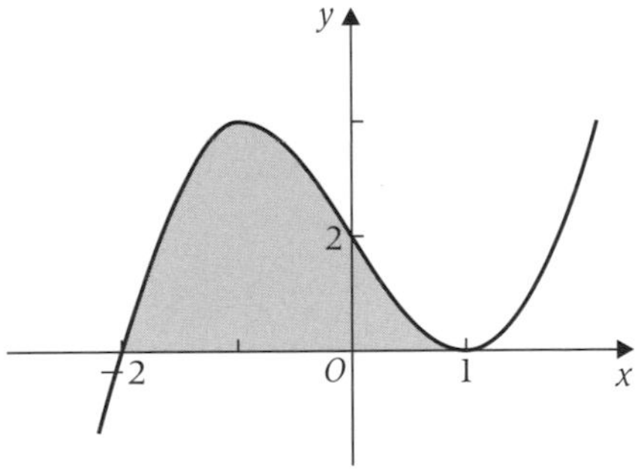

The diagram shows a sketch of a curve whose equation is $y = x^3 - 3x + 2$.
The region bounded by the curve and the x-axis is shaded.
The points $(-2, 0)$, $(1, 0)$ and $(0, 2)$ lie on the curve.

(a) Find $\frac{dy}{dx}$. (2 marks)

(b) Find the equation of the normal to the curve C at the point $(0, 2)$. Give your answer in the form $y = mx + c$. (3 marks)

(c) **(i)** Find $\int (x^3 - 3x + 2)\,dx$. (3 marks)

(ii) Hence show that the area of the shaded region is $6\frac{3}{4}$. (3 marks)

8 A curve C has equation $y = 4x^2 + 1$.

(a) Sketch the curve C. Indicate the coordinates of any point where the curve intersects the axes. (2 marks)

(b) A line L with equation $y = mx - 15$, where m is a constant, is drawn through the point $(0, -15)$. Show that the x-coordinates of any points of intersection of L and C must satisfy the equation $4x^2 - mx + 16 = 0$. (2 marks)

(c) Hence find the values of m for which L is a tangent to the curve C. (2 marks)

CHAPTER 1

C2: Indices

Learning objectives

After studying this chapter, you should be able to:

- use the language of indices
- simplify expressions involving indices
- know the laws of indices
- interpret negative, zero and fractional indices.

1.1 Introduction

At the end of the twentieth century, the English mathematician Andrew Wiles solved a problem involving powers of whole numbers that had perplexed people for hundreds of years. From his childhood he had learned that certain whole numbers could satisfy Pythagoras' relationship $a^2 + b^2 = c^2$, such as $a = 3$, $b = 4$ and $c = 5$ (or $a = 5$, $b = 12$ and $c = 13$, etc.). In 1637, the French mathematician Pierre de Fermat stated that, although this was true for squares, he had found a proof that this could never be done for cubes or any higher powers. Fermat wrote in the margin of his book that it was too small to contain the proof. This is not surprising because it took Andrew Wiles seven years research and 200 pages to present a proof of Fermat's Last Theorem, namely that there are no whole numbers a, b, and c that satisfy $a^n + b^n = c^n$ when the power n is greater than two.

During your mathematics course, you will combine some ideas discovered centuries ago with those that will inspire great minds of the twenty-first century. Some of the concepts in this chapter you may have met in your GCSE course but others will be new.

1.2 Index notation

Index is simply another word for **power**.
The plural of index is **indices**.

You should recall that $a \times a \times a$ can be written as a^3 in a shorthand way.

In this expression, a is called the **base** and 3 is called the power or **index**.

Read as a cubed or a to the power 3.

Similarly,

$a \times a \times a \times a$ can be written as a^4 (a to the power 4)

$\underbrace{a \times a \times \ldots a \times a}_{n \text{ of these}}$ can be written as a^n (a to the power n)

The name for the fourth power is biquadrate, but, whereas we regularly use the terms square and cube, we never really use this obscure word.

1.3 The laws of indices

There are three rules which will help us to simplify complicated expressions involving indices.

1 Consider the expression $a^4 \times a^5$

$$\begin{aligned} a^4 \times a^5 &= (a \times a \times a \times a) \times (a \times a \times a \times a \times a) \\ &= a \times a \times a \times a \times a \times a \times a \times a \times a \\ &= a^9 \end{aligned}$$

$a^4 = a \times a \times a \times a$

and

$a^5 = a \times a \times a \times a \times a$

So we have

$$a^4 \times a^5 = a^9$$

By now (probably long ago!) you will have spotted that we can simply add the powers in the expression to get the new power.

In general we have,

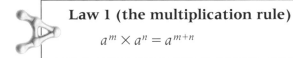

Law 1 (the multiplication rule)

$$a^m \times a^n = a^{m+n}$$

2 What about division?

Consider $a^5 \div a^2$

Now

$$a^5 \div a^2 = \frac{a \times a \times a \times a \times a}{a \times a} = a \times a \times a = a^3$$

$a \times a$ cancels in both the top and in the bottom.

So $a^5 \div a^2 = a^3$

Again there is a simple rule – when dividing expressions with indices, you can subtract the indices to find the new power.

In general we have,

Law 2 (the division rule)

$$a^m \div a^n = a^{m-n}$$

Note. The expressions $a^m + a^n$ and $a^m - a^n$ **cannot** be simplified to a single power of a.

3 Consider the expression $(a^3)^4$.

Now

$$\begin{aligned}(a^3)^4 &= a^3 \times a^3 \times a^3 \times a^3 \\ &= (a \times a \times a) \times (a \times a \times a) \times (a \times a \times a) \times (a \times a \times a) \\ &= a^{12}\end{aligned}$$

Again, a simple rule is that you can multiply the two indices to find the new index, i.e. $3 \times 4 = 12$.

In general we have,

Law 3 (the power rule)

$$(a^m)^n = a^{m \times n}$$

Worked example 1.1

Simplify: **1** $3x^4 \times 7x^6$ **2** $3x^5 \times 4y^3 \times 2y^2 \times 5x^3$

3 $(3p^4q)^3$ **4** $2x^4y^3 \times (x^2y^3)^4$

Solution

1 $3x^4 \times 7x^6 = 3 \times x^4 \times 7 \times x^6 = 3 \times 7 \times x^4 \times x^6 = 21x^{10}$

Add the indices $4 + 6 = 10$.

The order in which we multiply is not important.

2 $3x^5 \times 4y^3 \times 2y^2 \times 5x^3 = 3 \times 4 \times 2 \times 5 \times x^5 \times x^3 \times y^3 \times y^2$
$= 120x^8y^5$

Notice that there are two different bases and they must be handled separately.

3 $(3p^4q)^3 = 3^3 \times (p^4)^3 \times q^3$
$= 27 \times p^{12} \times q^3 = 27p^{12}q^3$

Each term is raised to the power 3.

4 $2x^4y^3 \times (x^2y^3)^4$
$= 2x^4y^3 \times x^8y^{12}$
$= 2x^{8+4}y^{3+12}$
$= 2x^{12}y^{15}$

Both x^2 and y^3 are raised to the power 4.

EXERCISE 1A

Simplify the following as far as possible.

Note that in general, expressions involving indices can only be simplified when the numbers involved have the **same base**. The expression $a^2 \times b^2$, for example, cannot be simplified since one base is a while the other is b.

1 $3 \times m \times m \times m \times m \times 2 \times m$

2 $2 \times b \times b \times b$

3 $6 \times 4 \times p \times p \times p \times q \times q$

4 $r^5 \times r^4$

5 $7t^2 \times 8t^6$

6 $4a^3 \times 9b^6$

7 $9a^4 \times 2a \times 3a^7$

8 $4a^4 \times 2b^3 \times a^8 \times 3b^2$

9 $10p^3 \times 3p^2 \times 2q^3 \times q$

10 $2pq^2 \times 5p^3q^4 \times p^6 \times q^2$

11 $6a^3p \times 2aq^3 \times 3p^2q$

12 $(m^3)^5$

13 $(m^5)^3$

14 $(a^2b^2)^5$

15 $(2x^2)^3$

16 $(a^2b^4)^3$

17 $(p^2q^2)^4 \times 3p$

18 $2m^2 \times 5n^3 \times (mn)^4$

19 $4r^2t \times (r^3t^2)^3$

20 $10p^2q^5 \times (q^3)^2 \times 2(p^4)^3$

Worked example 1.2

Simplify: **1** $8p^2 \div 4p^7$

2 $21x^3y^4 \div 3x^2y$

3 $\dfrac{(3s^3t^4)^2 \times s^2t^5}{(s^2t^3)^4}$

Solution

1 $8p^2 \div 4p^7 = \dfrac{8}{4}p^{2-7}$
$= 2p^{-5}$

First, we deal with the numbers 8 divided by 4, then we subtract the indices $2 - 7 = -5$.

We will explain what is meant by a negative index in the next section and show an alternative form in which the answer may be written.

2 $21x^3y^4 \div 3x^2y$
$= \dfrac{21}{3}x^{3-2}y^{4-1}$
$= 7xy^3$

The numbers give 21 divided by 3. Now we subtract the indices for each of the bases x and y separately.

Note that $3 - 2 = 1$ so we have x^1 but this is normally written simply as x.

3 $\dfrac{(3s^3t^4)^2 \times s^2t^5}{(s^2t^3)^4} = \dfrac{9s^6t^8 \times s^2t^5}{s^8t^{12}} = \dfrac{9s^8t^{13}}{s^8t^{12}} = 9t$

Note. It may be that you obtained an answer $9s^0t$. This is still correct as you will see in section 1.5.

EXERCISE 1B

Simplify the following as far as possible.

1 $a^5 \div a^2$ **2** $8p^3 \div 2p$ **3** $20t^6 \div 4t^5$

4 $60h^7 \div 6h^7$ **5** $(x^2)^4 \div x^5$ **6** $12(a^2)^7 \div 4a^{11}$

7 $5p^2q^3 \div 15pq^2$ **8** $10a^{15} \div 20a^7$ **9** $60p^2q^5 \div 30pq^2$

10 $25a^2b^9 \div 30a^2b^8$ **11** $\dfrac{5t^5 \times 2t^6}{t^7}$ **12** $\dfrac{4a^4 \times 5a^7}{2a^3}$

13 $\dfrac{8p^5q \times 3p^2q^3}{12p^3q}$ **14** $\dfrac{(x^2)^5 \times x^3y^4}{x^2y^5}$

15 $\dfrac{(m^3)^2 \times m^3n}{m^8n^3}$ **16** $\dfrac{5a^2c \times 2ac^3}{30a^3c^9}$

17 $\dfrac{9(mn)^3 \times 2(m^2n)^2}{24m^8n^2}$ **18** $\dfrac{(x^2y^3)^3 \times (xy^3)^2}{x^8y^{15}}$

19 $\dfrac{6x^3y^2z \times 5x^2yz^4}{(x^2yz^3)^2}$ **20** $\dfrac{4(ab)^4 \times (2ab^2)^3}{16(abc)^5}$

1.4 Negative indices

Consider the expression $a^4 \div a^6$

Now

$$a^4 \div a^6 = \frac{a \times a \times a \times a}{a \times a \times a \times a \times a \times a} = \frac{1}{a \times a} = \frac{1}{a^2},$$

We have simplified the expression using two different methods but the answers must be the same.

but if we use **the division rule** and subtract the powers we get

$$a^4 \div a^6 = a^{4-6} = a^{-2}.$$

Therefore $a^{-2} = \dfrac{1}{a^2}.$

Similarly, we can show that

$$a^{-1} = \frac{1}{a}, \quad a^{-3} = \frac{1}{a^3}, \quad a^{-4} = \frac{1}{a^4}, \quad \text{etc.}$$

In general, we have:

> The negative index rule
>
> $$a^{-n} = \frac{1}{a^n}$$

You often make use of the fact that when there is a negative power you take the reciprocal. For example, because $2^3 = 8$, we can write $2^{-3} = \dfrac{1}{8}$.

1.5 The zero index

Consider the expression $a^5 \div a^5$.

Now

$$a^5 \div a^5 = \frac{a \times a \times a \times a \times a}{a \times a \times a \times a \times a},$$

Everything cancels from top and bottom to give 1.

but using **the division law** we can simply subtract the indices to find the new index:

$$a^5 \div a^5 = a^{5-5} = a^0.$$

But $a^5 \div a^5 = 1$, hence:

$$a^0 = 1$$

In other words, any non-zero number raised to the power of 0 is equal to 1.

We can apply these rules to expressions involving numbers rather than letters.

Worked example 1.3

Without using a calculator, evaluate:

1 73^0 **2** $\left(\frac{3}{4}\right)^{-2}$ **3** $(-4)^{-3}$

Solution

1 $73^0 = 1$

Any non-zero number raised to the power of 0 is equal to 1.

2 Method 1 $\left(\frac{3}{4}\right)^{-2} = \frac{3^{-2}}{4^{-2}} = \left(\frac{1}{9}\right) \div \left(\frac{1}{16}\right) = \frac{16}{9}$

Method 2 Since $\left(\frac{3}{4}\right)^2 = \frac{9}{16}$ we can use the reciprocal idea

and so $\left(\frac{3}{4}\right)^{-2} = \frac{16}{9}$.

$a^{-2} = \frac{1}{a^2}$

3 $(-4)^{-3} = (-1)^{-3}(4)^{-3} = -1 \times \frac{1}{4^3} = -\frac{1}{64}$

Make sure you deal with the negative sign.

Worked example 1.4

Without using a calculator, simplify:

1 $\frac{2^7 \times 2^3}{4^3}$

2 $\frac{15^3}{5^2 \times 3^5}$

Solution

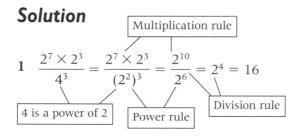

1 $\dfrac{2^7 \times 2^3}{4^3} = \dfrac{2^7 \times 2^3}{(2^2)^3} = \dfrac{2^{10}}{2^6} = 2^4 = 16$

Multiplication rule

4 is a power of 2

Power rule

Division rule

Notice that we have not had to use a calculator!

2 $\dfrac{15^3}{5^2 \times 3^5} = \dfrac{(5 \times 3)^3}{5^2 \times 3^5} = \dfrac{5^3 \times 3^3}{5^2 \times 3^5} = \dfrac{5^1}{3^2} = \dfrac{5}{9}$

You can check your answer using a calculator but be careful to use brackets for the denominator.

We could have worked with negative indices here and obtained $5^1 \times 3^{-2} = \dfrac{5}{9}$.

Back to algebra again.

Worked example 1.5

Simplify the following expressions as much as possible:

1 $\dfrac{(x^2)^{-2}}{x^{-7}}$ **2** $\dfrac{(p^{-2}q^3)^{\ 3}}{p^4q^{-7}}$

Solution

Power rule

Division rule

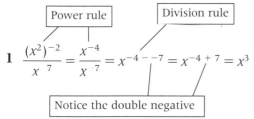

1 $\dfrac{(x^2)^{-2}}{x^{\ 7}} = \dfrac{x^{-4}}{x^{\ 7}} = x^{-4\,-\,-7} = x^{-4\,+\,7} = x^3$

Notice the double negative

This cannot be simplified any further.

2 $\dfrac{(p^{-2}q^3)^{-3}}{p^4q^{-7}} = \dfrac{p^6q^{-9}}{p^4q^{-7}} = p^{6\,-\,4}q^{-9\,-\,-7} = p^2q^{-9\,+\,7}$

$= p^2q^{-2}$

You may not like leaving your answer with negative indices and might prefer $\dfrac{p^2}{q^2}$.
Either form is acceptable for your final answer.

EXERCISE 1C

1 Write each of the following as a prime number raised to a particular power:

(a) 16, **(b)** 27, **(c)** 125, **(d)** 1,

(e) 49, **(f)** $\dfrac{1}{8}$, **(g)** $\dfrac{1}{81}$, **(h)** $\dfrac{1}{3}$,

(i) $\dfrac{1}{121}$, **(j)** $\dfrac{1}{289}$, **(k)** $\dfrac{1}{128}$.

2 Express each of the following in a form without indices, without using your calculator:

(a) 2^{-1}, **(b)** 3^0, **(c)** 5^4, **(d)** 7^{-2},
(e) $(-13)^0$, **(f)** $(-5)^{-3}$, **(g)** 12^{-1}, **(h)** $(-3)^{-2}$,
(i) $(5^2)^{-1}$, **(j)** $(3^{-2})^{-1}$.

3 Simplify each of the following without using a calculator:

(a) $\frac{2^3}{4^2}$, **(b)** $\frac{(3^2)^3}{9^5}$, **(c)** $\frac{2^3 \times 3^4}{6^3}$,
(d) $\frac{5^7}{125^3}$, **(e)** $\frac{12^3}{4^2 \times 3^5}$, **(f)** $\frac{20^7 \times 5^5}{4^{10} \times 25^6}$.

4 Simplify each of the following:

(a) $(x^5)^{-2}$, **(b)** $(a^{-3})^{-7}$,
(c) $(a^3b^2)^0$, **(d)** $(c^{-3}d^4)^5$,
(e) $x^{-9} \div x^3$, **(f)** $a^{-9} \div a^{-7}$,
(g) $p^{-8} \div p^3$, **(h)** $25p^5 \div (5p^{-7})$,
(i) $28a^{-3} \div 7a^{-5}$, **(j)** $13a^{14} \div (a^6)^3$,
(k) $14(a^{-3})^5 \div 2a^{-9}$, **(l)** $16(p^{-2})^4 \div 8p^{-8}$,
(m) $8(y^2)^3 \div 2y^{-6}$, **(n)** $28(x^2)^{-9} \div 4(x^{-2})^{10}$,
(o) $15p^{-2}q^{-7} \div (5pq^{-3})$, **(p)** $21a^3b^{-5}c \div (7a^{-4}bc^{-3})$,
(q) $15(a^3b)^{-3} \div (5ab^{-5})$, **(r)** $16a^{-2}b^2c^{-3} \times 2(abc)^{-2}$,
(s) $3a^{-2}bc^{-3} \times a^{-2}bc \div (2a^{-2}bc^{-5})$,
(t) $9xy^{-4} \times 4x^{-5}yz^{-4} \div [6(x^{-2}yz)^{-2}]$.

5 Simplify:

(a) $\frac{x^{-5} \times x^7}{x^{-3}}$, **(b)** $\frac{y^{-3} \times y \times y^2}{y^3}$, **(c)** $\frac{(p^{-2})^3}{p^{-5}}$,
(d) $\frac{r^5 \div r^8}{r^3}$, **(e)** $(t^5u^6)^{-3} \times (t^7u^{20})$, **(f)** $\frac{(pq)^7}{(pq^2)^4}$,
(g) $\frac{(a^2b^3)^{-2}}{(a^{-3} \div b^5)}$.

1.6 Fractional indices

It is very useful to be able to write $\sqrt{\ }$ and $\sqrt[3]{\ }$, and so on, as powers so that you can use the rules that we have already obtained.

Assume that $\sqrt{a}$ can be written using a power as a^x.
We need to find the value of x.

We know that

$$\sqrt{a} \times \sqrt{a} = a,$$

so

$$a^x \times a^x = a^1$$

Using the multiplication law for indices

$$a^{2x} = a^1.$$

This means that

$$2x = 1$$

and so

$$x = \frac{1}{2}.$$

Therefore, $\sqrt{a}$ can be written as $a^{\frac{1}{2}}$.

So, we know that $\quad a^{\frac{1}{2}} = \sqrt{a} \quad$ (square root).

Similarly $\quad a^{\frac{1}{3}} = \sqrt[3]{a} \quad$ (cube root).

In general

$$a^{\frac{1}{n}} = \sqrt[n]{a} \quad (n\text{th root})$$

Worked example 1.6

1 Write down the exact value of $27^{\frac{1}{3}}$ and $16^{\frac{1}{4}}$.

2 Find the value of the fifth root of 1200 giving your answer to three significant figures.

Solution

1 $27^{\frac{1}{3}} = \sqrt[3]{27} = 3 \qquad (\text{since } 3 \times 3 \times 3 = 27)$

$16^{\frac{1}{4}} = \sqrt[4]{16} = 2 \qquad (\text{since } 2 \times 2 \times 2 \times 2 = 16)$

Notice that we did not need a calculator to evaluate these expressions.

2 We need to express the fifth root as an index before we can work out its value on a calculator.

$$\sqrt[5]{1200} = 1200^{\frac{1}{5}} = 4.1289179\ldots$$

$$\approx 4.13 \text{ (to three significant figures)}$$

If you did not obtain this answer then you almost certainly forgot to put the power $\frac{1}{5}$ in brackets. Of course you could always use the index 0.2.

You can extend this idea to consider more complicated fractional indices such as $a^{\frac{2}{3}}$.

$$a^{\frac{2}{3}} = a^{2 \times \frac{1}{3}} = (a^2)^{\frac{1}{3}} \qquad \text{(using the fact that } a^{m \times n} = (a^m)^n\text{)}$$

$$\Rightarrow \qquad a^{\frac{2}{3}} = \sqrt[3]{a^2}$$

or

$$a^{\frac{2}{3}} = a^{\frac{1}{3} \times 2} = (a^{\frac{1}{3}})^2$$

$$\Rightarrow \qquad a^{\frac{2}{3}} = (\sqrt[3]{a})^2$$

In general

$$a^{\frac{m}{n}} = \sqrt[n]{a^m} \text{ or } (\sqrt[n]{a})^m$$

Worked example 1.7

Evaluate the following without using a calculator.

1 $64^{-\frac{1}{3}}$ **2** $\left(\frac{25}{9}\right)^{-\frac{3}{2}}$ **3** $\left(\frac{16}{81}\right)^{\frac{1}{4}}$

Solution

1 $64^{-\frac{1}{3}} = \frac{1}{64^{\frac{1}{3}}} = \frac{1}{\sqrt[3]{64}} = \frac{1}{4}$

2 $\left(\frac{25}{9}\right)^{-\frac{3}{2}} = \left(\frac{9}{25}\right)^{\frac{3}{2}} = \left(\sqrt{\frac{9}{25}}\right)^3 = \left(\frac{3}{5}\right)^3 = \frac{27}{125}$

3 $\left(\frac{16}{81}\right)^{\frac{1}{4}} = \frac{16^{\frac{1}{4}}}{81^{\frac{1}{4}}} = \frac{\sqrt[4]{16}}{\sqrt[4]{81}} = \frac{2}{3}$

Worked example 1.8

Simplify each of the following expressions as far as possible.

1 $\sqrt[3]{a^4b^5} \times \frac{b^{\frac{1}{3}}}{a}$ **2** $\frac{\sqrt[4]{xy^2} \times \sqrt[3]{xy}}{x^2y}$

Solution

1 $\sqrt[3]{a^4b^5} \times \frac{b^{\frac{1}{3}}}{a} = \sqrt[3]{a^4} \times \sqrt[3]{b^5} \times \frac{b^{\frac{1}{3}}}{a}$

$$= a^{\frac{4}{3}} \times b^{\frac{5}{3}} \times b^{\frac{1}{3}} \times a^{-1} = a^{\frac{1}{3}}b^2$$

2 $\dfrac{\sqrt[4]{xy^2} \times \sqrt[3]{xy}}{x^2y} = \dfrac{x^{\frac{1}{4}} \times y^{\frac{2}{4}} \times x^{\frac{1}{3}} \times y^{\frac{1}{3}}}{x^2y}$

$= \dfrac{x^{\frac{7}{12}} \times y^{\frac{5}{6}}}{x^2y} = x^{\frac{7}{12}-2} \times y^{\frac{5}{6}-1} = x^{-\frac{17}{12}}y^{-\frac{1}{6}}$

1.7 Solving equations with indices

We are now in a position to solve certain equations where the unknown quantity is either a power or forms part of the power.

Worked example 1.9

Solve the following equations.

1 $9^{x+1} = 3$ **2** $2^{3x} \times 4^{(x+1)} = 64$

Solution

1 $9^{x+1} = 3$

> Considering the RHS of the equation, 3 can be expressed as a power of 9, since $9^{\frac{1}{2}} = 3$, (or $\sqrt{9} = 3$).

So, $9^{x+1} = 9^{\frac{1}{2}}$

> Both sides of the equation now have the same **base**.

$\Rightarrow x + 1 = \dfrac{1}{2}$.

> Equating the powers.

Giving $x = -\dfrac{1}{2}$.

Alternative solution:

We could also express 9 as a power of 3 as follows:

$9^{x+1} = 3 \quad \Rightarrow \quad (3^2)^{(x+1)} = 3$

> Don't forget that 3 can be written as 3^1.

so that $2(x + 1) = 1$

and $x = -\dfrac{1}{2}$, as before.

2 $2^{3x} \times 4^{(x+1)} = 64$

> If you expressed every term as a power of $4(4^{\frac{1}{2}} = 2)$ then you would still get the same answer. Try it and see.

$2^{3x} \times 2^{2(x+1)} = 64$ ($4 = 2^2$)

and hence $2^{5x+2} = 64$.

But $64 = 2^6$

and so we can equate the indices

$5x + 2 = 6 \Rightarrow x = \dfrac{4}{5}$.

EXERCISE 1D

1 Work these out without using a calculator:

(a) $4^{\frac{3}{2}}$, **(b)** $100^{\frac{5}{2}}$,

(c) $27^{\frac{2}{3}}$, **(d)** $1000^{1\frac{1}{3}}$,

(e) $16^{\frac{5}{4}}$, **(f)** $32^{0.4}$.

2 Write these as fractions:

(a) $16^{-\frac{1}{2}}$, **(b)** $4^{-1\frac{1}{2}}$, **(c)** $125^{-\frac{4}{3}}$,

(d) $\left(\dfrac{4}{49}\right)^{-\frac{1}{2}}$, **(e)** $\left(\dfrac{125}{8}\right)^{\frac{2}{3}}$, **(f)** $\left(1\frac{7}{9}\right)^{\frac{1}{2}}$,

(g) $\left(\dfrac{16}{25}\right)^{-\frac{1}{2}}$, **(h)** $\left(\dfrac{27}{64}\right)^{\frac{2}{3}}$, **(i)** $\left(2\frac{7}{81}\right)^{\frac{1}{2}}$.

3 Simplify the following as far as possible:

(a) $\dfrac{\sqrt[3]{x} \times \sqrt[3]{x^4}}{x^{-\frac{1}{3}}}$, **(b)** $(\sqrt{a})^3 \times (\sqrt{a^7})$,

(c) $\dfrac{x^2 \times \sqrt{x^5}}{x^{-\frac{1}{2}}}$, **(d)** $\dfrac{(\sqrt[3]{y})^2 \times (y^2)^{\frac{1}{3}}}{\sqrt[3]{y}}$,

(e) $\dfrac{\sqrt[6]{a} \times a^{\frac{2}{3}}}{\sqrt[4]{a}}$, **(f)** $\dfrac{\sqrt{xy} \times \sqrt[3]{x} \times 2\sqrt[4]{y}}{(x^{10}y^9)^{\frac{1}{12}}}$.

4 Solve the following equations:

(a) $3^{x-2} = \dfrac{1}{27}$, **(b)** $5^{3x} \times 25^{x-2} = 1$,

(c) $7^{3x+1} = 49$, **(d)** $2^{3x-1} \times 8^{x-1} = 128$.

5 Solve the equation

$$\frac{81^x}{27^{2x+1}} = \frac{9^{2x-5}}{729}.$$

6 **(a)** Write each of the following as a power of 3:

(i) $\dfrac{1}{27}$, **(ii)** 9^x.

(b) Hence solve the equation $9^x \times 3^{1-x} = \dfrac{1}{27}$. [A]

7 **(a)** Write $\sqrt{2}$ as a power of 2.

(b) Hence write $4\sqrt{2}$ as a power of 2.

(c) Hence solve the equation $2^{3x+4} = 4\sqrt{2}$. [A]

8 It is given that $p = 8^{\frac{1}{2}}$ and $q = 4^{\frac{3}{4}}$.

(a) Show that $p = 2^{\frac{3}{2}}$.

(b) Similarly express q as a power of 2.

(c) Hence express pq as a power of 2. [A]

9 **(a)** Express each of the following as a power of 3:

(i) $\sqrt{3}$, **(ii)** $\dfrac{3^x}{\sqrt{3}}$.

(b) Hence, or otherwise, solve the equation

$$\frac{3^x}{\sqrt{3}} = \frac{1}{3}.$$ [A]

10 **(a)** Express each of the following as a power of 5:

(i) $5\sqrt{5}$, **(ii)** $\dfrac{1}{25}$.

(b) Hence, or otherwise, solve the equation

$$5^{x-3} \times 5\sqrt{5} = \frac{1}{25}.$$

11 **(a)** Show that the substitution $y - 2^x$ transforms the equation $2^{2x} - 5 \times 2^x + 4 = 0$ into the quadratic equation $y^2 - 5y + 4 = 0$.

(b) Solve the equation $y^2 - 5y + 4 = 0$.

(c) Hence solve the equation $2^{2x} - 5 \times 2^x + 4 = 0$.

12 Solve each of the following equations:

(a) $3^{2x} - 4 \times 3^x + 3 = 0$;

(b) $4 \times 2^{2x} - 33 \times 2^x + 8 = 0$;

(c) $25^x + 2 \times 5^x - 3 = 0$.

13 Fermat stated that every whole number could be expressed as the sum of no more than four squares, so that, for example

$$15 = 3^2 + 2^2 + 1^2 + 1^2$$
$$139 = 9^2 + 7^2 + 3^2$$
$$327 = 13^2 + 10^2 + 7^2 + 3^2$$

Do you think he was correct? Can you find a general formula for those numbers that require the sum of at least four squares to make up the number itself?

Key point summary

1 $a^m \times a^n = a^{m+n}$ p208

2 $a^m \div a^n = a^{m-n}$ p209

3 $(a^m)^n = a^{m \times n}$ p209

4 $a^{-n} = \dfrac{1}{a^n}$ p211

5 $a^0 = 1$ p212

6 $a^{\frac{1}{n}} = \sqrt[n]{a}$ p215

7 $a^{\frac{m}{n}} = \sqrt[n]{a^m}$ or $(\sqrt[n]{a})^m$ p216

Test yourself

Test yourself	What to review
1 Simplify each of the following: **(a)** $7 \times s \times s \times s \times s \times t \times t \times t$, **(b)** $5y^4 \times 4y^6 \times 3y^2$, **(c)** $3a^3 \times 4b^3 \times b \times 6a^2$, **(d)** $5p^2q^7 \times q^6 \times 2q^2 \times 3p^4$.	*Section 1.3*
2 Simplify each of the following: **(a)** $x^{-8} \times x^3$, **(b)** $y^7 \div y^{-3}$, **(c)** $\dfrac{p^7q^{-2}}{p^{-4}q^{-3}}$.	*Section 1.4*
3 Evaluate each of the following without using a calculator: **(a)** $27^{\frac{1}{3}}$, **(b)** $\left(\dfrac{5}{8}\right)^{-1}$, **(c)** $8^{-\frac{2}{3}}$, **(d)** $(-23)^{\circ}$.	*Sections 1.5 and 1.6*
4 Solve the equation: $3^{x+1} \times 9^{1-x} = \dfrac{1}{27}$.	*Section 1.7*

Test yourself ANSWERS

1 (a) $7s^4t^3$; **(b)** $60y^{12}$; **(c)** $72a^5b^4$; **(d)** $30p^6q^{15}$.

2 (a) x^{-5}; **(b)** y^{10}; **(c)** $p^{11}q$.

3 (a) 3; **(b)** $\dfrac{8}{5}$; **(c)** $\dfrac{1}{4}$; **(d)** 1.

4 $x = 6$.

CHAPTER 2

C2: Further differentiation

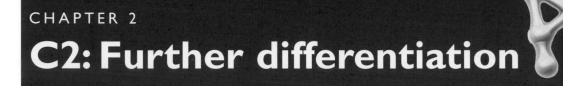

Learning objectives

After studying this chapter, you should be able to:

- differentiate x^n, where n is any rational number
- differentiate expressions which contain brackets and simple fractions
- extend the calculus techniques from the previous module to more complicated expressions than polynomials
- use the second derivative to establish the nature of stationary points for curves more advanced than polynomials
- solve more advanced optimisation problems using differentiation.

2.1 Review of differentiation techniques

You learned in C1 how to find the gradient of a curve using differentiation.

When differentiating, you used the following strategy:

1 If $y = x^n$, then $\dfrac{\mathrm{d}y}{\mathrm{d}x} = nx^{n-1}$.

2 If $y = cx^n$ (c is a constant), then $\dfrac{\mathrm{d}y}{\mathrm{d}x} = cnx^{n-1}$.

3 If $y = \mathrm{f}(x) \pm \mathrm{g}(x)$, then $\dfrac{\mathrm{d}y}{\mathrm{d}x} = \mathrm{f}'(x) \pm \mathrm{g}'(x)$.

2.2 Negative and fractional powers

The strategy for differentiation reviewed in the previous section still works if you have negative and/or fractional powers of x.

The use of the laws of indices from C2 chapter 1 will be needed for these questions.

Worked example 2.1

Differentiate the following functions with respect to x:

(a) $y = \dfrac{1}{x^3}$, **(b)** $y = \dfrac{7}{x^6}$, **(c)** $y = \dfrac{2}{3x^7}$,

(d) $y = \dfrac{5}{x^3} - \dfrac{4}{3x} + 8x^2$.

The expressions must be in the form $y = kx^n$ before you can differentiate.

Solution

(a) First you must rewrite y in an appropriate form:

$$y = \frac{1}{x^3} = x^{-3}$$

Using the negative power rule.

You can now use the quick method for differentiating:

$$\frac{dy}{dx} = -3x^{-4}$$

Bring the power to the front and reduce the power by 1.

This can be 'tidied-up' as

$$\frac{dy}{dx} = -\frac{3}{x^4}.$$

Again, using the negative power rule.

(b) First you need to rewrite it in an appropriate form:

$$\begin{aligned} y &= 7x^{-6} \\ \Rightarrow \frac{dy}{dx} &= 7(-6x^{-7}) \\ &= -42x^{-7} \\ &= -\frac{42}{x^7}. \end{aligned}$$

(c) Rewrite y using a power of x:

$$\begin{aligned} y &= \frac{2}{3}x^{-7} \\ \Rightarrow \frac{dy}{dx} &= \frac{2}{3}(-7x^{-8}) \\ &= -\frac{14}{3}x^{-8} \\ &= -\frac{14}{3x^8}. \end{aligned}$$

Notice that the 3 is left on the bottom since the power only applies to x.

(d) Rewrite y using powers of x:

$$\begin{aligned} y &= 5x^{-3} - \frac{4}{3}x^{-1} + 8x^2 \\ \Rightarrow \frac{dy}{dx} &= 5(-3x^{-4}) - \frac{4}{3}(-1x^{-2}) + 8(2x^1) \\ &= -15x^{-4} + \frac{4}{3}x^{-2} + 16x \\ &= -\frac{15}{x^4} + \frac{4}{3x^2} + 16x. \end{aligned}$$

Notice the double negative.

Worked example 2.2

Differentiate the following with respect to x:

(a) $y = 2x^{\frac{1}{3}}$, **(b)** $y = 7\sqrt{x}$, **(c)** $y = \frac{4}{\sqrt{x}}$.

Solution

$\frac{1}{3} - 1 = -\frac{2}{3}$

(a) $\frac{dy}{dx} = 2\left(\frac{1}{3}x^{-\frac{2}{3}}\right)$

$= \frac{2}{3}x^{-\frac{2}{3}}$

$= \frac{2}{3x^{\frac{2}{3}}}$

(b) Write y using a power of x:

Power of $\frac{1}{2}$ means square root.

$y = 7x^{\frac{1}{2}}$

$\Rightarrow \frac{dy}{dx} = 7\left(\frac{1}{2}x^{-\frac{1}{2}}\right)$

$\frac{1}{2} - 1 = -\frac{1}{2}$

$= \frac{7}{2}x^{-\frac{1}{2}}$

$= \frac{7}{2x^{\frac{1}{2}}}$

$= \frac{7}{2\sqrt{x}}$

(c) This can be written as

$y = 4x^{-\frac{1}{2}}$

$\Rightarrow \frac{dy}{dx} = 4\left(-\frac{1}{2}x^{-\frac{3}{2}}\right)$

$= -\frac{4}{2}x^{-\frac{3}{2}}$

$= -2x^{-\frac{3}{2}}$

$= -\frac{2}{\sqrt{x^3}}$

EXERCISE 2A

1 Find $\frac{dy}{dx}$ for each of the following:

(a) $y = 3x^3 - \frac{2}{x}$,

(b) $y = 5x - \frac{4}{3x^2} + 15$,

(c) $y = \frac{4}{7x} + \frac{3}{5x^2} - 8x^2$,

(d) $y = \frac{5}{x^7} + \frac{x^2}{5} - 27x$,

(e) $y = \frac{5}{x^5} + \frac{1}{x^6} - 7x$,

(f) $y = \frac{4}{5x^2} - \frac{2}{3x} + \frac{8x}{3}$,

(g) $y = \frac{3}{2x} + \frac{4}{3x^3} - 2x^3$,

(h) $y = \frac{10}{x^4} + \frac{x^7}{14} + 1$.

2 Differentiate the following functions with respect to x:

(a) $y = 5\sqrt{x} - 2$,

(b) $y = \sqrt[3]{x}$,

(c) $y = 3x^{\frac{1}{3}} + 2x^{\frac{1}{5}}$,

(d) $y = 7x^3 - \dfrac{1}{\sqrt{x}}$,

(e) $y = \dfrac{6}{5x^4} - \dfrac{x^3}{5}$,

(f) $y = 3x^{-\frac{1}{3}} - 10x^{-\frac{2}{5}}$,

(g) $y = 6x^2 - \sqrt{x^3} - \dfrac{2}{x^2}$,

(h) $y = \dfrac{4}{\sqrt[5]{x}} - 3\sqrt{x^3}$.

3 Find the derivative, $f'(x)$, for each of the following:

(a) $f(x) = \dfrac{6}{5x^3} + 4x^3$,

(b) $f(x) = 6\sqrt{x} - 3x + 12$,

(c) $f(x) = 23 - 4x - \dfrac{3}{x^2}$,

(d) $f(x) = \dfrac{5}{x^2} + \dfrac{3}{2x^5} - 3$,

(e) $f(x) = \sqrt[4]{x^3} + \sqrt[4]{x^5}$,

(f) $f(x) = 3x^4 - \dfrac{3}{\sqrt[5]{x}} + \dfrac{4}{5\sqrt{x^4}}$,

(g) $f(x) = 2 - 5\sqrt{x} - \dfrac{2}{x^2}$,

(h) $f(x) = \dfrac{6}{\sqrt[3]{x^2}} + 10\sqrt[5]{x} - 13$.

4 Differentiate each of the following with respect to the appropriate variable:

(a) $z = 3x^2 - \dfrac{2}{x^4}$,

(b) $v = 3t - \dfrac{7}{2t^2} + 3\sqrt{t}$,

(c) $p = \dfrac{4}{b^3} + \dfrac{3}{4b^{12}} - 8b^7$,

(d) $y = \dfrac{5}{z^2} + \dfrac{z^{\frac{3}{5}}}{5} - 27z$,

(e) $s = \dfrac{12}{t^6} + \dfrac{1}{t^2} - 7$,

(f) $h = \dfrac{4}{5c^2} - \dfrac{2}{3\sqrt{c}} + \dfrac{8\sqrt{c}}{3}$.

5 Find $f''(x)$ for each of the following:

(a) $f(x) = 8x - \dfrac{4}{x^2}$,

(b) $f(x) = 12\sqrt{x} - 4x\sqrt{x}$,

(c) $f(x) = \dfrac{2}{x} - 4\sqrt{x}$,

(d) $f(x) = \dfrac{6}{x} - \dfrac{1}{x^3}$.

6 Given that $y = 32x^2 + \dfrac{4}{x}$, find: **(a)** $\dfrac{dy}{dx}$ and **(b)** $\dfrac{d^2y}{dx^2}$.

7 Given that $f(x) = 3 - \dfrac{1}{x}$, show that f is an increasing function for $x > 0$.

Increasing and decreasing functions are introduced in C1 section 10.4.

8 Given that $y = 2 - \sqrt{x}$, $x \neq 0$, show that y is decreasing for $x > 0$.

2

2.3 Simplifying expressions before differentiating

Sometimes it is necessary to simplify expressions before you can differentiate them. You may have to expand brackets, simplify fractions, use the laws of indices, etc.

Your aim, regardless of how you have to simplify, is to obtain a line of expressions of the form kx^n.

Worked example 2.3

Find $\frac{dy}{dx}$ for each of the following:

(a) $y = \frac{x^6 - 8x^3 + 7}{x}$, **(b)** $y = \frac{4x^4 + 9x - 6}{2x^2}$.

Solution

(a) $y = \frac{x^6}{x} - \frac{8x^3}{x} + \frac{7}{x}$ ← With this type of expression, it is a simple matter to split the fraction into separate terms.

$y = x^5 - 8x^2 + 7x^{-1}$ ← Simplified using the laws of indices.

$\Rightarrow \frac{dy}{dx} = 5x^4 - 16x - 7x^{-2}$

$= 5x^4 - 16x - \frac{7}{x^2}$

(b) $y = \frac{4x^4}{2x^2} + \frac{9x}{2x^2} - \frac{6}{2x^2}$ — Again, split-up the fraction.

$= \frac{4}{2}x^2 + \frac{9}{2x} - \frac{6}{2x^2}$

$= 2x^2 + \frac{9}{2}x^{-1} - 3x^{-2}$ ← Rewrite terms in the form kx^n.

$\Rightarrow \frac{dy}{dx} = 4x - \frac{9}{2}x^{-2} + 6x^{-3}$ — Now differentiate.

$= 4x - \frac{9}{2x^2} + \frac{6}{x^3}$

The following worked example shows how the laws of indices can help you to differentiate some 'nasty-looking' expressions.

Worked example 2.4

Differentiate each of the following functions with respect to x:

(a) $y = \sqrt{x}(7\sqrt{x} - 4x^3)$, **(b)** $y = \frac{4x - 3\sqrt{x}}{x^3}$.

Solution

(a) $y = x^{\frac{1}{2}}(7x^{\frac{1}{2}} - 4x^3)$

Expand the brackets:

$$y = 7x^1 - 4x^{\frac{7}{2}}$$

$3 + \frac{1}{2} = \frac{7}{2}$

$$\Rightarrow \frac{dy}{dx} = 7 - 4\left(\frac{7}{2}x^{\frac{5}{2}}\right)$$

$$= 7 - 14x^{\frac{5}{2}}$$

$$= 7 - 14\sqrt{x^5}$$

(b) $$y = \frac{4x}{x^3} - \frac{3x^{\frac{1}{2}}}{x^3}$$

$$= 4x^{-2} - 3$$

$$\Rightarrow \frac{dy}{dx} = -8x^{-3} - 3\left(-\frac{5}{2}x^{-\frac{7}{2}}\right)$$

$$= -\frac{8}{x^3} + \frac{15}{2\sqrt{x^7}}$$

Before you attempt to differentiate, you must convert the square roots to powers and then simplify the expression.

Remember to add the powers of x when multiplying.

Either of these answers is acceptable.

You have to convert the root to a power.

Again, using the laws of indices.

EXERCISE 2B

1 Find $\frac{dy}{dx}$ for each of the following:

(a) $y = 4x(\sqrt{x} - 3)$, **(b)** $y = x^2(2\sqrt{x} - 3)$,
(c) $y = (3\sqrt{x} - 2)(4\sqrt{x} + 9)$, **(d)** $y = (\sqrt{x} - 3)^2$,
(e) $y = \sqrt{x}(x + 4)$, **(f)** $y = \sqrt{x}(\sqrt{x} - 2x)$.

2 Differentiate each of the following with respect to x:

(a) $y = x^3\left(3x + \frac{8}{x}\right)$, **(b)** $y = \frac{4x^3 - 3}{x^2}$,

(c) $y = \frac{3x^5 + 4x^3 + 3}{x^2}$, **(d)** $y = \frac{3(x - 4)}{2x^3}$,

(e) $y = \frac{(x + 4)(x - 3)}{x}$, **(f)** $y = \frac{4x^4 - 3x^3 + 1}{2x^4}$,

(g) $y = \frac{\sqrt{x} - 3x^8}{5x^3}$, **(h)** $y = \frac{(x - 2)(x + 3)}{x^3}$.

3 Find the gradient of the following curves at the given point:

(a) $y = 3\sqrt{x}(x - 2)^2$ at the point where $x = 3$,

(b) $y = \frac{x^2 + 6x^3 - 2}{x^4}$ at the point where $x = -1$,

(c) $y = \frac{\sqrt{x}(x^3 - \sqrt{x})}{x^2}$ at the point where $x = 1$.

4 Find $f'(x)$ for each of the following:

(a) $f(x) = \frac{(8 - x^4)}{x}$, **(b)** $f(x) = \left(\frac{4}{x}\right)\left(\frac{3}{\sqrt{x}} + \frac{\sqrt{x}}{4}\right)$,

(c) $f(x) = x(2\sqrt{x} + 9) - 7\sqrt[3]{x}$, **(d)** $f(x) = \frac{4x^2 - 7\sqrt{x}}{\sqrt[3]{x}} - \frac{3}{x}$.

2.4 Finding gradients of curves and equations of tangents and normals

In order to find the gradient at a given point (a, b) on a curve, you have already used the following procedure in the C1 module:

1 find $\frac{dy}{dx}$;

2 substitute the value $x = a$ into $\frac{dy}{dx}$.

When the gradient of the curve at the point (a, b) is m, then the equation of the tangent to the curve at this point is

$$y - b = m(x - a).$$

Worked example 2.5

A curve has equation $y = \frac{2}{x} - \sqrt{x}$ and P is the point on the curve where $x = 4$.

(a) Find the gradient of the curve at the point P.

(b) Hence find the equation of the tangent to the curve at the point P.

Solution

(a) You need to rewrite the expression so each term is of the form ax^n.

$$y = \frac{2}{x} - \sqrt{x} = 2x^{-1} - x^{\frac{1}{2}}$$

Hence, $\frac{dy}{dx} = (-2)x^{-2} - \frac{1}{2}x^{-\frac{1}{2}}$.

Using the properties of indices:

$4^{-2} = \frac{1}{4^2} = \frac{1}{16}$; $4^{-\frac{1}{2}} = \frac{1}{\sqrt{4}} = \frac{1}{2}$

Substituting $x = 4$

$$\Rightarrow \quad \text{gradient at } P = -2 \times \frac{1}{16} - \frac{1}{2} \times \frac{1}{2} = -\frac{3}{8}$$

(b) When $x = 4$, $y = \frac{2}{4} - \sqrt{4} = -1\frac{1}{2}$

Therefore the coordinates of P are $\left(4, -\frac{3}{2}\right)$.

Hence, the equation of the tangent is

$$y + \frac{3}{2} = -\frac{3}{8}(x - 4)$$

or $3x + 8y = 0$.

Unless the question asks for the answer in a particular form, ANY equivalent equation for the tangent is acceptable, e.g. $y = -\frac{3}{8}x$.

Worked example 2.6

(a) Find the points on the curve $y = \frac{9}{x} + 3x$, where the gradient is equal to 2?

(b) Find the equation of the normal to the curve $y = \frac{9}{x} + 3x$ at the point $(1, 12)$.

Solution

(a) The equation of the curve needs to be converted into an appropriate form, namely:

$$y = 9x^{-1} + 3x$$

$$\Rightarrow \quad \frac{dy}{dx} = -9x^{-2} + 3 = -\frac{9}{x^2} + 3$$

You need the points where the gradient = 2.

$$-\frac{9}{x^2} + 3 = 2$$

$$\Rightarrow \quad \frac{9}{x^2} = 1$$

$$x^2 = 9$$

Multiply through by x^2.

$$\Rightarrow \quad x = \pm 3$$

So there are two points where the gradient equals 2:

when $x = 3$, $y = \frac{9}{3} + 3(3) = 12$;

when $x = -3$, $y = \frac{9}{-3} + 3(-3) = -12$.

To find the corresponding *y*-coordinates, we substitute each *x*-coordinate into the original equation.

The points where the gradient is 2 are (3, 12) and (−3, −12).

(b) Since $\frac{dy}{dx} = -9x^{-2} + 3 = -\frac{9}{x^2} + 3$,

when $x = 1$, $\frac{dy}{dx} = -\frac{9}{1^2} + 3 = -6$.

Recall that the gradients of perpendicular lines satisfy $m_1 \times m_2 = -1$.

The gradient of the tangent is −6.

Hence, the gradient of the normal is $\frac{1}{6}$.

The equation of the normal is $y - 12 = \frac{1}{6}(x - 1)$ or any equivalent such as $x - 6y + 71 = 0$.

EXERCISE 2C

1 Find the gradient of $y = 3 - \frac{8}{x^2}$, when $x = 2$.

2 Calculate the gradient of $y = 6x + \frac{5}{x}$ at the point (1, 11).

3 Find the gradient of the following curves at the given point:

(a) $y = \frac{7}{x} - 12$, where $x = 1$,

(b) $y = 10\sqrt{x} + 10$, where $x = 9$.

4 The equation of a curve is $y = \dfrac{x^2 - 5x - 24}{x}$.

Find the gradient of the curve at each of the points where the curve crosses the x-axis.

5 Find the equation of the tangent to the curve $y = x^4 - \dfrac{8}{x}$ at the point where $x = 2$.

6 Find the equation of the normal to the curve $y = \sqrt{x} - \dfrac{3}{x^3}$ at the point where $x = 1$.

7 Find for each of the following curves:

(i) the equation of the tangent to the curve at the given point;

(ii) the equation of the normal to the curve at the given point.

(a) $y = \dfrac{2}{x^2} - 3x^2$ at the point $(1, -1)$,

(b) $y = 2\sqrt{x} + \dfrac{8}{x}$ at the point $(4, 6)$,

(c) $y = \dfrac{27}{x^2} - \dfrac{27}{x^3} + 5$ at the point $(3, 7)$.

2.5 Stationary points

You should recall that stationary points are the points on a curve where the gradient is zero.

You also learned in the C1 module how to use the second derivative to test whether a stationary point is a maximum point or a minimum point.

> If P is a point where $\dfrac{dy}{dx} = 0$ and $\dfrac{d^2y}{dx^2} < 0$, then P is a maximum point.
>
> If Q is a point where $\dfrac{dy}{dx} = 0$ and $\dfrac{d^2y}{dx^2} > 0$, then Q is a minimum point.

Worked example 2.7

Find the stationary point on the curve $y = 2x + \dfrac{8}{x^2}$ and determine whether it is a maximum or a minimum point.

Solution

$$y = 2x + \frac{8}{x^2} = 2x + 8x^{-2}$$

$$\Rightarrow \quad \frac{dy}{dx} = 2 - 16x^{-3} = 2 - \frac{16}{x^3}$$

At stationary points, the gradient = 0

$\Rightarrow \quad 2 - \dfrac{16}{x^3} = 0$

$\Rightarrow \quad x^3 = 8$

$\Rightarrow \quad x = 2$ and hence $y = 6$

To determine the nature of the stationary point at $x = 2$, use the second derivative test

$$\frac{d^2y}{dx^2} = 48x^{-4} = \frac{48}{x^4}$$

at $x = 2$, $\dfrac{d^2y}{dx^2} = \dfrac{48}{2^4} = \dfrac{48}{16} = 3$, which is positive.

$\Rightarrow$ (2, 6) is a minimum point.

Worked example 2.8

A function f is defined for $x > 0$ by $f(x) = x(3 - \sqrt{x})$.

(a) Find the value of x for which $f'(x) = 0$.

(b) Find $f''(x)$ and deduce the nature of the stationary value.

Solution

(a) $$f(x) = x(3 - \sqrt{x}) = 3x - x^{\frac{3}{2}}$$

$$\Rightarrow \quad f'(x) = 3 - \frac{3}{2}x^{\frac{1}{2}}$$

Hence, $f'(x) = 0$ when $x^{\frac{1}{2}} = 2$

$\Rightarrow \quad x = 4$

(b) $$f''(x) = \frac{-3}{4}x^{-\frac{1}{2}} = \frac{-3}{4\sqrt{x}}$$

$$f''(4) = \frac{-3}{4\sqrt{4}} = -\frac{3}{8} < 0$$

Hence, the stationary value is a maximum.

Worked examination question 2.9

A curve has equation $y = x^2 - \dfrac{54}{x}$.

Find $\dfrac{dy}{dx}$ and $\dfrac{d^2y}{dx^2}$ in terms of x.

Calculate the coordinates of the stationary point of the curve and determine whether it is a maximum or minimum point. [A]

2

Solution

$$y = x^2 - \frac{54}{x} = x^2 - 54x^{-1}$$

$$\Rightarrow \quad \frac{dy}{dx} = 2x + 54x^{-2} = 2x + \frac{54}{x^2}$$

$$\Rightarrow \quad \frac{d^2y}{dx^2} = 2 - 108x^{-3} = 2 - \frac{108}{x^3}$$

At stationary points, $\frac{dy}{dx} = 0$

$$2x + \frac{54}{x^2} = 0$$

$$\Rightarrow \quad 2x^3 + 54 = 0$$

$$\Rightarrow \quad x^3 = -27 \quad \Rightarrow \quad x = -3$$

The coordinates of the stationary point are $(-3, 27)$.

When $x = -3$, $\frac{d^2y}{dx^2} = 2 - \frac{108}{(-3)^3} = 2 - \frac{108}{(-27)} = 6$

$\Rightarrow \quad (-3, 27)$ is a minimum point since $\frac{d^2y}{dx^2}$ is positive.

EXERCISE 2D

1 A curve has equation

$$y = \frac{x^2 + 9}{x} \quad (x \neq 0).$$

(a) Show that $\frac{dy}{dx} = 0$ when $x = 3$ and find the x-coordinate of the other stationary point.

(b) Find the value of $\frac{d^2y}{dx^2}$ at each of the stationary points and hence determine the nature of each stationary point.

2 A curve has equation $y = 16x + \frac{1}{x}$.
Find the stationary points on the curve and determine whether they are maximum or minimum points.

3 A curve has equation $y = 8\sqrt{x} - 2x$, where $x > 0$. Find the stationary point on the curve and determine whether it is a maximum or a minimum point.

4 Find the maximum and minimum points of these curves:

(a) $y = 3\sqrt{x} - 5x + 1$, where $x > 0$,

(b) $y = x\sqrt{x} - 3x + 4$, where $x > 0$,

(c) $y = \frac{25}{x} + 4x$.

5 A curve has equation $y = \dfrac{x^4}{4} + \dfrac{32}{x^2}$ and is sketched opposite.

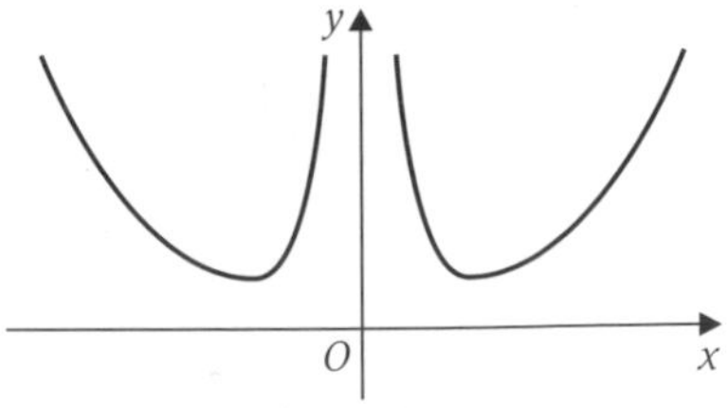

(a) Find $\dfrac{dy}{dx}$.

(b) Hence find the gradient of the curve at the point where $x = 1$.

(c) Show that the stationary points of the curve occur when $x^6 = 64$.

(d) Hence find the x-coordinates of the stationary points. [A]

6 The curve with equation $y = 2x + \dfrac{27}{x^2} - 7$ is defined for $x > 0$, and is sketched opposite.

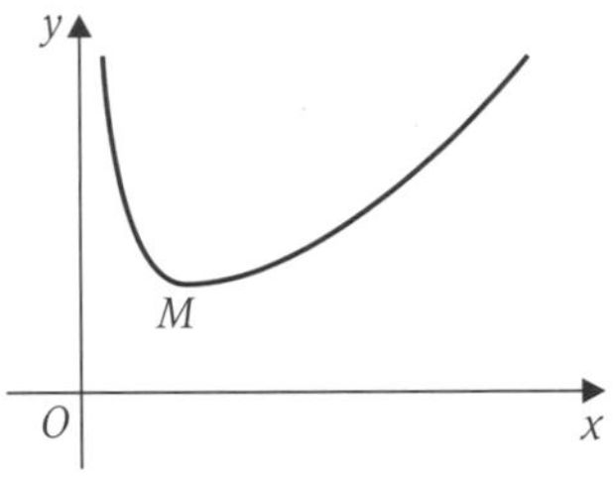

(a) Find $\dfrac{dy}{dx}$.

(b) The curve has a minimum point M. Find the x-coordinate of M.

(c) Find the value of $\dfrac{d^2y}{dx^2}$ at M and explain how this confirms that M is a minimum point. [A]

7 The curve with equation $y = 2x - x^{\frac{3}{2}}$ is defined for $x \geqslant 0$.

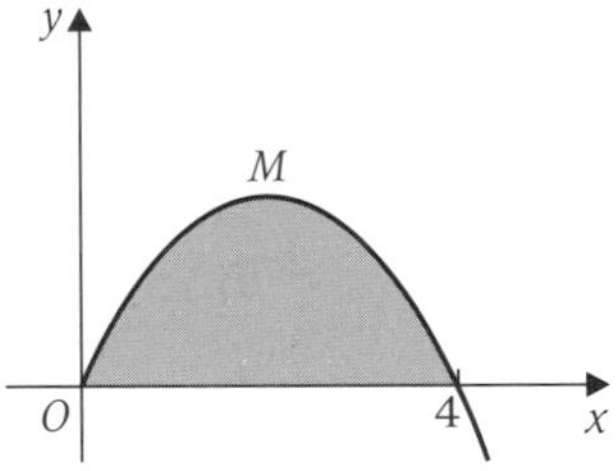

(a) Find $\dfrac{dy}{dx}$.

(b) The curve has a maximum point M. Show that the x-coordinate of M is $\dfrac{16}{9}$. [A]

8 The curve with equation $y = 2x^2 - 1 - \sqrt{x}$ is defined for $x \geqslant 0$ and is shown below.

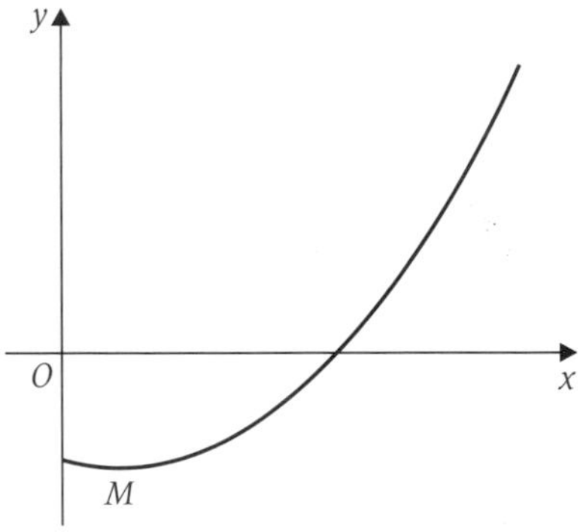

(a) Use differentiation to determine the coordinates of M, the minimum point of the curve.

(b) Find the value of $\dfrac{d^2y}{dx^2}$ at M.

Explain why this confirms that M is a minimum and not a maximum point.

9 Find the stationary points of the following curves:

(a) $y = 2x^2(x - 5)$, where $x > 0$,

(b) $y = \dfrac{x^2 + 1}{x}$,

(c) $y = \dfrac{108}{x} + 2x^2$.

10 For each of the following curves find:

(i) all the stationary points of the curve;

(ii) the value of the second derivative at each stationary point and hence determine their nature.

(a) $y = x^2 - 5x$; **(b)** $y = (5 - 2x)^2$;

(c) $y = x + \dfrac{4}{x^2}$; **(d)** $y = x(\sqrt{x} - 3)$.

11 Find the coordinates of the stationary points of the curve with equation $y = 8 - x^2 - \dfrac{16}{x^2}$. Explain why the curve has no minimum points but has two maximum points.

Use your graphics calculator to help you sketch its graph.

2.6 Optimisation

Many practical problems can be solved using differentiation. You have met some of these ideas in module C1. Previously, the differentiation only involved polynomials, but here you will need to differentiate terms with negative or fractional indices.

Worked example 2.10

An open tank is constructed from sheet metal to hold 13 500 cm^3 of water. It has vertical sides and has a horizontal square base of side x cm.

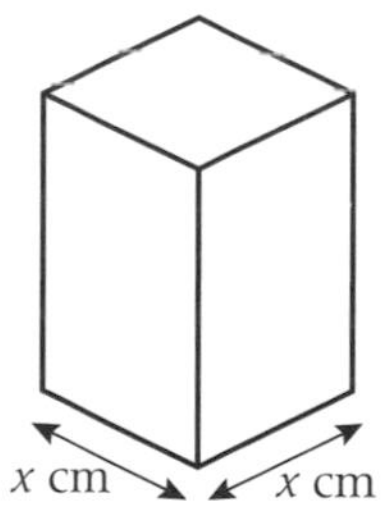

(a) Show that the surface area of metal, A cm^2, needed to make the tank is given by $A = x^2 + \dfrac{54\,000}{x}$.

(b) Find the value of x for which the area of metal used is least.

(c) Verify that this value does give a minimum.

(d) Calculate the minimum area of metal used to make the tank.

Solution

(a) Firstly, let the height of the tank be h cm.
Surface area of the tank, $A = x^2 + 4xh$.

Square base and the four rectangular sides.

In order to minimise this expression you need to differentiate. This is impossible at the moment since there are two variables, namely x and h. Before you can differentiate you need just one variable.

The expression in the printed answer contains just x so this suggests that you have to eliminate the variable h. This is done as follows:

You know that the volume of the tank is 13 500 cm^3

$$x^2h = 13\,500$$

$$\Rightarrow \quad h = \frac{13\,500}{x^2}$$

Substitute this into the surface area expression

$$A = x^2 + 4x\frac{13\,500}{x^2} = x^2 + \frac{54\,000}{x} \text{ as required.}$$

(b) $$\frac{dA}{dx} = 2x - \frac{54\,000}{x^2}$$

For a minimum value, $\frac{dA}{dx} = 0$

$$\Rightarrow \quad 2x - \frac{54\,000}{x^2} = 0$$

$$\Rightarrow \quad 2x^3 = 54\,000$$

$$\Rightarrow \quad x^3 = 27\,000$$

$$\Rightarrow \quad x = \sqrt[3]{27\,000} = 30.$$

(c) To verify the minimum point, use the second derivative test:

$$\frac{d^2A}{dx^2} = 2 + \frac{108\,000}{x^3}$$

When $x = 30$, $\frac{d^2A}{dx^2} = 2 + \frac{108\,000}{(30)^3} = 6$

$\frac{d^2A}{dx^2}$ is positive $\Rightarrow x = 30$ gives a minimum point.

(d) Minimum area used $= (30)^2 + \frac{54\,000}{30} = 2700$ cm^2.

Worked example 2.11

A closed cylindrical tin has a base radius r cm and height h cm. Its contents when full have a volume of 128π cm^3. Assuming that the stationary value gives a minimum value, find the minimum surface area of metal required to make the tin.

Solution

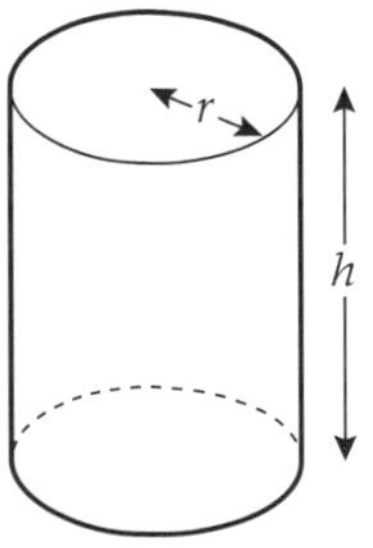

Surface area $A = 2\pi r^2 + 2\pi rh$

Top and bottom and curved surface.

You also know that the volume $= 128\pi$

$$\Rightarrow \quad \pi r^2 h = 128\pi$$

$$\Rightarrow \quad h = \frac{128\pi}{\pi r^2} = \frac{128}{r^2}$$

In order to eliminate one of the variables from the area expression.

Substituting this into the surface area expression gives

$$A = 2\pi r^2 + 2\pi r \frac{128}{r^2} = 2\pi r^2 + \frac{256\pi}{r}$$

Differentiating gives

$$\frac{dA}{dr} = 4\pi r - \frac{256\pi}{r^2}$$

At a stationary point

$$4\pi r - \frac{256\pi}{r^2} = 0$$

$$\Rightarrow \quad 4\pi r^3 - 256\pi = 0$$

$$\Rightarrow \quad r^3 = \frac{256\pi}{4\pi} = 64$$

$$\Rightarrow \quad r = 4$$

Multiply through by r^2.

When $r = 4$, $h = \dfrac{128}{4^2} = 8$

So the minimum surface area =

$$2\pi(4)^2 + 2\pi(4)(8) = 32\pi + 64\pi = 96\pi \text{ cm}^2 \approx 301.6 \text{ cm}^2$$

You are told that this stationary point gives the minimum surface area, so there is no need to do the second derivative test.

EXERCISE 2E

1 A closed rectangular box is made from cardboard. Its base has sides x and $2x$ and its height is h. The volume of the box is 1944 cm^3.

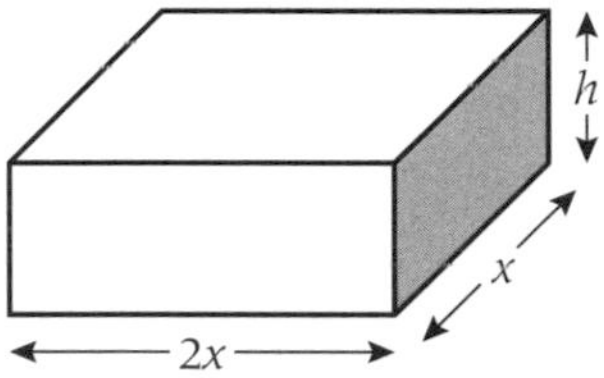

(a) Show that the surface area of the box is given by

$$A = 4x^2 + \frac{5832}{x}.$$

(b) Find the value of x that minimises A.

(c) Use the second derivative test to verify that your value of x gives a minimum.

(d) Write down the dimensions of the box that requires the least amount of cardboard and calculate the minimum surface area.

2 A closed cylindrical container has a volume of $16\pi \text{ cm}^3$. The base radius is r cm and the height is h cm.

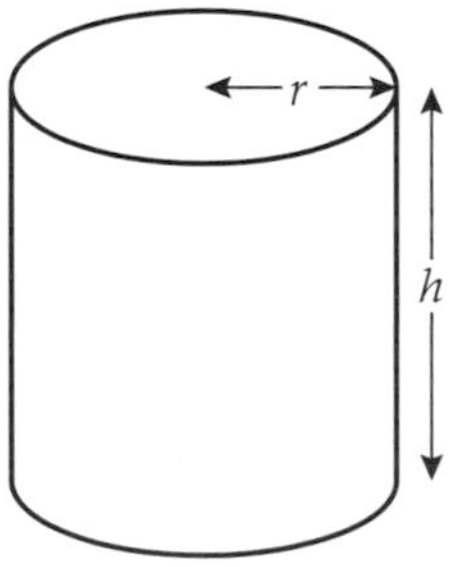

(a) Show that the total surface area, $S \text{ cm}^2$, of the outside of the cylinder is given by

$$S = 2\pi r^2 + \frac{32\pi}{r}.$$

(b) Given that r varies, find the positive value of r for which $\dfrac{dS}{dr} = 0$ and find the corresponding value of $\dfrac{d^2S}{dr^2}$.

(c) Hence find the minimum possible surface area of the outside of the container.

3 A glass window consists of a rectangle with sides of length $2r$ cm and h cm and a semicircle of radius r cm. The total area of one surface of the glass is 500 cm^2.

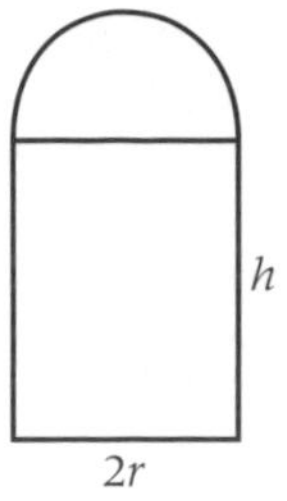

(a) (i) Write down a formula connecting h and r.

(ii) The perimeter of the window is p cm.
By eliminating h, show that

$$p = \left(2 + \frac{\pi}{2}\right)r + \frac{500}{r}.$$

(b) (i) Determine the positive value of r for which p has a stationary value, giving your answer correct to three significant figures.

(ii) Calculate $\frac{\mathrm{d}^2p}{\mathrm{d}r^2}$ and hence determine whether this stationary value is a maximum or minimum value. [A]

4 (a) The function f is defined for $x > 0$ by $\mathrm{f}(x) = x^2 + \frac{16}{x}$.

(i) Find the derivative $\mathrm{f}'(x)$.

(ii) Find the value of x for which $\mathrm{f}'(x) = 0$.

(iii) Find the value of $\mathrm{f}''(x)$ when $\mathrm{f}(x)$ is stationary.

(iv) Determine whether the stationary value of f is a maximum or minimum value.

(b) A right circular cone with base radius r cm and height h cm has volume $\frac{1}{3}\pi r^2 h$ cm^3 and curved surface area $\pi r(r^2 + h^2)^{\frac{1}{2}}$ cm^2.

Show that the curved surface area, S cm^2, of a cone with volume $\frac{4\pi}{3}$ cm^3 is given by

$$S^2 = \pi^2\left(r^4 + \frac{16}{r^2}\right).$$

Use results from **(a)** to find:

(i) the value of the radius, in centimetres, for which the cone has the minimum curved surface area, giving your answer correct to three significant figures;

(ii) the minimum curved surface area of the cone, in square centimetres, giving your answer correct to three significant figures. [A]

You can obtain the answers without any further use of calculus. You can assume that S^2 is a minimum when S is a minimum.

Key point summary

1 When differentiating, you can use the following strategy even when n is any rational number: *p221*

(a) if $y = x^n$, then $\frac{dy}{dx} = nx^{n-1}$;

(b) if $y = cx^n$ (c is a constant), then $\frac{dy}{dx} = cnx^{n-1}$;

(c) if $y = f(x) \pm g(x)$, then $\frac{dy}{dx} = f'(x) \pm g'(x)$.

2 The equation of the tangent to a curve at the point $P(x_1, y_1)$ is given by $y - y_1 = m(x - x_1)$, where m is the gradient of the curve at P. *p227*

3 The normal to a curve at the point P is a straight line that is perpendicular to the tangent at P. When the gradient of the tangent is m, the gradient of the normal is $-\frac{1}{m}$. *p227*

4 The nature of stationary points can be established using the second derivative test: *p229*

If P is a point where $\frac{dy}{dx} = 0$ and $\frac{d^2y}{dx^2}$ is negative then P is a maximum point.

If Q is a point where $\frac{dy}{dx} = 0$ and $\frac{d^2y}{dx^2}$ is positive then Q is a minimum point.

Test yourself

Test yourself	What to review
1 Differentiate with respect to x: **(a)** $y = \frac{2x^3 - 5x^2 + 7}{x^5}$, **(b)** $y = 3x(6x^3 + 2\sqrt{x})$.	*Sections 2.2 and 2.3*
2 Find the gradient of the following curves at the given points: **(a)** $y = \frac{(x-2)(x-5)}{x}$ at $x = -2$, **(b)** $y = 2\sqrt{x}(5x - 3)$ at $x = 9$.	*Section 2.4*
3 Find $\frac{d^2y}{dx^2}$ for the following functions: **(a)** $y = 5x^2 - \frac{3}{x}$, **(b)** $y = 2x(\sqrt{x} - 1)$.	*Sections 2.2 and 2.3*

Test yourself (continued)	What to review
4 Find the equation of the tangent to the curve $y = \dfrac{x^2 + 4}{x}$ at the point (1, 5).	*Section 2.4*
5 Find the equation of the normal to the curve $y = \dfrac{x^3 - x + 2}{x^2}$ at the point (−2, −1).	*Section 2.4*
6 Find the x-coordinates of the stationary points on the curve with equation $y = 49x + \dfrac{1}{x}$ and determine whether they are maximum or minimum points.	*Section 2.5*

Test yourself ANSWERS

1 (a) $\dfrac{15}{x^4} - \dfrac{4}{x^3} - \dfrac{35}{x^6}$; **(b)** $72x^3 + 9\sqrt{x}$.

2 (a) −1.5; **(b)** 44.

3 (a) $10 - \dfrac{6}{x^3}$; **(b)** $\dfrac{3}{2\sqrt{x}}$.

4 $y + 3x = 8$.

5 $4x + 7y + 15 = 0$.

6 Maximum when $x = -\frac{1}{7}$, minimum when $x = \frac{1}{7}$.

CHAPTER 3

C2: Further integration and the trapezium rule

Learning objectives

After studying this chapter, you should be able to:

- integrate expressions of the form kx^n, where n is any rational number, $n \neq -1$
- integrate expressions more difficult than polynomials to find both indefinite and definite integrals
- use definite integrals to find areas of regions bounded by curves where the curves involve expressions more difficult than polynomials
- use the trapezium rule with a given number of ordinates to find an approximation for the area under a curve
- realise situations where the trapezium rule gives an overestimate or underestimate of the area under a curve
- appreciate how the trapezium rule approximation can often be improved by increasing the number of ordinates used.

3.1 Basic rule for integration of x^n

You learned in the C1 module that integration is the reverse process of differentiation and that

> $\dfrac{dy}{dx} = x^n \quad (n \neq -1) \quad \Rightarrow \quad y = \dfrac{1}{n+1}x^{n+1} + c$, where c is an arbitrary constant.

Can you see why this formula would not work if you wanted to integrate $\dfrac{1}{x} = x^{-1}$? You will learn how to find $\int \dfrac{1}{x}\,dx$ in Core 3.

You should recall, for example, that

$$\int (6x^3 - 5x + 2)\,dx = 6 \times \frac{x^4}{4} - 5 \times \frac{x^2}{2} + 2x + c$$

$$= \frac{3x^4}{2} - \frac{5x^2}{2} + 2x + c,$$

where c is an arbitrary constant.

Many students get confused between differentiation and integration and so it is helpful to remember

To **I**ntegrate, you **I**ncrease the power by 1, whereas to **D**ifferentiate, you **D**ecrease the power by 1.

In this chapter you will learn to extend your ability to integrate more difficult expressions where the powers of x are negative numbers or fractions.

Worked example 3.1

Given that $\frac{dy}{dx} = \frac{3}{x^2} + \frac{8}{x^5} - 4$, find an expression for y.

Solution

$$\frac{dy}{dx} = \frac{3}{x^2} + \frac{8}{x^5} - 4 = 3x^{-2} + 8x^{-5} - 4x^0$$

Consider x^{-2} – this integrates to $\frac{x^{-2+1}}{-2+1} = \frac{x^{-1}}{-1}$.

$$\Rightarrow y = 3 \times \frac{x^{-1}}{-1} + 8 \times \frac{x^{-4}}{-4} - 4\frac{x^1}{1} + c$$

Increase the index by 1 and divide by the new index.

You could write the answer in a form without negative indices:

$$y = c - \frac{3}{x} - \frac{2}{x^4} - 4x,$$

where c is an arbitrary constant.

Worked example 3.2

Given that $f'(x) = x^4 - \sqrt{x}$, find an expression for $f(x)$.

Solution

Writing $\sqrt{x}$ as a power of x.

$$f'(x) = x^4 - \sqrt{x} = x^4 - x^{\frac{1}{2}}$$

Integrating: Increase the index by 1 and divide by the new index.

Don't forget the arbitrary constant.

$$\Rightarrow f(x) = \frac{1}{5}x^5 - \frac{1}{\left(\frac{3}{2}\right)}x^{\frac{3}{2}} + k = \frac{x^5}{5} - \frac{2}{3}x^{\frac{3}{2}} + k,$$

where k is an arbitrary constant.

Worked example 3.3

A curve passes through the point (4, 5) and $\frac{dy}{dx} = 3\sqrt{x} - \frac{16}{x^2}$.

Find its equation.

Solution

Rewriting as powers of x, $\frac{dy}{dx} = 3x^{\frac{1}{2}} - 16x^{-2}$.

Note the indices are increased by 1:
$\frac{1}{2} + 1 = \frac{3}{2}$; $-2 + 1 = -1$.

Integration gives $y = 3 \times \frac{x^{\frac{3}{2}}}{\left(\frac{3}{2}\right)} - 16 \times \frac{x^{-1}}{-1} + c$, where c is a constant.

$$\Rightarrow \quad y = 2x^{\frac{3}{2}} + \frac{16}{x} + c$$

Since the curve passes through (4, 5) you can find the value of c.

Putting $x = 4$ and $y = 5 \Rightarrow 5 = 2 \times 2^3 + \frac{16}{4} + c \Rightarrow 5 = 16 + 4 + c$

$\Rightarrow \quad c = -15$

The equation of the curve is $y = 2x^{\frac{3}{2}} + \frac{16}{x} - 15$

or $y = 2x\sqrt{x} + \frac{16}{x} - 15$.

EXERCISE 3A

3

Don't forget the arbitrary constant.

You can always check your final answer by differentiating and making sure you get the original expression for $\frac{dy}{dx}$ or f′(x).

1 Find y in terms of x for each of the following:

(a) $\frac{dy}{dx} = x^{-5}$, **(b)** $\frac{dy}{dx} = x^{-4}$,

(c) $\frac{dy}{dx} = \frac{4}{x^3}$, **(d)** $\frac{dy}{dx} = \frac{6}{x^2}$,

(e) $\frac{dy}{dx} = 3x^{-\frac{1}{4}}$, **(f)** $\frac{dy}{dx} = x^{\frac{1}{3}}$,

(g) $\frac{dy}{dx} = 6\sqrt{x}$, **(h)** $\frac{dy}{dx} = \frac{2}{\sqrt{x}}$,

(i) $\frac{dy}{dx} = x\sqrt{x} - \frac{1}{x^2}$, **(j)** $\frac{dy}{dx} = \frac{3}{x^4} - \frac{1}{x^2}$.

2 Find an expression for f(x) in each of the following cases:

(a) $f'(x) = 2x^{-2} - 3x^{-4}$, **(b)** $f'(x) = 10x^{-6} - 12x^{-6} + 7x^{-2}$,

(c) $f'(x) = \frac{3}{x^4} + \frac{10}{x^6}$, **(d)** $f'(x) = \frac{2}{x^3} - \frac{12}{x^5}$,

(e) $f'(x) = \frac{18}{x^{10}} - \frac{14}{x^8}$ **(f)** $f'(x) = 6\sqrt{x} + \frac{9}{x^2\sqrt{x}}$.

3 A curve passes through the point (−2, 7) and is such that $\frac{dy}{dx} = \frac{4}{x^2} - 8x + 5$. Find the equation of the curve.

4 A curve passes through the point (4, 11) and is such that $\frac{dy}{dx} = 3\sqrt{x} + 10x\sqrt{x}$. Find the equation of the curve.

5 A curve passes through the point (27, 4) and is such that $\frac{dy}{dx} = 8\sqrt[3]{x} - \frac{2}{\sqrt[3]{x}}$. Find the equation of the curve.

3.2 Simplifying expressions before integrating

Sometimes it is necessary to obtain a series of terms of the form kx^n before integrating.

Worked example 3.4

Given that $\frac{dy}{dx} = \frac{x^3 - 3x + 4}{x^5}$, find an expression for y.

Solution

$$\frac{dy}{dx} = \frac{x^3 - 3x + 4}{x^5} = \frac{x^3}{x^5} - \frac{3x}{x^5} + \frac{4}{x^5}$$

Writing as separate fractions.

$$\Rightarrow \frac{dy}{dx} = x^{-2} - 3x^{-4} + 4x^{-5}$$

Each term is now of the form kx^n.

$$\Rightarrow y = \left(\frac{x^{-1}}{-1}\right) - \left(3 \times \frac{x^{-3}}{-3}\right) + \left(4 \times \frac{x^{-4}}{-4}\right) + k$$

k is the arbitrary constant.

$$\Rightarrow y = -x^{-1} + x^{-3} - x^{-4} + k$$

or if you prefer this can be written as

$$y = \frac{1}{x^3} - \frac{1}{x} - \frac{1}{x^4} + k, \text{ where } k \text{ is a constant.}$$

EXERCISE 3B

1 Find an expression for y in terms of x and a constant in each of the following cases:

(a) $\frac{dy}{dx} = x^{-4}(3x - 2)$, **(b)** $\frac{dy}{dx} = x^{-3}(10x + 6)$,

(c) $\frac{dy}{dx} = \sqrt{x}(2x^2 + 3)$, **(d)** $\frac{dy}{dx} = x^{-5}(3x + 8)$,

(e) $\frac{dy}{dx} = x\sqrt{x}(5 + 7x)$, **(f)** $\frac{dy}{dx} = \sqrt{x}(x - 2)(x + 3)$,

(g) $\frac{dy}{dx} = x^{\frac{2}{3}}(16x + 5)$, **(h)** $\frac{dy}{dx} = x^{-6}(x + 3)(x + 5)$,

(i) $\frac{dy}{dx} = \frac{(2x + 3)(x + 4)}{x^4}$, **(j)** $\frac{dy}{dx} = \frac{(3x - 4)(x + 2)}{\sqrt{x}}$.

2 Find f(x) in terms of x and an arbitrary constant for each of the following where:

(a) $f'(x) = x^{-5}(12 - 3x)$, **(b)** $f'(x) = \sqrt{x}(5x - 3)$,

(c) $f'(x) = 12x^3(x + 1)(x - 1)$, **(d)** $f'(x) = \sqrt{x}(6 - 2\sqrt{x})$,

(e) $f'(x) = \frac{x^4 - 3x^2 + 7}{x^2}$, **(f)** $f'(x) = \frac{x^2 - 3x + 5}{\sqrt{x}}$,

(g) $f'(x) = \frac{4x - 5}{\sqrt[3]{x}}$, **(h)** $f'(x) = \frac{\sqrt{x} - 4x + 6}{\sqrt{x}}$.

3 A curve passes through the point $(-1, 3)$ and is such that $\frac{dy}{dx} = x^{-3}(25x - 16)$. Find the equation of the curve.

4 A curve passes through the point (2, −2) and is such that $\frac{dy}{dx} = \frac{x^5 - 3x^2 + 2}{x^2}$. Find the equation of the curve.

5 A curve passes through the point (1, 2) and is such that $\frac{dy}{dx} = \frac{(x-3)(5-2x)}{x^4}$. Find the equation of the curve.

6 A curve passes through the point (4, 25) and is such that $\frac{dy}{dx} = \frac{x^2 - 3x + 7}{\sqrt{x}}$. Find the equation of the curve.

7 A curve passes through the point (1, −5) and is such that $\frac{dy}{dx} = \frac{(x-3)(2x+1)}{\sqrt[3]{x}}$. Find the equation of the curve.

3.3 Indefinite integrals

Whenever you are given $\frac{dy}{dx} = f(x)$ you can find y by integration and you can write $y = \int f(x)\,dx$.

This is called an indefinite integral.

When you find an indefinite integral you should always include '$+ c$', where c is an arbitrary constant.

Worked example 3.5

Find **(a)** $\int \sqrt{x}(10x - 3)\,dx$, **(b)** $\int \frac{(x+2)}{\sqrt{x}}\,dx$.

Solution

(a) $\int \sqrt{x}(10x-3)\,dx = \int (10x^{\frac{3}{2}} - 3x^{\frac{1}{2}})\,dx$

$$= 10 \times \frac{x^{\frac{5}{2}}}{\left(\frac{5}{2}\right)} - 3 \times \frac{x^{\frac{3}{2}}}{\left(\frac{3}{2}\right)} + c$$

$$= 4x^{\frac{5}{2}} - 2x^{\frac{3}{2}} + c, \text{ where } c \text{ is a constant}$$

Notice that you keep the integral sign in your working while you simplify the expression.

(b) $\int \frac{(x+2)}{\sqrt{x}}\,dx = \int \left(\frac{x}{\sqrt{x}} + \frac{2}{\sqrt{x}}\right)dx$

$$= \int \left(x^{\frac{1}{2}} + 2x^{-\frac{1}{2}}\right)dx$$

$$= \frac{1}{\left(\frac{3}{2}\right)}x^{\frac{3}{2}} + \frac{2}{\left(\frac{1}{2}\right)}x^{\frac{1}{2}} + k$$

$$= \frac{2}{3}x^{\frac{3}{2}} + 4x^{\frac{1}{2}} + k, \text{ where } k \text{ is a constant.}$$

EXERCISE 3C

Find each of the following:

1 $\int 12x^{\frac{1}{2}}\,dx$

2 $\int 20x^{\frac{3}{2}}\,dx$

3 $\int 2x^{-3}\,dx$

4 $\int x^{-7}(4x + 6)\,dx$

5 $\int 88x^{-12}\,dx$

6 $\int 20x^{-4}\,dx$

7 $\int 3x^{-2}\,dx$

8 $\int x^{-2}(4 + 6x^7)\,dx$

9 $\int \frac{6}{x^4}\,dx$

10 $\int 3\sqrt{x}\,dx$

11 $\int \frac{1}{x^3}\,dx$

12 $\int \frac{4}{x^2} + \frac{6}{x^3}\,dx$

13 $\int \sqrt{x}(3x + 5)\,dx$

14 $\int \frac{(6x - 5)}{\sqrt{x}}\,dx$

15 $\int x^2\sqrt{x}\,dx$

16 $\int \frac{x^2 - 1}{\sqrt{x}}\,dx$

17 $\int \frac{(3 + 2x)^2}{x^4}\,dx$

18 $\int \frac{(2 - 3x)^2}{\sqrt{x}}\,dx$

19 $\int (\sqrt{x} + 3)^2\,dx$

20 $\int (\sqrt{x} - 1)^2\,dx$

21 $\int \sqrt{x}(3 - x)(x + 3)\,dx$

22 $\int (x^3 - 1)\sqrt[3]{x}\,dx$

23 $\int (5 - 6x)^2\,dx$

24 $\int \frac{2(x^4 + 1)}{x^3}\,dx$

25 $\int 18x^3\sqrt{x}\,dx$

26 $\int \frac{x^4 - 3}{x\sqrt{x}}\,dx$

27 $\int \sqrt{x}(x^3 - 1)^2\,dx$

28 $\int \frac{x^5 + 5}{x^3\sqrt{x}}\,dx$

29 $\int x(\sqrt{x} - 3)^2\,dx$

3.4 Definite integrals

A definite integral is of the form $\int_a^b f(x)\,dx$.

The function f(x) is called the integrand.

The values $x = a$ and $x = b$ at the bottom and top of the integral sign are called **limits**.

There is no need to add '+c' when finding definite integrals.

So if $\int f(x)\,dx = F(x)$, then $\int_a^b f(x)\,dx = F(b) - F(a)$.

Worked example 3.6

Find the value of $\int_1^4 \sqrt{x}(x+1)\,dx$.

Solution

$$\int_1^4 \sqrt{x}(x+1)\,dx = \int_1^4 x\sqrt{x} + \sqrt{x}\,dx = \int_1^4 x^{\frac{3}{2}} + x^{\frac{1}{2}}\,dx$$

Write each term as a power of x.

$$= \left[\frac{2}{5}x^{\frac{5}{2}} + \frac{2}{3}x^{\frac{3}{2}}\right]_1^4$$

Integrating each term.

$$= \left[\frac{2}{5}\times 4^{\frac{5}{2}} + \frac{2}{3}\times 4^{\frac{3}{2}}\right] - \left[\frac{2}{5}\times 1^{\frac{5}{2}} + \frac{2}{3}\times 1^{\frac{3}{2}}\right]$$

'top limit value' minus 'bottom limit value'.

$$= \left(\frac{64}{5} + \frac{16}{3}\right) - \left(\frac{2}{5} + \frac{2}{3}\right) = \frac{256}{15}$$

3

3.5 Interpretation as an area

Suppose the portion of the curve with equation $y = f(x)$ between $x = a$ and $x = b$ lies above the x-axis, then the area of the shaded region bounded by the curve, the x-axis and the lines $x = a$ and $x = b$ is given by

$$\text{Area} = \int_a^b f(x)\,dx.$$

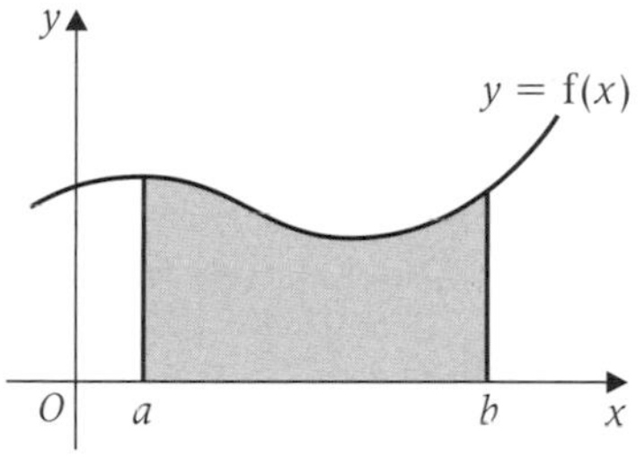

Worked example 3.7

Find the area of the region bounded by the curve with equation $y = x + \frac{1}{x^2}$, the lines $x = 1$, $x = 2$ and the x-axis from $x = 1$ to $x = 2$.

Your graphics calculator could be used to get a sketch of the region and give you a check on your answer.

Solution

$$\text{Area} = \int_1^2 x + \frac{1}{x^2}\,dx = \int_1^2 x + x^{-2}\,dx$$

$$= \left[\frac{x^2}{2} - \frac{1}{x}\right]_1^2 = \left(\frac{4}{2} - \frac{1}{2}\right) - \left(\frac{1}{2} - \frac{1}{1}\right) = 2$$

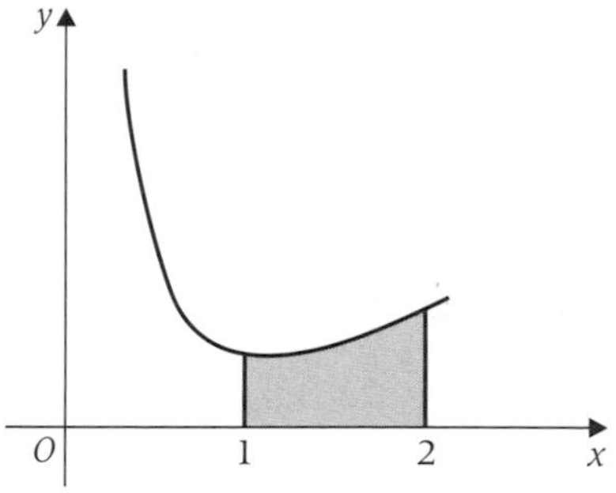

EXERCISE 3D

Evaluate each of the definite integrals **1** to **11**.

Don't worry if some of the final answers are negative.

1 $\int_0^1 6x\,dx$

2 $\int_1^3 9x^{-2}\,dx$

3 $\int_0^2 12x^3\,dx$

4 $\int_1^4 5x\sqrt{x}\,dx$

5 $\int_1^4 (3\sqrt{x} - 6x)\,dx$

6 $\int_1^2 \frac{(x-1)(x+1)}{x^2}\,dx$

7 $\int_0^1 \sqrt{x}(10x - 3)\,dx$

8 $\int_1^4 3\sqrt{x} - x^{-3}\,dx$

9 $\int_0^1 \sqrt{x^3}\,dx$

10 $\int_1^4 (2\sqrt{x} - 3)(\sqrt{x} - 1)\,dx$

You may wish to sketch the curves and regions using a graphic calculator.

11 $\int_1^2 \frac{(2x-3)(x+1)}{x^4}\,dx$

12 **(a)** Express $\frac{x^5 + 1}{x^2}$ in the form $x^p + x^q$, where p and q are integers.

(b) Hence, find $\int_1^2 \frac{x^5 + 1}{x^2}\,dx$. [A]

13 Calculate the area of the region bounded by the curve with equation $y = 3\sqrt{x} + \frac{2}{\sqrt{x}}$, the x-axis from $x = 1$ to $x = 4$ and the lines $x = 1$ and $x = 4$.

14 Calculate the area of the region bounded by the curve with equation $y = x^2 + \frac{4}{x^3}$, the x-axis from $x = 1$ to $x = 2$ and the lines $x = 1$ and $x = 2$.

You may wish to sketch the curves and regions using a graphics calculator.

15 Calculate the area of the region bounded by the curve $y = x^{\frac{3}{2}}$, the x-axis from the origin to $x = 4$ and the line $x = 4$.

16 Calculate the area of the region bounded by the curve with equation $y = 3\sqrt{x} + \frac{12}{x^2}$, the x-axis from $x = 1$ to $x = 4$ and the lines $x = 1$ and $x = 4$.

17 The curve with equation $y = 2x - x^{\frac{3}{2}}$ is defined for $x \geqslant 0$, and is sketched below.

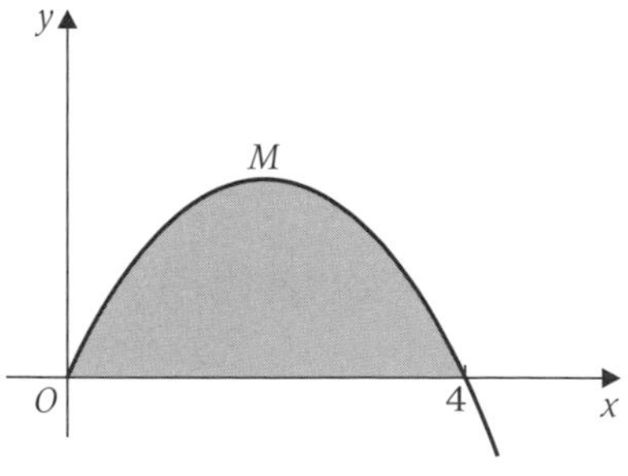

Calculate the area of the shaded region bounded by the curve and the x-axis. [A]

18 The curve with equation $y = 2x + \dfrac{27}{x^2} - 7$ is defined for $x > 0$, and is sketched opposite.

(a) Find $\int \left(2x + \dfrac{27}{x^2} - 7\right) dx$.

(b) Hence determine the area of the region bounded by the curve, the lines $x = 1$, $x = 2$ and the x-axis. [A]

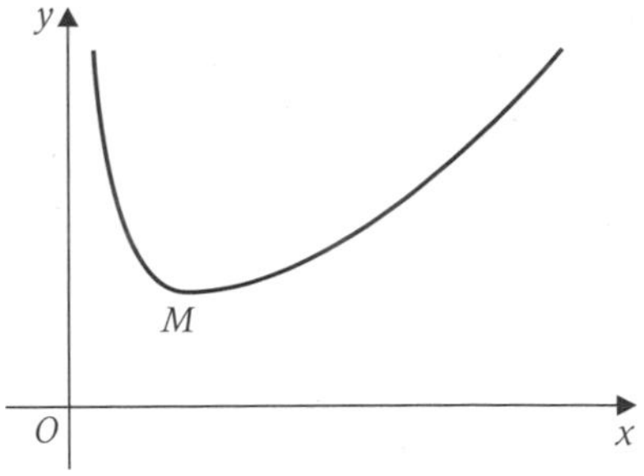

19 The graph of $y = \sqrt{x}$ (defined for $x > 0$) is sketched together with the line $y = \dfrac{1}{2}x$. Find the coordinates of the point of intersection A, other than the origin O.
Calculate the area of the finite region bounded by the line and the curve.

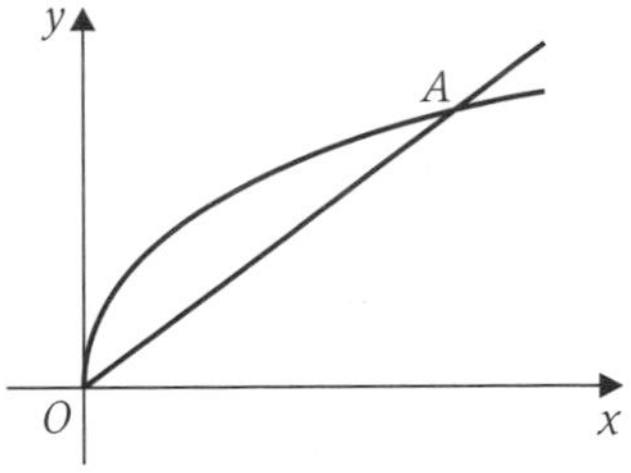

20 **(a)** Find the positive roots of the equation $x^2 + \dfrac{4}{x^2} = 5$.

(b) Find the area of the region in the first quadrant, bounded by the curve $y = x^2 + \dfrac{4}{x^2}$ and the line $y = 5$.

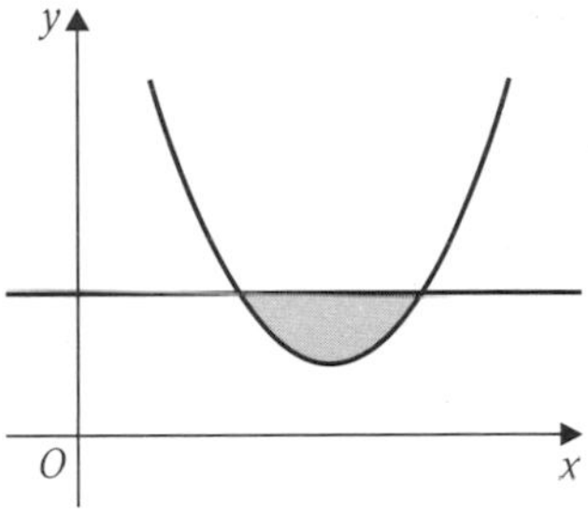

3.6 Estimation of area using the trapezium rule

Sometimes, it is not easy (or even possible) to find the exact value of a definite integral. It is then necessary to use an approximation method.

One such method is called the trapezium rule. The area under the curve is approximated using a number of trapezia.

Consider the area represented by $\int_a^b f(x)\, dx$ to be split up into n vertical strips of equal width, h, so that $h = \dfrac{(b - a)}{n}$.

> Let $x_0 = a$, $x_1 = a + h$, $x_2 = a + 2h$, etc., so that $b = x_n = a + nh$.

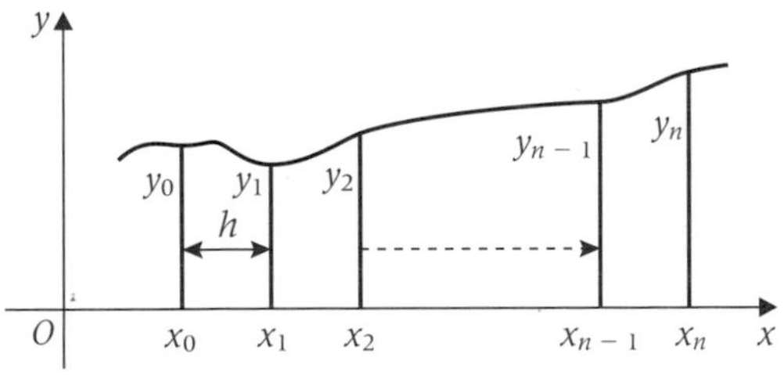

The vertical distances $y_0, y_1, \ldots, y_n$ are called the **ordinates** and are given by $y_0 = f(x_0)$, $y_1 = f(x_1)$, …, $f(x_n)$.

Approximating the first strip as a trapezium, it has area $\dfrac{1}{2}h(y_0 + y_1)$.

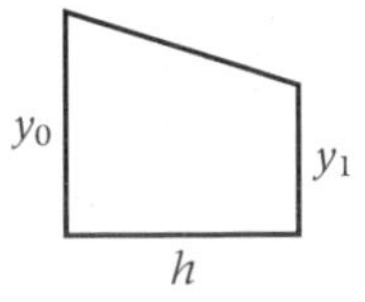

Repeating for all n strips gives the total area of the trapezia as

$$\frac{1}{2}h(y_0 + y_1) + \frac{1}{2}h(y_1 + y_2) + \frac{1}{2}h(y_2 + y_3) + \ldots + \frac{1}{2}h(y_{n-1} + y_n)$$

which can be simplified by taking out a common factor to

$$\frac{1}{2}h(y_0 + 2y_1 + 2y_2 + 2y_3 + \ldots + 2y_{n-1} + y_n).$$

Hence,

$$\int_a^b \mathrm{f}(x)\,\mathrm{d}x \approx \frac{1}{2}h\{(y_0 + y_n) + 2(y_1 + y_2 + \ldots + y_{n-1})\},$$

where $h = \dfrac{b-a}{n}$

This formula is in the formulae book.

Worked example 3.8

Use the trapezium rule with six ordinates (five strips) to find a four significant figure approximation to $\int_1^2 \frac{1}{x}\,\mathrm{d}x$.

Solution I

Using the notation as above:

$$a = 1,\ b = 2,\ h = \frac{(b-a)}{n} = \frac{(2-1)}{5} = 0.2$$

Also $\mathrm{f}(x) = \frac{1}{x}$, and

$$a = x_0 = 1.0,$$
$$\text{so } x_1 = 1.2,\ x_2 = 1.4,\quad x_3 = 1.6,\quad x_4 = 1.8,\quad x_5 = 2.0 = b.$$

Hence, since $y_0 = \mathrm{f}(x_0),\ y_1 = \mathrm{f}(x_1),\ \ldots,\ y_5 = \mathrm{f}(x_5)$ you have

It is good practice to work to at least one (usually two) more significant figures than required for the final answer because rounding can take place.

$$y_0 = 1.0,\ y_1 = \frac{1}{1.2} \approx 0.833\,333,\ y_2 = \frac{1}{1.4} = 0.714\,286,$$

$$y_3 = \frac{1}{1.6} = 0.625,\ y_4 = \frac{1}{1.8} \approx 0.555\,556,\ y_5 = \frac{1}{2.0} = 0.5.$$

Substituting in the formula in the box above gives:

$$\frac{1}{2} \times 0.2\{(1.0 + 0.5) + 2(0.833\,333 + 0.714\,286 + 0.625 + 0.555\,556)\}$$

$$= 0.695\,635$$

Hence, the trapezium rule with five strips gives the approximation

$$\int_1^2 \frac{1}{x}\,\mathrm{d}x = 0.6956 \text{ (to four significant figures).}$$

It is neater to set the working out in tabular form so you are likely to make fewer errors.

Solution 2

The expression $\frac{1}{2}h\{(y_0 + y_n) + 2(y_1 + y_2 + \dots + y_{n-1})\}$ can be thought of as $\frac{1}{2}h(\text{'ends'} + 2 \times \text{'middles'})$ and the previous working could be set out in a table as follows:

x	y		
	ends	**middles**	
$x_0 = 1.0$	1.000 000	←	y_0
$x_1 = 1.2$		0.833 333 ←	y_1
$x_2 = 1.4$		0.714 286	
$x_3 = 1.6$		0.625 000	
$x_4 = 1.8$		0.555 556	
$x_5 = 2.0$	0.500 000		
	1.500 000	2.728 175	**Totals**

The trapezium rule gives $\frac{1}{2}h(1.500\,000 + 2 \times 2.728\,175)$

$$= 0.1 \times 6.956\,350 = 0.695\,635$$

Don't forget the '2 times the middles'.

Hence, to four significant figures, the integral ≈ 0.6956.

Worked example 3.9

(a) Use the trapezium rule with five ordinates (four strips) to find a three significant figure approximation to $\int_0^2 3^x \, dx$.

(b) Sketch the graph of $y = 3^x$ for $x > 0$ and comment on whether your estimate is likely to be an overestimate or underestimate of the true value.

(c) Comment on how you could obtain a better approximation to the value of the integral using the trapezium rule.

Solution

(a) Using the formula $\int_a^b y \, dx \approx \frac{1}{2}h\{y_0 + y_n) + 2(y_1 + y_2 + \dots + y_{n-1})\}$

$$h = \frac{(b-a)}{n} = \frac{2-0}{4} = 0.5.$$

Setting this out in tabular form where

$$y_0 = 3^0 = 1, \quad y_1 = 3^{0.5} = 1.732\,05, \text{ etc.}$$

x	y ends	y middles	
$x_0 = 0.0$	1.000 00		
$x_1 = 0.5$		1.732 05	
$x_2 = 1.0$		3.000 00	
$x_3 = 1.5$		5.196 15	
$x_4 = 2.0$	9.000 00		
	10.000 00	9.928 20	**Totals**

The trapezium rule gives $\frac{1}{2}h(10.000\,00 + 2 \times 9.928\,20)$, where $h = 0.5$.

Don't forget the '2 times the middles'.

Hence, the estimate is $0.25 \times 29.8564 = 7.4641$
The approximate value is 7.46 to three significant figures.

(b) The graph is an increasing function and its gradient is increasing. The trapezia will therefore have an area slightly greater than the corresponding actual area of each strip.

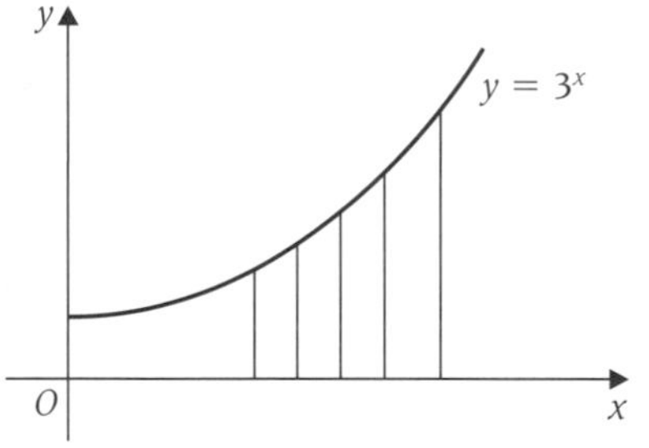

Using the trapezium rule will therefore give an overestimate of the true area and, therefore, of the integral.

(c) By increasing the number of strips the trapezia will more closely approximate to the true area and, therefore, if you want to find a better approximation to the integral you need to increase the number of ordinates.

The true value of this integral is 7.2819 … correct to four decimal places.
The error using the trapezium rule with only four strips is approximately 0.18 which is $\frac{0.18}{7.28} \times 100\%$ or about 2.5%.

The diagram shows why the estimate using the trapezium rule is an overestimate in this case.

For a region such as that below, the trapezium rule would give an underestimate of the area. Can you see why?

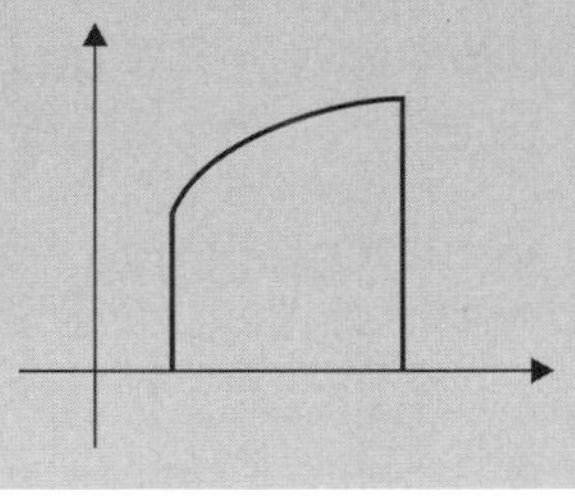

EXERCISE 3E

For each of the definite integrals in questions **1** to **5**, use the trapezium rule with six ordinates (five strips) to estimate its value, giving your answers to four significant figures.

1 $\int_0^1 \sqrt{x + 1}\, dx$ **2** $\int_1^3 2^x\, dx$ **3** $\int_0^2 \sqrt{1 + x^3}\, dx$

4 $\int_1^4 \frac{\sqrt{x}}{1 + x}\, dx$ **5** $\int_{-1}^0 \sqrt{4 - x^2}\, dx$

6 The integral in **1** has the exact value $\frac{2}{3}(2\sqrt{2} - 1)$. Find the percentage error in using the trapezium rule with six ordinates. How could you have obtained a better approximation using the trapezium rule?

7 Sketch the graph of $y = 2^x$ for $1 \leqslant x \leqslant 3$. Hence explain whether your estimate of the integral in **2** will be an overestimate or an underestimate of the exact value.

8 Repeat **4** using seven ordinates (six strips). Which answer is likely to be the most accurate?

9 For each of the following regions, determine whether the use of the trapezium rule to estimate the area would give an overestimate or an underestimate of the true area.

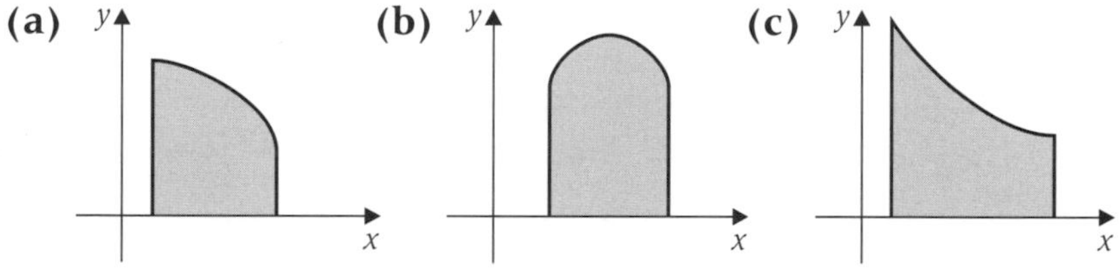

10 Sketch the circle with equation $x^2 + y^2 = 4$. Indicate on your sketch the region whose area is represented by the integral in **5**. Hence, explain whether your trapezium rule estimate of the integral in **5** will be an overestimate or an underestimate of the exact value.

Key point summary

1 $\frac{dy}{dx} = x^n \ (n \neq -1) \Rightarrow y = \frac{1}{n+1}x^{n+1} + c$, where c is an arbitrary constant. *p239*

2 An indefinite integral has no limits and is of the form $\int f(x)\,dx$. You must remember to include an arbitrary constant in your answer. *p243*

3 A definite integral is of the form $\int_a^b f(x)\,dx$, where a and b are the limits. After integrating a definite integral, you obtain $[F(x)]_a^b$ and this is evaluated as $F(b) - F(a)$. *p244*

4 The area under a curve $y = f(x)$ from $x = a$ to $x = b$ is given by *p245*

$$\int_a^b f(x)\,dx.$$

5 The trapezium rule can be used to find an approximation to a definite integral: p248

$$\int_a^b f(x)\,dx \approx \frac{1}{2}h\{(y_0 + y_n) + 2(y_1 + y_2 + \ldots + y_{n-1})\},$$

where $h = \dfrac{b-a}{n}$

Test yourself	What to review
1 Obtain the equation of the curve passing through $(-1, 5)$ which has gradient $\dfrac{dy}{dx} = 6x^2 - \dfrac{2}{x^3}$.	*Section 3.1*
2 Find $f(x)$ when $f'(x) = \dfrac{(6x+5)}{\sqrt{x}}$.	*Section 3.2*
3 Find $\int \sqrt{x}(7x^2 + 6)\,dx$.	*Section 3.3*
4 Evaluate $\int_1^2 \left(\dfrac{x^3 - 3}{x^2}\right) dx$.	*Section 3.4*
5 Find the area bounded by the curve with equation $y = \sqrt[3]{x}$, the x-axis and the lines $x = 1$ and $x = 8$.	*Section 3.5*
6 Use the trapezium rule with five ordinates (four strips) to estimate the value of $\int_2^3 2^{-x}\,dx$, giving your answer to four significant figures. By reference to a sketch of $y = 2^{-x}$, state whether your estimate is an overestimate or underestimate of the exact value.	*Section 3.6*

Test yourself ANSWERS

1 $y = 2x^3 + \dfrac{1}{x^2} + 6$.

2 $f(x) = 4x^{\frac{3}{2}} + 10x^{\frac{1}{2}} + c = 2\sqrt{x}(2x + 5) + c$.

3 $2x^{\frac{7}{2}} + 4x^{\frac{3}{2}} + c$

4 0

5 $\dfrac{45}{4}$.

6 0.1808, overestimate.

CHAPTER 4

C2: Basic trigonometry

Learning objectives

After studying this chapter, you should:

- be familiar with the sine, cosine and tangent functions
- recognise the graphs of the sine, cosine and tangent functions, their symmetries and their periodicity
- be able to solve trigonometrical equations of the form $\sin x = k$, $\cos x = k$ and $\tan x = c$, where k and c are constants and $-1 \leqslant k \leqslant 1$
- recall and use the sine rule to solve problems
- use the cosine rule to solve problems
- recall and use the formula: area of a triangle $= \frac{1}{2}ab \sin C$.

4.1 Introduction

Trigonometry originally meant the study of triangles, or three points. You have no doubt used trigonometry before, considering mainly right-angled triangles. It does, however, have a wide variety of applications including surveying, navigation and engineering. The sine, cosine and related functions can model the behaviour of natural phenomena such as sound, music and many aspects of our physical environment including planetary motion.

It is usual to use variables such as the Greek letter θ (pronounced 'theta') or x for an angle, although any variable can be used.

4.2 The sine function – an informal approach

You will recall that $\sin \theta = \dfrac{\text{side opposite } \theta}{\text{hypotenuse}} = \dfrac{o}{h}$ in a right-angled triangle where the angle θ is less than 90°.

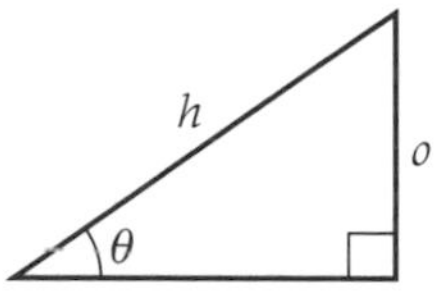

From a calculator in degree mode, $\sin 140° = 0.642\ldots$, $\sin(-55)° = -0.819\ldots$, $\sin 370° = 0.173\ldots$, etc.
It is possible to calculate the sine of any angle and so a definition of $\sin \theta$ is required that is valid for any angle.

The angle can be greater than 360° or less than zero. Try it and see – any angle!

If you have a graphics calculator and you plot $y = \sin x$ in degree mode between -360 and 360 degrees you will get:

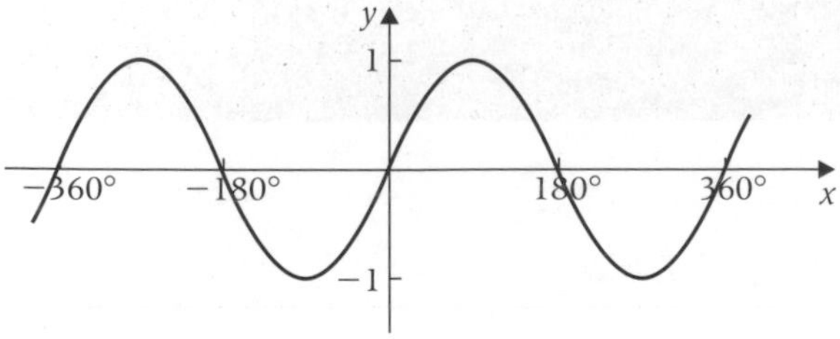

> This is not surprising because the word **sine** comes from the Latin word 'sinus' meaning a curve or bay.

Next a diversion to the funfair, where you are going to take a ride on a special type of Ferris Wheel. You have to enter from a platform that is level with the horizontal diameter of the wheel. You travel in an anticlockwise direction and consider your distance from the platform as the angle you turn through varies.

> Imagine that you are very small so that you are always on the circumference of the wheel.

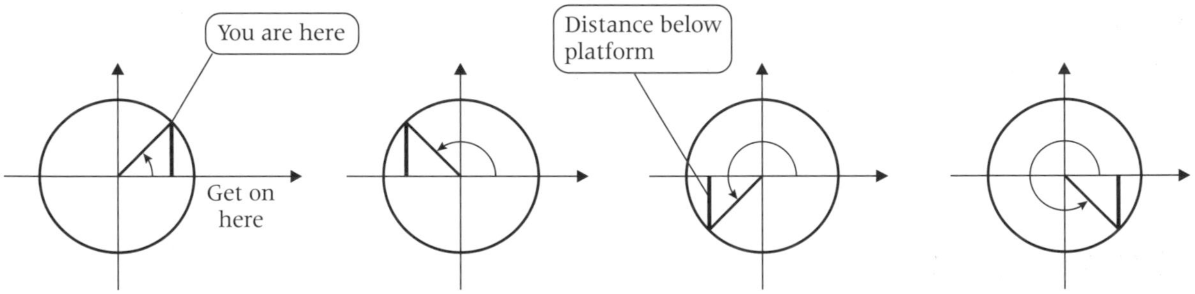

(a) Travel in an anticlockwise direction. Here you have turned through about 50°

(b) Here you have travelled through about 140° and you are still above the platform.

(c) Here the angle is about 240° and you are below the platform.

(d) About 330° travelled. No need to get off after one revolution (360°). You can go round as often as you like.

Various methods can be used to find your position, such as estimation or scale drawings. You could say that when you are at the maximum distance above the platform, then this is one unit.

This distance is equal to the radius of the wheel.

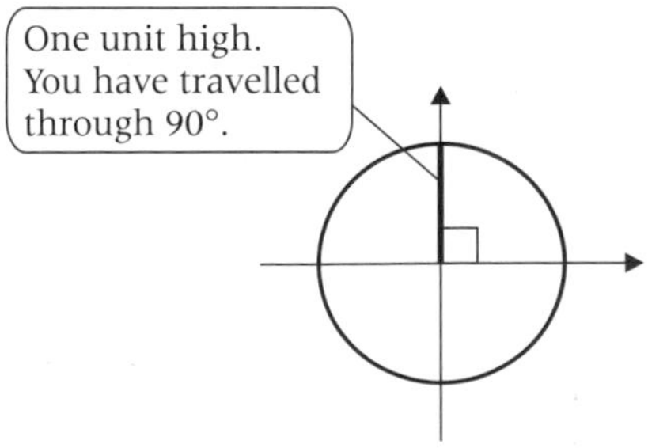

Worked example 4.1

How far above or below the platform are you in each of the diagrams above?

Solution

The sine of the angle that you have travelled through gives your height above or below the platform.

Let y be the vertical distance in each particular case.

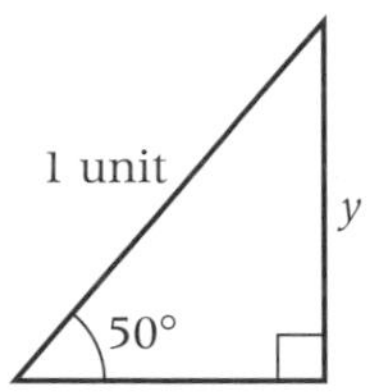

(a) You can probably see that y can be found using

$$\sin \theta = \frac{\text{side opposite } \theta}{\text{hypotenuse}}$$

Since the hypotenuse is one unit long, then $y = \sin 50° = 0.766$ (three decimal places). This is your distance above the platform.

0.766 of a unit.

(b) You do not really need to draw a right-angled triangle for each particular case.

Since $y = \sin 140° = 0.623$ units (three decimal places), this is your distance above the platform.

(c) In this case, $y = \sin 240° = -0.866$ units (three decimal places). The negative sign indicates 'below'.

(d) $y = \sin 330° = -0.5$ units. Again below the platform.

If you write sin 50°, sin 140°, sin 240° then these are exact values, just as $\sqrt{2}$ is exact.

Generally, these values are rounded when you work them out on a calculator. A few angles do have a sine that has an exact decimal form. For example, $\sin 330° = -0.5$ exactly.

4.3 The cosine function – an informal approach

You are still at the funfair, riding on the Ferris Wheel. The wheel has a radius of 1 unit.
You travel in an anticlockwise direction and consider your distance from the **vertical diameter**.
Call this distance x.
To the right of the vertical diameter is positive, to the left is negative.

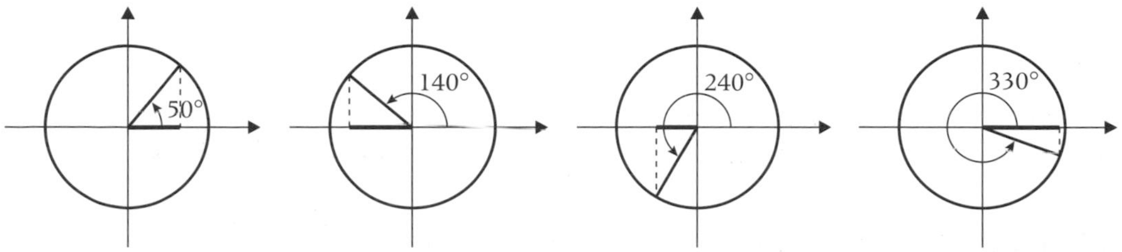

The **cosine function** will give this distance, depending again on the angle that you have turned through.

The sine function and the cosine function are closely related.

Worked example 4.2

Work out your distance from the vertical diameter for each of the above (as a fraction of 1 unit):

Solution

(a) $\cos 50° = 0.643$ units (three decimal places)

Positive, so to the right.

(b) $\cos 140° = -0.766$ units (three decimal places)

Negative, so to the left.

(c) $\cos 240° = -0.5$ units

(d) $\cos 330° = 0.866$ units (three decimal places)

4.4 Definitions of the sine and cosine functions

The sine and cosine functions are called **circular functions** for the following reason.

Take a circle with radius one unit, centred at the origin with the x-axis and the y-axis positioned in the usual way. This is called a **unit circle**.

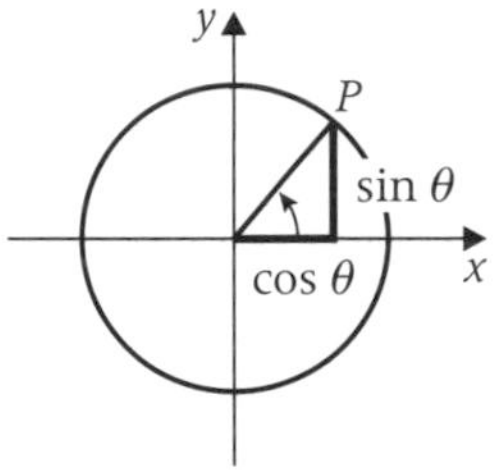

Positive angles are measured in an **anticlockwise** direction.

Negative angles are measured in a **clockwise** direction.

Sine

The **sine** of an angle is the **vertical** distance from the x-axis of a point P as it travels around a unit circle.

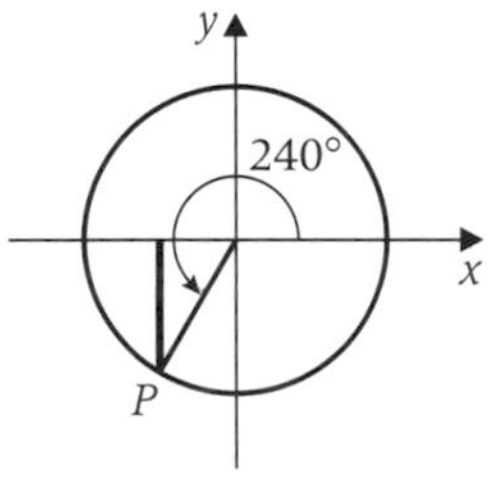

It can be defined as the y-coordinate of P.

Locate a journey of 240°. Does this look right?

For example, $\sin 240° = -0.866$ (three decimal places)

A journey of $-120°$ takes you to the same position as P, so $\sin(-120°) = -0.866$ (three decimal places).

Cosine

The **cosine** of an angle is the **horizontal** distance from the y-axis of a point P as it travels around a unit circle.

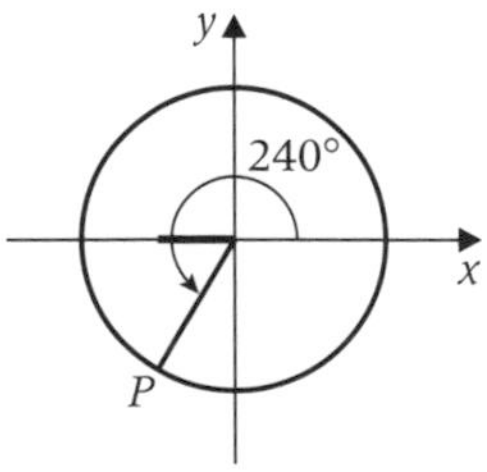

It can be defined as the x-coordinate of P.

For example, $\cos 240° = -0.5$. Also, $\cos(-120°) = -0.5$.

Point P has coordinates $(\cos\theta, \sin\theta)$

The point P is uniquely defined by the coordinates $(\cos\theta, \sin\theta)$

4.5 The graph of the sine function

For this section, you should make good use of your graphics calculator or graph plotting software to plot the graphs.

The graph of $y = \sin\theta$

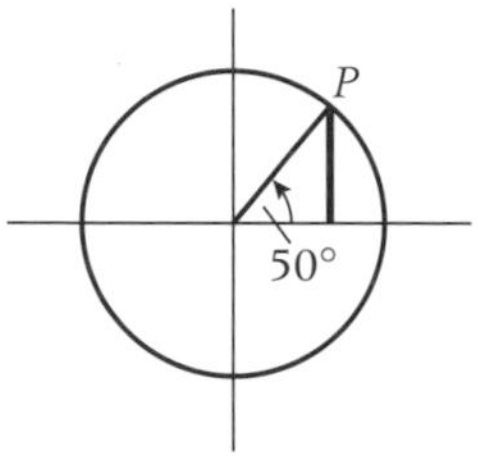

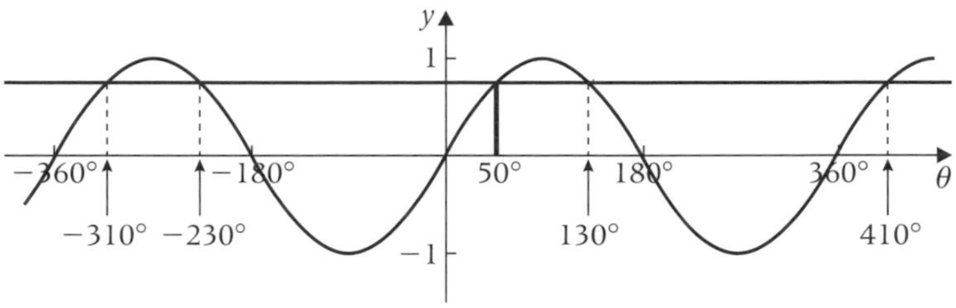

From the graph

$\sin 50° = \sin 130° = \sin 410° = \sin(-230°)$, etc.

Key features of the graph of $y = \sin\theta$

(a) The graph is **periodic**, The graph of $y = \sin\theta$ has a **period** of 360°. It repeats itself in every 360° interval.

(b) The maximum value of $\sin\theta$ is 1.
The minimum value of $\sin\theta$ is -1.

(c) The graph has rotational symmetry, of order 2 about the origin.

(d) Vertical lines of symmetry occur at $\theta = 90°$, $\theta = 270°$, $\theta = -90°$..., etc.

Just as you repeat your position on the Ferris Wheel every 360° no matter what your starting position.

$\sin(360° + \theta) = \sin\theta$ $\quad$ $\sin(360° - \theta) = -\sin\theta$
$\sin(180° - \theta) = \sin\theta$ $\quad$ $\sin(180° + \theta) = -\sin\theta$
$\sin(-\theta) = -\sin\theta$

From the graph of $y = \sin\theta$ you can see that two solutions of the equation $\sin\theta = \sin\alpha$, in a 360° cycle, are $\theta = \alpha$ and $\theta = 180° - \alpha$.

Since the graph of $y = \sin\theta$ has period 360°, the solutions of $\sin\theta = \sin\alpha$ are

$$\theta = \alpha,\ \alpha + 360°,\ \alpha - 360°,\ \alpha + 720°,\ \alpha - 720°,\ \ldots$$

and $\theta = 180° - \alpha,\ 180° - \alpha + 360°,\ 180° - \alpha - 360°,\ 180° - \alpha + 720°,\ 180° - \alpha - 720°,\ \ldots.$

4.6 The graph of the cosine function

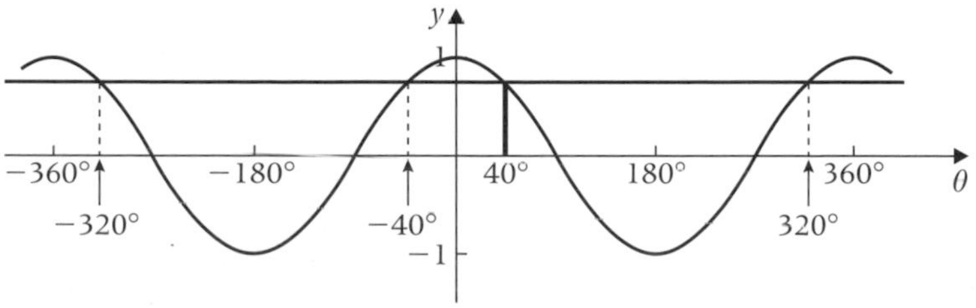

From the graph of $y = \cos\theta$

$$\cos 40° = \cos(-40°) = \cos 320° \ldots, \text{etc.}$$

Key features of the graph of y = cos θ

(a) The graph is periodic, with period 360°.

(b) The maximum value of $\cos\theta$ is 1.
The minimum value of $\cos\theta$ is -1.

(c) The graph of $y = \cos\theta$ has reflective symmetry in the y-axis.

(d) Vertical lines of symmetry occur at $\theta = 0°$, $\theta = 180°$, $\theta = -360°$, …, etc.

> $\cos(360° + \theta) = \cos\theta$ $\quad$ $\cos(180° - \theta) = -\cos\theta$
> $\cos(360° - \theta) = \cos\theta$ $\quad$ $\cos(180° + \theta) = -\cos\theta$
> $\cos(-\theta) = \cos\theta$

> From the graph of $y = \cos\theta$ you can see that two solutions of the equation $\cos\theta = \cos\alpha$, in a 360° cycle, are $\theta = \alpha$ and $\theta = -\alpha$.
>
> Since the graph of $y = \cos\theta$ has period 360°, the solutions of $\cos\theta = \cos\alpha$ are
>
> $\theta = \alpha, \alpha + 360°, \alpha - 360°, \alpha + 720°, \alpha - 720°, \ldots$
>
> and $\theta = -\alpha, -\alpha + 360°, -\alpha - 360°, -\alpha + 720°, -\alpha - 720°, \ldots.$

EXERCISE 4A

Only use your calculator to ***check*** your answers.

1 Make a copy of the unit circle and for each angle state:
(i) whether the sine is positive or negative,
(ii) whether the cosine is positive or negative:
(a) 70°,
(b) 130°,
(c) 300°,
(d) 200°,
(e) $-60°$,
(f) $-100°$.

2 Make a copy of the sketch of the unit circle and the graph of $y = \sin\theta$.

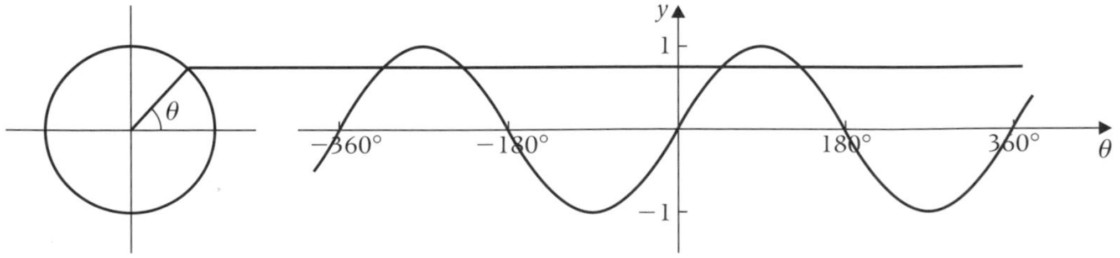

(a) On your sketch, mark the point $(\theta, \sin\theta)$ for each of the values of θ given:

(i) $\theta = 20°$, **(ii)** $\theta = 160°$, **(iii)** $\theta = 50°$,
(iv) $\theta = -200°$, **(v)** $\theta = 230°$, **(vi)** $\theta = 140°$,
(vii) $\theta = -340°$, **(viii)** $\theta = 130°$, **(ix)** $\theta = -20°$,
(x) $\theta = -230°$, **(xi)** $\theta = 200°$, **(xii)** $\theta = -130°$.

(b) Identify values of θ above for which:

(i) $\sin\theta = \sin 20°$, **(ii)** $\sin\theta = \sin 50°$.

3 Make a copy of the sketch of the graph of $y = \cos\theta$.

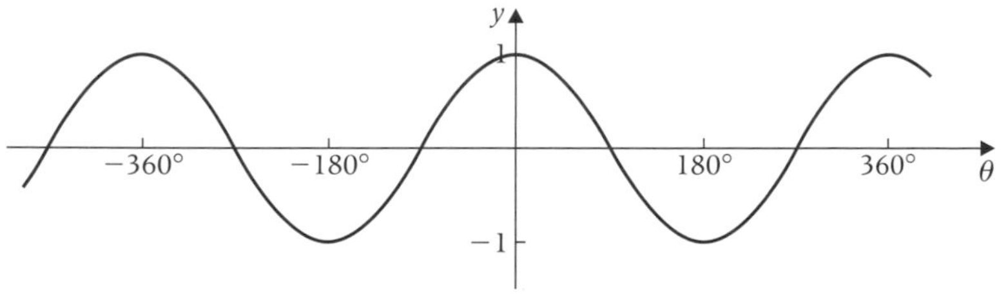

(a) On your sketch, mark the point $(\theta, \cos\theta)$ for each of the values of θ given:

(i) $\theta = 50°$, **(ii)** $\theta = 40°$, **(iii)** $\theta = -50°$,
(iv) $\theta = 310°$, **(v)** $\theta = 220°$, **(vi)** $\theta = 140°$,
(vii) $\theta = 320°$, **(viii)** $\theta = -40°$, **(ix)** $\theta = 230°$,
(x) $\theta = -410°$, **(xi)** $\theta = 400°$, **(xii)** $\theta = -220°$.

(b) Identify values of θ above for which

(i) $\cos\theta = \cos 50°$, **(ii)** $\cos\theta = \cos 140°$.

4.7 Finding angles with the same sine or cosine

You can work out on your calculator that $\sin 30° = 0.5$. There are, however, other angles that have the same sine as 30°, as you have already seen. For example, $\sin 150° = 0.5$, $\sin 390° = 0.5$, etc.

If no interval were specified, there would be an infinite number of solutions.

Sometimes it is necessary to find all of the angles within a given interval that have the same sine or cosine.

Worked example 4.3

Find all the values of θ in the interval $-360° \leqslant \theta \leqslant 360°$ for which $\sin\theta = \sin 60°$.

θ can take any value between $-360°$ to $360°$ inclusive.

Solution

Sketch a graph of $y = \sin\theta$.

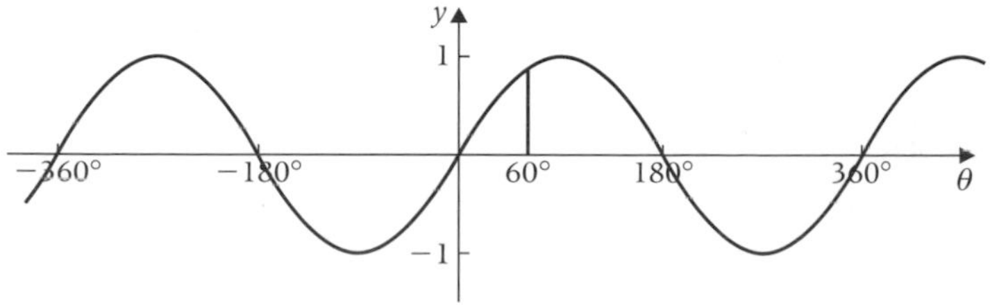

It is now quite an easy task to use the symmetry and periodicity properties of the function to locate the other angles.

The angle 60° is located on the θ-axis 60° to the right of the origin.

You could also say that 60° is 30° to the left of 90° so another angle is 30° to the right of 90°.

Since $\theta = 90°$ is a vertical line of symmetry, another angle with the same sine as 60° in the given interval is $(180 - 60)° = 120°$. Since the graph is periodic with period 360° there are two more solutions in the interval $-360° \leqslant \theta \leqslant 360°$, namely $(60 - 360)°$ and $(120 - 360)°$.

Check using a calculator that the sine of each angle does equal sin 60°

The values of θ in the interval $-360° \leqslant \theta \leqslant 360°$ for which $\sin\theta = \sin 60°$ are $\theta = 60°, 120°, -300°$ and $-240°$.

Worked example 4.4

Find all values of θ within the interval $-360° \leqslant \theta \leqslant 360°$ for which $\cos\theta = \cos 130°$.

Solution

Sketch the graph of $y = \cos\theta$.

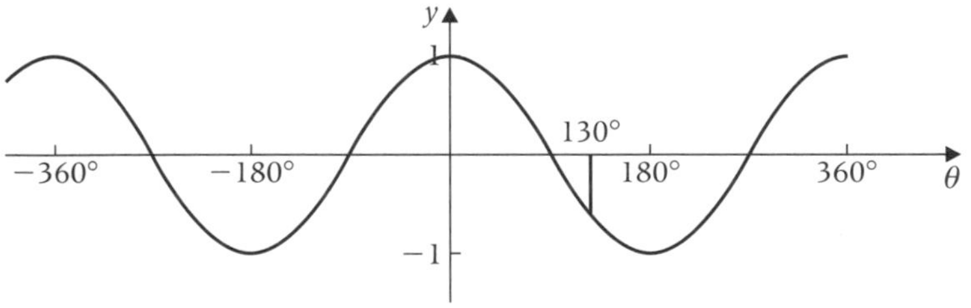

One solution is $\theta = 130°$.

Locate 130° on the θ-axis. This is $(180 - 50)°$

Since $\theta = 180°$ is a vertical line of symmetry, another angle is $\theta = (180 + 50)° = 230°$.

You could have used the fact that $\theta = 0°$ is a vertical line of symmetry to get the other angle as $(0 - 130)° = -130°$ and then use $(-130 + 360)°$ to get the $\theta = 230°$ solution.

Since the graph is periodic with period 360° there are two more solutions in the interval $-360° \leqslant \theta \leqslant 360°$, namely $(130 - 360)°$ and $(230 - 360)°$.

The values of θ in the interval $-360° \leqslant \theta \leqslant 360°$ for which $\cos\theta = \cos 130°$ are $\theta = 130°, 230°, -230°$ and $-130°$.

EXERCISE 4B

1 Find all possible angles θ for which

(a) $\sin\theta = \sin 50°$, in the interval $-360° \leqslant \theta \leqslant 180°$,

(b) $\sin\theta = \sin(-20°)$, in the interval $-180° \leqslant \theta \leqslant 180°$,

(c) $\sin\theta = \sin 90°$, in the interval $-360° \leqslant \theta \leqslant 360°$,

(d) $\sin\theta = \sin 0°$, in the interval $-360° < \theta < 360°$,

(e) $\sin\theta = \sin 170°$, in the interval $-360° \leqslant \theta \leqslant 180°$,

(f) $\sin\theta = \sin(-130°)$, in the interval $-180° \leqslant \theta \leqslant 360°$.

2 Find all possible angles θ for which

(a) $\cos\theta = \cos(-40°)$, in the interval $-180° \leqslant \theta \leqslant 180°$,

(b) $\cos\theta = \cos 70°$, in the interval $0° \leqslant \theta \leqslant 360°$,

(c) $\cos\theta = \cos 110°$, in the interval $0° \leqslant \theta \leqslant 360°$,

(d) $\cos\theta = \cos 0°$, in the interval $-360° \leqslant \theta < 360°$,

(e) $\cos\theta = \cos 90°$, in the interval $-90° < \theta \leqslant 360°$,

(f) $\cos\theta = \cos 20°$, in the interval $-180° \leqslant \theta \leqslant 180°$.

3 Use the graphs of $y = \cos\theta$ and $y = \sin\theta$ to state the smallest **positive** angle θ for which

(a) $\cos(-35°) = \cos\theta$, **(b)** $\cos 380° = \cos\theta$,

(c) $\sin -180° = \sin\theta$, **(d)** $\sin(-45°) = \sin\theta$,

(e) $\sin 310° = \sin\theta$, **(f)** $\cos(-225°) = \cos\theta$.

4.8 The tangent function

Draw a unit circle with a vertical tangent on the right of the circle. The radius travels through an angle θ in an anticlockwise (positive) direction starting at the x-axis. Think of the radius as a double-ended torch with a ray of light shining from both ends. Think of the tangent as a screen that the light from the torch hits in different places, depending on the angle that it travels through.

What is the distance of the beam of light, as it hits the screen, from the x-axis?

Euclid said that to draw a tangent to a circle, you should draw a line at right angles to a diameter so that no other line can be inserted between the line drawn and the circle. In other words, the tangent line touches the circle at one point.

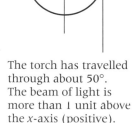

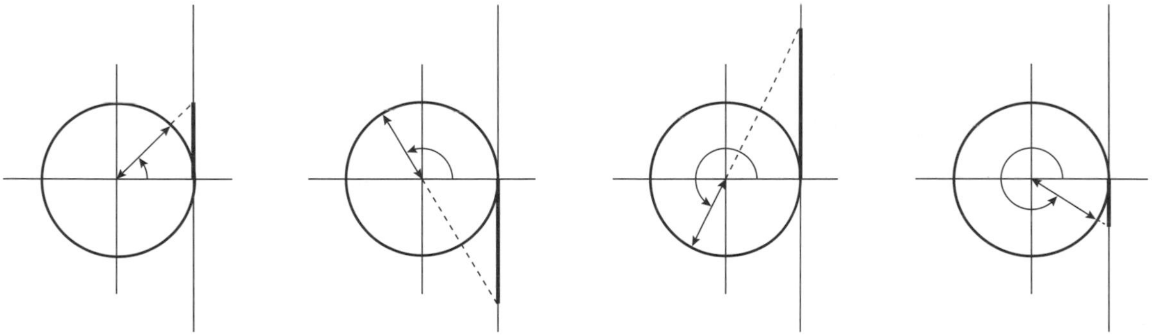

The torch has travelled through about 50°. The beam of light is more than 1 unit above the x-axis (positive).

The torch has travelled through about 110°. The beam of light is about 2.5 units below the x-axis (negative).

The torch has travelled through about 240°. The beam of light is about 2 units above the x-axis (positive).

The torch has travelled through about 330°. The beam of light is about half a unit below the x-axis (negative).

The exact distance of the beam of light from the x-axis is found using the tangent function, again with positive values for 'above' and negative values for 'below'. Try working out the tangent for these four angles using a calculator.

A more formal definition of the tangent function is given later.

Use your calculator to work out the tangent of any angle between 0° and 360°, considering the idea outlined above. Then try angles outside this interval.

'Tangent' is abbreviated to 'tan'.

If you tried to work out the tangent of 90° or 270° you would receive an 'error' message. It is obvious that, when the torch is at 90° or 270° (180 + 90)°, for example, the beam of light will not hit the screen and so for these angles, the tangent is not defined.

Also for (360 + 90)° (540 + 90)°, −90° (−90 − 180)° etc.

Below is a table of values of θ and approximate values for $\tan\theta$ in the region of 90°

θ	89.00°	89.10°	89.20°	89.30°	89.40°	89.50°	89.60°	89.70°	89.80°	89.90°	89.91°	89.92°	89.93°	89.94°	89.95°	89.96°	89.97°	89.98°	89.99°
$\tan\theta$	57.3	63.7	71.6	81.8	95.5	115	143	191	286	573	637	716	819	955	1146	1432	1910	2865	5730
θ	90.01°	90.02°	90.03°	90.04°	90.05°	90.06°	90.07°	90.08°	90.09°	90.10°	90.20°	90.30°	90.40°	90.50°	90.60°	90.70°	90.80°	90.90°	91.00°
$\tan\theta$	−5730	−2865	−1910	−1432	−1146	−955	−819	−716	−637	−573	−286	−191	−143	−115	−95.5	−81.8	−71.6	−63.7	−57.3

Notice how quickly $\tan\theta$ becomes very large and positive as the angle θ approaches 90° and how large and negative $\tan\theta$ is when θ is just over 90°.

4.9 The graph of the tangent function

The graph of $y = \tan\theta$

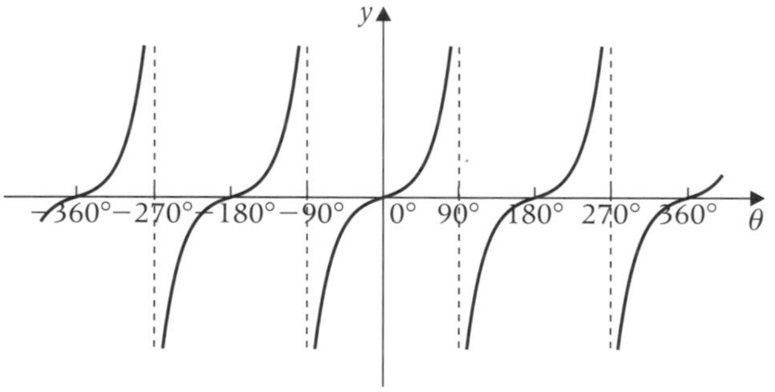

The dotted vertical lines are called asymptotes. You can see from the table that $\tan\theta$ becomes 'large positive' or 'large negative' close to either side of the asymptotes. The value of $\tan\theta$ 'tends towards plus or minus infinity' or is said to be undefined wherever there is an asymptote.

Key features of the graph of $y = \tan\theta$

(a) The graph is periodic with period 180°.

(b) $\tan\theta$ can take any value, it is not restricted in the same way that $\sin\theta$ and $\cos\theta$ are.

(c) The graph has rotational symmetry of order 2 about the origin.

$\tan(180° + \theta) = \tan\theta$
$\tan(-\theta) = -\tan\theta$

$\tan(360° + \theta) = \tan\theta$ $\quad$ $\tan(360° - \theta) = -\tan\theta$
$\tan(180° + \theta) = \tan\theta$ $\quad$ $\tan(180° - \theta) = -\tan\theta$
$\tan(-\theta) = -\tan\theta$

Since the graph of $y = \tan\theta$ has period of 180°, the solutions of $\tan\theta = \tan\alpha$ are $\theta = \alpha, \alpha + 180°, \alpha - 180°, \alpha + 360°, \alpha - 360°, \alpha + 540°, \alpha - 540°, \ldots$.

Worked example 4.5

Find all angles in the interval $-270° \leqslant \theta \leqslant 360°$ for which $\tan\theta = \tan(-35°)$.

Solution

The graph of $y = \tan\theta$ repeats itself every 180°, so once you have sketched $\tan\theta$ and found one solution, it is quite a simple task to identify all of the angles.

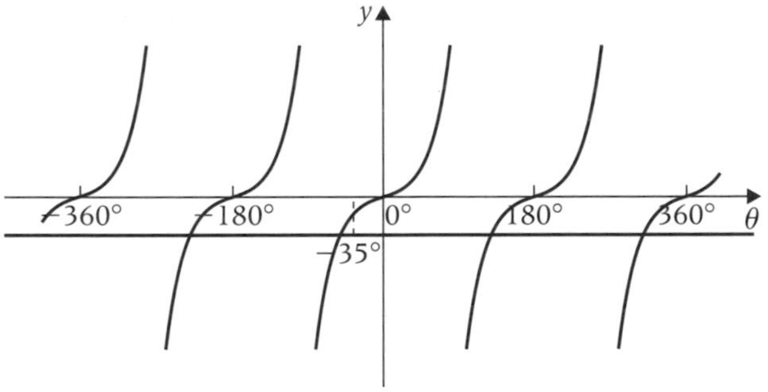

One solution is $\theta = -35°$.

Since the graph is periodic with period 180°, the other angles in the given interval are $\theta = (-35 + 180)°, (-35 - 180)°$ and $(-35 + 360)°$.

The values of θ in the interval $-270° \leqslant \theta \leqslant 360°$ for which $\tan\theta = \tan(-35°)$ are $\theta = -35°, 145°, -215°$ and $325°$.

EXERCISE 4C

1 Find all angles θ for which

(a) $\tan\theta = \tan 50°$, in the interval $-180° \leqslant \theta \leqslant 360°$,

(b) $\tan\theta = \tan(-20)°$, in the interval $0° \leqslant \theta \leqslant 360°$,

(c) $\tan\theta = \tan 120°$, in the interval $-360° \leqslant \theta \leqslant 180°$,

(d) $\tan\theta = \tan 20°$, in the interval $-360° \leqslant \theta \leqslant 0°$,

(e) $\tan\theta = \tan 230°$, in the interval $-180° \leqslant \theta \leqslant 360°$,

(f) $\tan\theta = \tan(-60)°$, in the interval $-360° \leqslant \theta \leqslant 180°$.

2 State the smallest **positive** angle θ for which

(a) $\tan(-35°) = \tan\theta$, **(b)** $\tan 380° = \tan\theta$,

(c) $\tan(-280°) = \tan\theta$, **(d)** $\tan(-45°) = \tan\theta$,

(e) $\tan 310° = \tan\theta$, **(f)** $\tan(-225°) = \tan\theta$.

4.10 Solving simple trigonometrical equations

Suppose you are told that the cosine of an angle θ is 0.7. How could a value of the angle θ be found that would give 0.7 as its cosine?

The equation is $\cos\theta = 0.7$ and

θ is 'the angle whose cosine is 0.7'.

Using the idea of a flow diagram

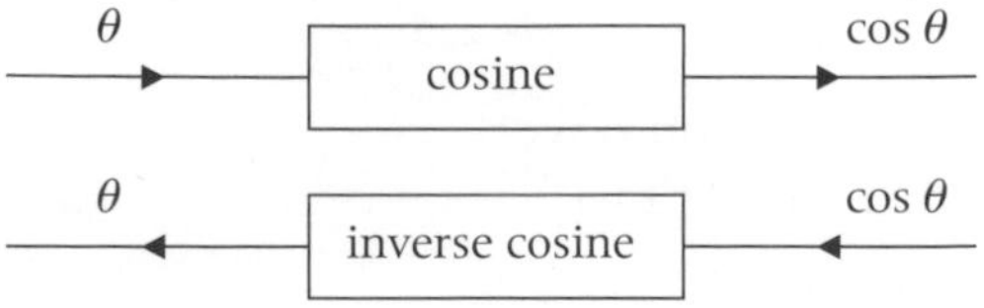

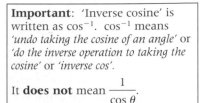

Important: 'Inverse cosine' is written as $\cos^{-1}$. $\cos^{-1}$ means *'undo taking the cosine of an angle'* or *'do the inverse operation to taking the cosine'* or *'inverse cos'*.

It **does not** mean $\dfrac{1}{\cos\theta}$.

While a calculator will give the cosine of any angle, reversing the process does not give all of the angles with a given cosine.

A calculator considers a specific interval and returns one angle. This is called the **principal value**.

Because there is an infinite number of angles having the same cosine!

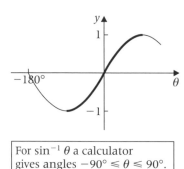

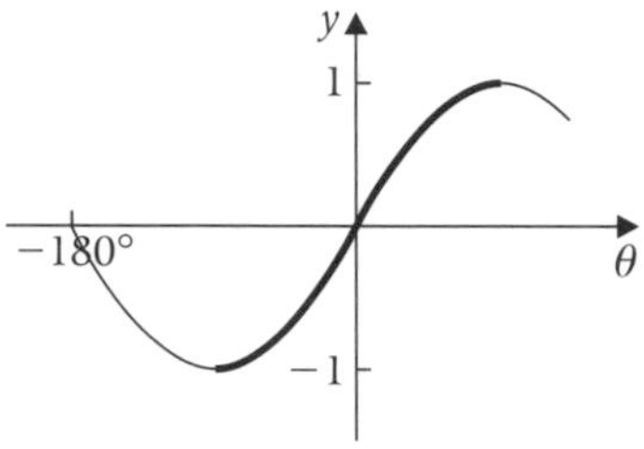

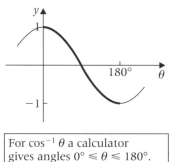

For $\sin^{-1}\theta$ a calculator gives angles $-90° \leqslant \theta \leqslant 90°$.

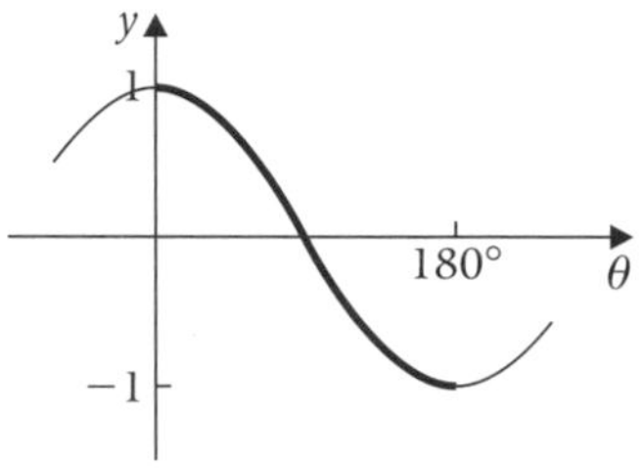

For $\cos^{-1}\theta$ a calculator gives angles $0° \leqslant \theta \leqslant 180°$.

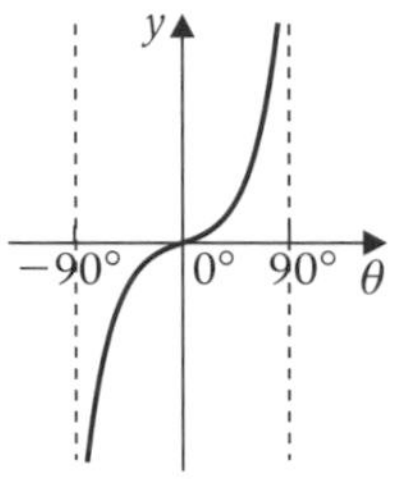

For $\tan^{-1}\theta$ a calculator gives angles $-90° \leqslant \theta \leqslant 90°$.

To solve equations of the form $\sin x = k$, $\cos x = k$, where $-1 \leqslant k \leqslant 1$:

Step 1: Use a calculator to find one solution.

Step 2: Sketch the graphs of $y = \sin x$ (or $y = \cos x$) and $y = k$ and use the symmetry of the graph to find a second solution within a 360° cycle.

Step 3: Use the fact that $y = \sin x$ and $y = \cos x$ are periodic with period 360° to find all other solutions in the given interval by adding and subtracting multiples of 360° to the two angles found in steps **1** and **2**.

To solve equations of the form $\tan x = c$, where c is a constant:

Step 1: Use a calculator to find one solution α, ($\alpha = \tan^{-1}c$).

Step 2: Use the fact that $y = \tan x$ is periodic, with period 180° to find all other solutions in the given interval ($x = \alpha$, $\alpha \pm 180°$, $\alpha \pm 360°$, $\alpha \pm 540°$, $\alpha \pm 720°$, etc.)

Worked example 4.6

Solve the equation $\cos\theta = 0.5$, in the interval $0° \leqslant \theta \leqslant 360°$.

Which angles within the given interval have a cosine of 0.5?

Solution

Using the 'inverse cos' key on a calculator gives $\theta = 60°$.

One angle that has a cosine of 0.5.

So $\cos 60° = 0.5$.

Other solutions within the interval can be found by considering the graph of $y = \cos\theta$ and this should now be a familiar process.

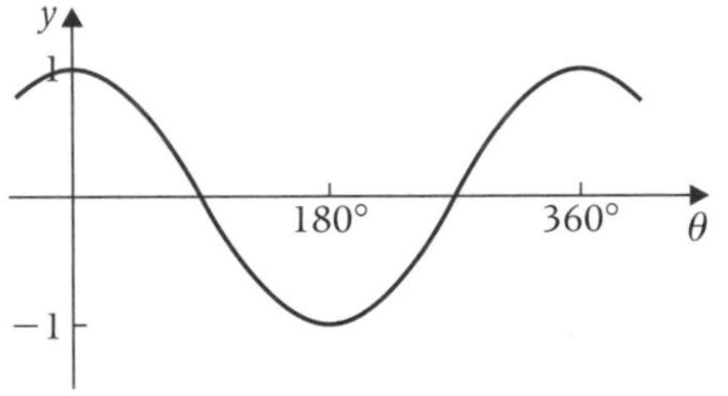

Using the symmetry properties of the graph, $(360 - 60)°$ is also a solution. There are two solutions within the given interval and these are $\theta = 60°$ and $\theta = 300°$.

Check using your calculator that $\cos 300° = 0.5$.

For a larger interval, for example, $-720° \leqslant \theta \leqslant 720°$, you would use the fact that the graph of $y = \cos\theta$ has period 360° to get the extra solutions: $\theta = 420°, -300°, -60°$ and $-420°$.

Worked example 4.7

Solve the equation $5\sin x = 1$, in the interval $-360° \leqslant x \leqslant 0°$ giving your answers to the nearest 0.1°.

Solution

By now you should be quite familiar with the graph of $\sin x$ so rearrange to give

$$\sin x = \frac{1}{5} \quad \text{or} \quad \sin x = 0.2$$

If you input $\sin^{-1}\left(\frac{1}{5}\right)$ remember the brackets.

Solving $\sin x = 0.2$ using a calculator gives $x = 11.536\ldots°$ which is outside the given interval since $-360° \leqslant x \leqslant 0°$.

Sketch the graphs of $y = \sin x$ and $y = 0.2$.

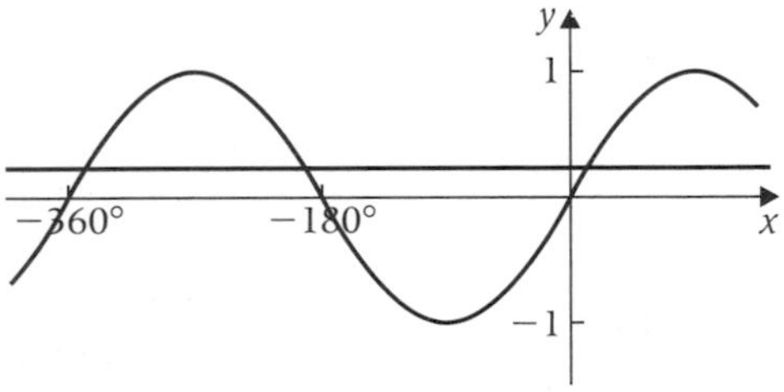

If you have a graphics calculator you can use the zoom and trace facilities to find approximate values for the solutions.

The graph shows that x can be found using the symmetry and periodic properties of $\sin x$ and so in this case

either $\quad x = -180° - 11.536\ldots° \quad \Rightarrow x = -191.536\ldots°$

or $\quad x = 11.536\ldots° - 360° \quad \Rightarrow x = -348.463\ldots°$

Always use your full calculator display to evaluate inverse cosine and inverse sine or you may lose accuracy.

There are two solutions, $x = -191.5°$

$x = -348.5°$ both to the nearest 0.1°

Worked example 4.8

Solve the equation $2 \tan \theta + 11 = 0$ in the interval $-720° \leqslant \theta \leqslant 720°$ giving your answers to one decimal place.

Solution

Rearrange $2 \tan \theta + 11 = 0$ to give $\tan \theta = -5.5$.
Solving $\tan \theta = -5.5$ using a calculator gives $\theta = -79.695\ldots°$.
Sketch the graphs of $y = \tan \theta$ and $y = -5.5$.

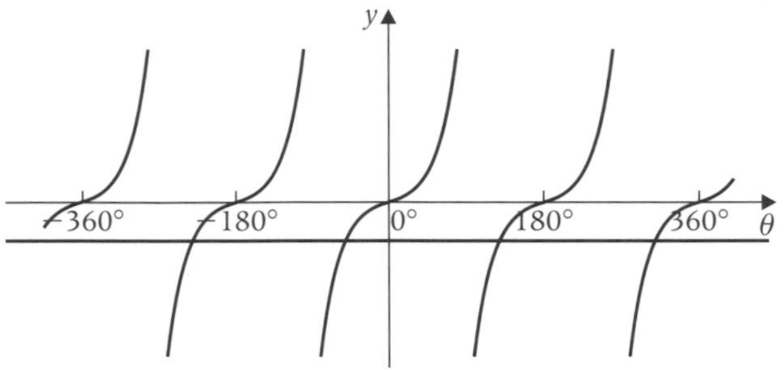

Using the fact that $y = \tan \theta$ is periodic with period 180°, the other angles in the given interval are $\theta = -79.695\ldots° + 180°$, $-79.695\ldots° - 180°$, $-79.695\ldots° + 360°$, $-79.695\ldots° - 360°$, $-79.695\ldots° + 540°$, $-79.695\ldots° - 540°$, $-79.695\ldots° + 720°$, $-79.695\ldots° - 720°$.

The final expression is outside the given interval. In the exam you will not be penalised for giving extra answers outside the interval.

The eight solutions of the equation $2 \tan \theta + 11 = 0$ in the interval $-720° \leqslant \theta \leqslant 720°$, to one decimal place are:

$$\theta = -79.7°, 100.3°, -259.7°, 280.3°, -439.7°, 460.3°, -619.7° \text{ and } 640.3°.$$

Checking the solution 100.3° gives $2 \times \tan 100.3° + 11 = -0.005\,28\ldots$ which is not exactly 0. This is not surprising as θ has been rounded to one decimal place and $y = \tan \theta$ has a 'steep' gradient at the points of intersection with the line $y = -5.5$.

It is worth noting that:

> The equation $\sin x = k$, where $-1 \leqslant k \leqslant 1$, normally has two solutions for each 360° cycle within the given interval.

> The equation $\cos x = k$, where $-1 \leqslant k \leqslant 1$, normally has two solutions for each 360° cycle within the given interval.

Exceptions can come from stationary points on the graphs or end points of intervals, e.g. in the interval $0° \leqslant x \leqslant 360°$, $\sin x = 1$ has one solution, $\cos x = -1$ has one solution, $\sin x = 0$ has three solutions.

> The equation $\tan x = c$, where c is a constant, has two solutions for each 360° cycle within the given interval.

EXERCISE 4D

1 Solve the following equations for the given interval, giving answers to the nearest 0.1°:

(a) $\sin\theta = 0.25$, in the interval $0° \leqslant \theta \leqslant 360°$,

(b) $\cos x = \frac{4}{5}$, in the interval $-360° \leqslant x \leqslant 90°$,

(c) $\sin x = 0.4$, in the interval $-360° \leqslant x \leqslant 360°$,

(d) $\cos\theta = -0.7$, in the interval $-360° \leqslant \theta \leqslant 360°$,

(e) $\cos x = 0.3$, in the interval $-360° \leqslant x \leqslant 0°$,

(f) $\sin\theta = -0.5$, in the interval $0° \leqslant \theta \leqslant 360°$,

(g) $\sin x = 0$, in the interval $0° < x < 360°$,

(h) $\cos\theta = 1$, in the interval $-360° \leqslant \theta \leqslant 360°$.

2 Find the first four positive solutions of the following equations, giving your answers to the nearest 0.1°:

(a) $10\sin\theta = 5$,
(b) $7\sin x + 4 = 0$,

(c) $5\cos x + 2 = 0$
(d) $\frac{\sin\theta}{2} + \frac{3}{10} = \frac{1}{5}$,

(e) $4\cos\theta - \cos\theta = 2$,
(f) $2\sin x - \frac{1}{2}\sin x = 1$.

3 Solve the following equations, giving your answers to the nearest 0.1°:

(a) $\tan x = 0.5$, in the interval $0° \leqslant x \leqslant 360°$,

(b) $4\tan\theta = 4$, in the interval $0° \leqslant \theta \leqslant 360°$,

(c) $\tan x = 0$, in the interval $0° \leqslant x \leqslant 360°$,

(d) $6\tan x = 1$, in the interval $0° \leqslant x \leqslant 180°$,

(e) $10\tan\theta + 3 = 0$, in the interval $-180° \leqslant \theta \leqslant 180°$.

4 Solve the equation $p\tan x + q = 0$, giving your answers in the given interval to one decimal place for the following values of p and q:

(a) $p = 2, q = -3$, in the interval $0° \leqslant x \leqslant 360°$,

(b) $p = 5, q = 2$, in the interval $-180° \leqslant x \leqslant 720°$,

(c) $p = 8, q = 7$, in the interval $360° \leqslant x \leqslant 720°$,

(d) $p = -4, q = 3$, in the interval $-900° \leqslant x \leqslant 0°$.

When the answer is an exact whole number of degrees, such as 30°, there is no need to write 30.0°.

The idea of asking for answers to the nearest 0.1°, which will appear in exam questions, is to prevent you from leaving your answer as a full calculator display.

4

4.11 Sine rule

You will all have used trigonometry to calculate unknown lengths and angles in a right-angled triangle. Some of you will have also used the sine rule and cosine rule to solve triangles which are not right-angled. Here we shall revise the work but at the same time introduce an extra part to the sine rule which is useful when a question involves the **circumcircle** of a triangle.

It is always possible to draw a circle through the vertices of a triangle. Such a circle is called the **circumcircle** of the triangle.

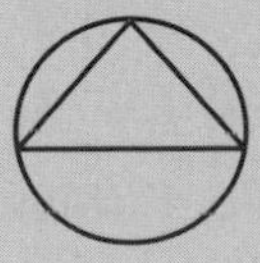

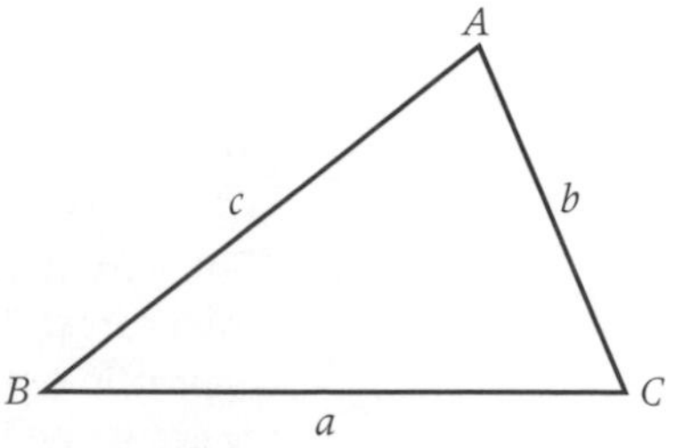

The diagram above shows a general triangle ABC with sides of length a, b and c. The length of the side opposite angle A is a, b is opposite angle B and c is opposite angle C.

To prove the sine rule we shall use the fact that the angle subtended by an arc at the centre of a circle is twice the angle subtended at any point on the circumference.

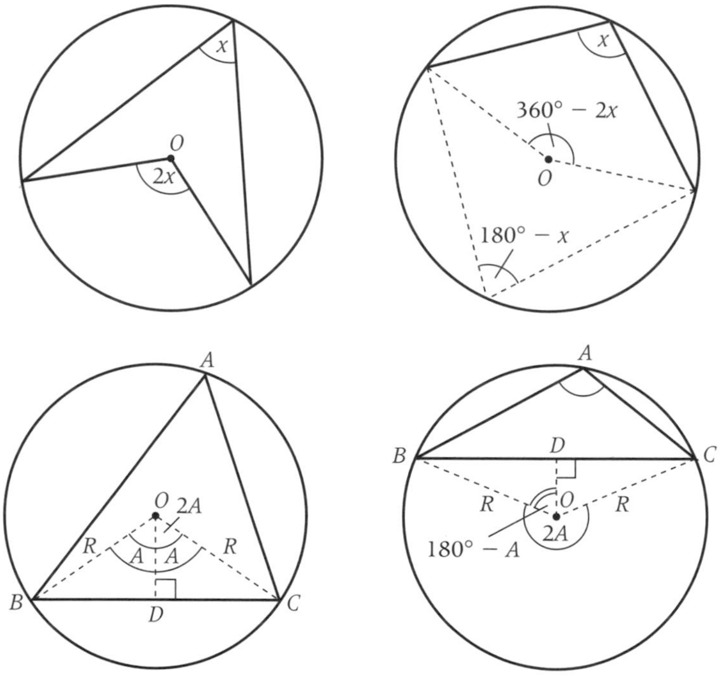

Angle A acute

Angle A obtuse

We shall prove the sine rule for a general triangle ABC by considering the two cases, A acute and A obtuse as shown in the diagrams above.
Point O is the centre and R is the radius of the circumcircle of triangle ABC.

If $A = 90°$, the results are consistent with the trigonometry of a right-angled triangle.

Since OBC is an isosceles triangle the perpendicular, OD, bisects both BC and angle BOC.

Therefore $DC = \frac{1}{2}a$.

When angle A is acute, angle $DOC = A$ and when angle A is obtuse, angle $DOC = 180° - A$.

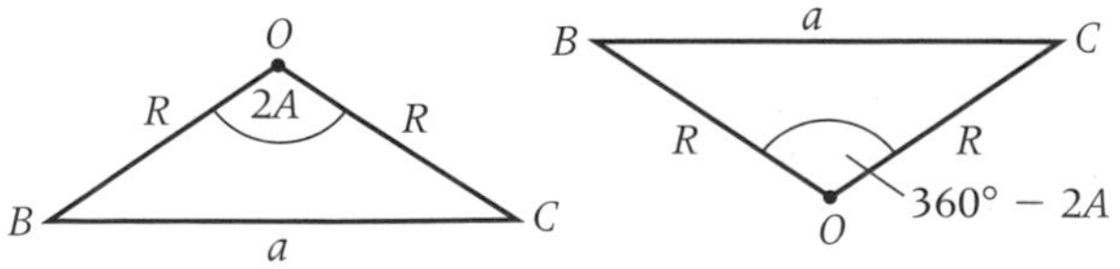

Using the fact, from section 4.5, that $\sin(180° - A) = \sin A$, you can see from triangle ODC in each diagram that

$$\sin A = \frac{DC}{OC} = \frac{\frac{1}{2}a}{R} = \frac{a}{2R} \Rightarrow 2R = \frac{a}{\sin A}.$$

Using a similar approach for triangle OAC and then for triangle OAB we can deduce that

$$2R = \frac{b}{\sin B} \text{ and } 2R = \frac{c}{\sin C}.$$

The lettering of the triangle does not matter. For triangle DEF, the sine rule is

$$\frac{d}{\sin D} = \frac{e}{\sin E} = \frac{f}{\sin F} = 2R$$

Hence we have the 'extended' sine rule for triangle ABC

$$\frac{a}{\sin A} = \frac{b}{\sin B} = \frac{c}{\sin C} = 2R,$$

where R is the radius of the circumcircle of triangle ABC.

$$\frac{a}{\sin A} = \frac{b}{\sin B} \text{ or } \frac{a}{\sin A} = 2R$$
$$\text{or } \frac{a}{\sin A} = \frac{c}{\sin C} \text{ or } \frac{b}{\sin B} = 2R$$
$$\text{or } \frac{b}{\sin B} = \frac{c}{\sin C} \text{ or } \frac{c}{\sin C} = 2R$$

This sine rule is really a combination of 6 formulae. When solving problems choose the ratio which is completely known and the one which involves the length or angle you are finding.

Worked example 4.9

In the diagram, the three points A, B and C, lie on the circumference of a circle.

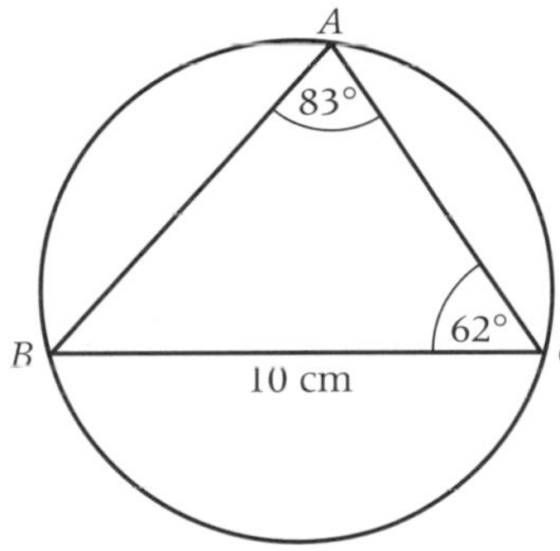

Calculate:

(a) the length of AC,

(b) the circumference of the circle.

(Give your answers to three significant figures.)

Solution

Angle $B = 180° - (83° + 62°) = 35°$.

Sine rule $\frac{a \checkmark}{\sin A \checkmark} = \frac{b}{\sin B \checkmark} = \frac{c}{\sin C \checkmark} = 2R$

Tick the 'knowns'.

(a) $\frac{b}{\sin B} = \frac{a}{\sin A} \Rightarrow \frac{b}{\sin 35°} = \frac{10}{\sin 83°}$

$$\Rightarrow b = \frac{10 \times \sin 35°}{\sin 83°} = 5.778 \ldots$$

The length of $AC = 5.78$ cm (to 3 sf).

We could use $2R = \frac{b}{\sin B}$ but if possible it is always safer to use given data rather than calculated data.

(b) $2R = \frac{a}{\sin A}$

$$2R = \frac{10}{\sin 83°} = 10.075 \ldots$$

To avoid rounding errors keep the full value on your calculator.

The circumference of the circle $= \pi \times 2R = \pi \times 10.075\ \ldots$
$= 31.65\ldots$

The circumference of the circle = 31.7 cm (to 3 sf).

Write down a more accurate answer before rounding to the required degree of accuracy.

Ambiguous case

It is sometimes possible to find two different triangles which satisfy the given data. This is known as 'the ambiguous case'. Such cases will only arise when the given angle is acute and the unknown side is not opposite the given angle. The next worked example shows how to deal with an ambiguous case.

Worked example 4.10

In triangle ABC, $A = 30°$, $a = 7$ cm and $b = 12$ cm.

Calculate the lengths of the unknown side and the sizes of the unknown angles in the triangle.

Give all your answers to three significant figures.

Solution

Tick the 'knowns'.

Sine rule $\dfrac{a\checkmark}{\sin A\checkmark} = \dfrac{b\checkmark}{\sin B} = \dfrac{c}{\sin C} = 2R$

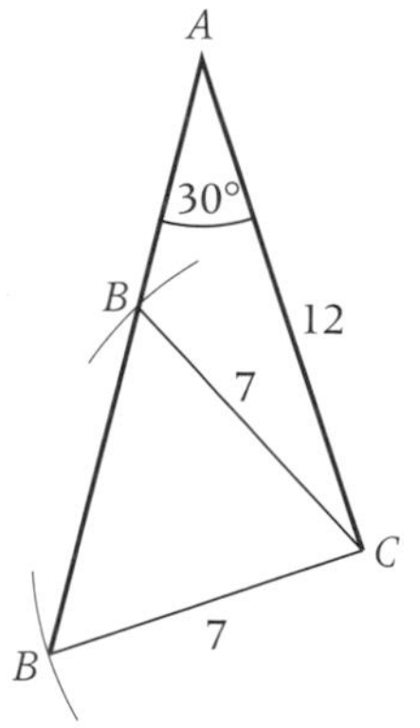

Rearrange $\dfrac{b}{\sin B} = \dfrac{a}{\sin A}$

$\Rightarrow \dfrac{\sin B}{b} = \dfrac{\sin A}{a}$

$\Rightarrow \dfrac{\sin B}{12} = \dfrac{\sin 30°}{7} \Rightarrow \sin B = \dfrac{12 \times \sin 30°}{7} = 0.8571\ \ldots$

$\Rightarrow B = 58.997\ldots°$ **or** $B = (180 - 58.997\ldots)°$

so $B = 59.0°$ **or** $B = 121°$ (to 3 sf).

Case **(i)**, when $B = 58.997\ \ldots°$,
$C = 180° - (30° + 58.997\ \ldots°) = 91.002\ \ldots°$

Using $\dfrac{c}{\sin C} = \dfrac{a}{\sin A}$ leads to $c = \dfrac{7 \times \sin 91.002\ \ldots°}{\sin 30°} = 13.997\ \ldots$

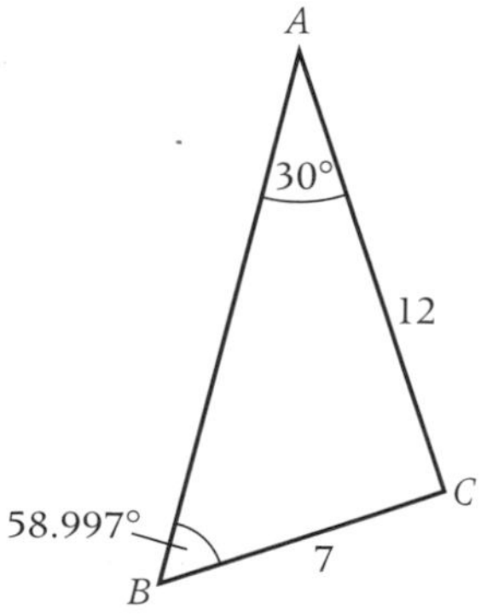

Case **(ii)**, when $B = 121.002\ldots°$,
$C = 180° - (30° + 121.002\ldots°) = 28.997\ldots°$

Using $\dfrac{c}{\sin C} = \dfrac{a}{\sin A}$ leads to $c = \dfrac{7 \times \sin 28.997\ldots°}{\sin 30°} = 6.78675\ldots$

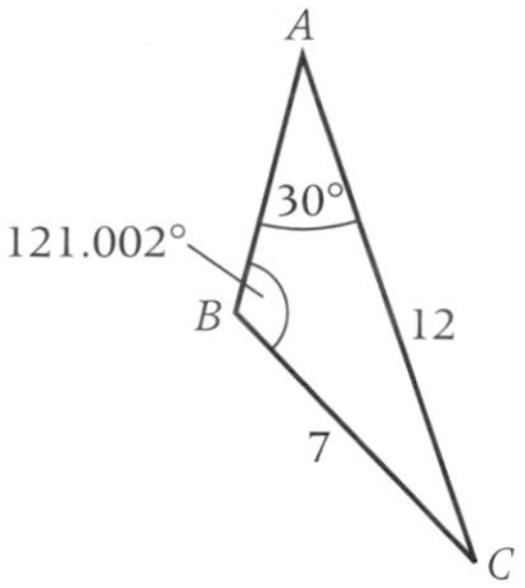

So the two possible triangles have, to three significant figures,

(i) $B = 59.0°, C = 91.0°, c = 14.0$ cm

(ii) $B = 121°, C = 29.0°, c = 6.79$ cm

4

4.12 Cosine rule

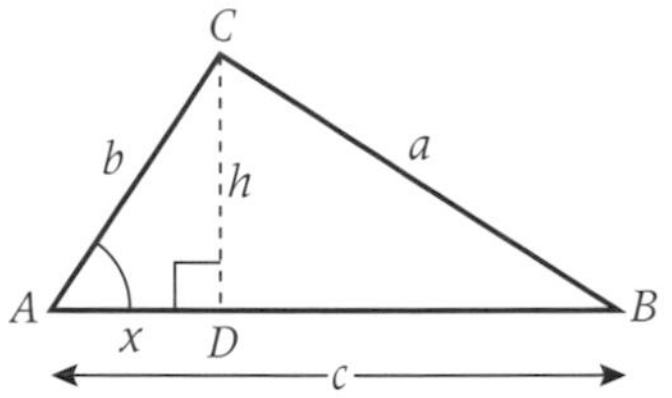

Case (i)
Angle A acute

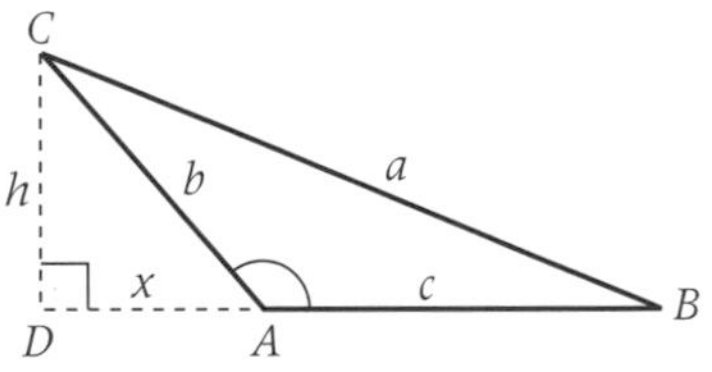

Case (ii)
Angle A obtuse

By applying Pythagoras' theorem to triangle ADC and to triangle CDB and then replacing x by $b\cos A$ in case **(i)** and by $b\cos(180° - A)$ in case **(ii)** you can show that

$\cos(180° - A) = -\cos A$, see section 4.5.

$$a^2 = b^2 + c^2 - 2\,b\,c\cos A$$

This result is known as the cosine formula and, by symmetry, it can be written in different forms.

$b^2 = a^2 + c^2 - 2ac\cos B$
$c^2 = a^2 + b^2 - 2ab\cos C$

To find the angles when all three sides are known, the formula can be rearranged into the form $\cos A = \dfrac{b^2 + c^2 - a^2}{2bc}$.

Worked example 4.11

Find the length of AC in this triangle.
Give your answer to three significant figures.

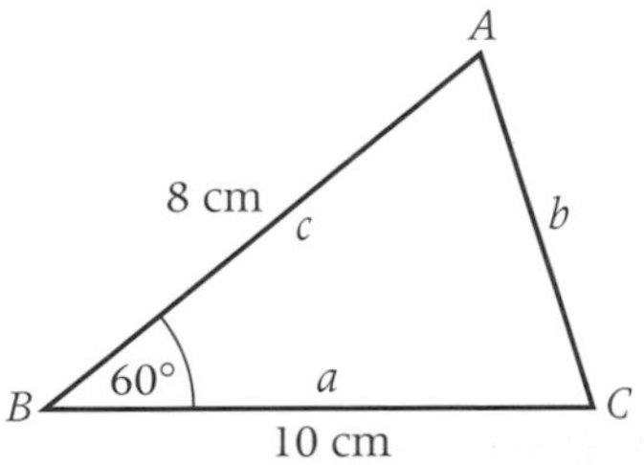

Solution

Using the cosine rule in the form
$b^2 = a^2 + c^2 - 2\,a\,c\cos B$
leads to

$$\begin{aligned} b^2 &= 10^2 + 8^2 - 2 \times 10 \times 8\cos 60° \\ &= 164 - 160\cos 60° \\ &= 164 - 80 \\ b^2 &= 84 \\ b &= \sqrt{84} = 9.1651\ldots \end{aligned}$$

A common error is to write $4\cos 60°$, so always work out the final term before combining.

Don't forget to take the square root.

$AC = 9.17$ cm (to 3 sf)

Worked example 4.12

Calculate the size of the largest angle in this triangle

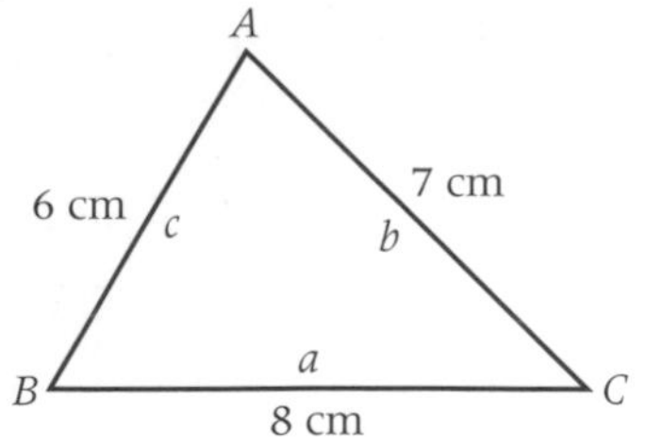

> The largest angle is opposite the longest side; the smallest angle is opposite the shortest side. You can use these two facts to check that your answers are sensible when solving triangles. For example in worked example 4.9, *AC* is opposite the smallest angle so it is the shortest side. We can then deduce that *AC* has to be less than 10 cm.

Solution

The longest side is *BC* so angle *A* is the largest angle. Using the cosine rule in the form $a^2 = b^2 + c^2 - 2\,b\,c\cos A$ leads to

$$\cos A = \frac{b^2 + c^2 - a^2}{2bc}$$

$$= \frac{7^2 + 6^2 - 8^2}{2 \times 7 \times 6}$$

$$\Rightarrow \quad \cos A = \frac{21}{84} = 0.25$$

$$\Rightarrow \quad A = 75.52\ldots^\circ$$

The largest angle is 75.5° (to 3 sf).

4.13 The area of a triangle

$$\sin C = \frac{AD}{AC} \Rightarrow AD = b \sin C$$

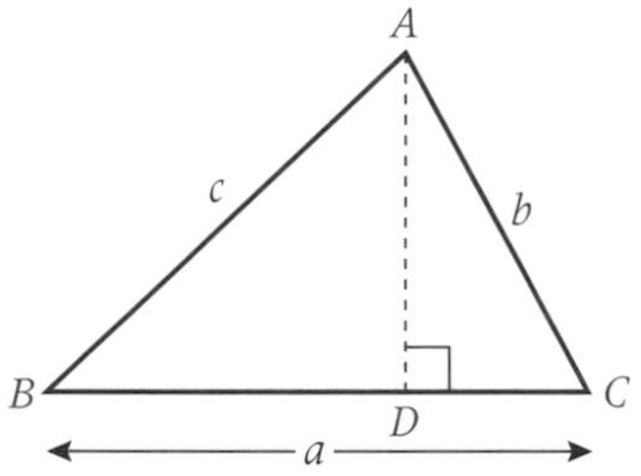

> Area of triangle $ABC = \frac{1}{2} \times BC \times AD$
>
> $= \frac{1}{2}ab\sin C$

By symmetry, $\frac{1}{2}ac\sin B$ and $\frac{1}{2}bc\sin A$ are also equal to the area of triangle *ABC*.

> In words:
> Area of a triangle equals half the product of two of the sides times the sine of the angle between them.

Worked example 4.13

These two triangles have the same area. Calculate the size of the acute angle *R*.

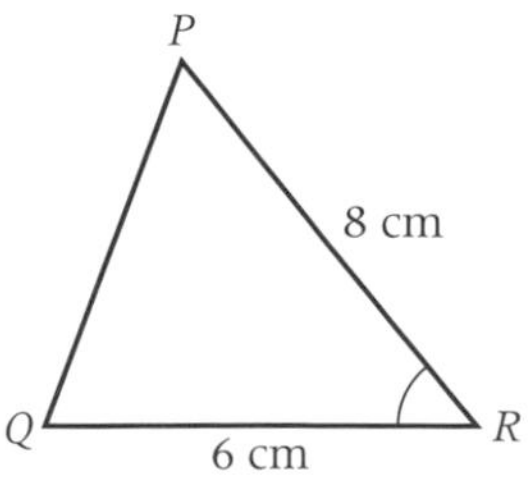

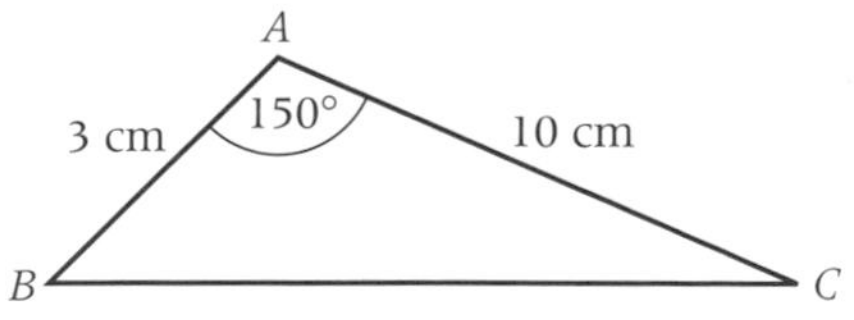

Solution

Area of triangle PQR = Area of triangle ABC

$$\frac{1}{2} \times 8 \times 6 \times \sin R = \frac{1}{2} \times 3 \times 10 \times \sin 150^\circ$$

$$24\sin R = 15\sin 150^\circ \Rightarrow \sin R = \frac{7.5}{24}$$

$$\Rightarrow R = 18.209\ldots^\circ = 18.2^\circ \text{ to 3 sf.}$$

You are told that R is acute so there is no ambiguous case to consider.

EXERCISE 4E

[In this exercise, give all final numerical answers to three significant figures.]

1 Find the values of θ, x and the radius of the circumcircle for each of the following triangles.

(a)

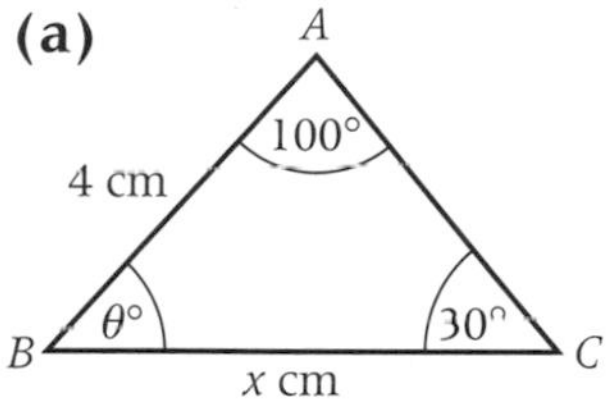

(b)

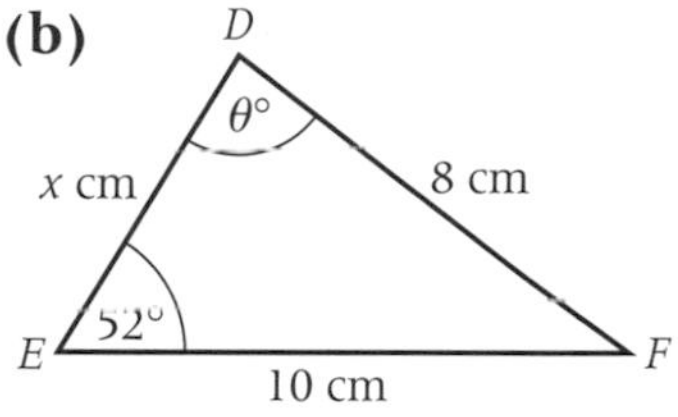

(c)

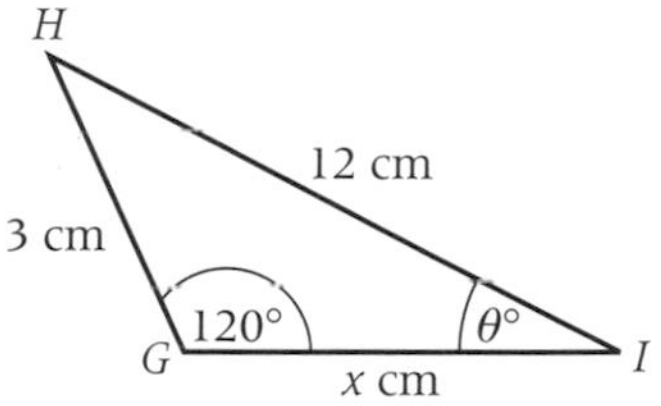

2 Find the length of BC for each of these triangles.

(a)

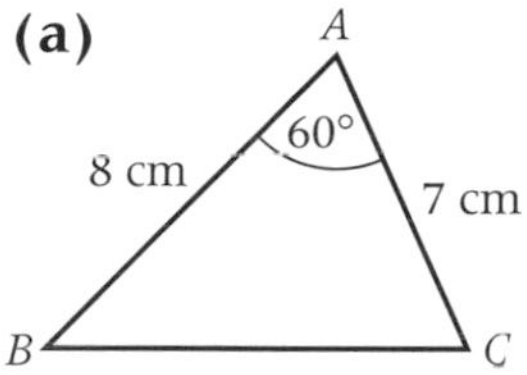

(b)

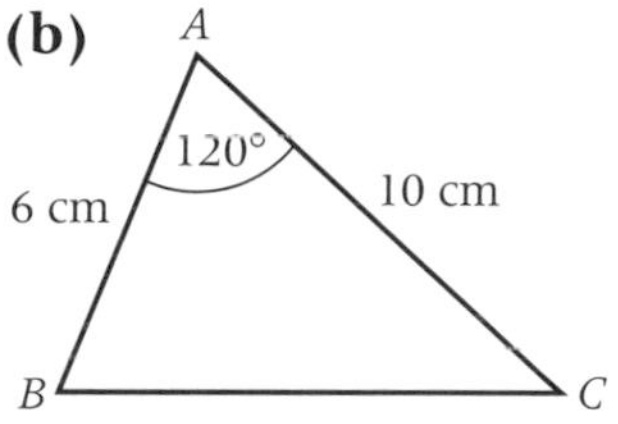

(c)

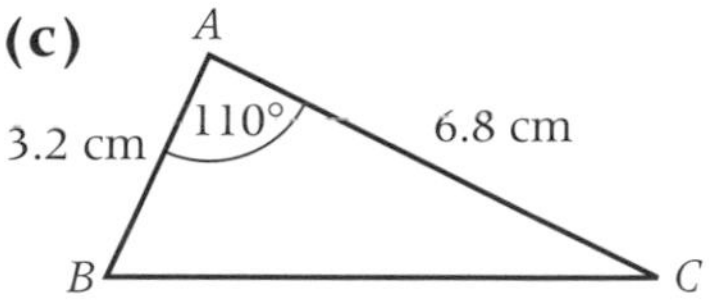

3 Find the area of each of these triangles.

(a)

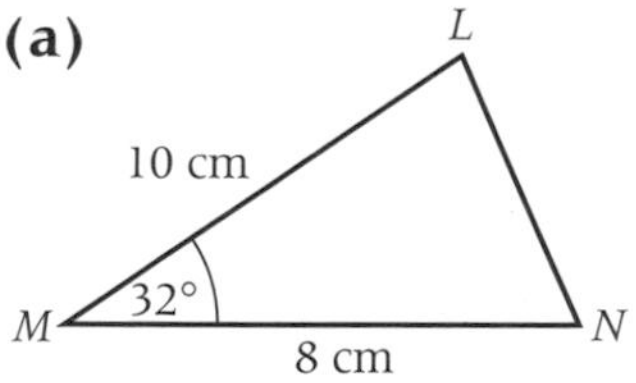

(b)

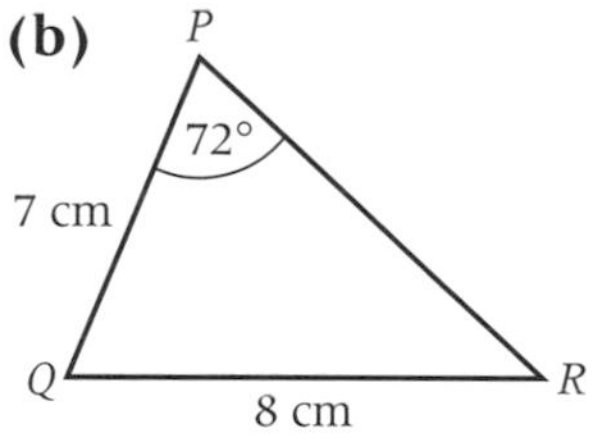

(c)

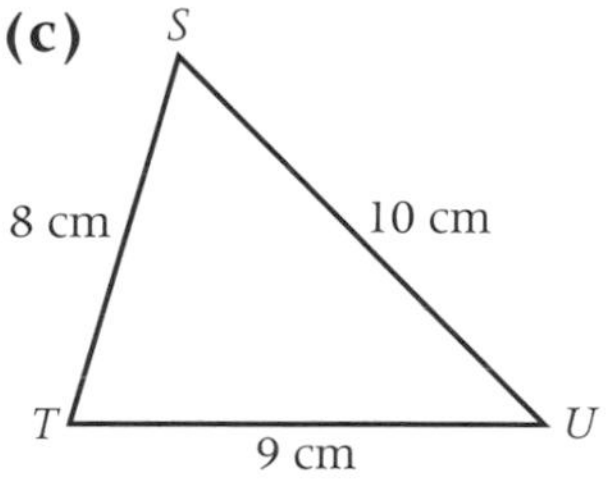

4 Find the size of the largest angle in each of these triangles.

(a)

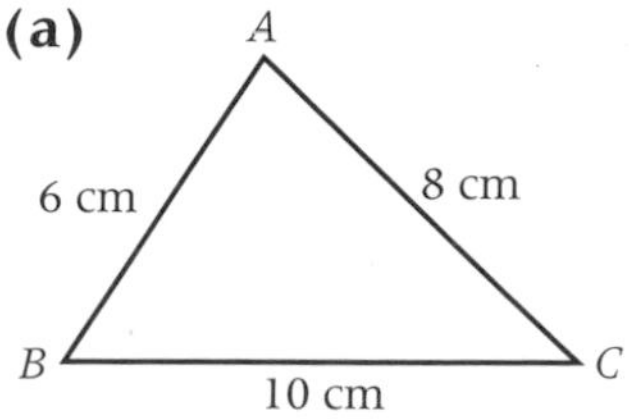

(b)

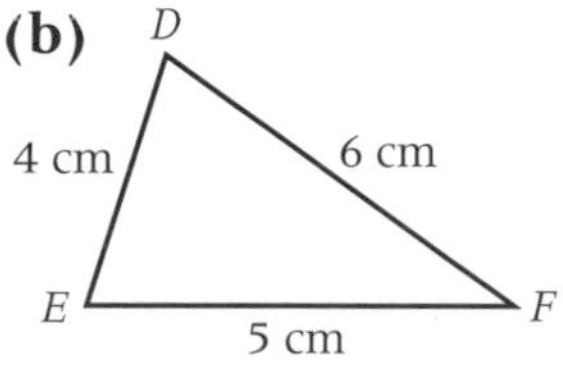

(c)

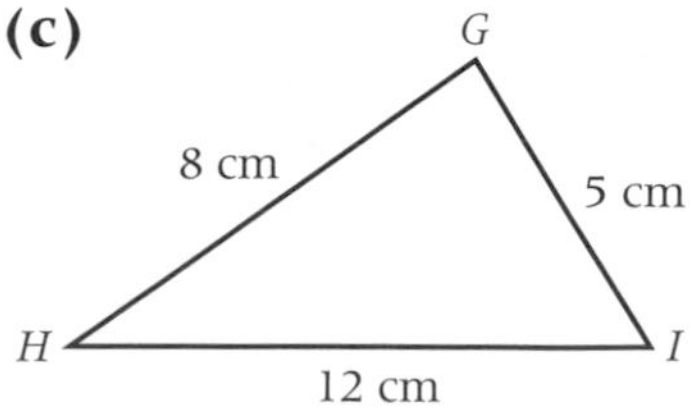

5 Find the circumference of the circumcircle for each of the triangles in Question **4**.

6 Calculate the lengths of the unknown sides and the sizes of the unknown angles in these triangles. [Where it is possible to find two different triangles with the same given data you should give both possibilities as your answer.]

(a)

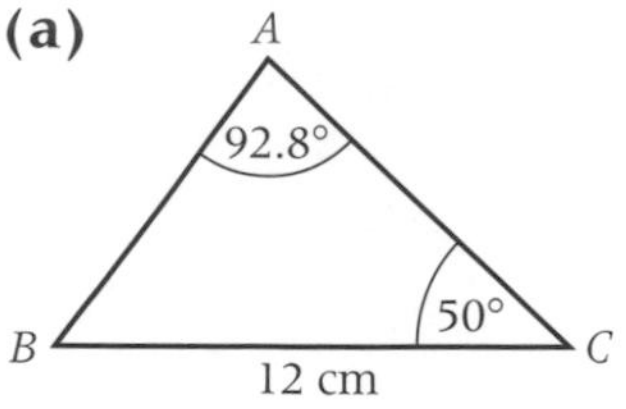

(b)

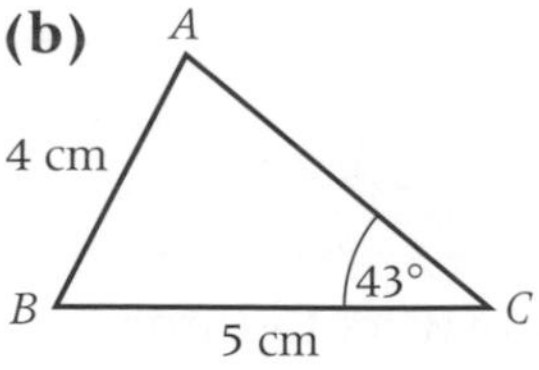

(c)

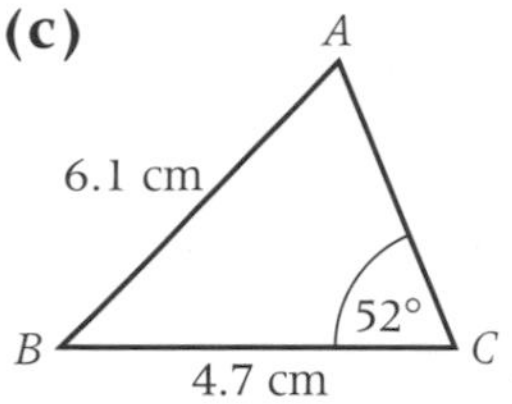

(d)

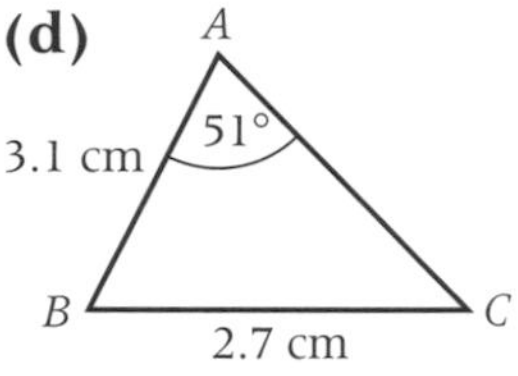

7 Find **(a)** the length of AB,
(b) the area of triangle ABC.

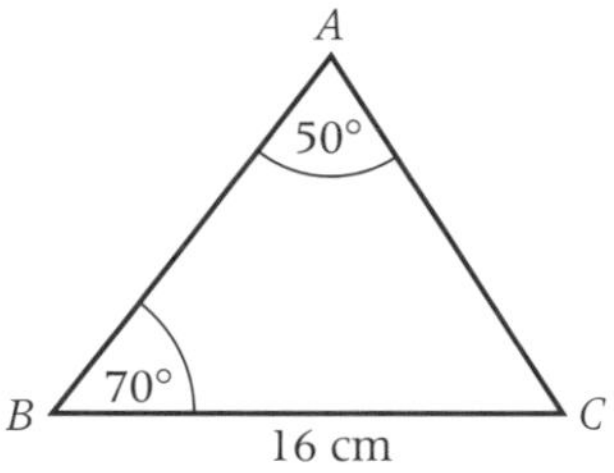

8 Find **(a)** the radius of the circumcircle of triangle XYZ,
(b) the area of the circumcircle.

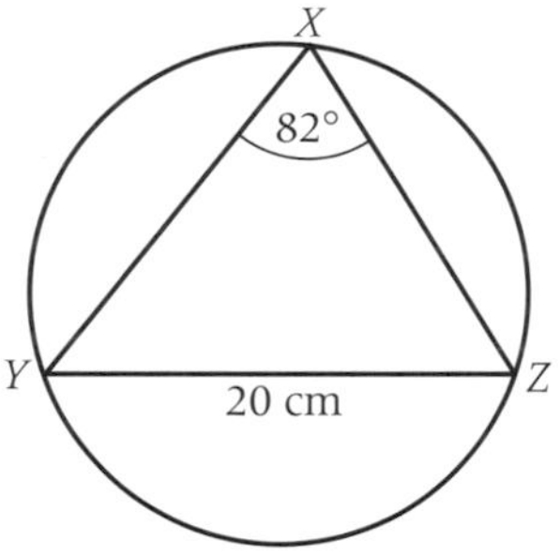

9 The diagram shows three points D, E and F on the circumference of a circle. Find the area of the shaded part of the circle.

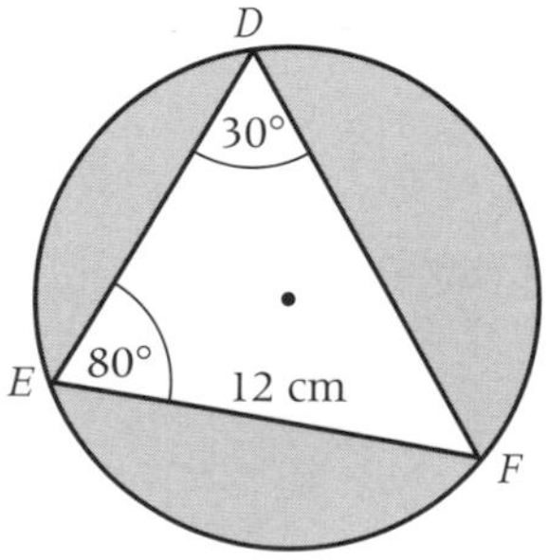

10 An equilateral triangle of side exactly 10 cm is cut from a circular piece of card. The card left over is wasted. Calculate the least area of card wasted.

11 Prove, with the usual notation, that

the area of triangle $ABC = \dfrac{abc}{4R}$,

where R is the radius of the circumcircle of triangle ABC.

Key point summary

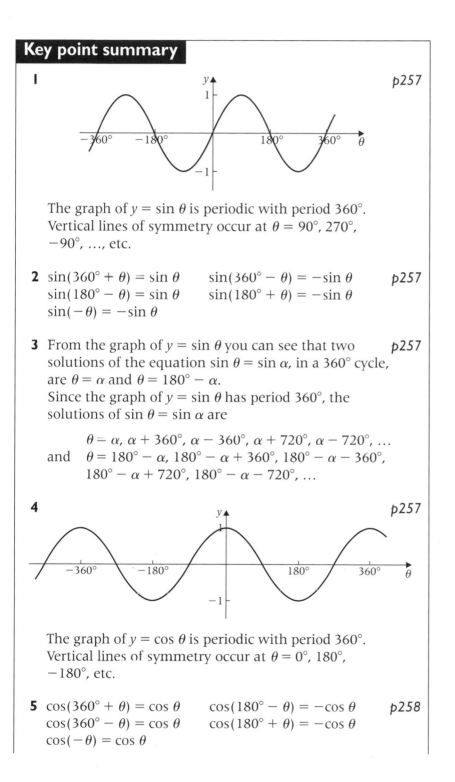

1 p257

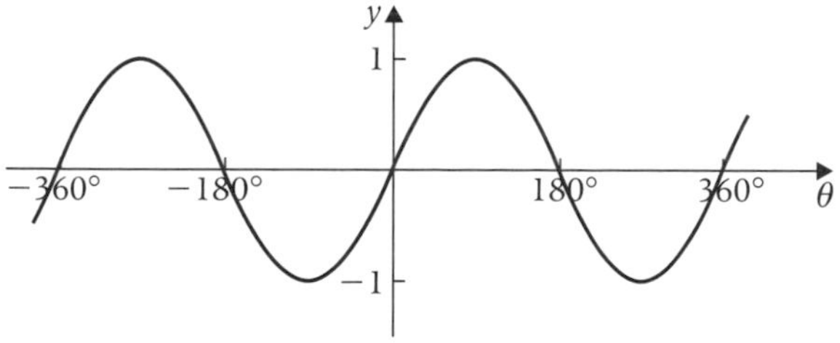

The graph of $y = \sin\theta$ is periodic with period 360°. Vertical lines of symmetry occur at $\theta = 90°, 270°, -90°, \ldots$, etc.

2 $\sin(360° + \theta) = \sin\theta$ $\quad \sin(360° - \theta) = -\sin\theta$ p257
$\sin(180° - \theta) = \sin\theta$ $\quad \sin(180° + \theta) = -\sin\theta$
$\sin(-\theta) = -\sin\theta$

3 From the graph of $y = \sin\theta$ you can see that two solutions of the equation $\sin\theta = \sin\alpha$, in a 360° cycle, are $\theta = \alpha$ and $\theta = 180° - \alpha$. p257
Since the graph of $y = \sin\theta$ has period 360°, the solutions of $\sin\theta = \sin\alpha$ are

$\theta = \alpha, \alpha + 360°, \alpha - 360°, \alpha + 720°, \alpha - 720°, \ldots$
and $\theta = 180° - \alpha, 180° - \alpha + 360°, 180° - \alpha - 360°, 180° - \alpha + 720°, 180° - \alpha - 720°, \ldots$

4 p257

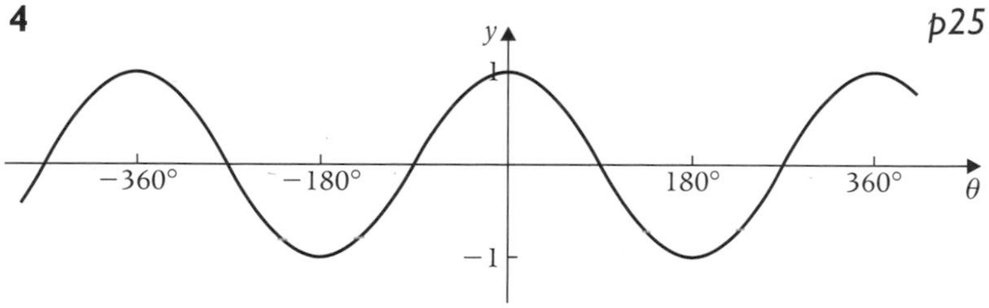

The graph of $y = \cos\theta$ is periodic with period 360°. Vertical lines of symmetry occur at $\theta = 0°, 180°, -180°$, etc.

5 $\cos(360° + \theta) = \cos\theta$ $\quad \cos(180° - \theta) = -\cos\theta$ p258
$\cos(360° - \theta) = \cos\theta$ $\quad \cos(180° + \theta) = -\cos\theta$
$\cos(-\theta) = \cos\theta$

6 From the graph of $y = \cos\theta$ you can see that two solutions of the equation $\cos\theta = \cos\alpha$, in a 360° cycle, are $\theta = \alpha$ and $\theta = -\alpha$. *p258*
Since the graph of $y = \cos\theta$ has period 360°, the solutions of $\cos\theta = \cos\alpha$ are

$\theta = \alpha, \alpha + 360°, \alpha - 360°, \alpha + 720°, \alpha - 720°, \ldots$
and $\theta = -\alpha, -\alpha + 360°, -\alpha - 360°, -\alpha + 720°, -\alpha - 720°, \ldots$

7 *p262*

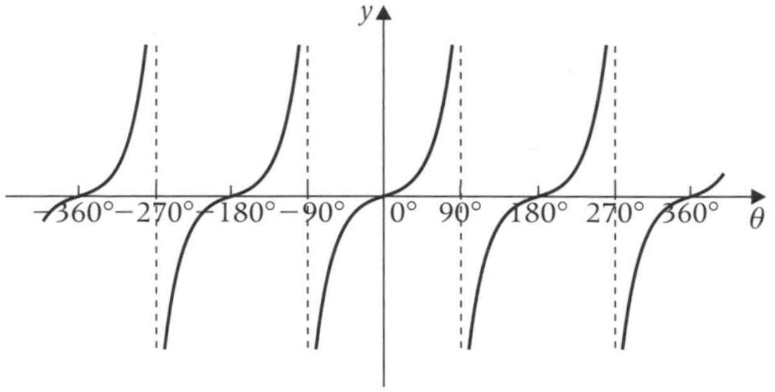

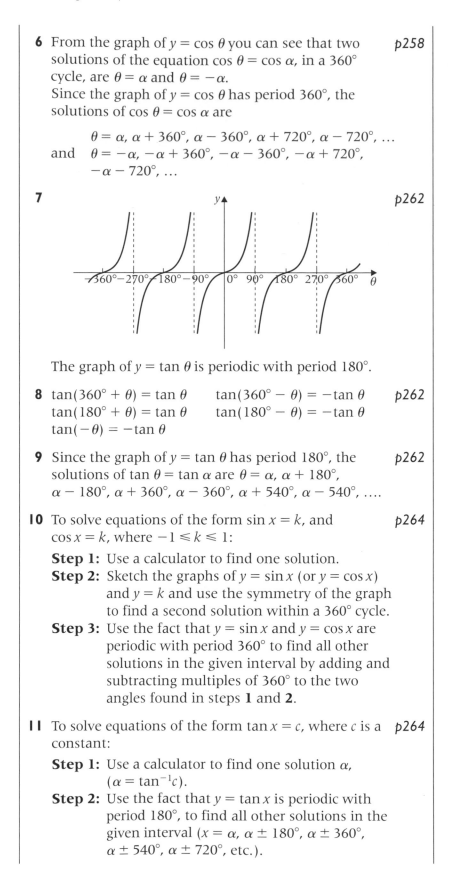

The graph of $y = \tan\theta$ is periodic with period 180°.

8 $\tan(360° + \theta) = \tan\theta$ $\tan(360° - \theta) = -\tan\theta$ *p262*
$\tan(180° + \theta) = \tan\theta$ $\tan(180° - \theta) = -\tan\theta$
$\tan(-\theta) = -\tan\theta$

9 Since the graph of $y = \tan\theta$ has period 180°, the solutions of $\tan\theta = \tan\alpha$ are $\theta = \alpha, \alpha + 180°, \alpha - 180°, \alpha + 360°, \alpha - 360°, \alpha + 540°, \alpha - 540°, \ldots$. *p262*

10 To solve equations of the form $\sin x = k$, and $\cos x = k$, where $-1 \leqslant k \leqslant 1$: *p264*

Step 1: Use a calculator to find one solution.
Step 2: Sketch the graphs of $y = \sin x$ (or $y = \cos x$) and $y = k$ and use the symmetry of the graph to find a second solution within a 360° cycle.
Step 3: Use the fact that $y = \sin x$ and $y = \cos x$ are periodic with period 360° to find all other solutions in the given interval by adding and subtracting multiples of 360° to the two angles found in steps **1** and **2**.

11 To solve equations of the form $\tan x = c$, where c is a constant: *p264*

Step 1: Use a calculator to find one solution α, ($\alpha = \tan^{-1}c$).
Step 2: Use the fact that $y = \tan x$ is periodic with period 180°, to find all other solutions in the given interval ($x = \alpha, \alpha \pm 180°, \alpha \pm 360°, \alpha \pm 540°, \alpha \pm 720°$, etc.).

12 The equation $\sin x = k$, where $-1 \leqslant k \leqslant 1$, normally has two solutions for each 360° cycle within the given interval. *p266*

13 The equation $\cos x = k$, where $-1 \leqslant k \leqslant 1$, normally has two solutions for each 360° cycle within the given interval. *p266*

14 The equation $\tan x = c$, where c is a constant, has two solutions for each 360° cycle within the given interval. *p266*

15 Sine rule for triangle ABC: *p269*

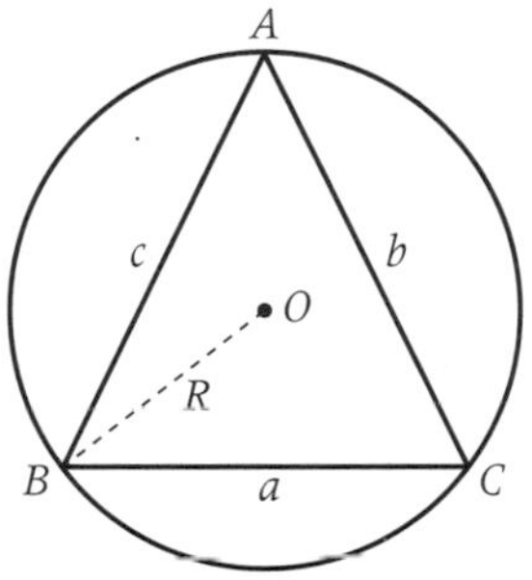

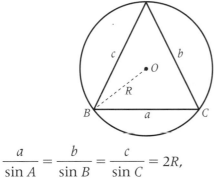

$$\frac{a}{\sin A} = \frac{b}{\sin B} = \frac{c}{\sin C} = 2R,$$

where R is the radius of the circumcircle of triangle ABC.

16 Cosine rule for triangle ABC: *p271*

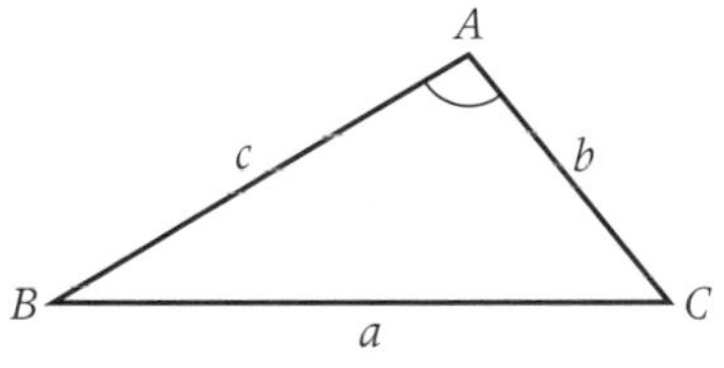

$$a^2 = b^2 + c^2 - 2\,b\,c \cos A$$

To find the angles when all three sides are known use

$$\cos A = \frac{b^2 + c^2 - a^2}{2bc}.$$

17 Area of triangle $ABC = \frac{1}{2}ab \sin C$. *p272*

In words:
Area of a triangle equals half the product of two of the sides times the sine of the angle between them.

Test yourself	What to review
1 Find all the angles θ for which:	*Section 4.7*
(a) $\sin\theta = \sin(-45)°$, in the interval $-180° \leqslant \theta \leqslant 360°$,	
(b) $\cos\theta = \cos 100°$, in the interval $-270° \leqslant \theta \leqslant 0°$,	
(c) $\tan\theta = \tan 150°$, in the interval $-90° \leqslant \theta \leqslant 540°$.	*Section 4.9*
2 Find all the solutions in the given interval, giving answers to the nearest 0.1°:	*Section 4.10*
(a) $\sin\theta = 0.3$, in the interval $-360° \leqslant \theta \leqslant 360°$,	
(b) $\cos x = -\frac{2}{7}$, in the interval $0° \leqslant x \leqslant 360°$,	
(c) $\tan\theta = -1$, in the interval $-540° \leqslant \theta \leqslant 540°$.	
3 In triangle *ABC*, $A = 100°$, $B = 50°$ and $AB = 10$ cm.	
(a) Find the length of *BC* to three significant figures.	*Section 4.11*
(b) Find the area of triangle *ABC*, giving your answer to three significant figures.	*Section 4.13*
4 The sides of a triangle are 5 cm, 7 cm and 8 cm.	
(a) Calculate the size of the smallest angle of the triangle, giving your answer to one decimal place.	*Section 4.12*
(b) Calculate the radius of the circumcircle of the triangle, giving your answer to three significant figures.	*Section 4.11*

Test yourself ANSWERS

1 **(a)** −45°, −135°, 225°, 315°;
(b) −260°, −100°;
(c) −30°, 150°, 330°, 510°.

2 **(a)** −342.5°, −197.5°, 17.5°, 162.5°;
(b) 106.6°, 253.4°;
(c) −405°, −225°, −45°, 135°, 315°, 495°.

3 **(a)** 19.7 cm; **(b)** 75.4 cm^2.

4 **(a)** 38.2°; **(b)** 4.04 cm.

CHAPTER 5

C2: Simple transformations of graphs

Learning objectives

After studying this chapter, you should be able to:

- use correct terms for simple transformations such as translation, reflection, stretch in the x- or y-direction
- transform graphs using translations
- reflect graphs in the coordinate axes to produce new graphs
- transform graphs using stretches in the x- or y-direction
- find the effect of simple transformations on the graph of $y = \mathrm{f}(x)$ as represented by $y = \mathrm{f}(x - a)$, $y = \mathrm{f}(x) + b$, $y = \mathrm{f}\left(\frac{x}{c}\right)$, $y = d\,\mathrm{f}(x)$.

5.1 Translations parallel to the y-axis

You have already applied translations to quadratic graphs in Chapter 4 of C1.

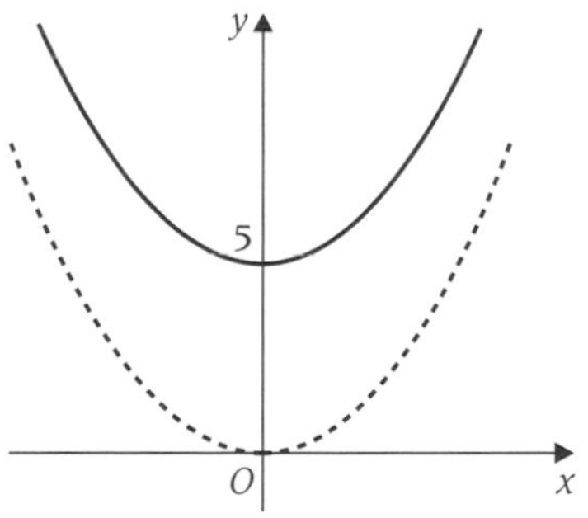

The graph of $y = x^2 + 5$ is easily obtained from the graph of $y = x^2$. Each y-coordinate is increased by 5 and so the graph of $y = x^2$ is translated by the vector $\begin{bmatrix} 0 \\ 5 \end{bmatrix}$ to obtain the graph of $y = x^2 + 5$.

A **translation** moves every point of an object to the corresponding point on its image through the same vector.

Try to use the correct term 'translation' rather than words like 'shift' or 'move'. Remember that a **translation** is a special type of transformation so avoid confusing these two words.

In general, a translation of $\begin{bmatrix} 0 \\ b \end{bmatrix}$ transforms the graph of $y = \mathrm{f}(x)$ into the graph of $y = \mathrm{f}(x) + b$.

Worked example 5.1

Describe geometrically the transformation that maps the graph of $y = (x + 2)^2$ onto the graph of $y = x^2 + 4x - 3$.

Solution

The first graph has equation $y = (x + 2)^2 = x^2 + 4x + 4$.
By subtracting 7 from each y-value you obtain the graph of $y = x^2 + 4x - 3$.
Each point has been translated by -7 units in the y-direction.

The transformation is a translation of $\begin{bmatrix} 0 \\ -7 \end{bmatrix}$.

5.2 Translations parallel to the *x*-axis

In C1, you applied translations in the x-direction to quadratic graphs using the result below:

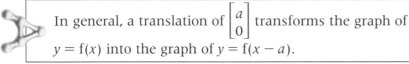

> In general, a translation of $\begin{bmatrix} a \\ 0 \end{bmatrix}$ transforms the graph of $y = \mathrm{f}(x)$ into the graph of $y = \mathrm{f}(x - a)$.

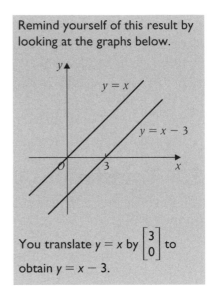

Remind yourself of this result by looking at the graphs below.

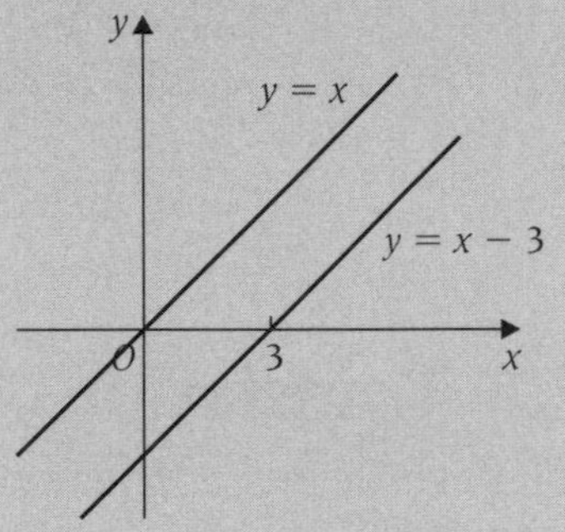

You translate $y = x$ by $\begin{bmatrix} 3 \\ 0 \end{bmatrix}$ to obtain $y = x - 3$.

This formula can be applied to any graph of the form $y = \mathrm{f}(x)$.

Worked example 5.2

The graph of $y = (x - 5)^3$ is translated by $\begin{bmatrix} -4 \\ 0 \end{bmatrix}$. Find the equation of the new graph.

Solution

Use the result from the previous Key Point box.

Replace the variable x by $x - (-4) = x + 4$.

The new graph has equation $y = (\{x + 4\} - 5)^3 = (x - 1)^3$.

Worked example 5.3

Describe the geometrical transformation that maps the graph of $y = \sin x$ into the graph of $y = \sin(x + 30°)$.

Solution

The first graph has been mapped onto the second graph by a **translation** with vector $\begin{bmatrix} -30° \\ 0 \end{bmatrix}$.

5.3 General translations of graphs

You can combine the two results from the previous sections to find the effect of a general translation. You have already done this with quadratic graphs in C1.

> In general, a translation of $\begin{bmatrix} a \\ b \end{bmatrix}$ transforms the graph of $y = f(x)$ into the graph of $y - b = f(x - a)$.

EXERCISE 5A

A useful strategy is to find the point on the second graph that corresponds to the origin in the first graph.

If you need to, sketch the graphs on your graphics calculator.

1 The second graph has been transformed from the first by a translation. State the vector of the translation in each case.

(a) (i) $y = x^2$, **(ii)** $y = x^2 - 5$,

(b) (i) $y = 2^x$, **(ii)** $y = 2^{x-3}$,

(c) (i) $y = x^5$, **(ii)** $y = (x - 6)^5$,

(d) (i) $y = 3x^7$, **(ii)** $y = 3x^7 + 2$,

(e) (i) $y = 5x^4$, **(ii)** $y = 5(x - 1)^4$,

(f) (i) $y = x^2$, **(ii)** $y = 3 + (x + 2)^2$,

(g) (i) $y = 3^x - 7$, **(ii)** $y = 3^x - 1$,

(h) (i) $y = 16x^4$, **(ii)** $y = (2x + 1)^4$.

2 Find the equation of the graph after it has been translated by the vector given in each case.

(a) $y = 3x$, $\begin{bmatrix} 4 \\ 0 \end{bmatrix}$, **(b)** $y = x^6$, $\begin{bmatrix} 0 \\ 3 \end{bmatrix}$, **(c)** $y = x^2$, $\begin{bmatrix} 2 \\ 5 \end{bmatrix}$,

(d) $y = 5x^2$, $\begin{bmatrix} 0 \\ -1 \end{bmatrix}$, **(e)** $y = 2x^3$, $\begin{bmatrix} 3 \\ 0 \end{bmatrix}$, **(f)** $y = x^5$, $\begin{bmatrix} -2 \\ 0 \end{bmatrix}$,

(g) $y = x^7$, $\begin{bmatrix} -3 \\ 0 \end{bmatrix}$, **(h)** $y = x^3 - 5$, $\begin{bmatrix} 5 \\ 0 \end{bmatrix}$ **(i)** $y = \dfrac{2}{x + 1}$, $\begin{bmatrix} 3 \\ 0 \end{bmatrix}$.

3 State the transformation that has taken place in mapping the graph of $y = \sin x$ onto each of the following graphs:

(a) $y = 4 + \sin x$,

(b) $y = \sin(x - 10°)$,

(c) $y = \sin(x + 70°)$,

(d) $y = 2 + \sin(x - 30°)$.

4 State the transformation that maps the graph of $y = \cos x$ onto each of the following:

(a) $y = 1 + \cos x$, **(b)** $y = \cos(x + 120°)$,
(c) $y = \cos(x - 50°)$, **(d)** $y = \cos(x - 80°) - 3$.

5 The graph of $y = \mathrm{f}(x)$ for $0 \leqslant x \leqslant 3$ is sketched below and $\mathrm{f}(x) = 0$ outside this interval.

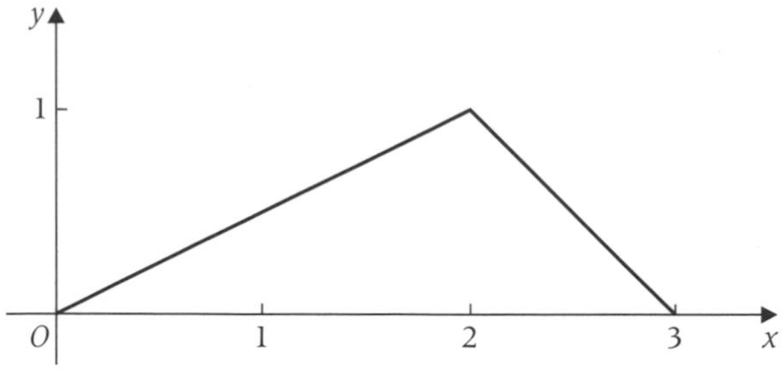

On separate sets of axes, sketch, for $-2 \leqslant x \leqslant 5$, the graphs of

(a) $y = \mathrm{f}(x) + 2$, **(b)** $y = \mathrm{f}(x - 1)$,
(c) $y = \mathrm{f}(x + 2)$, **(d)** $y = \mathrm{f}(x - 1) - 3$.

6 Find the equation of the resulting curve when $y = \tan x$ is translated by the given vector:

(a) $\begin{bmatrix} 0 \\ 3 \end{bmatrix}$, **(b)** $\begin{bmatrix} 30° \\ 0 \end{bmatrix}$ **(c)** $\begin{bmatrix} -45° \\ 0 \end{bmatrix}$ **(d)** $\begin{bmatrix} 36° \\ -4 \end{bmatrix}$.

7 (a) Express **(i)** 16×2^x and **(ii)** $\dfrac{2^x}{8}$ as powers of 2.

(b) Hence determine translations that would map the graph of $y = 2^x$ onto the graph of **(i)** $y = 16 \times 2^x$ and **(ii)** $y = \dfrac{2^x}{8}$.

(c) Use your answer to **(b)** to find a transformation that maps the graph of $y = 16 \times 2^x$ onto the graph of $y = \dfrac{2^x}{8}$.

8 State all the possible values of k between 0 and 1000 for which $\sin(x - k°) \equiv \cos x$.

Explain the result geometrically with reference to the graphs of $y = \sin x$ and $y = \cos x$.

5.4 Reflections

Worked example 5.4

Draw the graphs of $y = x^4$ and $y = -x^4$ on the same axes and state a transformation that maps one graph onto the other.

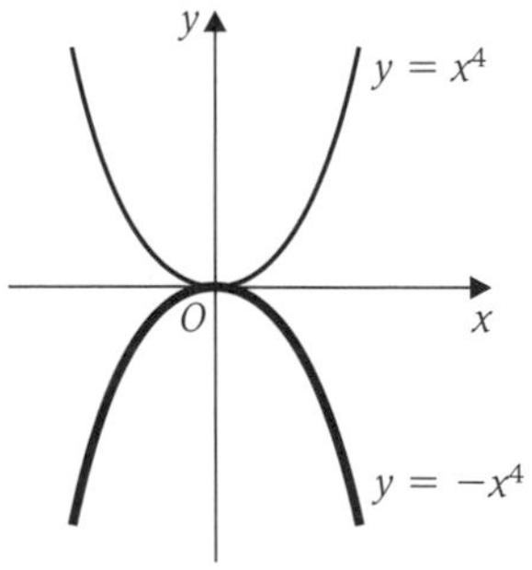

Solution

Each graph is mapped onto the other graph by a reflection in the line $y = 0$ (or the x-axis).

In general, the graph of $y = f(x)$ is transformed into the graph of $y = -f(x)$ by a reflection in the line $y = 0$ (the x-axis).

Worked example 5.5

Sketch the graphs of $y = 1 + x^3$ and $y = 1 - x^3$ on the same axes and state a transformation that maps one graph onto the other.

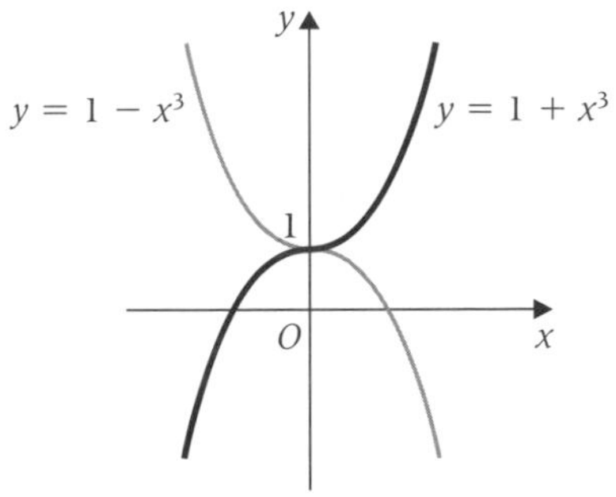

Solution

Each graph is mapped onto the other graph by a reflection in the line $x = 0$ (the y-axis).

In general, the graph of $y = f(x)$ is transformed into the graph of $y = f(-x)$ by a reflection in the line $x = 0$ (the y-axis).

5

5.5 Stretches in the y-direction

The graph of $y = 5x^2$ is similar to the graph of $y = x^2$, but each y-coordinate corresponding to a given value of x is five times as large. The transformation mapping $y = x^2$ onto $y = 5x^2$ is a **one-way stretch** with factor 5 in the y-direction.

The term **one-way stretch** is slightly more precise than **stretch** because you will meet **two-way stretches** in C3.

In general, the graph of $y = f(x)$ is transformed into the graph of $y = d\,f(x)$ by a stretch of scale factor d in the y-direction.

Worked example 5.6

(a) Find the equation of the resulting graph when the graph of $y = x^3$ is stretched by scale factor 4 in the y-direction.

(b) The graph of $y = x^5$ is mapped onto the graph of $y = (2x)^5$. Describe the geometrical transformation that has taken place.

Solution

(a) Each y-coordinate is now 4 times its previous value. The new equation is $y = 4x^3$.

(b) You need to express $y = (2x)^5$ in the form $y = 32x^5$, since the original curve had equation $y = x^5$.

Therefore, the transformation can be described geometrically as a one-way stretch in the y-direction with scale factor 32.

You will see in the next section that it can also be described as a one-way stretch in the x-direction but with a different scale factor.

5.6 Stretches in the x-direction

Worked example 5.7

Draw the graphs $y = x^3$ and $y = \left(\frac{x}{2}\right)^3$ on the same axes and state a transformation that maps the first graph onto the second.

Solution

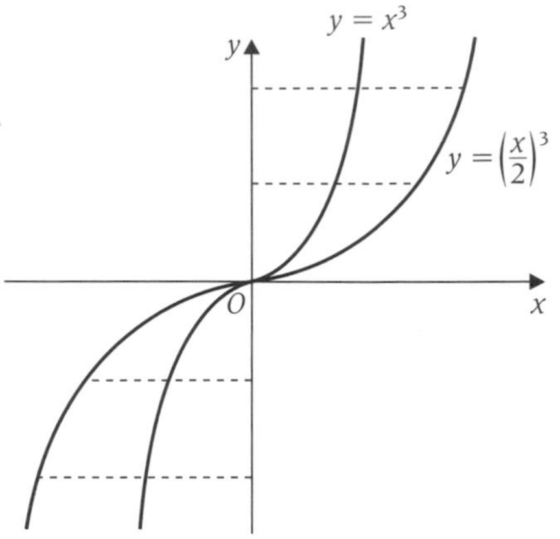

As you consider the graphs, you can see that in the x-direction, the second graph is stretched 2 times as much as the first graph.

A rearrangement of the equation in the form $x = 2\sqrt[3]{y}$ perhaps makes this clearer.

The transformation is a one-way stretch of scale factor 2 in the x-direction.

Note. Because $y = \left(\frac{x}{2}\right)^3 = \frac{x^3}{8} = \frac{1}{8}x^3$

You could also describe this as a one-way stretch with scale factor $\frac{1}{8}$ in the y-direction.

> In general, the graph of $y = \text{f}(x)$ is transformed into the graph of $y = \text{f}\left(\frac{x}{c}\right)$ by a stretch of scale factor c in the x-direction.

5.7 Scale factors

It is possible for a scale factor to be a fraction. Suppose the stretch in the x-direction has scale factor $\frac{1}{2}$. Each distance from the y-axis will now only be half of the previous distance. However the transformation is still called a **stretch**.

> Some students have used terms such as 'shrink' or 'reduction' because the scale factor is smaller than 1. This is incorrect and the term stretch should be used even for fractional or negative scale factors.

You may also find that a scale factor is negative. Suppose a stretch has scale factor -3 in the y-direction. This means that the points on one side of the x-axis are transformed to points on the opposite side of the x-axis and three times as far from the x-axis as the original points. The transformation is still called a **stretch**.

EXERCISE 5B

1 Describe in detail the geometrical transformation that maps the curve with the first equation onto the second.

(a) (i) $y = x^2$, **(ii)** $y = -x^2$,

(b) (i) $y = 2^x$, **(ii)** $y = 5 \times 2^x$,

(c) (i) $y = x^5$, **(ii)** $y = (x - 3)^5$,

(d) (i) $y = x^7$, **(ii)** $y = \left(\frac{x}{3}\right)^7$,

> You may find it useful to plot the graphs on a computer or on a graphics calculator to confirm your answers.

(e) (i) $y = 5 + x^4$, (ii) $y = 15 + 3x^4$,
(f) (i) $y = x^2 + 7$, (ii) $y = 3 + (x + 2)^2$,
(g) (i) $y = 3^x$, (ii) $y = 3^{\frac{x}{2}}$,
(h) (i) $y = 16 + x^5$, (ii) $y = 16 - x^5$.

2 Find the equation of each graph after it had been transformed as indicated:

(a) $y = x^3$, stretch with scale factor 5 in x-direction,
(b) $y = 4 - x^6$, reflection in the y-axis,
(c) $y = x^2 + 3x - 1$, reflection in the x-axis,
(d) $y = 9x^2 + 4$, stretch with scale factor 3 in y-direction,
(e) $y = 2x^3 - x^2 + 7$, stretch with scale factor -3 in x-direction,
(f) $y = x^2$, stretch with scale factor $\frac{1}{3}$ in x-direction,
(g) $y = x^7$, stretch with scale factor -2 in y-direction,
(h) $y = x^3 - 5$, stretch with scale factor $\frac{1}{2}$ in x-direction,
(i) $y = \frac{3}{x + 2}$, reflection in the y-axis.

3 State the transformation that has taken place in mapping the graph of $y = \tan x$ onto each of the following graphs:

(a) $y = 3 \tan x$, **(b)** $y = \tan(2x)$, **(c)** $y = \tan\left(\frac{x}{5}\right)$.

4 State the transformation that maps the graph of $y = 2 + \cos x$ onto each of the following:

(a) $y = -2 - \cos x$,
(b) $y = 2 + \cos(3x)$,
(c) $y = 10 + 5 \cos x$.

5 The graph of $y = f(x)$ for $0 \leqslant x \leqslant 3$ is sketched below and $f(x) = 0$ outside this interval.

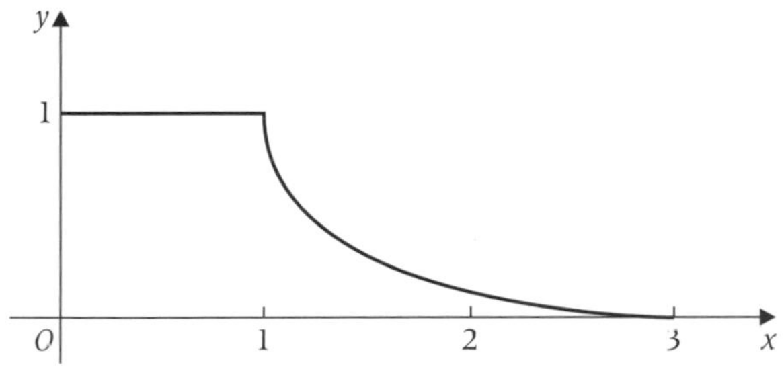

On separate sets of axes, sketch for $-6 \leqslant x \leqslant 6$ the graphs of

(a) $y = f(-x)$, **(b)** $y = f(3x)$,
(c) $y = -f(x)$, **(d)** $y = f\left(\frac{x}{2}\right)$.

5

6 Find the equation of the resulting curve when the curve with equation $y = \sin 6x$ has been

(a) stretched by a factor 3 in the x-direction,

(b) stretched by a factor 2 in the y-direction,

(c) reflected in the y-axis.

7 **(a)** The graph of $y = 2^x$ can be transformed into the graph of $y = \frac{2^x}{32}$ by a stretch. State the scale factor and direction of the stretch.

(b) The graph of $y = 2^x$ can be transformed into the graph of $y = \frac{2^x}{32}$ by a translation. By expressing $\frac{1}{32}$ as a power of 2, find the vector of the translation.

Key point summary

1 A translation of $\begin{bmatrix} 0 \\ b \end{bmatrix}$ transforms the graph of $y = f(x)$ into the graph of $y = f(x) + b$. *p279*

2 A translation of $\begin{bmatrix} a \\ 0 \end{bmatrix}$ transforms the graph of $y = f(x)$ into the graph of $y = f(x - a)$. *p280*

3 A translation of $\begin{bmatrix} a \\ b \end{bmatrix}$ transforms the graph of $y = f(x)$ into the graph of $y = f(x - a) + b$. *p281*

4 The graph of $y = f(x)$ is transformed into the graph of $y = -f(x)$ by a reflection in the line $y = 0$ (the x-axis). *p282*

5 The graph of $y = f(x)$ is transformed into the graph of $y = f(-x)$ by a reflection in the line $x = 0$ (the y-axis). *p283*

6 The graph of $y = f(x)$ is transformed into the graph of $y = d\,f(x)$ by a stretch of scale factor d in the y-direction. *p283*

7 The curve $y = f(x)$ is transformed into the graph $y = f\left(\frac{x}{c}\right)$ by a stretch of scale factor c in the x-direction. *p284*

Test yourself	What to review
1 Find the equation of the curve resulting from translating the graph of $y = x^7$ through $\begin{bmatrix} 2 \\ 0 \end{bmatrix}$.	*Section 5.2*
2 Determine the transformation that maps $y = 4 - x^3$ onto $y = x^3 - 4$.	*Section 5.4*
3 The graph of $y = x^3$ can be mapped onto the graph of $y = 8x^3$ by means of a one-way stretch. State the scale factor of the stretch when it is parallel to the **(a)** y-axis, **(b)** x-axis.	*Sections 5.5 and 5.6*
4 Describe the transformations that maps the curve $y = \sin x$ onto **(a)** $y = \sin(x - 20°)$, **(b)** $y = \sin 2x$.	*Sections 5.2 and 5.6*

5

Test yourself ANSWERS

1 $y = (x - 2)^7$.

2 Reflection in the line $y = 0$ (x-axis).

3 **(a)** 8; **(b)** $\frac{1}{2}$.

4 **(a)** Translation of $\begin{bmatrix} 20° \\ 0 \end{bmatrix}$;

(b) Stretch in the x-direction with scale factor $\frac{1}{2}$.

CHAPTER 6

C2: Solving trigonometrical equations

Learning objectives

After studying this chapter, you should:

- be able to solve trigonometrical equations of the form $\sin bx = k$, $\cos bx = k$ and $\tan bx = c$, where b, k and c are constants and $-1 \leqslant k \leqslant 1$
- be able to solve trigonometrical equations of the form $\sin(x + b) = k$, $\cos(x + b) = k$, and $\tan(x + b) = c$, where b, k and c are constants and $-1 \leqslant k \leqslant 1$
- be able to solve quadratic equations in $\sin x$, $\cos x$ and $\tan x$
- know and be able to use the identities $\cos^2\theta + \sin^2\theta \equiv 1$ and $\tan\theta \equiv \dfrac{\sin\theta}{\cos\theta}$ to simplify trigonometrical equations and to prove other identities

6.1 Solving equations of the form $\sin bx = k$, $\cos bx = k$ and $\tan bx = c$

In chapter 5 you were shown how to sketch graphs of the form $y = \sin bx$ by stretching the graph of $y = \sin x$ parallel to the x-axis by a scale factor $\frac{1}{b}$. In this section you will be shown how to solve equations of the form $\sin bx = k$ by applying a simple substitution although such equations can be solved by sketching and using the graph of $y = \sin bx$ in a similar way to that used in section 4.10.

Worked example 6.1

Solve the equation $\cos 2x = \frac{1}{2}$, in the interval $0° \leqslant x \leqslant 360°$.

Solution

Let $u = 2x$, then $\cos 2x = \frac{1}{2} \Rightarrow \cos u = \frac{1}{2}$.

You now have to find the interval for u.

Multiplying each part of the inequality $0° \leqslant x \leqslant 360°$ by 2 gives $0° \leqslant 2x \leqslant 720°$ so the interval for u is $0° \leqslant u \leqslant 720°$.

Using a calculator, $u = \cos^{-1}\left(\frac{1}{2}\right) = 60°$.

Since the graph of $y = \cos u$ is symmetrical about the vertical line $u = 0°$, another solution is $u = -60°$.
Since the graph of $y = \cos u$ is periodic with period 360°, solutions of $\cos u = \frac{1}{2}$ are

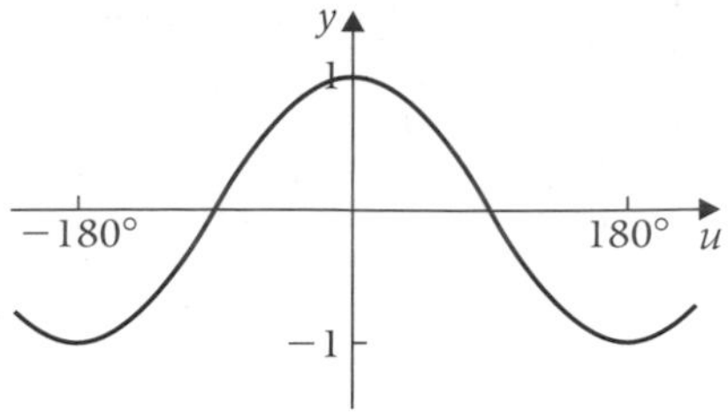

$$u = 60°, 60° \pm 360°, 60° \pm 720°, \ldots$$
$$-60°, -60° \pm 360°, -60° \pm 720°, \ldots.$$

Picking out the values of u that lie in the interval $0° \leqslant u \leqslant 720°$ gives:

$$u = 60°, 300°, 420°, 660°.$$

> We would expect four solutions of $\cos u = \frac{1}{2}$ in a 720° cycle. (See Key Point 13 in chapter 4.)

Since you were asked to solve $\cos 2x = \frac{1}{2}$ you need values of x, so replacing u by $2x$ gives

$$2x = 60°, 300°, 420°, 660°.$$
$$\Rightarrow \quad x = 30°, 150°, 210°, 330°.$$

Worked example 6.2

Solve the equation $4 \sin 4x + 3 = 0$, in the interval $-180° \leqslant x \leqslant 180°$, giving your answers to one decimal place.

Solution

Rearrange $4 \sin 4x + 3 = 0$ to get $\sin 4x = -0.75$.
Let $u = 4x \Rightarrow \sin u = -0.75$.
Finding the interval for u, multiply each part of the inequality $-180° \leqslant x \leqslant 180°$ by 4 to give $-720° \leqslant 4x \leqslant 720°$ so the interval for u is $-720° \leqslant u \leqslant 720°$.
Using a calculator, $u = \sin^{-1}(-0.75) = -48.590\ldots°$.
Since the graph of $y = \sin u$ is symmetrical about the vertical line $u = 90°$, another solution is $u = 180° + 48.590\ldots° = 228.590\ldots°$.
Since the graph of $y = \sin u$ is periodic with period 360°, solutions of $\sin u = -0.75$ are

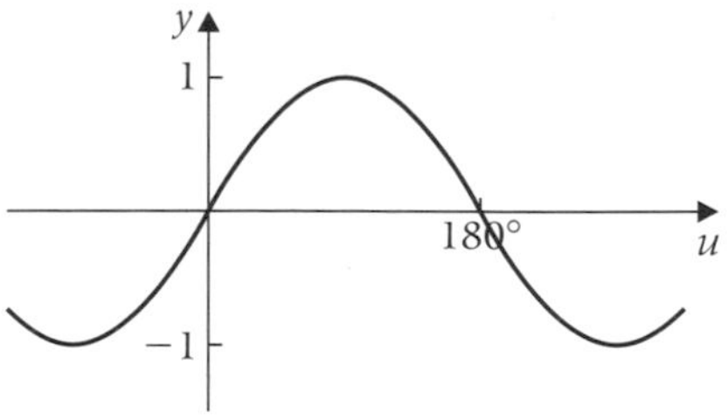

$$u = -48.590\ldots°, -48.590\ldots° \pm 360°, -48.590\ldots° \pm 720°, \ldots$$
$$228.590\ldots°, 228.590\ldots° \pm 360°, 228.590\ldots° \pm 720°, \ldots$$

Picking out the values of u that lie in the interval $-720° \leqslant u \leqslant 720°$ gives

$$u = -48.590\ldots°, 311.409\ldots°, -408.590\ldots°, 671.409\ldots°,$$
$$228.590\ldots°, -131.409\ldots°, 588.590\ldots°, -491.409\ldots°.$$
$$4x = -491.409\ldots°, -408.590\ldots°, -131.409\ldots°, -48.590\ldots°,$$
$$228.590\ldots°, 311.409\ldots°, 588.590\ldots°, 671.409\ldots°$$
$$\Rightarrow \quad x = -122.9°, -102.1°, -32.9°, -12.1°, 57.1°, 77.9°, 147.1°, 167.9°.$$

> We would expect eight solutions of $\sin u = -0.75$ in a 1440° cycle. (See Key Point 12 in chapter 4.)

Worked example 6.3

Solve the equation $4\tan\frac{\theta}{2} - 10 = 0$, in the interval $-360° \leqslant \theta \leqslant 540°$, giving your answers to one decimal place.

Solution

Rearrange $4\tan\frac{\theta}{2} - 10 = 0$ to get $\tan\frac{\theta}{2} = 2.5$.

Let $u = \frac{\theta}{2} \Rightarrow \tan u = 2.5$.

Finding the interval for u, divide each part of the inequality $-360° \leqslant \theta \leqslant 540°$ by 2 to give $-180° \leqslant \frac{\theta}{2} \leqslant 270°$, so the interval for u is $-180° \leqslant u \leqslant 270°$.

Using a calculator, $u = \tan^{-1}(2.5) = 68.198...°$.

Since the graph of $y = \tan u$ is periodic with period 180°, solutions of $\tan u = 2.5$ are

$$u = 68.198...°,\ 68.198...° \pm 180°,\ 68.198...° \pm 360°,\ ....$$

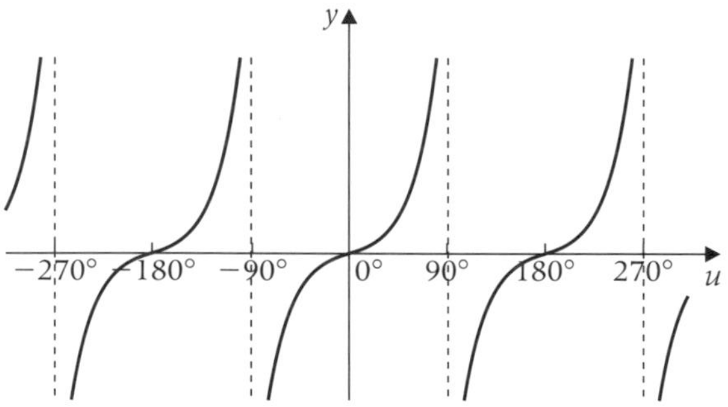

Picking out the values of u that lie in the interval $-180° \leqslant u \leqslant 270°$ gives

$$u = 68.198...°,\ -111.802...°,\ 248.198...°.$$

$$\frac{\theta}{2} = -111.802...°,\ 68.198...°,\ 248.198...°$$

$$\Rightarrow \quad \theta = -223.6°,\ 136.4°,\ 496.4°.$$

In summary:

> To solve equations of the form $\sin bx = k$, $\cos bx = k$ and $\tan bx = c$ in a given interval for x
>
> **Step 1:** Let $u = bx$ to get an equation of the form $\sin u = k$, $\cos u = k$ and $\tan u = c$.
>
> **Step 2:** Find the interval for u.
>
> **Step 3:** Solve $\sin u = k$, $\cos u = k$ and $\tan u = c$ using the symmetries of their graphs, as in section 4.10, to get the values of u in the interval for u.
>
> **Step 4:** Substitute bx for u and hence obtain the values of x.

EXERCISE 6A

1 Given the interval for x is $-360° \leqslant x \leqslant 360°$, find the interval for:

(a) $2x$, **(b)** $4x$, **(c)** $3x$, **(d)** $\frac{x}{2}$, **(e)** $\frac{x}{4}$.

2 Solve the equations for the given interval giving your answers to one decimal place where appropriate:

(a) $\tan 3x = 1$, $-180° \leqslant x \leqslant 180°$,

(b) $\sin 5x = 0.5$, $0° \leqslant x \leqslant 90°$,

(c) $\sin 2x = 0.42$, $0° \leqslant x \leqslant 360°$,

(d) $\cos 3\theta = 0.6$, $-180° \leqslant \theta \leqslant 90°$,

(e) $4 \sin 2\theta = 1$, $-180° \leqslant \theta \leqslant 180°$,

(f) $\tan 3\theta = -8$, $0° \leqslant \theta \leqslant 180°$,

(g) $\tan 2x - 1 = 0$, $0° \leqslant x \leqslant 180°$,

(h) $3 \cos 3\theta + 2 = 0$, $-90° \leqslant \theta \leqslant 90°$,

(i) $2 \tan 2\theta - 3 = 0$, $-180° \leqslant \theta \leqslant 180°$,

(j) $4 \cos \frac{\theta}{3} - 2 = 0$, $-180° \leqslant \theta \leqslant 720°$,

(k) $2 \sin \frac{\theta}{2} = 0.6$, $0° \leqslant \theta \leqslant 360°$,

(l) $10 \sin 3x - 4 = 0$, $0° < x < 180°$,

(m) $\tan(-x) = 1$, $0° \leqslant x \leqslant 360°$.

6.2 Solving equations of the form $\sin(x + b) = k$, $\cos(x + b) = k$ and $\tan(x + b) = c$

In chapter 5 you were shown how to sketch graphs of the form $y = \sin(x + b)$ by translating the graph of $y = \sin x$ parallel to the x-axis. In this section you will be shown how to solve equations of the form $\sin(x + b) = k$ by again applying a simple substitution, although such equations can be solved by sketching and using the graph of $y = \sin(x + b)$ in a similar way to that used in section 4.10.

Worked example 6.4

Solve the equation $\sin(x - 60°) = 0.4$, in the interval $-180° \leqslant x \leqslant 180°$, giving your answer to one decimal place.

Solution

Let $u = x - 60° \quad \Rightarrow \quad \sin u = 0.4$.
Finding the interval for u, subtract $60°$ from each part of the inequality $-180° \leqslant x \leqslant 180°$ to give
$-180° - 60° \leqslant x - 60° \leqslant 180° - 60°$, so the interval for u is
$-240° \leqslant u \leqslant 120°$.

6

Using a calculator, $u = \sin^{-1}(0.4) = 23.578\ldots°$.
Since the graph of $y = \sin u$ is symmetrical about the vertical line $u = 90°$, another solution is $u = 180° - 23.578\ldots° = 156.42\ldots°$.
Since the graph of $y = \sin u$ is periodic with period 360°, solutions of $\sin u = 0.4$ are

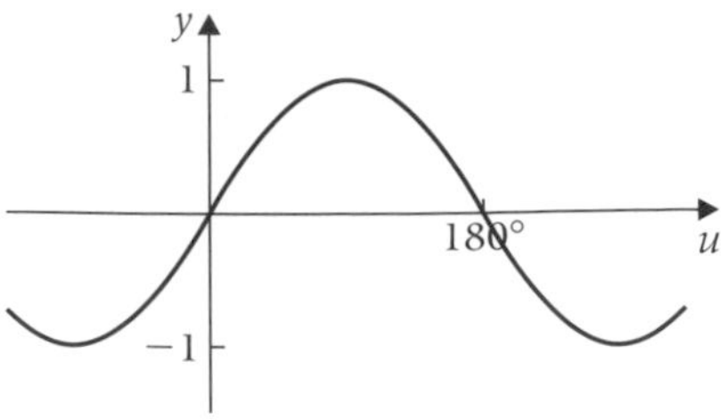

$$u = 23.57\ldots°,\ 23.57\ldots° \pm 360°,\ \ldots$$
$$156.42\ldots°,\ 156.42\ldots° \pm 360°,\ \ldots.$$

Picking out the values of u that lie in the interval $-240° \leqslant u \leqslant 120°$ gives

$$u = 23.57\ldots°,\ -203.58\ldots°.$$
$$x - 60° = 23.57\ldots°,\ -203.58\ldots°$$
$$\Rightarrow \quad x = 83.6°,\ -143.6° \text{ (to one decimal place).}$$

You would expect two solutions of $\sin u = 0.4$ in a 360° cycle. (See Key Point 12 in chapter 4.)

In summary:

To solve equations of the form $\sin(x + b) = k$, $\cos(x + b) = k$ and $\tan(x + b) = c$ in a given interval for x:

Step 1: Let $u = (x + b)$, to get an equation of the form $\sin u = k$, $\cos u = k$ and $\tan u = c$.
Step 2: Find the interval for u.
Step 3: Solve $\sin u = k$, $\cos u = k$ and $\tan u = c$ using the symmetries of their graphs as in section 4.10 to get the values of u in the interval for u.
Step 4: Substitute $x + b$ for u and hence obtain the values of x.

EXERCISE 6B

1 Given the interval for x is $-360° \leqslant x \leqslant 360°$, find the interval for
(a) $x - 90°$, **(b)** $x + 60°$, **(c)** $x - 300°$, **(d)** $x + 230°$.

2 Solve the equations for the given interval giving your answers to one decimal place where appropriate:
(a) $\sin(x + 45°) = 0.5$, $-180° \leqslant x \leqslant 360°$,
(b) $\tan(x - 60°) = -1$, $-360° \leqslant x \leqslant 360°$,
(c) $2\cos(x + 10°) = \sqrt{3}$, $-540° \leqslant x \leqslant 360°$,
(d) $10\sin(x - 160°) - 7 = 0$, $-180° \leqslant x \leqslant 270°$,
(e) $\tan(x - 100°) = -\sqrt{3}$, $-270° \leqslant x \leqslant 360°$,
(f) $5\cos(90° + x) + 3 = 0$, $-180° < x < 270°$.

6.3 Solving quadratic equations in sin x, cos x and tan x

In this section you will be shown how to solve equations of the form:

$$a\sin^2 x + b\sin x + c = 0$$
$$a\cos^2 x + b\cos x + c = 0$$
$$a\tan^2 x + b\tan x + c = 0.$$

$(\sin x)^2$ is usually written as $\sin^2 x$ (similarly $(\cos x)^2$ as $\cos^2 x$ and $(\tan x)^2$ as $\tan^2 x$). This is done to avoid confusion with $\sin x^2$ which means 'square x and then take the sine'.
Think: $\sin^2 x$ is $\sin x$ squared or $\sin x \times \sin x$.

Worked example 6.5

Solve the equation $2\sin^2 x + 3\sin x - 2 = 0$, in the interval $-270° \leq x \leq 180°$.

Solution

Factorising $2\sin^2 x + 3\sin x - 2 = 0$ gives $(2\sin x - 1)(\sin x + 2) = 0$

$\Rightarrow$ either $2\sin x - 1 = 0$ or $\sin x + 2 = 0$

$\Rightarrow$ $\sin x = 0.5$ or $\sin x = -2$.

Consider solutions from the equation $\sin x = 0.5$

$x = \sin^{-1}(0.5) = 30°$

Using the symmetry of the $y = \sin x$ graph, another solution is $x = 180° - 30° = 150°$ and since the graph has period 360°, the solutions of $\sin x = 0.5$ are

$x = 30°, 30° \pm 360°, 150°, 150° \pm 360°, \ldots$

so in the interval $-270° \leq x \leq 180°$, $\sin x = 0.5$ gives $x = 30°$, 150° and −210°.

Consider $\sin x = -2$. Since $-1 \leq \sin x \leq 1$, $\sin x$ cannot equal −2, so $\sin x = -2$ has no solutions.
The only solutions of the equation $2\sin^2 x + 3\sin x - 2 = 0$ in the interval $-270° \leq x \leq 180°$ are $x = 30°$, 150° and −210°.

If you cannot spot the factors, let $u = \sin x$ then write $2\sin^2 x + 3\sin x - 2 = 0$ as $2u^2 + 3u - 2 = 0$ which factorises and leads to $(2u - 1)(u + 2) = 0$.

Worked example 6.6

Solve the equation $3\cos^2\theta - 5\cos\theta + 2 = 0$, in the interval $-360° \leq \theta \leq 360°$ giving your answers to the nearest 0.1°

Solution

Factorising gives $(3\cos\theta - 2)(\cos\theta - 1) = 0$

either

$3\cos\theta - 2 = 0 \Rightarrow \cos\theta = \frac{2}{3} \Rightarrow \cos^{-1}\frac{2}{3} = \theta = 48.189\ldots°$

from which the other solutions are

$\theta = -48.189\ldots°$

$\theta = (360 - 48.189\ldots)° = 311.810\ldots°$

$\theta = (-360 + 48.189\ldots)° = -311.810\ldots°$

or

$\cos\theta - 1 = 0 \Rightarrow \cos\theta - 1 \Rightarrow \theta = 0°$

Other solutions are

$\theta = 360°$ or $\theta = -360°$

The solutions are (to the nearest 0.1°),

$\theta = -360°, -311.8°, -48.2°, 0°, 48.2°, 311.8°, 360°.$

You may not at first glance notice that this is a quadratic equation. If you write the equation as

$3(\cos\theta)^2 - 5\cos\theta + 2 = 0$

and compare with

$3x^2 - 5x + 2 = 0$

then this is seen more easily, with 'x' replaced by '$\cos\theta$'

If $-360° < \theta < 360°$ then these two solutions would be outside the range.

6

Worked example 6.7

Solve the equation $2\tan^2\theta - 3\tan\theta - 1 = 0$, in the interval $0° \leqslant \theta \leqslant 360°$, giving your answer to one decimal place.

Solution

$x = \tan\theta \Rightarrow 2x^2 - 3x - 1 = 0$, which is solved using the quadratic formula $x = \dfrac{-b \pm \sqrt{b^2 - 4ac}}{2a}$.

$a = 2,\ b = -3,\ c = -1$

Since $b^2 - 4ac = 17$, which is not a perfect square, the quadratic in $\tan\theta$ cannot be factorised, so you use the quadratic formula to get:

$$\tan\theta = \frac{-(-3) \pm \sqrt{(-3)^2 - 4(2)(-1)}}{2(2)} = \frac{3 \pm \sqrt{17}}{4}$$

$$\tan\theta = \frac{3 + \sqrt{17}}{4} = 1.7807\ldots \text{ or } \tan\theta = \frac{3 - \sqrt{17}}{4} = -0.2807\ldots$$

When $\tan\theta = 1.7807\ldots$, $\theta = \tan^{-1}(1.7807\ldots) = 60.68\ldots°$.
Since $y = \tan\theta$ is periodic with period 180°, other solutions are $60.68\ldots° \pm 180°$, $60.68\ldots° \pm 360°$, …. In the interval $0° \leqslant \theta \leqslant 360°$, solutions of $\tan\theta = 1.7807\ldots$ are $\theta = 60.7°$ and $240.7°$.

When $\tan\theta = -0.2807\ldots$, $\theta = \tan^{-1}(-0.2807\ldots) = -15.68\ldots°$.
Other solutions are $-15.68\ldots° \pm 180°$, $-15.68\ldots° \pm 360°$, …
In the interval $0° \leqslant \theta \leqslant 360°$, solutions of $\tan\theta = -0.2807\ldots$ are $\theta = 164.3°$ and $344.3°$.

The four solutions of the equation $2\tan^2\theta - 3\tan\theta - 1 = 0$, in the interval $0° \leqslant \theta \leqslant 360°$, are $\theta = 60.7°$, $164.3°$, $240.7°$ and $344.3°$.

EXERCISE 6C

1 Solve the following equations in the interval $-180° \leqslant x \leqslant 180°$:

(a) $\cos^2 x - 1 = 0$, **(b)** $\tan^2 x = 3$,

(c) $4\sin^2 x = 1$, **(d)** $3\tan^2 x = 1$.

2 Solve the following equations in the interval $0° \leqslant \theta < 360°$, to the nearest 0.1°.

(a) $2\cos^2\theta - \cos\theta - 1 = 0$, **(b)** $3\sin^2\theta - \sin\theta = 0$,

(c) $\cos^2\theta - 1 = 0$, **(d)** $5\cos^2\theta + \cos\theta = 0$,

(e) $6\sin^2\theta - \sin\theta = 1$, **(f)** $4\cos^2\theta + 7\cos\theta = 2$,

(g) $6\cos^2\theta + \cos\theta - 1 = 0$, **(h)** $4\sin^2\theta + 3\sin\theta = 1$.

3 Use the quadratic equation formula to solve the following equations in the interval $0° \leqslant x \leqslant 360°$ to the nearest 0.1°.

(a) $\cos^2 x + \cos x - 1 = 0$, **(b)** $\sin^2 x + 6\sin x + 1 = 0$,

(c) $\sin^2 x + 2\sin x = 2$, **(d)** $3\cos^2 x + 6\cos x - 2 = 0$,

(e) $2\sin^2 x + 4\sin x = 3$, **(f)** $4\cos^2 x + 24\cos x + 17 = 0$.

4 Solve the following equations, giving your answers to the nearest 0.1°.

(a) $\tan^2 \theta = \frac{1}{4}$, in the interval $-180° \leqslant \theta \leqslant 270°$,

(b) $\tan^2 x + 14 \tan x + 40 = 0$, in the interval $-180° \leqslant x \leqslant 0°$,

(c) $\tan^2 \theta - \tan \theta - 2 = 0$, in the interval $-180° \leqslant \theta \leqslant 180°$,

(d) $2 \tan^2 \theta - 7 \tan \theta + 2 = 0$, in the interval $0° \leqslant \theta \leqslant 360°$,

(e) $3 \tan^2 x + 3 \tan x = 1$, in the interval $-180° \leqslant x \leqslant 180°$.

6.4 The relationship $\cos^2\theta + \sin^2\theta = 1$

P has coordinates $(\cos \theta, \sin \theta)$ on a unit circle

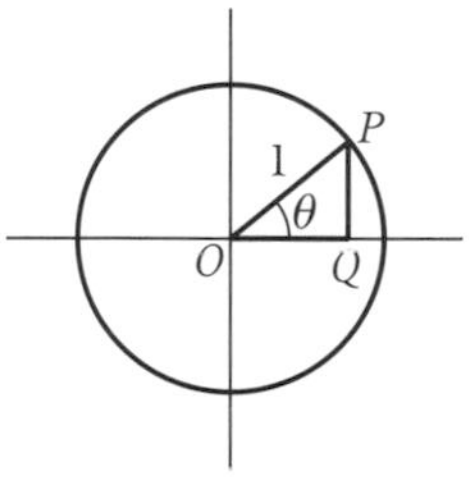

Taking the right-angled triangle OPQ gives

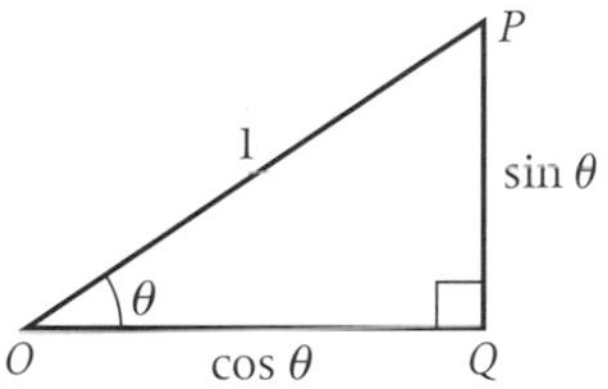

Applying Pythagoras' theorem to the triangle gives

$$(OQ)^2 + (PQ)^2 = 1^2$$
$$(\cos \theta)^2 + (\sin \theta)^2 = 1^2$$
$$\cos^2 \theta + \sin^2 \theta = 1$$

This is an identity (because it is true for all values of θ) and so

$$\cos^2 \theta + \sin^2 \theta \equiv 1$$

You can also write
$\cos^2 \theta \equiv 1 - \sin^2 \theta$
$\sin^2 \theta \equiv 1 - \cos^2 \theta$.

6.5 The relationship between $\sin \theta$, $\cos \theta$ and $\tan \theta$

Taking the right-angled triangles OAB and OPQ and using the idea of similar triangles

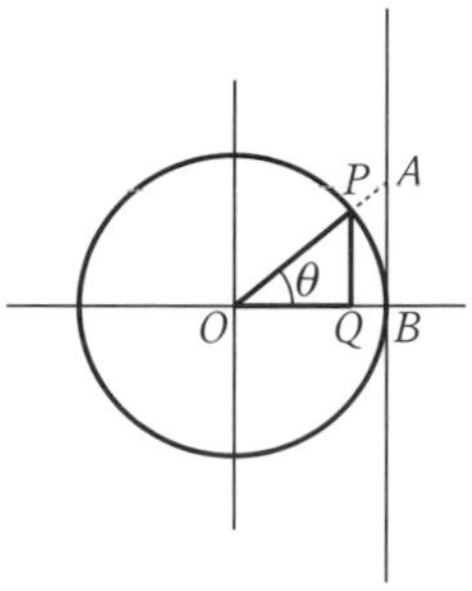

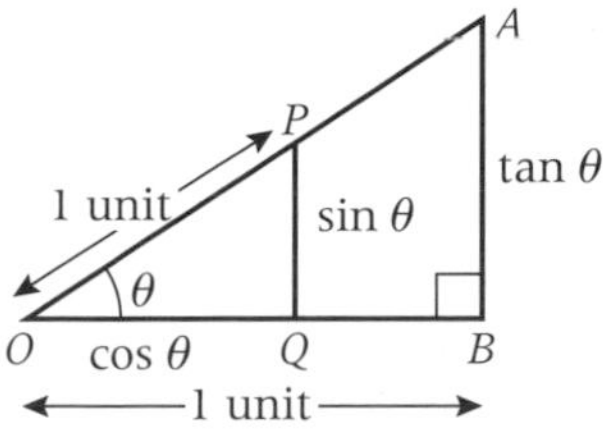

$$\frac{AB}{OB} = \frac{PQ}{OQ} \text{ so that } \frac{\tan\theta}{1} = \frac{\sin\theta}{\cos\theta},$$

since P has coordinates $(\cos\theta, \sin\theta)$.

More simply $\tan\theta = \dfrac{\sin\theta}{\cos\theta}$

The relationship $\tan\theta = \dfrac{\sin\theta}{\cos\theta}$, is also an identity.

An identity is true for all values of the variable, in this case θ, whereas an equation is possibly satisfied by only a few values of the variable.

Hence,

$$\tan\theta \equiv \frac{\sin\theta}{\cos\theta}$$

Provided $\cos\theta \neq 0$.

This is generally given as the definition of $\tan\theta$.

Knowing $\tan\theta \equiv \dfrac{\sin\theta}{\cos\theta}$ means that given any two of these values, the third value can be found.

Worked example 6.8

Find the value of $\tan\theta$ when $\sin\theta = \frac{3}{5}$ and $\cos\theta = -\frac{4}{5}$,

If you use your calculator to work with fractions, always use the fraction button. Do not be tempted to convert fractions to decimals since some fractions do not have an exact decimal representation. For example,

$$\frac{5}{13} = 0.384615384615$$

↑ exact ↑ approximate

Solution

$$\tan\theta \equiv \frac{\sin\theta}{\cos\theta} = \frac{3}{5} \div -\frac{4}{5} = \frac{3}{\cancel{5}} \times -\frac{\cancel{5}}{4} = -\frac{3}{4}$$

It is also possible to create further identities using $\tan\theta \equiv \dfrac{\sin\theta}{\cos\theta}$.

Worked example 6.9

Show that $\dfrac{1-\sin\theta}{\cos\theta} \equiv \dfrac{1}{\cos\theta} - \tan\theta$.

If you are asked to 'show that' or 'prove that' then consider one side of the identity and, step by step, reduce it to the same form as the other side of the identity.

Solution

LHS: $\dfrac{1-\sin\theta}{\cos\theta} \equiv \dfrac{1}{\cos\theta} - \dfrac{\sin\theta}{\cos\theta} \equiv \dfrac{1}{\cos\theta} - \tan\theta$, since

$$\tan\theta \equiv \frac{\sin\theta}{\cos\theta}.$$

You could just as easily have started with the RHS, in this case, and reduced it to the LHS.

The identity $\cos^2\theta + \sin^2\theta \equiv 1$ can be used to simplify other identities.

Worked example 6.10

Show that $\dfrac{2 - \cos^2 x}{1 + \sin^2 x} \equiv 1$.

> Since the numerator and denominator contain $\cos^2 x$ and $\sin^2 x$ it makes sense to try the identity $\cos^2 \theta + \sin^2 \theta \equiv 1$.
>
> For this example, start with the LHS and reduce it to 1.

Solution

$\cos^2 x = 1 - \sin^2 x$

$$\text{LHS: } \frac{2 - \cos^2 x}{1 + \sin^2 x} \equiv \frac{2 - (1 - \sin^2 x)}{1 + \sin^2 x} \equiv \frac{1 + \sin^2 x}{1 + \sin^2 x} \equiv 1$$

> You could also consider the denominator using
>
> $\sin^2 x \equiv 1 - \cos^2 x$

The identity $\cos^2 \theta + \sin^2 \theta \equiv 1$ can also be used to manipulate equations into a form whereby they can be solved more easily.

Worked example 6.11

Solve the equation $\sin^2 x + \cos x + 1 = 0$, giving all answers in degrees in the interval $0° < x < 360°$.

> Contains both sine and cosine terms **but** $\sin^2 x \equiv 1 - \cos^2 x$ so this equation can be written in terms of cosine only by replacing $\sin^2 x$ with $1 - \cos^2 x$.

Solution

Using $\cos^2 x + \sin^2 x \equiv 1$, so that $\sin^2 x \equiv 1 - \cos^2 x$ enables the equation to be written as a quadratic in $\cos x$, so that

$$\underbrace{1 - \cos^2 x}_{\sin^2 x} + \cos x + 1 = 0$$

Rearrange to give

$$\cos^2 x - \cos x - 2 = 0.$$

Factorising

$$(\cos x - 2)(\cos x + 1) = 0.$$

Either $\cos x - 2 = 0$, which does not give any solutions.

or $\cos x = -1 \Rightarrow x = \cos^{-1}(-1) = 180°$

which is the only solution for the given interval as the graph of $y = \sin^2 x + \cos x + 1$ would confirm.

Worked example 6.12

Given that $4 \sin 2x = 3 \cos 2x$, find the value of $\tan 2x$ and hence solve the equation, giving all solutions in degrees to the nearest 0.1° in the interval $-180° \leqslant x \leqslant 180°$.

Solution

Dividing both sides by $\cos 2x$

$$\Rightarrow 4\frac{\sin 2x}{\cos 2x} = 3 \Rightarrow \frac{\sin 2x}{\cos 2x} = \frac{3}{4}$$

$$\Rightarrow \tan 2x = \frac{3}{4}$$

> Use the identity $\tan \theta \equiv \dfrac{\sin \theta}{\cos \theta}$.

Find one value for $2x = \tan^{-1} 0.75 = 36.8699...^\circ$

The interval to be considered is $-360^\circ \leqslant 2x \leqslant 360^\circ$.

Other values are $2x = 180^\circ + 36.87^\circ = 216.87^\circ$

and $2x = -180^\circ + 36.87^\circ = -143.13^\circ$

and $2x = -360^\circ + 36.87^\circ = -323.13^\circ$

Hence the values of x are given by dividing by 2

$\Rightarrow x = -161.6^\circ, -71.6^\circ, 18.4^\circ, 108.4^\circ$.

Other values of $2x$ are obtained by adding and subtracting 180° using the symmetries of the graph of $y = \tan\theta$.

It is customary to present the final answers in ascending order, but provided you have found all the solutions the order does not matter.

EXERCISE 6D

1 Given that x is an acute angle and $\sin x = \frac{12}{13}$, use the identity $\cos^2 x + \sin^2 x \equiv 1$ to show that $13 \cos x = 5$.

2 Find the value of:

(a) $\tan\theta$, when $\sin\theta = -\frac{4}{5}$ and $\cos\theta = \frac{3}{5}$,

(b) $\tan x$, when $\cos x = \frac{5}{13}$ and $\sin x = \frac{12}{13}$,

(c) $\sin\theta$, when $\tan\theta = -\frac{5}{12}$ and $\cos\theta = -\frac{12}{13}$,

(d) $\sin x$, when $\cos x = -\frac{3}{5}$ and $\tan x = \frac{4}{3}$,

(e) $\tan x$, when $\cos x = \frac{24}{25}$ and $\sin x = -\frac{7}{25}$,

(f) $\cos\theta$, when $\tan\theta = \frac{24}{7}$ and $\sin\theta = \frac{24}{25}$,

(g) $\cos x$, when $\sin x = \frac{3}{5}$ and $\tan x = -\frac{3}{4}$.

3 Use the identities $\tan\theta \equiv \frac{\sin\theta}{\cos\theta}$ and $\cos^2 x + \sin^2 x \equiv 1$ to show that:

(a) $(\cos x + \sin x)^2 \equiv 1 + 2\cos x \sin x$,

(b) $1 + \cos^2 x \equiv 2 - \sin^2 x$,

(c) $\sin^2\theta + 2\cos^2\theta \equiv 2 - \sin^2\theta$,

(d) $\sin^2\theta - \cos^2\theta \equiv 2\sin^2\theta - 1$,

(e) $\frac{1 - \sin x}{\tan x} \equiv \frac{1}{\tan x} - \cos x$, **(f)** $\frac{\cos x \sin x}{\tan x} \equiv 1 - \sin^2 x$,

(g) $\frac{\tan x \cos^2 x}{\sin x \cos x} \equiv 1$, **(h)** $\frac{5 - \sin^2\theta}{4 + \cos^2\theta} \equiv 1$,

(i) $\frac{3 + \sin^2\theta}{2 + \cos\theta} \equiv 2 - \cos\theta$, **(j)** $\frac{\tan\theta \sin\theta}{1 - \cos\theta} \equiv 1 + \frac{1}{\cos\theta}$.

Although these are identities and the symbol $\equiv$ has been used, it is very common to switch to an equals sign (=) instead. It is important to realise that these relations are valid for all values of the variable. You are not solving an equation.

However, you rarely write $2x + 3x \equiv 5x$, but usually write $2x + 3x = 5x$ and so in a similar way, you can write

$\cos^2 x + \sin^2 x = 1$.

LHS: put numerator in terms of cosine, then factorise

4 Solve the following equations giving your answers in degrees to the nearest 0.1°:

(a) $2\sin^2 x + 3\cos x + 1 = 0$, in the interval $0° \leqslant x \leqslant 360°$,

(b) $\sin^2\theta + 3\cos\theta = 2$, in the interval $0° \leqslant \theta \leqslant 360°$,

(c) $2\cos^2 x + 4\sin x - 3 = 0$, in the interval $-360° \leqslant x \leqslant 180°$,

(d) $\tan x + \cos x = 0$, in the interval $-180° \leqslant x \leqslant 180°$,

(e) $\tan\theta\sin\theta - \cos\theta = 0$, in the interval $0° \leqslant \theta \leqslant 360°$.

5 Find the value of $\tan x$ given that:

(a) $2\sin x = 5\cos x$,

(b) $4\sin x = \cos x - 2\sin x$,

(c) $\frac{1}{3}\sin x = 4\cos x$,

(d) $\frac{2}{3}\cos x = \frac{1}{2}\sin x$.

6 Solve the following equations giving your answers to one decimal place where appropriate:

(a) $\sin x = \cos x$, in the interval $-360° \leqslant x \leqslant 360°$,

(b) $\sin x - 2\cos x = 0$, in the interval $0° \leqslant x \leqslant 360°$,

(c) $8\sin x + 5\cos x = 0$, in the interval $-180° \leqslant x \leqslant 180°$.

7 Solve each of the following equations, giving all solutions between 0° and 180° to the nearest 0.1°:

(a) $5\sin 3x = 7\cos 3x$,

(b) $2\sin 2x + 5\cos 2x = 0$.

8 **(a)** Express $\frac{8 + \cos^2\theta}{3 - \sin\theta}$ in the form $a + b\sin\theta$, whcrc a and b are integers.

(b) Hence find all solutions of the equation

$$\frac{8 + \cos^2 2x}{3 - \sin 2x} = \frac{5}{2}$$

in the interval $0° \leqslant x \leqslant 360°$.

9 **(a)** Prove the identity $\frac{\sin\theta}{1 + \cos\theta} + \frac{1 + \cos\theta}{\sin\theta} \equiv \frac{2}{\sin\theta}$.

(b) Hence find all solutions of the equation

$$\frac{\sin 3x}{1 + \cos 3x} + \frac{1 + \cos 3x}{\sin 3x} = 5$$

in the interval $0° \leqslant x \leqslant 360°$, giving your answers to the nearest 0.1°.

10 **(a)** Given that $3\cos^2\theta - 4\sin\theta = 4$, show that $3x^2 + 4x + 1 = 0$, where $x = \sin\theta$.

(b) **(i)** Solve $3x^2 + 4x + 1 = 0$.

(ii) Hence find all solutions of the equation $3\cos^2\theta - 4\sin\theta = 4$, in the interval $0° < \theta < 360°$, giving your answers to the nearest degree. [A]

11 (a) Given that $7\sin^2x - 2\cos^2x + 6\sin x = 1$, show that $3\sin^2x + 2\sin x - 1 = 0$.

(b) Hence, find all solutions of the equation $7\sin^2x - 2\cos^2x + 6\sin x = 1$ in the interval $0° \leqslant x \leqslant 360°$, giving your answers to the nearest $0.1°$. [A]

12 (a) Given that $3\cos 5x = 4\sin 5x$, write down the value of $\tan 5x$.

(b) Hence, find all the solutions of the equation $3\cos 5x = 4\sin 5x$ in the interval $0° \leqslant x \leqslant 90°$, giving your answers correct to the nearest $0.1°$. [A]

13 (a) On one diagram sketch the graphs of $y = \cos x$ and $y = \tan x$ for values of x between $0°$ and $360°$.

(b) Show that the x-coordinates of the points of intersection of the graphs of $y = \cos x$ and $y = \tan x$ satisfy the equation

$$\sin^2 x + \sin x - 1 = 0.$$

(c) Hence find the x-coordinates of the points of intersection of the two graphs in the interval $0° \leqslant x \leqslant 360°$, giving your answers to the nearest $0.1°$. [A]

14 Find all solutions in the interval $0° < x < 360°$ of the equation

$$3\tan x + 2\cos x = 0.$$ [A]

Key point summary

1 To solve equations of the form $\sin bx = k$, $\cos bx = k$ and $\tan bx = c$ in a given interval for x: *p290*

Step 1: Let $u = bx$, to get an equation of the form $\sin u = k$, $\cos u = k$ and $\tan u = c$
Step 2: Find the interval for u.
Step 3: Solve $\sin u = k$, $\cos u = k$ and $\tan u = c$ using the symmetries of their graphs as in section 4.10 to get the values of u in the interval for u.
Step 4: Substitute bx for u and hence obtain the values of x.

2 To solve equations of the form $\sin(x + b) = k$, $\cos(x + b) = k$ and $\tan(x + b) = c$ in a given interval for x: *p292*

Step 1: Let $u = (x + b)$, to get an equation of the form $\sin u = k$, $\cos u = k$ and $\tan u = c$.
Step 2: Find the interval for u.

Step 3: Solve $\sin u = k$, $\cos u = k$ and $\tan u = c$ using the symmetries of their graphs as in section 4.10 to get the values of u in the interval for u.

Step 4: Substitute $x + b$ for u and hence obtain the values of x.

3 To solve a quadratic equation in $\sin x$ in a given interval for x: *p292*

Step 1: Factorise or use the quadratic formula to find the value(s) (normally two different values) of $\sin x$, for example, $\sin x = p$ or $\sin x = q$.

Step 2: Consider each of the equations $\sin x = p$ and $\sin x = q$ separately to find all the solutions for x in the given interval.

[The same method is used to solve quadratic equations in $\cos x$ and in $\tan x$.]

4 The identity $\cos^2\theta + \sin^2\theta \equiv 1$ is frequently used to simplify equations into quadratics in either $\sin\theta$ or in $\cos\theta$. *p295*

5 The identity $\tan\theta = \frac{\sin\theta}{\cos\theta}$, where $\cos\theta \neq 0$, is frequently used to solve equations of the form $p\sin b\theta = q\cos b\theta$, where p, q and b are constants. *p296*

6 To prove an identity, consider the expression on one side and try to reduce it to the expression on the other side. If this proves to be too difficult then try to show that the RHS and the LHS are equivalent to a third expression. *p296*

Test yourself	**What to review**
1 Solve the following equations, giving your answers to the nearest 0.1°: **(a)** $2\sin 2\theta = 0.5$, in the interval $0° \leqslant \theta \leqslant 90°$, **(b)** $2\cos 3\theta + 1 = 0$, in the interval $-90° \leqslant \theta \leqslant 90°$. **(c)** $3\tan 4\theta = 4$, in the interval $-90° \leqslant \theta \leqslant 90°$.	*Section 6.1*
2 Solve the equation $\sin(x - 150°) = -0.5$, in the interval $-360° \leqslant x < 360°$.	*Section 6.2*
3 Solve $\tan^2 x - \tan x - 6 = 0$, giving your answers to the nearest 0.1° in the interval $0° \leqslant x \leqslant 360°$.	*Section 6.3*
4 Given that $\sin\theta = \frac{24}{25}$ and that $\cos\theta = -\frac{7}{25}$, find $\tan\theta$.	*Section 6.4*

Test yourself (continued)	What to review
5 Prove the identity $\dfrac{1}{1+\cos\theta}+\dfrac{1}{1-\cos\theta}\equiv\dfrac{2}{\sin^2\theta}$.	*Section 6.4*
6 (a) Given that $\cos x+\sin x=0$, find the value of $\tan x$. **(b)** Hence solve the equation $\cos x+\sin x=0$ in the interval $-180°\leqslant x\leqslant 180°$.	*Section 6.4*

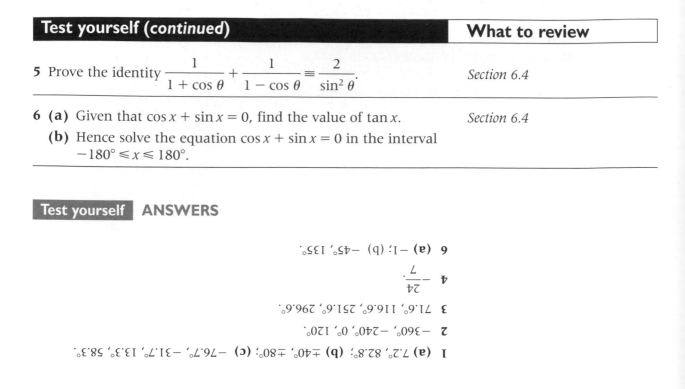

Test yourself ANSWERS

1 (a) 7.2°, 82.8°; **(b)** ±40°, ±80°; **(c)** −76.7°, −31.7°, 13.3°, 58.3°.

2 −360°, −240°, 0°, 120°.

3 71.6°, 116.6°, 251.6°, 296.6°.

4 $-\dfrac{24}{7}$.

6 (a) −1; (b) −45°, 135°.

CHAPTER 7

C2: Factorials and binomial expansions

Learning objectives

After studying this chapter, you should be able to:

- evaluate factorial expressions
- understand the notations $n!$ and $\binom{n}{r}$
- construct the rows of Pascal's Triangle
- use the binomial expansion to expand $(a + b)^n$ for a positive integer n
- use the binomial coefficients $\binom{n}{r}$ to find terms in the expansion of $(a + b)^n$.

7.1 Factorial notation

Suppose you had ten different CDs and you wanted to arrange them in a rack with ten slots.

How many ways could you do it?

You have a choice of ten different CDs to put into the first position. Once this slot is filled, you have nine different ways of filling the second slot, then eight for the third slot. So altogether, the number of different ways of arranging your ten CDs is

$$10 \times 9 \times 8 \times 7 \times 6 \times 5 \times 4 \times 3 \times 2 \times 1.$$

This product is called ten factorial and is denoted by the special symbol 10!. You might wish to check that your calculator has a factorial button. You should then be able to evaluate 10! showing that its value is 3 628 800.

The factorial function grows very rapidly, so that 50! is approximately 3.04×10^{64}. Try to find the largest value of n for which your calculator can evaluate $n!$ Don' be surprised if your calculator cannot evaluate 100! for instance.

> $x!$
>
> You may find the factorial function is in a special statistics or probability menu. Find where it is on your calculator so you can use it in this chapter.

The expression $n!$ is read as 'n factorial' and its value is given by the product

$$n \times (n-1) \times (n-2) \times \ldots \times 3 \times 2 \times 1.$$

Worked example 7.1

Find the values of each of the following:

(a) $5!$, **(b)** $\frac{12!}{10!}$, **(c)** $\frac{6! \times 4!}{(3!)^2}$.

Solution

(a) $5! = 5 \times 4 \times 3 \times 2 \times 1 = 120$;

(b) $\frac{12!}{10!} = \frac{12 \times 11 \times 10 \times 9 \times 8 \times 7 \times 6 \times 5 \times 4 \times 3 \times 2 \times 1}{10 \times 9 \times 8 \times 7 \times 6 \times 5 \times 4 \times 3 \times 2 \times 1}$

$= 12 \times 11 = 132$;

With experience, you will realise without writing down the full product that all the terms from 10 down to 1 cancel out.

(c) $\frac{6! \times 4!}{(3!)^2} = \frac{(6 \times 5 \times 4 \times 3 \times 2 \times 1) \times (4 \times 3 \times 2 \times 1)}{(3 \times 2 \times 1) \times (3 \times 2 \times 1)}$

$= (6 \times 5 \times 4) \times 4$

$= 480.$

Cancel terms rather than evaluate 6! and 4! separately.

7.2 Inductive definition of factorial and 0!

Because $6! = 6 \times 5 \times 4 \times 3 \times 2 \times 1$ and $5! = 5 \times 4 \times 3 \times 2 \times 1$, it follows that

$$6! = 6 \times 5!$$

and similarly

$$10! = 10 \times 9!.$$

In other words, if you know that the value of 9! is 362 880 you can find 10! by simply multiplying the previous answer by 10. Hence 10! = 3 628 800.

So that, in general, the inductive definition of factorial is given by

$$n! = n \times (n-1)!.$$

Using this definition, substituting $n = 1$, gives

$$1! = 1 \times 0!.$$

But since $1! = 1$, it follows that

$$0! = 1.$$

7.3 The notation $\binom{n}{r}$

You will often need to evaluate expressions such as $\dfrac{10!}{7! \times 3!}$.

Notice that $7 + 3 = 10$.

If you have ten objects and need to select exactly seven of them, the number of ways of doing this is denoted by the symbol $\binom{10}{7}$ and its value is $\dfrac{10!}{7! \times 3!} = \dfrac{10 \times 9 \times 8}{3 \times 2 \times 1} = 120$.

In general $\binom{n}{r} = \dfrac{n!}{r!(n-r)!}$ and represents the number of ways that r different objects can be chosen from n objects.

Worked example 7.2

Find the number of different ways that six people can be chosen to go in a lift from a total of nine people who are waiting for the lift.

Solution

Number of ways is $\binom{9}{6} = \dfrac{9!}{6! \times 3!} = \dfrac{9 \times 8 \times 7}{3 \times 2 \times 1} = 84$.

Notice that $\binom{9}{3} = \dfrac{9!}{3! \times 6!} = \dfrac{9 \times 8 \times 7}{3 \times 2 \times 1} = 84$ also.

This is not surprising, because if you choose six people to go in the lift, it is exactly the same as if you choose three people who do **not** go in the lift.

In general $\binom{n}{r} = \binom{n}{n-r}$.

EXERCISE 7A

1 Find the value of:

(a) $3!$, **(b)** $6!$, **(c)** $8!$, **(d)** $(4!)^2$, **(e)** $0!$.

2 Simplify each of the following:

(a) $\frac{5!}{3!}$, **(b)** $\frac{7!}{4!}$, **(c)** $\frac{14!}{13!}$,

(d) $\frac{19!}{18!}$, **(e)** $\frac{8!}{5!}$, **(f)** $\frac{4!}{0!}$.

3 Find the value of each of the following:

(a) $\frac{7!}{(3!)^2}$, **(b)** $\frac{9!}{3! \times 4!}$, **(c)** $\frac{12!}{8! \times 3!}$,

(d) $\frac{18!}{14! \times 2!}$, **(e)** $\frac{20!}{17! \times 4!}$.

4 Evaluate:

(a) $\binom{5}{2}$, **(b)** $\binom{7}{4}$, **(c)** $\binom{3}{3}$, **(d)** $\binom{4}{0}$,

(e) $\binom{8}{6}$, **(f)** $\binom{8}{2}$, **(g)** $\binom{9}{7}$, **(h)** $\binom{25}{24}$,

(i) $\binom{50}{50}$, **(j)** $\binom{13}{3}$, **(k)** $\binom{17}{14}$, **(l)** $\binom{100}{99}$,

(m) $\binom{80}{2}$.

5 There are 20 people at a party and a taxi arrives that can hold five people. In how many different ways could the taxi be filled?

6 There are four tickets for a concert and 30 students each put their name on a piece of paper and four names are selected to get the tickets. In how many different ways could the four be chosen?

7.4 Pascal's Triangle

An expression with two terms and having the form $a + b$ is called a binomial expression and you are often required to find powers of $(a + b)$ such as $(a + b)^5$.

No doubt you can write down straight away that

$$(a + b)^2 = a^2 + 2ab + b^2.$$

Hence
$$\begin{aligned}(a + b)^3 &= (a + b)(a^2 + 2ab + b^2)\\ &= a^3 + 2a^2b + ab^2\\ &\quad + a^2b + 2ab^2 + b^3\\ &= a^3 + 3a^2b + 3ab^2 + b^3\end{aligned}$$

Multiply everything in the second bracket by a then multiply everything by b, and collect like terms.

In a similar way, you can expand $(a+b)^4$ to give

$$(a+b)^4 = a^4 + 4a^3b + 6a^2b^2 + 4ab^3 + b^4$$

but it is quite tedious.

An interesting pattern is produced when you expand $(a+b)^n$ for different integers n.

$$\begin{aligned}
(a+b)^0 &= 1\\
(a+b)^1 &= a+b\\
(a+b)^2 &= a^2 + 2ab + b^2\\
(a+b)^3 &= a^3 + 3a^2b + 3ab^2 + b^3\\
(a+b)^4 &= a^4 + 4a^3b + 6a^2b^2 + 4ab^3 + b^4
\end{aligned}$$

The triangular array of numbers forming the coefficients is called Pascal's Triangle. Each term is formed by adding together the two terms immediately above.

$$\begin{array}{ccccccccccccc}
 & & & & & & 1 & & & & & & \\
 & & & & & 1 & & 1 & & & & & \\
 & & & & 1 & & 2 & & 1 & & & & \\
 & & & 1 & & 3 & & 3 & & 1 & & & \\
 & & 1 & & 4 & & 6 & & 4 & & 1 & & \\
 & 1 & & 5 & & 10 & & 10 & & 5 & & 1 & \\
1 & & 6 & & 15 & & 20 & & 15 & & 6 & & 1
\end{array}$$

Notice that the coefficients are the numbers in the row of Pascal's Triangle that starts 1 4 ....

Worked example 7.3

Write down the next two rows of Pascal's Triangle.

Solution

Since $1 + 6 = 7$, $6 + 15 = 21$, $15 + 20 = 35$, etc. and the row begins with 1 (which can be thought of as $0 + 1$ from the previous row), the next row of Pascal's Triangle is

1 7 21 35 35 21 7 1.

The row after that is obtained in a similar way and is

1 8 28 56 70 56 28 8 1.

Pascal's Triangle can be used to find the expansion of $(a+b)^n$ for small positive integers n.

You have already seen that

$$(a+b)^4 = a^4 + 4a^3b + 6a^2b^2 + 4ab^3 + b^4.$$

This can be written as

$$(a+b)^4 = \mathbf{1}a^4b^0 + \mathbf{4}a^3b^1 + \mathbf{6}a^2b^2 + \mathbf{4}a^1b^3 + \mathbf{1}a^0b^4.$$

Also, the terms consist of a product of the form a^pb^q, where $p+q=4$.

Worked example 7.4

Use Pascal's Triangle to find the expansion of $(x+y)^6$.

Solution

Since the index is 6, you need the row of Pascal's Triangle with values

1 6 15 20 15 6 1

Hence $(x+y)^6 = 1x^6y^0 + 6x^5y^1 + 15x^4y^2 + 20x^3y^3 + 15x^2y^4$
$+ 6x^1y^5 + 1x^0y^6$

which can be written in simplified form as

$$(x+y)^6 = x^6 + 6x^5y + 15x^4y^2 + 20x^3y^3 + 15x^2y^4 + 6xy^5 + y^6.$$

Since $x^0 = 1$ and $y^0 = 1$, etc.

Worked example 7.5

Use Pascal's Triangle to find the expansions of:

(a) $(2+q)^4$, **(b)** $(3x+1)^5$, **(c)** $(1-2x)^6$.

Solution

(a) Since the index is 4, you need the row of Pascal's Triangle with values 1 4 6 4 1.

Although the first term in the bracket is 2, you treat it in the same way as in the previous example.

$$\begin{aligned}(2+q)^4 &= 1\times 2^4q^0 + 4\times 2^3q^1 + 6\times 2^2q^2 + 4\times 2^1q^3 + 1\times 2^0q^4\\ &= 16q^0 + 4\times 8q^1 + 6\times 4q^2 + 4\times 2q^3 + 1\times 1q^4\\ &= 16 + 32q + 24q^2 + 8q^3 + q^4\end{aligned}$$

This expansion is said to be written in ascending powers of q.

(b) The power this time is 5, so you need the row of Pascal's Triangle with values 1 5 10 10 5 1.

$$\begin{aligned}(3x+1)^5 = {} & 1\times(3x)^5\times 1^0 + 5\times(3x)^4\times 1^1\\ & + 10\times(3x)^3\times 1^2 + 10\times(3x)^2\times 1^3\\ & + 5\times(3x)^1\times 1^4 + 1\times(3x)^0\times 1^5\end{aligned}$$

Hence,

$$\begin{aligned}(3x+1)^5 = {} & 3^5x^5 + 5\times 3^4x^4 + 10\times 3^3x^3 + 10\times 3^2x^2\\ & + 5\times 3x + 1\end{aligned}$$

which becomes

$$(3x+1)^5 = 243x^5 + 405x^4 + 270x^3 + 90x^2 + 15x + 1$$

This expansion is said to be written in descending powers of x.

(c) Because the power is 6, you need to use the coefficients

1 6 15 20 15 6 1.

An added difficulty here is the minus sign and so it is best to write $(1-2x)^6$ as $[1+(-2x)]^6$.

$$\begin{aligned}[1+(-2x)]^6 &= 1\times 1^6\times(-2x)^0 + 6\times 1^5\times(-2x)^1\\ &\quad +15\times 1^4\times(-2x)^2 + 20\times 1^3\times(-2x)^3\\ &\quad + 15\times 1^2\times(-2x)^4 + 6\times 1^1\times(-2x)^5 + 1\times 1^0\times(-2x)^6\\ &= 1 + 6\times(-2x) + 15\times(-2)^2x^2 + 20(-2)^3x^3\\ &\quad +15\times(-2)^4x^4 + 6\times(-2)^5x^5 + (-2)^6x^6\end{aligned}$$

This expansion simplifies to

$$1 - 12x + 60x^2 - 160x^3 + 240x^4 - 192x^5 + 64x^6.$$

Notice how the signs alternate when a binomial expansion contains a minus sign.

EXERCISE 7B

1 Write down the first ten rows of Pascal's Triangle.

2 One of the rows of Pascal's Triangle begins

1 15 105 455 1365.

Find the first five terms of each of the next two rows.

3 One of the rows of Pascal's Triangle begins

1 20 190 1140 4845.

Find the first five terms of each of the next two rows.

4 Use Pascal's Triangle to find the expansions of:

(a) $(a+b)^3$, **(b)** $(c+d)^5$,

(c) $(x+y)^7$, **(d)** $(r+s)^8$.

5 Expand each of the following:

(a) $(1+x)^5$, **(b)** $(1+2y)^3$,

(c) $\left(1+\frac{x}{3}\right)^6$, **(d)** $\left(1+\frac{p}{2}\right)^4$.

6 Obtain the expansion of:

(a) $(1-a)^3$, **(b)** $(1-3b)^4$,

(c) $\left(1-\frac{x}{2}\right)^5$, **(d)** $\left(1-\frac{2x}{3}\right)^3$.

7 Find the expansion of each of the following:

(a) $(3+p)^5$, **(b)** $(5+x)^4$,

(c) $(2-m)^3$, **(d)** $(3-2t)^6$.

8 Find the first four terms in increasing powers of x of:

(a) $(1 - x)^5$,

(b) $(1 + 3x)^4$,

(c) $\left(1 - \frac{x}{3}\right)^6$,

(d) $\left(1 + \frac{2x}{5}\right)^5$.

9 Find the first four terms in decreasing powers of y of:

(a) $(y + 2)^5$, **(b)** $(2y - 1)^6$,

(c) $(4y + 3)^4$, **(d)** $(3y - 1)^7$.

10 Simplify each of the following:

(a) $(1 + 3x)^4 - (1 - 3x)^4$,

(b) $(3 - 2y)^5 + (3 + 2y)^5$.

11 Show that the first four rows of Pascal's Triangle can be written as

$$\binom{0}{0}$$

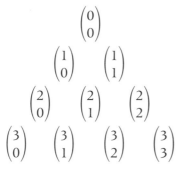

$$\binom{1}{0} \quad \binom{1}{1}$$

$$\binom{2}{0} \quad \binom{2}{1} \quad \binom{2}{2}$$

$$\binom{3}{0} \quad \binom{3}{1} \quad \binom{3}{2} \quad \binom{3}{3}$$

Write the next two rows of Pascal's triangle in a similar form.

7.5 Binomial coefficients

Expressions of the form $\binom{n}{r}$ are called the binomial coefficients.

Read as binomial coefficient n r.

As seen in the previous exercise, the binomial coefficients are in fact the same as the numbers that appeared in Pascal's Triangle.

When n is a small positive integer, it is a good idea to use Pascal's Triangle to expand $(a + b)^n$.

However, there is a more general expression that can be used without needing Pascal's Triangle.

The problem with using Pascal's Triangle is that if you want to know the numbers in row thirteen, you need to know the numbers in row twelve, and so on.

Imagine you are trying to expand $(a+b)^7$. This could be written out as

$$(a+b)(a+b)(a+b)(a+b)(a+b)(a+b)(a+b).$$

You could reason equally well that you are selecting the three brackets which have the letter a in the product, but you will get the same answer since $\binom{7}{4}=\binom{7}{3}$.

You know, for instance, that one of the terms is going to be a multiple of a^3b^4. You can select as from some brackets and bs from others. Since there are seven brackets, you can choose the bs in $\binom{7}{4}$ ways. This term in the expansion is therefore

$$\binom{7}{4}a^3b^4.$$

Another term is going to be $\binom{7}{2}a^5b^2$, and the leading term, when you select the a from each bracket, is a^7.

The full expansion of $(a+b)^7$ is

$$a^7+\binom{7}{1}a^6b+\binom{7}{2}a^5b^2+\binom{7}{3}a^4b^3+\binom{7}{4}a^3b^4$$
$$+\binom{7}{5}a^2b^5+\binom{7}{6}ab^6+b^7.$$

This result can now be generalised for $(a+b)^n$.

> The general binomial expansion can be written as
>
> $$(a+b)^n=a^n+\binom{n}{1}a^{n-1}b+\binom{n}{2}a^{n-2}b^2+\dots$$
> $$+\binom{n}{r}a^{n-r}b^r+\dots+b^n$$
>
> where n is a positive integer.

This formula appears in the formula booklet under the heading *Binomial Series*.

Worked example 7.6

Find the first four terms, in ascending powers of x, in the expansion of:

(a) $(1+3x)^{12}$, **(b)** $(2-x)^{20}$.

In **(b)**, leave your answer in terms of powers of 2.

Solution

(a) Using the formula above, with $a=1$, $b=3x$ and $n=12$ gives

$$(1+3x)^{12}=1^{12}+\binom{12}{1}\times 1^{11}\times(3x)$$
$$+\binom{12}{2}\times 1^{10}\times(3x)^2+\binom{12}{3}\times 1^{9}\times(3x)^3+\dots$$

You only need the first four terms.

Since $\binom{12}{1} = \frac{12!}{11! \times 1!} = 12,$

$\binom{12}{2} = \frac{12!}{10! \times 2!} = \frac{12 \times 11}{2 \times 1} = 66$ and

$\binom{12}{3} = \frac{12!}{9! \times 3!} = \frac{12 \times 11 \times 10}{3 \times 2 \times 1} = 220,$

$$(1 + 3x)^{12} = 1 + 12 \times (3x) + 66 \times 3^2x^2 + 220 \times 3^3x^3 + \ldots$$

Hence, the first four terms of $(1 + 3x)^{12}$ in ascending powers of x are

$$1 + 36x + 594x^2 + 5940x^3.$$

(b) This time, the values to substitute into the formula are $a = 2$, $b = -x$ and $n = 20$, so you get

$$(2 - x)^{20} = 2^{20} + \binom{20}{1} \times 2^{19} \times (-x) + \binom{20}{2} \times 2^{18} \times (-x)^2 + \binom{20}{3} \times 2^{17} \times (-x)^3 + \ldots$$

Since $\binom{20}{1} = \frac{20!}{19! \times 1!} = 20,$

$\binom{20}{2} = \frac{20!}{18! \times 2!} = \frac{20 \times 19}{2 \times 1} = 190$ and

$\binom{20}{3} = \frac{20!}{17! \times 3!} = \frac{20 \times 19 \times 18}{3 \times 2 \times 1} = 1140,$

$$\begin{aligned}(2 - x)^{20} &= 2^{20} - 20 \times 2^{19} \times x + 190 \times 2^{18} \times x^2 - 1140 \times 2^{17} \times x^3 + \ldots \\ &= 2^{20} - 5 \times 2^2 \times 2^{19} \times x + 95 \times 2 \times 2^{18} \times x^2 - 285 \times 2^2 \times 2^{17} \times x^3 + \ldots\end{aligned}$$

Notice how the terms alternate between plus and minus.

Hence, the first four terms of $(2 - x)^{20}$ in ascending powers of x are

$$2^{20} - 5 \times 2^{21}x + 95 \times 2^{19}x^2 - 285 \times 2^{19}x^3.$$

Worked example 7.7

Find the coefficient of x^{15} in the expansion of $(3 - 2x)^{18}$, leaving your answer as the product of prime numbers.

Solution

The term required is of the form $\binom{18}{15} \times 3^3 \times (-2x)^{15}$.

$$\binom{18}{15} = \frac{18!}{15! \times 3!} = \frac{18 \times 17 \times 16}{3 \times 2 \times 1} = 3 \times 17 \times 2^4.$$

So $\binom{18}{15} \times 3^3 \times (-2x)^{15} = -3 \times 17 \times 2^4 \times 3^3 \times 2^{15}x^{15}$

Hence the coefficient of x^{15} in the expansion of $(3 - 2x)^{18}$ is

$$-2^{19} \times 3^4 \times 17$$

Worked examination question 7.8

Find the binomial expansion of $\left(1 + \frac{1}{2}x\right)^{16}$ in ascending powers of x up to and including the term in x^3.
Hence determine the coefficient of x^3 in the expansion of $(23 + 13x)\left(1 + \frac{1}{2}x\right)^{16}$. [A]

Solution

$$\left(1 + \tfrac{1}{2}x\right)^{16} = 1^{16} + \binom{16}{1} \times 1^{15} \times \left(\tfrac{1}{2}x\right) + \binom{16}{2} \times 1^{14} \times \left(\tfrac{1}{2}x\right)^2 + \binom{16}{3} \times 1^{13} \times \left(\tfrac{1}{2}x\right)^3 + \ldots$$

Since $\binom{16}{1} = \dfrac{16!}{15! \times 1!} = 16,$

$$\binom{16}{2} = \frac{16!}{14! \times 2!} = \frac{16 \times 15}{2 \times 1} = 120 \quad \text{and}$$

$$\binom{16}{3} = \frac{16!}{13! \times 3!} = \frac{16 \times 15 \times 14}{3 \times 2 \times 1} = 560.$$

Hence, $\left(1 + \frac{1}{2}x\right)^{16} = 1 + 16 \times \left(\frac{1}{2}x\right) + 120 \times \left(\frac{1}{2}x\right)^2 + 560 \times \left(\frac{1}{2}x\right)^3 + \ldots$
$= 1 + 8x + 30x^2 + 70x^3 + \ldots$

In order to find the coefficient of x^3 in the expansion of

$$(23 + 13x)\left(1 + \tfrac{1}{2}x\right)^{16}$$

it is necessary to multiply 23 by the term in x^3 from the previous expansion and to multiply $13x$ by the term in x^2 from the expansion of $\left(1 + \frac{1}{2}x\right)^{16}$.

This gives $23 \times 70x^3 + 13x \times 30x^2 = 2000x^3$.

The coefficient of x^3 is therefore 2000.

EXERCISE 7C

1 Find the first four terms in ascending powers of x for each of the following:

(a) $(1 + x)^{11}$, **(b)** $(1 + x)^{15}$, **(c)** $(1 - x)^8$,
(d) $(1 + 2x)^7$, **(e)** $(1 - x)^{12}$, **(f)** $\left(1 + \frac{1}{3}x\right)^9$,
(g) $\left(1 - \frac{1}{2}x\right)^{10}$, **(h)** $(2 - x)^9$.

7

2 Obtain the first four terms in descending powers of x for each of the following:

(a) $(x + 1)^{12}$, **(b)** $(x + 2)^{17}$,

(c) $(x - 3)^{10}$, **(d)** $(2x - 1)^7$.

3 For each of the following binomial expansions, find the term indicated:

(a) $(1 + 2x)^{15}$ term in x^4, **(b)** $(1 + 3x)^{17}$ term in x^2,

(c) $(1 - 4x)^{13}$ term in x^3, **(d)** $(2 - x)^{14}$ term in x^{12},

(e) $\left(1 + \frac{2}{3}x\right)^{12}$ term in x^4, **(f)** $(3 - x)^{18}$ term in x^{14}.

4 For each of the following binomial expansions, find the coefficients indicated:

(a) $(1 - x)^{11}$ coefficient of x^7,

(b) $(1 + 2x)^{13}$ coefficient of x^4,

(c) $(2 - 3x)^{14}$ coefficient of x^3,

(d) $(2 - x)^{15}$ coefficient of x^{11},

(e) $\left(1 + \frac{1}{3}x\right)^9$ coefficient of x^4,

(f) $(5 - x)^{17}$ coefficient of x^{15}.

5 Find the binomial expansion of $\left(1 + \frac{1}{2}x\right)^8$ in ascending powers of x up to and including the term in x^3. Simplify the coefficients as much as possible. [A]

6 Write down the binomial expansion of $(1 + x)^7$ in ascending powers of x up to and including the term in x^3. Hence determine the value of $1.000\,01^7$ correct to 15 decimal places. [A]

7 Find the coefficient of the term in x^4 in the binomial expansion of $(3 + 2x)^7$. [A]

8 **(a)** Write down the first four terms in ascending powers of x in the expansion of $(1 + x)^8$, simplifying your coefficients as much as possible.

(b) Find the coefficient of x^3 in the expansion of $(3 - 2x)(1 + x)^8$. [A]

9 A polynomial can be expressed in the form $p(x) = (x - 2)^4 - (x + 1)^3$.

(a) Find the remainder when $p(x)$ is divided by $x - 3$.

(b) Use binomial expansions to express $p(x)$ in the form $x^4 + ax^3 + bx^2 + cx + d$, where a, b, c and d are integers. [A]

10 **(a)** Write down the first four terms in ascending powers of x in the expansion of $(1 + x)^{10}$, simplifying your coefficients as much as possible.

(b) Find the coefficient of x^4 in the expansion of $(5 - 2x)(1 + x)^{10}$.

11 A polynomial can be expressed in the form $p(x) = (x + 1)^5 - (x - 1)^4$. Use binomial expansions to express $p(x)$ in the form $x^5 + ax^4 + bx^3 + cx^2 + dx$, where a, b, c and d are integers.

12 Find the binomial expansion of $(1 - 3x)^{11}$ in ascending powers of x up to and including the term in x^3. Hence determine the coefficient of x^3 in the expansion of $(2 + 7x)(1 - 3x)^{11}$.

13 **(a)** Simplify $\dfrac{(2 + h)^5 - 32}{h}$.

(b) A curve has equation $y = x^5$ and the points P and Q lie on the curve and have x-coordinates equal to 2 and $2 + h$, respectively. By considering the gradient of the chord PQ and letting h tend to zero, deduce the gradient of the curve at P.

This question has been included for those students who might be taking the FPI module.

14 Find the first four terms in ascending powers of x in the expansion of:

(a) $(1 + 3x)^{12}(5 - 4x)$, **(b)** $\left(1 - \frac{1}{4}x\right)^{16}(2 + 3x)$.

15 Prove that $\binom{n+1}{k} = \binom{n}{k} + \binom{n}{k-1}$.

7

Key point summary

1 The expression $n!$ is read as 'n factorial' and its value is given by the product $n \times (n - 1) \times (n - 2) \times \ldots \times 3 \times 2 \times 1$. *p304*

2 $n! = n \times (n - 1)!$. *p304*

3 $0! = 1$. *p305*

4 In general $\binom{n}{r} = \dfrac{n!}{r!(n - r)!}$ and represents the number of ways that r different objects can be chosen from n objects. *p305*

5 In general $\binom{n}{r} = \binom{n}{n-r}$. *p305*

6 The first rows of Pascal's Triangle are *p307*

$$
\begin{array}{ccccccccccccc}
 & & & & & & 1 & & & & & & \\
 & & & & & 1 & & 1 & & & & & \\
 & & & & 1 & & 2 & & 1 & & & & \\
 & & & 1 & & 3 & & 3 & & 1 & & & \\
 & & 1 & & 4 & & 6 & & 4 & & 1 & & \\
 & 1 & & 5 & & 10 & & 10 & & 5 & & 1 & \\
1 & & 6 & & 15 & & 20 & & 15 & & 6 & & 1
\end{array}
$$

7 Pascal's Triangle is useful for expanding $(a + b)^n$ when n is a small positive integer. p308

8 The general binomial expansion can be written as p311

$$(a + b)^n = a^n + \binom{n}{1}a^{n-1}b + \binom{n}{2}a^{n-2}b^2 + \dots + \binom{n}{r}a^{n-r}b^r + \dots + b^n$$

where n is a positive integer.

Test yourself	What to review
1 Find the value of each of the following: **(a)** $3!$, **(b)** $0!$, **(c)** $\frac{5!}{3!}$.	*Sections 7.1 and 7.2*
2 Evaluate: **(a)** $\binom{7}{2}$, (b) $\binom{8}{8}$, **(c)** $\binom{47}{46}$.	*Section 7.3*
3 One row of Pascal's Triangle begins 1 19 171 969 … Find the first four terms of the next row.	*Section 7.4*
4 Find the expansion of $(x - w)^8$.	*Section 7.4*
5 Find the first four terms in ascending powers of x in the binomial expansion of $(1 - 2x)^{13}$.	*Section 7.5*
6 Find the coefficient of x^4 in the binomial expansion of $(2 + 3x)^{11}$.	*Section 7.5*

Test yourself ANSWERS

1 **(a)** 6; **(b)** 1; **(c)** 20.

2 **(a)** 21; **(b)** 1; **(c)** 47.

3 1 20 190 1140.

4 $x^8 - 8wx^7 + 28w^2x^6 - 56w^3x^5 + 70w^4x^4 - 56w^5x^3 + 28w^6x^2 - 8w^7x + w^8$.

5 $1 - 26x + 312x^2 - 2288x^3$.

6 3 421 440.

CHAPTER 8

C2: Sequences and series

Learning objectives

After studying this chapter, you should be able to:

- generate sequences using formulae or inductive definitions
- understand the difference between sequences and series
- understand what is meant by an arithmetic series
- find the nth term and the sum of the first n terms of an arithmetic series
- find the sum of the first n natural numbers.

8.1 Sequences

Probably from an early age you will have been presented with puzzles such as the following.

Find the next term in each of the following

(a) 1, 8, 27, 64, ____

(b) 0, 3, 8, 15, ____

These patterns of numbers, separated by commas, are called **sequences**.

Provided you could justify your answer, you could insert almost any number in the missing space.

For instance, although most people might argue that 125 is ***the*** answer to puzzle sequence **(a)** above, you might have been using the formula

$$n^3 + 7(n-1)(n-2)(n-3)(n-4)$$

to generate the terms 1, 8, 27, 64 in which case substituting $n = 5$ gives 293.

No doubt you can invent many more formulae which would give a different value for the 5th term.

Try to make up some formulae which would lead to an answer different from **24** for the missing term in sequence **(b)** above.

8.2 Suffix notation

> The most common notation for the terms of a sequence is the suffix notation.
>
> The first term is often written as t_1, the second term t_2, the third t_3 and so on, with the nth term being t_n.

This is very similar to function notation, so that if

$$t_n = n^2 + 1$$

then $t_1 = 1^2+1 = 2, \quad t_2 = 2^2 + 1 = 5$, etc.

Rather like $f(n) = n^2 + 1$, so that $f(1) = 1^2 + 1 = 2$ and $f(2) = 2^2 + 1 = 5$.

Any letter with suffices can be used to represent the terms of a sequence, but the most common letter used is u and the terms of the sequence are $u_1, u_2, u_3, \ldots$

This can be thought of as a function u where you calculate $u_1, u_2, u_3, \ldots$ as the next example illustrates.

Worked example 8.1

Given that $u_n = n(n + 2)$, find u_3 and u_5.

Find u_{n-1}.

Hence, find an expression for $u_n - u_{n-1}$ and simplify your answer.

Solution

$u_3 = 3(3 + 2) = 15$, $u_5 = 5(5 + 2) = 35$

Substituting $n = 3$, $n = 5$ in turn into the formula for u_n.

To find u_{n-1} it is necessary to replace n by $n - 1$ in the formula.

$$u_{n-1} = (n - 1)\ (n - 1 + 2) = (n - 1)\ (n + 1) = n^2 - 1.$$

Also, $u_n = n(n + 2) = n^2 + 2n$

Hence $u_n - u_{n-1} = n^2 + 2n - (n^2 - 1) = 2n + 1$.

8.3 Inductive definition

The previous worked example showed that $u_n - u_{n-1} = 2n + 1$.

This can be rewritten as $u_n = u_{n-1} + 2n + 1$.

Provided you know the first term, and in this case $u_1 = 3$, it is now possible to find all the terms in the sequence.

$$u_1 = 3 \Rightarrow u_2 = u_1 + 2 \times 2 + 1 = 3 + 4 + 1 = 8$$
$$u_2 = 8 \Rightarrow u_3 = u_2 + 2 \times 3 + 1 = 8 + 6 + 1 = 15$$

Continue this process and verify that $u_5 = 35$.

This agrees with the answer for u_3 in Worked example 8.1 when you were given the formula.

The **inductive definition** for this sequence is

$u_1 = 3, \quad u_n = u_{n-1} + 2n + 1.$

Notice you need two things: the first term and an expression relating one term to the previous term.

Worked example 8.2

Find the first four terms of the sequence defined inductively by $u_1 = 2, u_n = u_{n-1} + 3n^2$.

Solution

$u_1 = 2 \Rightarrow u_2 = u_1 + 3 \times 2^2 = 2 + 12 = 14$

$u_2 = 14 \Rightarrow u_3 = u_2 + 3 \times 3^2 = 14 + 27 = 41$

$u_3 = 41 \Rightarrow u_4 = u_3 + 3 \times 4^2 = 41 + 48 = 89$

The first four terms are 2, 14, 41, 89

You can produce sequences of this type very easily on a spreadsheet.

	A	B
1	1	2
2	2	14
3	3	41
4	4	89
5	5	164
6	6	272
7	7	419
8	8	611
9	9	854
10	10	1154

	A	B
1	1	2
2	=A1+1	=B1+3*A2^2
3	=A2+1	=B2+3*A3^2
4	=A3+1	=B3+3*A4^2
5	=A4+1	=B4+3*A5^2
6	=A5+1	=B5+3*A6^2
7	=A6+1	=B6+3*A7^2
8	=A7+1	=B7+3*A8^2
9	=A8+1	=B8+3*A9^2
10	=A9+1	=B9+3*A10^2

8.4 Limit of a sequence

Suppose a sequence defined inductively is given by

$$u_{n+1} + \frac{1}{2}u_n + \frac{1}{4}, \text{ where } u_1 = 1,$$

then $u_2 = \frac{1}{2}u_1 + \frac{1}{4} = \frac{1}{2} + \frac{1}{4} = \frac{3}{4}$.

Also, $u_3 = \frac{1}{2}u_2 + \frac{1}{4} = \frac{3}{8} + \frac{1}{4} = \frac{5}{8}$.

This sequence can be continued.

By converting the answers into decimals, you should start to see that the sequence is settling down to a particular value.

The sequence is 1, 0.75, 0.625, 0.5625, 0.531 25, 0.515, 625, ... Why not try using a spreadsheet to see what value the sequence seems to be approaching?

You can find this limit algebraically.
Suppose u_n is tending to the limit L. Then u_{n+1} must be tending to the same limit L.

You can form an equation:

$$L = \frac{1}{2}L + \frac{1}{4}$$

Hence, $\frac{1}{2}L = \frac{1}{4} \Rightarrow L = \frac{1}{2}$.

The limit of the sequence is $\frac{1}{2}$.

Worked example 8.3

A sequence is defined inductively by

$$u_{n+1} = 2 - \frac{1}{3}u_n,\ u_1 = 3.$$

(a) Find the values of u_2, u_3, u_4 and u_5.

(b) Find the limiting value of u_n as n tends to infinity.

Solution

(a) $u_2 = 2 - \frac{1}{3}u_1 = 2 - 1 = 1$

Hence, $u_3 = 2 - \frac{1}{3}u_2 = 2 - \frac{1}{3} = \frac{5}{3}$.

Also $u_4 = 2 - \frac{1}{3}u_3 = 2 - \frac{5}{9} = \frac{13}{9}$.

And therefore $u_5 = 2 - \frac{1}{3}u_4 = 2 - \frac{13}{27} = \frac{41}{27}$.

On a calculator, the values of the sequence, to three decimal places are

1, 1.667, 1.444, 1.519, ...

which seem to be oscillating above and below 1.5.

(b) In order to find the limit of the sequence, let u_n tend to the limit L. Then u_{n+1} must also tend to L.
You can form an equation

$$L = 2 - \frac{1}{3}L$$

Hence, $\frac{4}{3}L = 2 \Rightarrow L = \frac{3}{2}$.

The limit is therefore 1.5.

EXERCISE 8A

1 Use the formula $s_n = n^2$, to find s_1, s_2 and s_3.

2 Use the formula $t_n = 3^n - 1$, to find t_1, t_3 and t_5.

3 Write down the first five terms of each of the following sequences:

(a) $u_n = 2n + 1$, **(b)** $u_n = n^2 - 1$, **(c)** $u_n = n(n - 1)$,
(d) $u_n = n^3 + 1$, **(e)** $u_n = 2^n$, **(f)** $u_n = 3^n - 1$.

4 Find the first five terms of the following sequences which are defined inductively:

(a) $u_1 = 4$, $u_n = u_{n-1} + 3$, **(b)** $u_1 = 20$, $u_n = u_{n-1} - 2$,

(c) $u_1 = 3$, $u_n = u_{n-1} + n^2$, **(d)** $u_1 = 2$, $u_n = u_{n-1} + 3n - 2$,

(e) $u_1 = 1$, $u_n = u_{n-1} + 5n^2$, **(f)** $u_1 = 100$, $u_n = u_{n-1} - 2n^2$.

5 For each of the following sequences, find $u_n - u_{n-1}$ simplifying your answer where possible.

(a) $u_n = 3n + 2$, **(b)** $u_n = 5n - 3$,

(c) $u_n = 3 - 2n$, **(d)** $u_n = n^2$,

(e) $u_n = n^2 - 3$, **(f)** $u_n = 3n^2 + 2$,

(g) $u_n = n(3n + 1)$.

6 Find a possible formula for the *n*th term of each of the following sequences:

(a) 1, 3, 5, 7, 9, …, **(b)** 2, 4, 8, 16, 32, …,

(c) 6, 5, 4, 3, 2, …, **(d)** $\frac{1}{2}, \frac{2}{3}, \frac{3}{4}, \frac{4}{5}, \ldots$.

7 Write down the first four terms of each of the following sequences and try to find a formula for u_n.

(a) $u_1 = 5$, $u_n = u_{n-1} + 2$, **(b)** $u_1 = 40$, $u_n = u_{n-1} - 3$,

(c) $u_1 = 4$, $u_n = u_{n-1} + 2n - 1$, **(d)** $u_1 = 2$, $u_n = u_{n-1} + n$.

8 A sequence is defined inductively by $u_n = 5 - \frac{1}{4}u_{n-1}$, $u_1 = 2$.

(a) Find the values of u_2, u_3, u_4 and u_5.

(b) Find the limiting value of u_n as n tends to infinity.

9 A sequence is defined inductively by $u_{n+1} = \frac{1}{5}u_n + 1$, $u_1 = 5$.

(a) Find the values of u_2, u_3, u_4 and u_5.

(b) Find the limiting value of u_n as n tends to infinity.

10 A sequence is defined inductively by $u_{n+1} = 6 - \frac{1}{2}u_n$, $u_1 = 1$.

(a) Find the values of u_2, u_3, u_4 and u_5.

(b) Find the limiting value of u_n as n tends to infinity.

8.5 Arithmetic sequences

> A sequence with an inductive relation of the form
>
> $$u_n = u_{n-1} + d$$
>
> is called an **arithmetic sequence**.

The sequence increases by a constant amount from one term to the next (or decreases by a constant amount if d is negative).

The quantity d is called the **common difference**.

Examples of arithmetic sequences are

3, 7, 11, 15, ... where the common difference is 4,

20, 17, 14, 11, ... where the common difference is −3.

Sometimes an arithmetic sequence is called an **arithmetic progression**.

The first term is usually denoted by a and so the full inductive definition for an arithmetic sequence is

$$u_1 = a, \ u_n = u_{n-1} + d.$$

Writing out the first few terms of the sequence gives

$$a, a + d, a + 2d, a + 3d, \ldots$$

The tenth term, for example would be $a + 9d$.

> The nth term of an arithmetic sequence is $a + (n - 1)d$.

Worked example 8.4

The third term of an arithmetic sequence is 11 and the seventh term is 23. Find the first term and the common difference.

Solution

Let the first term be a and let the common difference be d.

The nth term would be $a + (n - 1)d$.

The third term is $a + 2d$ and this must equal 11.

$\Rightarrow a + 2d = 11.$ [A]

Similarly since the seventh term is equal to 23

$\Rightarrow a + 6d = 23.$ [B]

Solving [A] and [B] simultaneously, [B] − [A] $\Rightarrow 4d = 12 \Rightarrow d = 3$.

Substituting $d = 3$ into [A] $\Rightarrow a + 6 = 11 \Rightarrow a = 5$

The first term is 5 and the common difference is 3.

Checking in [B]
$5 + 6 \times 3 = 23$. ✓

8.6 Arithmetic series

When the terms of a sequence are added together you produce a **series**.

Whereas 1, 4, 9, 16, … is a sequence
1 + 4 + 9 + 16 + … is a series.

You can use the formula for the nth term to find the number of terms in an **arithmetic series**.

Worked example 8.5

Find the number of terms in the arithmetic series

$13 + 17 + 21 + \ldots + 93.$

Solution

The formula for the nth term is $a + (n - 1)d$.

In this series, the first term $a = 13$
and the common difference $d = 4$

Because each term is obtained from the previous one by adding 4.

Hence $13 + (n - 1)4 = 93$

$\Rightarrow 4(n - 1) = 80 \Rightarrow n - 1 = 20$

$\Rightarrow n = 21$ so the series has 21 terms.

An arithmetic series is of the form

$$a + [a + d] + [a + 2d] + [a + 3d] + \ldots + [a + (n - 1)d].$$

Worked example 8.6

An arithmetic series has first term 6 and common difference $2\frac{1}{2}$. Find the least value of n for which the nth term exceeds 1000.

Solution

The formula for the nth term is $a + (n - 1)d = 6 + 2.5(n - 1)$.

You need to solve the inequality $6 + 2.5(n - 1) > 1000$

$$\Rightarrow 2.5(n - 1) > 994$$

$$\Rightarrow (n - 1) > \frac{994}{2.5} \Rightarrow (n - 1) > 397.6$$

$$\Rightarrow n > 398.6$$

Since n must be a whole number, the number of terms must be at least 399 for the nth term to exceed 1000.

8.7 Sum of the first n natural numbers

In 1786 Carl Friedrich Gauss living in Brunswick Germany was just 9 years old. His teacher wanted to keep the class occupied and asked them to calculate the sum $1 + 2 + 3 + \ldots + 100$. The teacher had barely finished explaining the assignment when Gauss wrote the single number 5050 on his slate and deposited it on the teacher's desk. How did he get the answer so quickly?

He realised that you could pair off numbers from the beginning and end, $1 + 100$, $2 + 99$, $3 + 98$ and so on. Each pair sums to 101 and there are 50 pairs. The total is $50 \times 101 = 5050$.

This method can be extended to find the sum of

$1 + 2 + 3 + 4 + \ldots + n$, where n is any positive integer.

Let $S = 1 + 2 + 3 + 4 + \ldots + n$

and writing in reverse order

$$S = n + (n - 1) + (n - 2) + (n - 3) + \ldots + 1$$

Adding the two series together as pairs

$$2S = [n + 1] + [(n - 1) + 2] + [(n - 2)+3] + [(n - 3) + 4] + \ldots [n + 1]$$

But the total inside each square bracket is $n + 1$ and there are n terms in square brackets.

$$\Rightarrow 2S = n(n + 1)$$

$$\Rightarrow S = \tfrac{1}{2}n(n + 1)$$

The sum of the first n natural numbers is $\frac{1}{2}n(n + 1)$.

Use this formula to check that the sum $1 + 2 + 3 \ldots + 100 = 5050$.

Worked example 8.7

Find the sum of all the positive integers from 1000 to 2000 inclusive.

Solution

You can find the sum from 1 to 2000 and subtract the sum from 1 to 999 using the given formula.

$$1 + 2 + 3 + \dots + 2000 = \tfrac{1}{2} \times 2000 \times 2001 = 2\,001\,000$$
$$1 + 2 + 3 + \dots + 999 = \tfrac{1}{2} \times 999 \times 1000 = 499\,500$$

Hence

$$1000 + 1001 + 1002 + \dots + 2000 = 2\,001\,000 - 499\,500 = 1\,501\,500$$

EXERCISE 8B

1 Calculate the sum of the first **(a)** 10, **(b)** 20, **(c)** 70 natural numbers.

2 Determine the sum of the series:

(a) $20 + 21 + 22 + \dots + 100$,

(b) $100 + 101 + 102 + \dots + 1000$,

(c) $201 + 202 + 203 + \dots + 400$,

(d) $97 + 98 + 99 + \dots + 700$.

3 The first term of an arithmetic sequence is 17 and the common difference is 5.
Find **(a)** the third, **(b)** the seventh, **(c)** the twenty-fifth term.

4 The second term of an arithmetic sequence is 12 and the common difference is 9.
Find **(a)** the first, **(b)** the tenth, **(c)** the twenty-first term.

5 The first term of an arithmetic sequence is 9 and the common difference is 7. Find the value of n if the nth term is 380.

6 The 17th term of an arithmetic sequence is 50 and the first term is 2. Find the common difference.

7 Find the first term of an arithmetic sequence with common difference 6 and tenth term equal to 50.

8 The fifth term of an arithmetic sequence is 10 and the eighth term is 19. Find the first term and the common difference.

9 The twelfth term of an arithmetic sequence is 60 and the sixteenth term is 70. Find the first term and the common difference.

10 The sixth term of an arithmetic sequence is equal to twice the fourth term and the tenth term is equal to 48. Find the first term and the common difference.

11 The first term of an arithmetic series is 8 and the common difference is 13. Find the value of n for which the nth term first exceeds 1000.

12 The first term of an arithmetic series is -3 and the common difference is 7. Find the value of n for which the nth term first exceeds 650.

8.8 Sum of the first *n* terms of an arithmetic series

Let S_n represent the sum of the first n terms of an arithmetic series with first term a and common difference d.

It is sometimes convenient to represent the nth term or last term of the series by l, so that $l = a + (n - 1)d$.

Using a method similar to that in the last section

$$S_n = a + (a + d) + (a + 2d) + (a + 3d) + \ldots + (l - d) + l.$$
$$S_n = l + (l - d) + (l - 2d) + (l - 3d) + \ldots + (a + d) + a.$$

Writing the terms in reverse order.

Adding the two equations gives

$$2S_n = [a + l] + [(a + d) + (l - d)] + [(a + 2d) + (l - 2d)] + \ldots + [a + l].$$

Each of the square brackets has value $a + l$ and there are n terms so $2S_n = n(a + l)$.

> The sum of the first n terms of an arithmetic series with first term a and last term l is $\frac{1}{2}n(a + l)$. This formula is in the formulae booklet.

Worked example 8.8

Find the sum of the arithmetic series
$23 + 26 + 29 + \ldots + 113$.

Solution

You know that $a = 23$ and that $l = 113$.

You need to find n but the formula for the nth term is $a + (n - 1)d$.

Therefore, since $d = 3$, you can write
$23 + 3(n - 1) = 113 \Rightarrow 3(n - 1) = 90 \Rightarrow (n - 1) = 30 \Rightarrow n = 31$.

The sum is equal to

$$\frac{1}{2}n(a + l) = \frac{1}{2} \times 31 \times (23 + 113) = 31 \times 68 = 2108$$

Alternative formula for S_n

An alternative formula may be used for $S_n = \frac{1}{2}n(a + l)$.

Since $l = a + (n - 1)d \Rightarrow a + l = 2a + (n - 1)d$.

> The sum of the first n terms of an arithmetic series with first term a, common difference d is $\frac{1}{2}n[2a + (n - 1)d]$. This formula is in the formulae booklet.

Worked example 8.9

The sum of the first 25 terms of an arithmetic series is 500. The tenth term of the series is 19. Find the first term and the common difference.

Solution

Let the first term be a and the common difference d.

$$\frac{25}{2}(2a + 24d) = 500$$

Using $S_n = \frac{1}{2}n[2a + (n - 1)d]$.

$$a + 12d = 20 \qquad [A]$$

Tenth term is $19 \Rightarrow a + 9d = 19 \qquad [B]$

Solving the simultaneous equations [A] − [B] gives

$$3d = 1$$

$$\Rightarrow d = \frac{1}{3}$$

substituting into [A] gives $a + 4 = 20 \Rightarrow a = 16$

Therefore the first term is 16 and the common difference is $\frac{1}{3}$.

EXERCISE 8C

1 Find the sum of the 20 terms of the arithmetic series with first term 3 and last term 17. Find also the common difference of this series.

2 The first term of an arithmetic series is 6 and the common difference is 5. Find the sum of the first **(a)** ten, **(b)** twenty terms of this series.

3 The twentieth term of an arithmetic series is 50. The sum of the first 20 terms is 200. Find the first term of the series.

4 The sum of the first 30 terms of an arithmetic series with first term 7 is equal to 450. Find the thirtieth term.

5 Find the sum of the following arithmetic series:

(a) $3 + 5 + 7 + 9 + \ldots + 47$,

(b) $22 + 26 + 30 + \ldots + 422$,

(c) $17 + 20 + 23 + \ldots + 167$,

(d) $41 + 38 + 35 + \ldots + (-19)$,

(e) $123 + 107 + 91 + \ldots + (-197)$.

6 Find the sum of the first 50 terms of the arithmetic series

$4 + 7 + 10 + \ldots$ [A]

7 An arithmetic series has sixth term 28 and tenth term 44.

(a) Find the first term and the common difference.

(b) Find the sum of the first 50 terms of the series. [A]

8 The fourth term of an arithmetic series is 25. The common difference is 3.

(a) Find the first term.

(b) Find the sum of the first twenty terms of the series. [A]

9 The first term of an arithmetic series is -11 and the ninth term of the series is 1.

(a) Find the common difference and the sum of the first twenty terms of the series.

(b) Find the value of n for which the nth term of the series is 100. [A]

10 The first term of an arithmetic series is 7. The tenth term is 43.

(a) Find the common difference.

(b) Find the sum of the first fifty terms of the series.

(c) The kth term has a value greater than 1000.

(i) Show that $4k > 997$.

(ii) Find the least possible value of k. [A]

11 An arithmetic series has first term a and common difference d. The sum of the first 19 terms is 266.

(a) Show that $a + 9d = 14$.

(b) The sum of the fifth and eighth terms is 7. Find the values of a and d. [A]

12 Using the formula for the sum of the first N positive integers, write down a formula for $1 + 2 + 3 + 4 + 5 + \ldots + (n - 1)$.

Hence, by rewriting

$S_n = a + [a + d] + [a + 2d] + \ldots + [a + (n - 1)d]$ in the form

$S_n = a + a + \ldots + a + d[1 + 2 + 3 + \ldots + (n - 1)]$,

prove that $S_n = \frac{1}{2}n[2a + (n - 1)d]$.

8.9 Sigma notation

A shorthand notation for finding sums is to use the Greek capital letter sigma, Σ. This is particularly useful in statistics where there are lots of data values to be added together.

At the top of the sigma sign is the final value the variable can take.

$$\sum_{n=2}^{10} n^3$$

At the bottom of the sigma sign is the variable which serves as a counter and the initial value it takes.

Shorthand for $2^3 + 3^3 + 4^3 + \ldots + 10^3$.

Worked example 8.10

Write each of the following in full and hence find the value of:

(a) $\sum_{k=2}^{7} k^2$, **(b)** $\sum_{r=4}^{8} (2r + 1)$, **(c)** $\sum_{n=1}^{4} \frac{n}{n+1}$.

Solution

(a) The expression to be summed here is k^2.
You need to substitute $k = 2$ then $k = 3$ ... up to $k = 7$ adding the terms together.

$$\sum_{k=2}^{7} k^2 = 2^2 + 3^2 + 4^2 + 5^2 + 6^2 + 7^2$$
$$= 139$$

(b) This time the summation expression is $2r + 1$ with r ranging from 4 to 8.

$$\sum_{r=4}^{8} (2r + 1) = 9 + 11 + 13 + 15 + 17$$
$$= 65$$

(c) Substituting $n = 1, 2, 3, 4$ and summing gives

$$\sum_{n=1}^{4} \frac{n}{n+1} = \frac{1}{2} + \frac{2}{3} + \frac{3}{4} + \frac{4}{5} = \frac{30 + 40 + 45 + 48}{60} = \frac{163}{60}$$

Note that different letters may be used *within* the summation such as k, r, n etc and yet the final answer does not involve any of these variables.

8.10 Use of the sigma notation for arithmetic series

Suppose the nth term of a series is u_n. Then the sum of the first 20 terms could be expressed as $\sum_{n=1}^{20} u_n$.

It is possible to express the sum of an arithmetic series in sigma notation.

For instance, consider the arithmetic series

$5 + 8 + 11 + 14 + \ldots$.

The nth term of this series is $5 + (n - 1)3$,
or in a simpler form $3n + 2$.

Using $u_n = a + (n - 1)d$.

Hence the sum of the first 20 terms of the given series can be written as $\sum_{n=1}^{20} 3n + 2$.

Worked example 8.11

The nth term of an arithmetic sequence is u_n, where $u_n = 70 - 3n$.

(a) Find the values of u_1 and u_2.

(b) Write down the common difference of the arithmetic sequence.

(c) Evaluate $\sum_{n=1}^{30} u_n$

Solution

(a) $u_1 = 70 - 3 = 67$

$u_2 = 70 - 6 = 64$

(b) The common difference is therefore -3.

(c) The arithmetic series is $67 + 64 + 61 + \ldots$ and you need to find the sum of the first 30 terms.

Using the formula $S_n = \frac{n}{2}[2a + (n - 1)d]$

with $a = 67$, $d = -3$ and $n = 30$ gives

$S_{30} = \frac{30}{2}[134 + 29 \times (-3)] = 15 \times 47 = 705$

Worked example 8.12

Use appropriate formulae to evaluate **(a)** $\sum_{k=5}^{500} k$, **(b)** $\sum_{r=1}^{50} 3r - 5$.

Solution

(a) $\sum_{k=5}^{500} k = 5 + 6 + 7 + \ldots + 500$

You can use the formula for the sum of the first n natural numbers

$$1 + 2 + 3 + \ldots + n = \frac{1}{2}n(n + 1)$$

$\Rightarrow 1 + 2 + 3 + \ldots + 500 = 250 \times 501 = 125\,250$

$\Rightarrow 5 + 6 + 7 + \ldots + 500 = 125\,250 - (1 + 2 + 3 + 4) = 125\,240$

(b) $\sum_{r=1}^{50} 3r - 5 = -2 + 1 + 4 + \dots + 145$

You should recognise this as an arithmetic series with first term $a = -2$, common difference $d = 3$ and last term $l = 145$.

Because the 'counter' r starts at 1 and goes up in steps of 1 to 50, there are 50 terms so $n = 50$.

Using the formula for the sum of the first n terms of an arithmetic series $\frac{1}{2}n(a + l)$ gives

$$25 \times (-2 + 145) = 3575$$

Again, you may find it interesting to set up some of these problems on a spreadsheet to check your answers.

EXERCISE 8D

1 Write each of the following in full and hence find the value of:

(a) $\sum_{k=3}^{7} k$, **(b)** $\sum_{r=6}^{9} (3r - 11)$,

(c) $\sum_{n=1}^{5} n^2$, **(d)** $\sum_{r=4}^{7} (5r - 19)$,

(e) $\sum_{k=2}^{4} k^3$, **(f)** $\sum_{n=2}^{6} (n - 2)(n - 3)$.

2 Find the sum of each of the following by using appropriate formulae:

(a) $\sum_{k=13}^{100} k$, **(b)** $\sum_{r=3}^{102} (2r - 5)$,

(c) $\sum_{n=11}^{299} n$, **(d)** $\sum_{k=10}^{99} 2 + 3k$,

(e) $\sum_{r=13}^{122} 2r$, **(f)** $\sum_{n=10}^{509} 4n - 37$.

3 Find the nth term of each of the following arithmetic series and hence express each series in sigma notation:

(a) $1 + 4 + 7 + 10 + \dots + 91$

(b) $25 + 34 + 43 + 52 + \dots + 916$

(c) $75 + 71 + 67 + 63 + \dots + 7$

(d) $11 + 9 + 7 + 5 + \dots -31$

4 The nth term of an arithmetic sequence is u_n, where $u_n = 5n - 23$.

(a) Find the values of u_1 and u_2.

(b) Write down the common difference of the arithmetic sequence.

(c) Evaluate $\sum_{n=1}^{40} u_n$.

5 The nth term of an arithmetic sequence is u_n, where $u_n = 20 - 3n$.

(a) Find the values of u_1 and u_2.

(b) Write down the common difference of the arithmetic sequence.

(c) Evaluate $\sum_{n=1}^{40} u_n$.

8.11 Review of arithmetic series

You will find it helpful to check the formulae in the booklet to be used in the examination and to make sure you are familiar with the notation:

a is used for the first term
d is used for the common difference
S_n is the sum of the first n terms of the series

Always read questions carefully to check whether you are being given information about the nth term or the sum of the first n terms.

For instance, if you are told that the 7th term is equal to 53, you can write $a + 6d = 53$.

However, if you are told that the sum of the first 11 terms is 123, then you need to use the formula $S_n = \frac{n}{2}[2a + (n - 1)d]$ and you will write

$$\frac{11}{2}[2a + 10d] = 123.$$

You will often need to solve simultaneous equations to find a and d.

Worked examination question 8.13

The tenth term of an arithmetic series is 23 and the sum of the first seven terms is 56. Find the first term and the common difference.

Solution

Let the first term be a and the common difference be d.

The tenth term is $23 \Rightarrow a + 9d = 23$.

The sum of first seven terms is 56 so $\frac{7}{2}[2a + 6d] = 56$.

This equation can be simplified to $a + 3d = 8$.

Solving simultaneously $a + 9d = 23$ and $a + 3d = 8$.

Subtracting gives $6d = 15$. Hence, $d = 2.5$.

Substituting into either equation gives $a = 0.5$.

The first term is $\frac{1}{2}$ and the common difference is $2\frac{1}{2}$.

Worked examination question 8.14

The sum of the first five terms of an arithmetic series is 5 and the sum of the **next** five terms is 105. Find the first term and the common difference. [A]

Solution

$$S_5 = 5 \Rightarrow 5 = \frac{5}{2}(2a + 4d)$$

Using $S_n = \frac{1}{2}n[2a + (n - 1)d]$.

$$S_{10} = 110 \Rightarrow 110 = \frac{10}{2}(2a + 9d)$$

There are several ways of dealing with the second condition, e.g. $[a + 5d] + [a + 6d] + [a + 7d] + [a + 8d] + [a + 9d] = 105$

You now have two simultaneous equations equivalent to $a + 2d = 1$ and $2a + 9d = 22$, which can be solved to give $a = -7$, $d = 4$.

First term is -7; common difference is 4.

EXERCISE 8E

1 An arithmetic sequence has fifth term 10 and seventeenth term 31. Find the first term and the common difference of the sequence. [A]

2 In an arithmetic series, the ninth term is 7, and the twenty-ninth term is equal to twice the fifth term.

(a) Determine the first term and the common difference of the series.

(b) Calculate the sum of the first 200 terms of the series. [A]

3 The ninth term of an arithmetic series is 17 and the sum of the first five terms is 10. Determine the first term and the common difference of the series. [A]

4 The first term of an arithmetic series is 3. The seventh term is twice the third term.

(a) Find the common difference.

(b) Calculate the sum of the first 20 terms of the series. [A]

5 An arithmetic sequence with first term 37 has common difference 29. Find the value of n for which the nth term first exceeds 5000.

6 An arithmetic series has first term 12. The sum of the first 15 terms is equal to four times the twenty-eighth term. Find the common difference.

7 Find the sum of the three hundred integers from 101 to 400 inclusive. [A]

8 The nth term of an arithmetic sequence is u_n, where $u_n = 10 + 0.5n$.

(a) Find the values of u_1 and u_2.

(b) Write down the common difference of the arithmetic sequence.

(c) Find the value of n for which $u_n = 25$.

(d) Evaluate $\sum_{n=1}^{30} u_n$. [A]

9 (a) Find the sum of the first 16 terms of the arithmetic series $2 + 5 + 8 + \ldots + 47$.

(b) An arithmetic sequence $u_1, u_2, u_3, \ldots$ has rth term u_r, where

$$u_r = 50 - 3r.$$

(i) Write down the values of u_1, u_2, u_3 and u_4.

(ii) Show that the sequence has exactly 16 positive terms. [A]

10 A pipeline is to be constructed under a lake. It is calculated that the first mile will take 15 days to construct. Each further mile will take 3 days longer than the one before, so the 1st, 2nd and 3rd miles will take 15, 18 and 21 days, respectively, and so on.

(a) Find the nth term of the arithmetic series 15, 18, 21, … .

(b) Show that the total time taken to complete the first n miles of the pipeline is $\frac{3}{2}n(n + 9)$ days.

(c) Calculate the total length of pipeline that can be constructed in 600 days. [A]

Key point summary

1 A sequence is a set of numbers separated by commas following a particular rule. *p317*

An example of a sequence is 1, 4, 9, 16, … .

2 A sequence may be defined by a formula such as $u_n = 3n^2 - 7$. *p318*

3 A sequence may be defined inductively by giving one of the terms and a rule which relates one term to the previous term such as $u_1 = 5$; $u_n = u_{n-1} + 4n^3$. *p319*

4 An arithmetic sequence with common difference d has the inductive definition $u_1 = a$; $u_n = u_{n-1} + d$. *p320*

5 The nth term of an arithmetic sequence is given by $u_n = a + (n - 1)d$ and this formula is in the formulae booklet.	*p320*
6 When the terms of a sequence are added together they form a series. For example, $1 + 8 + 27 + \ldots + 2197$ is a series.	*p323*
7 An arithmetic series is of the form $a + [a + d] + [a + 2d] + [a + 3d] + \ldots + [a + (n - 1)d]$.	*p323*
8 The sum of the first n terms of an arithmetic series is $S_n = \frac{1}{2}n[2a + (n - 1)d]$. This formula is in the formulae booklet.	*p325*
9 The sum of the first n terms of an arithmetic series with first term a and last term l is $\frac{1}{2}n(a + l)$. This formula is in the formulae booklet.	*p326*
10 Sigma notation can be used as a shorthand for series. The sum of the first n natural numbers is given by $\sum_{r=1}^{n} r = \frac{1}{2}n(n + 1)$.	*p329*

Test yourself	What to review
1 Write down the first five terms of the sequence defined by $u_n = 2n^3 - 1$.	*Section 8.2*
2 A sequence has the following inductive definition: $u_1 = 3;\ u_n = 3u_{n-1} - 7$. Find the first five terms.	*Section 8.3*
3 The twentieth term of an arithmetic sequence is 100 and the first term is 43. Find the common difference.	*Section 8.4*
4 The sum of the first 16 terms of an arithmetic series is 48. The tenth term is 6. Find the first term and the common difference.	*Section 8.7*
5 Calculate the sum of all the natural numbers from 150 to 250 inclusive.	*Section 8.6*
6 Find the nth term of the arithmetic series $3 + 8 + 13 + \ldots + 38$ and express the series in sigma notation.	*Section 8.9*

Test yourself ANSWERS

1 1, 15, 53, 127, 249.

2 3, 2, -1, -10, -37.

3 $d = 3$.

4 $a = -12$, $d = 2$.

5 20 200

6 $5n - 2$, $\sum_{n=1}^{8} 5n - 2$.

CHAPTER 9

C2: Radian measure

Learning objectives

After studying this chapter, you should be able to:

- understand what is meant by a radian
- convert between degrees and radians
- recall and use the formula for the length of an arc of a circle
- recall and use the formula for the area of a sector of a circle.

9.1 Radians as a unit of measure of angles

You will have measured angles in degrees, where 1 degree is $\frac{1}{360}$ of a complete turn. In higher level mathematics a degree is not the most appropriate unit to use to measure angles. It is more useful to measure an angle in units called radians.

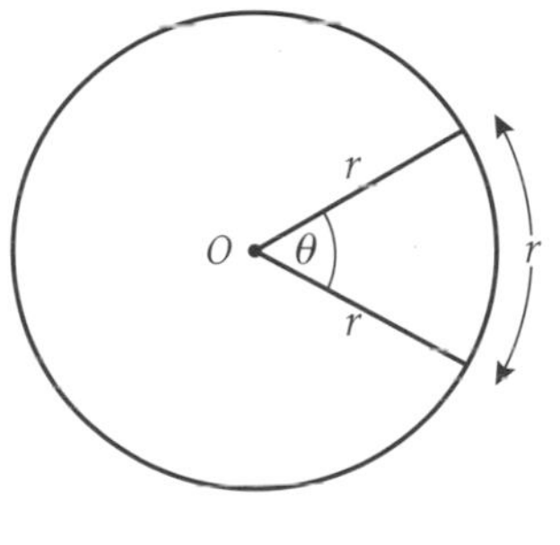

One radian is the angle subtended at the centre of a circle by an arc whose length is equal to the radius of the circle. In the diagram on the right, $\theta = 1$ radian.

1 radian is written as 1 rad or as 1^c

1 radian $\approx 57°$

9.2 Changing between degrees and radians

An arc of length $2r$ subtends an angle of 2 radians at the centre of the circle.

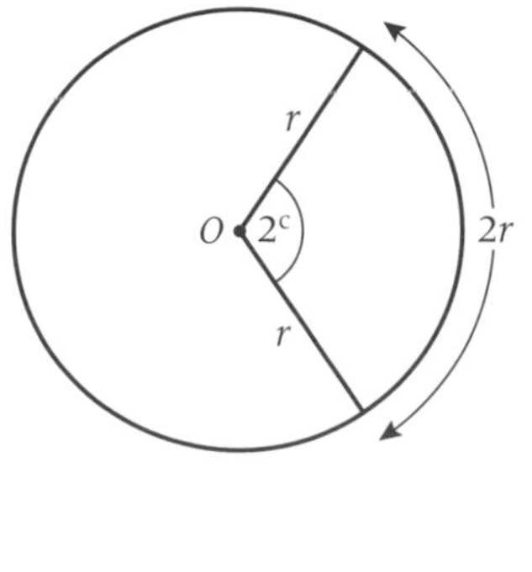

An arc of length πr subtends an angle of π radians at the centre of the circle.

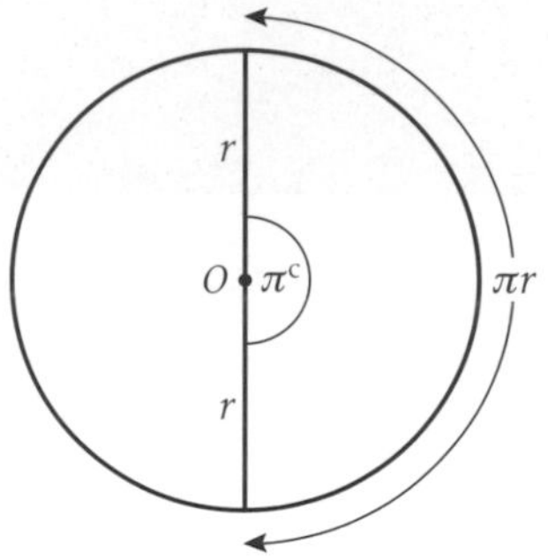

An arc of length $2\pi r$ subtends an angle of 2π radians at the centre of the circle. The circumference of the circle is $2\pi r$ which subtends an angle of 360° at the centre of the circle.
So $360° = 2\pi$ rads.

Worked example 9.1

Convert 20° to radians.

> Since 20° is $\frac{1}{18}$ of 360° we could have written $\frac{1}{18}$ of 2π rads $= \frac{\pi}{9}$.

Solution

$360° = 2\pi$ rads.

$$1° = \frac{2\pi}{360} \text{ rads.}$$

$$20° = 20 \times \frac{2\pi}{360} \text{ rads} = \frac{\pi}{9} \text{ rads } \{= 0.34906 \ldots\text{rads}\}$$

> If an exact answer is needed, or an answer in terms of π, give $\frac{\pi}{9}$ rads as the final answer.

Note: The expression $\frac{\pi}{9}$ is usually read as 'pi by nine' – no doubt as an abbreviation for 'pi divided by nine'.

Worked example 9.2

Convert $\frac{3\pi}{5}$ radians to degrees.

Solution

$$2\pi \text{ rads} = 360°$$

$$1 \text{ rad} = \frac{360°}{2\pi} \{= 57.2957 \ldots°\}$$

$$\frac{3\pi}{5} \text{ rads} = \frac{3\pi}{5} \times \frac{360°}{2\pi} = 108°$$

EXERCISE 9A

1 Change the following to degrees:

(a) $\frac{\pi}{2}$ rads, **(b)** 4π rads, **(c)** $\frac{\pi}{3}$ rads, **(d)** $\frac{\pi}{4}$ rads,

(e) $\frac{5\pi}{6}$ rads, **(f)** $\frac{3\pi}{4}$ rads, **(g)** $\frac{2\pi}{3}$ rads, **(h)** $\frac{11\pi}{6}$ rads,

(i) $\frac{7\pi}{4}$ rads, **(j)** $\frac{5\pi}{2}$ rads.

2 Convert the following to radians. Give your answers in terms of π.

(a) 180°, **(b)** 120°, **(c)** 36°, **(d)** 24°,

(e) 108°, **(f)** 540°, **(g)** 80°, **(h)** 225°,

(i) 405°, **(j)** 15°.

3 Convert the following to degrees, giving your answers to 3 sf.

(a) 2 rads, **(b)** 0.5 rads, **(c)** 1.8 rads, **(d)** 3 rads,

(e) 0.3 rads, **(f)** 2.3 rads, **(g)** 1.28 rads, **(h)** 1.6 rads,

(i) $\frac{7\pi}{11}$ rads, **(j)** $\frac{3\pi}{7}$ rads.

4 Convert the following to radians, giving your answers to 3 sf.

(a) 60°, **(b)** 150°, **(c)** 25°, **(d)** 305°,

(e) 96°, **(f)** 78°, **(g)** 14°, **(h)** 82°,

(i) 38°, **(j)** 500°.

5 Use your calculator to find the value, to 3 sf, of
(i) $\sin\theta$, **(ii)** $\cos\theta$, **(iii)** $\tan\theta$
for the following values of θ in radians:

(a) 0.4, **(b)** 1.2, **(c)** 2, **(d)** $\frac{5\pi}{8}$.

Remember to set your calculator to radian mode.

6 Find the angle in radians turned in exactly 1 day by:

(a) the hour hand, **(b)** the minute hand

of a clock. Leave your answers in terms of π.

7 Find the obtuse angle, in radians, between the hands of a clock showing

(a) 4 p.m, **(b)** 2:30 a.m, **(c)** 3:45 p.m.

Leave your answers in terms of π.

8 By considering a right-angled triangle, show that:

(a) $\sin\left(\frac{\pi}{2} - \theta\right) = \cos\theta$,

(b) $\cos\left(\frac{\pi}{2} - \theta\right) = \sin\theta$.

9.3 Arc length of a circle

Let arc PQ of a circle subtend an angle θ at the centre of the circle. Suppose the length of the arc is l then l is directly proportional to θ. (As θ doubles the arc length doubles.)

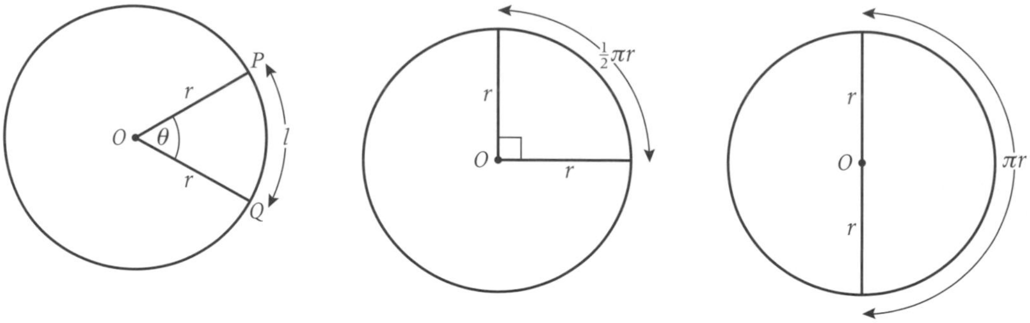

$$l \propto \theta \Rightarrow \quad l = k\theta$$

When $\theta = 2\pi$, $l = 2\pi r \Rightarrow 2\pi r = k2\pi$

$$\Rightarrow \quad k = r$$

$$\Rightarrow \quad l = r\theta, \text{ where } \theta \text{ is in radians}$$

> The length l of an arc of a circle is given by $l = r\theta$, where r is the radius and θ is the angle, in radians, subtended by the arc at the centre of the circle.

Worked example 9.3

An arc PQ of a circle of radius 15 cm subtends an angle of 1.6 radians at the centre, O, of the circle.

(a) Find the perimeter of the sector OPQ.

(b) Find the length of the chord PQ.

A sector of a circle is the region bounded by two radii and an arc. The larger region is called the major sector, the smaller region is the minor sector.

Solution

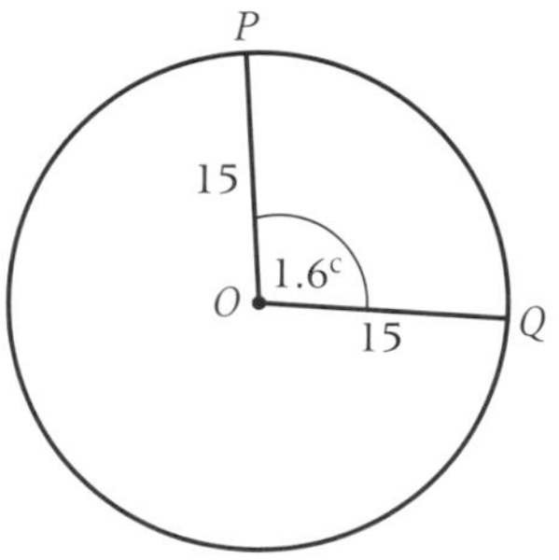

(a) Using $l = r\theta$,

length of arc $PQ = 15 \times 1.6 = 24$ cm.

Perimeter of sector $OPQ = 15 + 15 + 24$

$= 54$ cm

Forgetting to include the lengths of the two radii is a common error.

(b) OPQ is an isosceles triangle.

Chord $PQ = 2 \times PM = 2 \times r \sin(\theta/2)$

$= 2 \times 15 \sin 0.8^c$

$= 30 \times 0.717\,35 \ldots$

$= 21.52 \ldots = 21.5$ cm (to 3 sf)

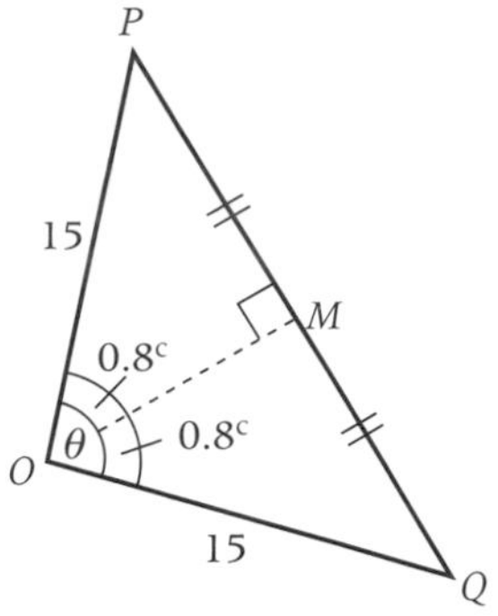

Worked example 9.4

OA and OB are radii of a circle. Arc AB subtends an angle of θ radians at O. P and Q are the mid-points of OA and OB, respectively. PQ is an arc of a circle with centre O. Given that the perimeter of sector OAB = perimeter of $ABQP$, show that $\theta = 2$.

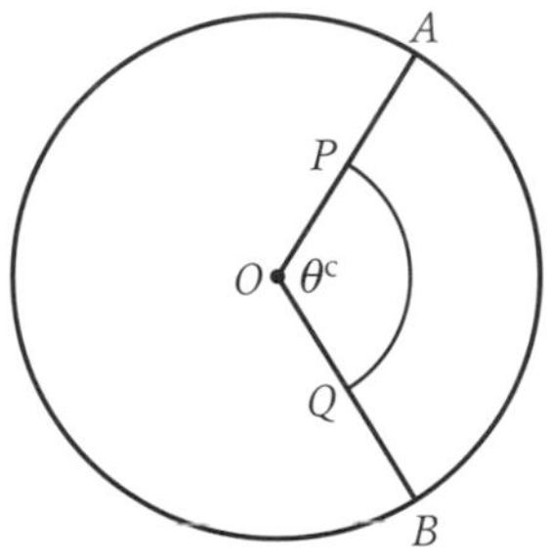

Solution

Let $OA = OB = 2x$

Perimeter of sector $OAB = OA + \text{arc } AB + BO$

$= 2x + 2x\theta + 2x$

Perimeter of $ABQP$ $= \text{arc } AB + QB + \text{arc } QP + PA$

$= 2x\theta + x + x\theta + x$

Perimeter of sector OAB = perimeter of $ABQP$

$\Rightarrow 4x + 2x\theta = 2x + 3x\theta$

$\Rightarrow 2x = x\theta \Rightarrow \theta = 2.$

EXERCISE 9B

[In this exercise, M and N are points on the circumference of a circle with centre O and θ is the angle subtended by the minor arc, MN, at O.]

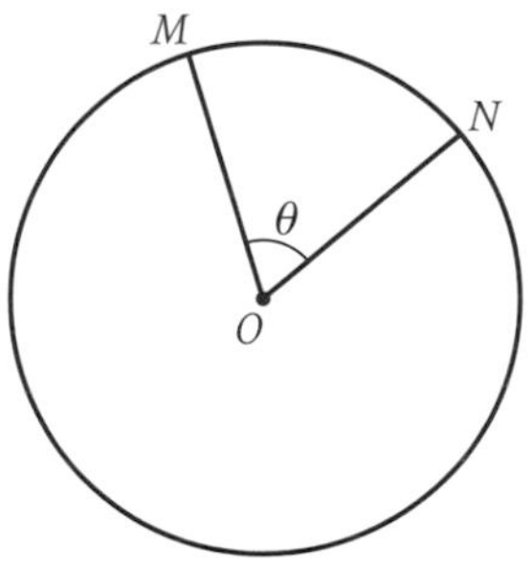

1 Find the length of the minor arc MN in each of the following cases:

(a) $OM = 18$ cm, $\theta = 1.5$ rads,

(b) $ON = 20$ cm, $\theta = 2.5$ rads,

(c) $ON = 7$ cm, $\theta = 0.3$ rads,

(d) $OM = 5.5$ cm, $\theta = 1.4$ rads.

9

2 Find the perimeter of the **major** sector OMN for each part of Question 1.

3 Find the angle subtended by the minor arc MN in each of the following:

(a) arc $MN = 21$ cm, $OM = 10$ cm,

(b) arc $MN = 30$ cm, $ON = 12$ cm,

(c) arc $MN = 10$ cm, $ON = 16$ cm,

(d) arc $MN = 16$ cm, $OM = 6$ cm,

(e) $OM = 10$ cm, perimeter of minor sector $OMN = 30$ cm,

(f) $OM = 4$ cm, perimeter of minor sector $OMN = 10$ cm,

(g) $OM = 12$ cm, perimeter of minor sector $OMN = 50$ cm,

(h) $OM = 5$ cm, perimeter of minor sector $OMN = 18$ cm.

4 Find the radius of a circle whose minor arc, MN, subtends an angle θ at the centre of the circle in each of the following:

(a) arc $MN = 10$ cm, $\theta = 2$ rads,

(b) arc $MN = 8$ cm, $\theta = 2.5$ rads,

(c) arc $MN = 16$ cm, $\theta = 10$ rads,

(d) arc $MN = 18$ cm, $\theta = 30°$,

(e) arc $MN = 20$ cm, $\theta = 80°$,

(f) arc $MN = 8\pi$ cm, $\theta = \frac{\pi}{4}$ rads.

5 The perimeter of the minor sector OMN is the same as the length of the major arc MN. Show that the minor arc MN subtends an angle $(\pi - 1)$ radians at O.

6 Given that $OM = 6$ m and $\theta = \frac{\pi}{3}$ radians, show that the perimeter of the minor sector OMN is about 283 mm longer than the perimeter of triangle OMN.

7 Find the length of the chord, MN, to the nearest millimetre, in each of the following cases:

(a) arc $MN = 20$ cm, $OM = 10$ cm,

(b) arc $MN = 8$ cm, $ON = 8$ cm,

(c) arc $MN = 18$ cm, $OM = 12$ cm,

(d) arc $MN = 45$ cm, $ON = 15$ cm.

9.4 Area of a sector of a circle

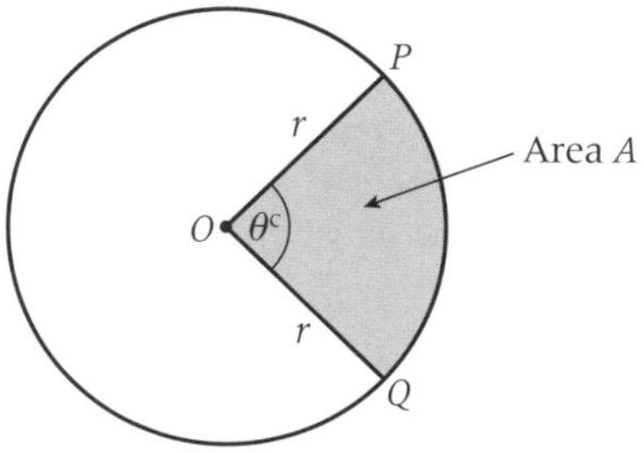

Arc PQ, which subtends an angle θ at the centre, O, of the circle and radii OP and OQ form a sector of the circle. Suppose the area of the sector is A, then A is directly proportional to θ.

$$A \propto \theta \Rightarrow A = k\,\theta$$

When $\theta = 2\pi$, $A = \pi r^2 \Rightarrow \pi r^2 = k2\pi$

$$\Rightarrow k = \frac{1}{2}r^2$$

$$\Rightarrow A = \frac{1}{2}r^2\,\theta, \text{ where } \theta \text{ is in radians}$$

> The area A of a sector of a circle is given by $\frac{1}{2}r^2\,\theta$,
>
> where r is the radius and θ is the angle, in radians, subtended by the arc at the centre of the circle.

Worked example 9.5

Find the area of the shaded segment.

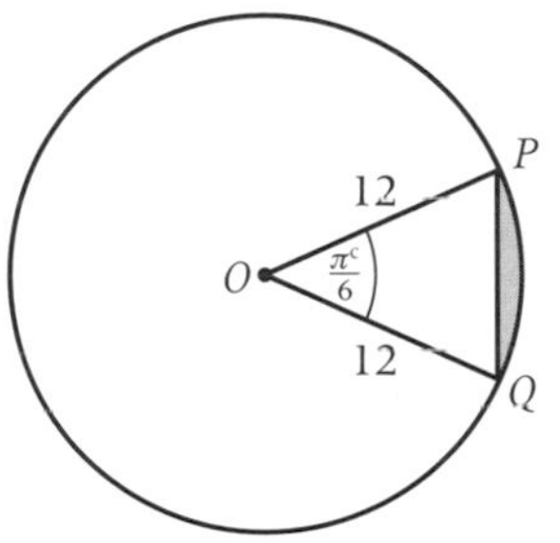

A segment of a circle is the region bounded by an arc and its chord.

Solution

Area of shaded segment = area of sector POQ − area of triangle POQ

$$= \frac{1}{2}r^2\theta - \frac{1}{2}r^2\sin\theta$$

$$= \frac{1}{2}(12)^2\frac{\pi}{6} - \frac{1}{2}(12)^2\sin\frac{\pi}{6}$$

$$= 12\pi - 36 = 1.70 \text{ cm}^2 \text{ (to 3 sf)}$$

Use $\frac{1}{2}ab\sin C$ for area of triangle.

Worked example 9.6

The area of the minor sector OAB is 24 cm^2. The radius of the circle is 4 cm. Find the length of the **major** arc AB.

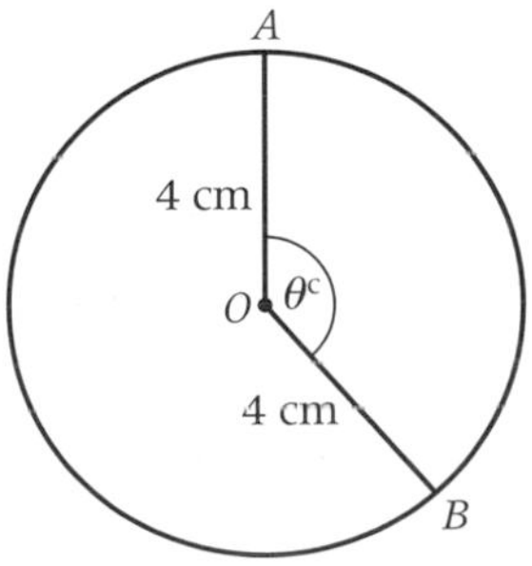

Solution

Using $A = \frac{1}{2}r^2\theta$,

$$24 = \frac{1}{2}(4)^2\theta \Rightarrow \theta = 3$$

Length of minor arc $AB = r\theta = 4 \times 3 = 12$.

Length of **major** arc $AB = 2\pi r - 12 = 8\pi - 12 = 13.1$ cm (to 3 sf).

Worked examination question 9.7

A circular sector, of area A cm^2, has bounding radii, each of length x cm, and the angle between these radii is θ radians. Given that the perimeter of the sector is 12 cm,

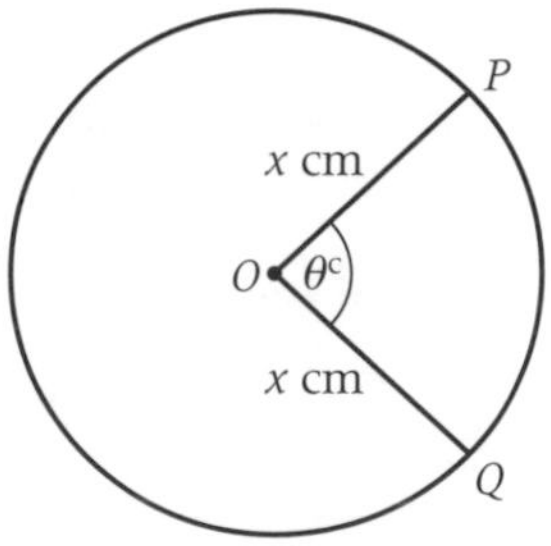

(a) express θ in terms of x,

(b) show that $A = 6x - x^2$. [A]

Solution

(a) Perimeter of sector $= OP + \text{arc } PQ + OQ$

$$12 = x + x\theta + x$$

$$\Rightarrow \quad 12 - 2x = x\theta$$

$$\Rightarrow \quad \theta = \frac{12 - 2x}{x}$$

(b) $A = \frac{1}{2}r^2\theta = \frac{1}{2}x^2\theta = \frac{1}{2}x(x\theta) = \frac{1}{2}x(12 - 2x) = 6x - x^2$.

EXERCISE 9C

[In Questions **1–4**, E and F are points on the circumference of a circle with centre O and θ is the angle subtended at O by the minor arc EF.]

1 Find the area of the minor sector, OEF, in each of the following:

(a) $OE = 4$ cm, $\theta = 2$ rads,

(b) $OF = 6$ cm, $\theta = 1.5$ rads,

(c) $OE = 1.5$ cm, $\theta = 0.75$ rads,

(d) $OF = 1$ cm, $\theta = 60°$.

2 The area of the minor sector, OEF, is A cm^2. Find the length of OE in each of the following:

(a) $A = 16$, $\theta = 2$ rads,

(b) $A = 25$, $\theta = 0.5$ rads,

(c) $A = 2.25$, $\theta = 0.125$ rads,

(d) $A = 5$, $\theta = 30°$.

3 Find the area of the **major** sector, OEF, in each of the following:

(a) $OE = 10$ cm, $\theta = \frac{\pi}{6}$ rads,

(b) $OE = 6$ cm, $\theta = \frac{\pi}{3}$ rads,

(c) $OE = 2.4$ cm, $\theta = 2$ rads,

(d) $OF = 4.5$ cm, $\theta = 72°$.

4 Find the length of the major arc EF in each of the following:

(a) $OF = 8$ cm, area of minor sector $OEF = 40$ cm^2,

(b) $OE = 6$ cm, area of minor sector $OEF = 18$ cm^2,

(c) $OF = 10$ cm, area of minor sector $OEF = 42$ cm^2,

(d) $OE = 12$ cm, area of minor sector $OEF = 30$ cm^2.

5 The diagram shows two concentric circles with centre O and radii 4 cm and 6 cm. Find the area of the shaded region in each of the following:

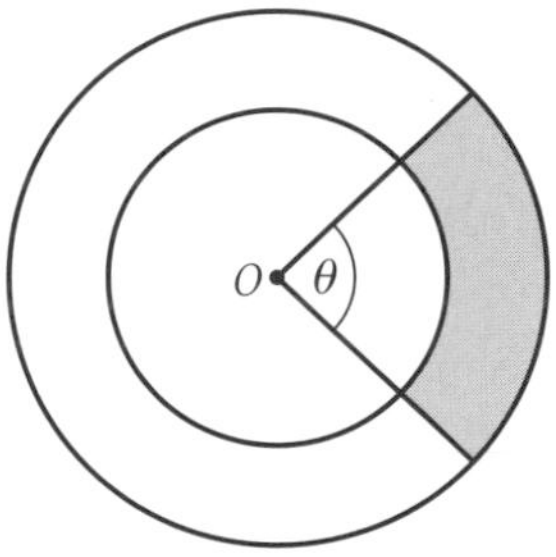

(a) $\theta = \frac{\pi}{2}$ rads, **(b)** $\theta = \frac{\pi}{3}$ rads,

(c) $\theta = 2$ rads, **(d)** $\theta = 120°$.

6 The diagram shows two concentric circles with centre O and radii 4 cm and 6 cm. The areas of the two shaded regions are equal. Show that $\theta = \frac{8\pi}{9}$ radians.

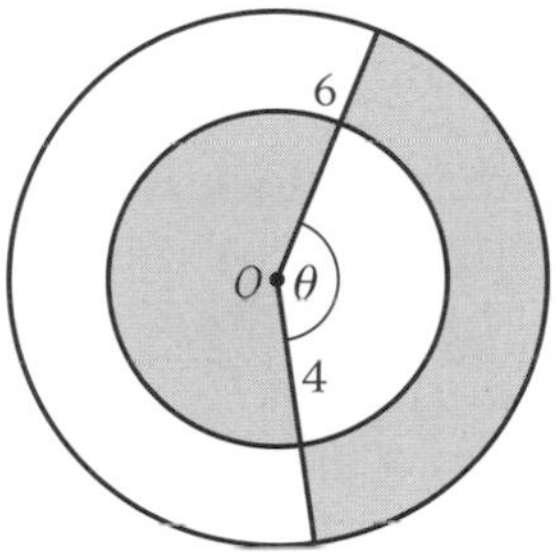

7 Points A and B lie on the circumference of a circle of radius 9 cm and centre O such that angle $AOB = \frac{2\pi}{3}$ radians. PA and PB are tangents to the circle. Find the area of the region bounded by the arc AB and the tangents PA and PB.

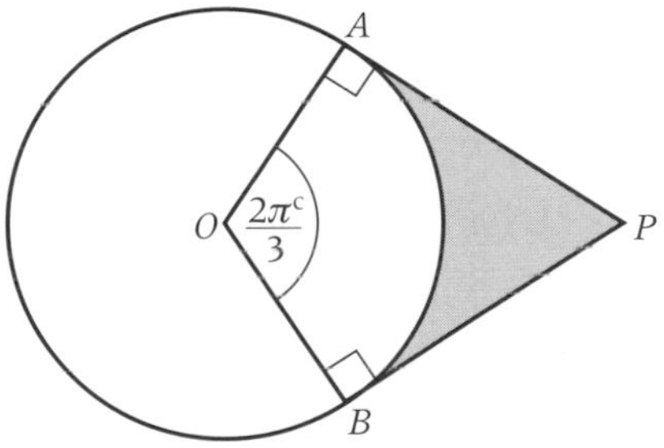

8 The perimeter of a sector of a circle of radius r and angle θ is the same as the perimeter of a square of side r.

(a) Show that $\theta = 2$ radians.

(b) Show that the area of the sector is the same as the area of the square. [A]

MIXED EXERCISE

1 The area of a sector of a circle of radius 10 cm is 75 cm^2. Find the arc length of this sector. [A]

2 A circle with centre O and radius 8 cm passes through points A and B such that OAB is an equilateral triangle. The point P is on the major arc AB. Find, in terms of π,

(a) the size of angle APB,

(b) the perimeter of the major segment ABP.

3 The perimeter of a sector OAB of a circle of centre O and radius 25 cm is 80 cm.

(a) Calculate the area of the sector OAB.

(b) AB cuts the circle into two segments. Calculate the area of the **major** segment.

4 The diagram shows a circular path with centre O and radius r, together with two other paths along the radii AO and OB. The size of the angle AOB is θ radians, where $\theta < \pi$. The widths of the paths may be neglected in the calculations. Peter runs along the radii AO and OB, then along the minor arc BA. Mary runs along the major arc AB.

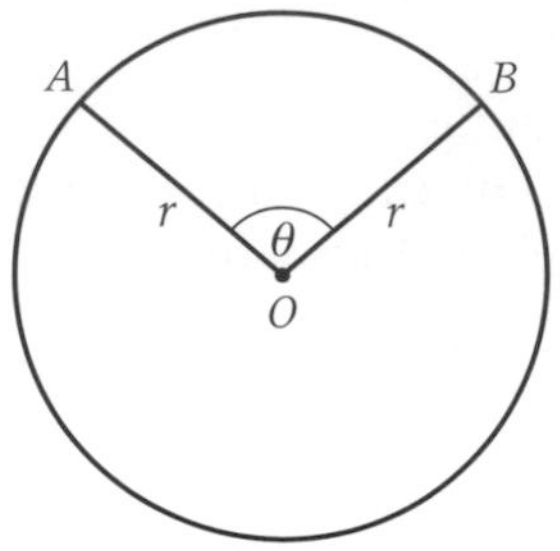

(a) Given that Peter and Mary run the same distance, show that

$$\theta = \pi - 1.$$

(b) Given that they each run 410 metres, find the radius of the circular path correct to the nearest metre. [A]

5 On a circular clock face, with centre O, the minute hand OA is of length 10 cm and the hour hand OB is of length 6 cm. Prove that the difference between the distances that A and B travel during the period of 1 hour from 12 o'clock to 1 o'clock is 19π cm. Calculate, correct to the nearest millimetre, how far A is from B at 1 o'clock. [A]

6 The diagram shows an equilateral triangle ABC with sides of length 6 cm and an arc BC of a circle with centre A.

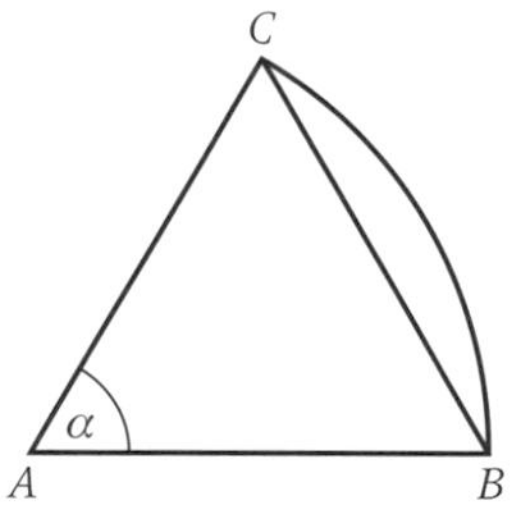

(a) Write down, in radians, the value of the angle α.

(b) Find the length of the arc BC.

(c) Show that the area of the triangle ABC is $9\sqrt{3}$ cm^2.

(d) Show that the area of the sector ABC is 6π cm^2. [A]

7 The diagram shows a circle with centre O and radius 3 cm. The points A and B on the circle are such that the angle AOB is 1.5 radians.

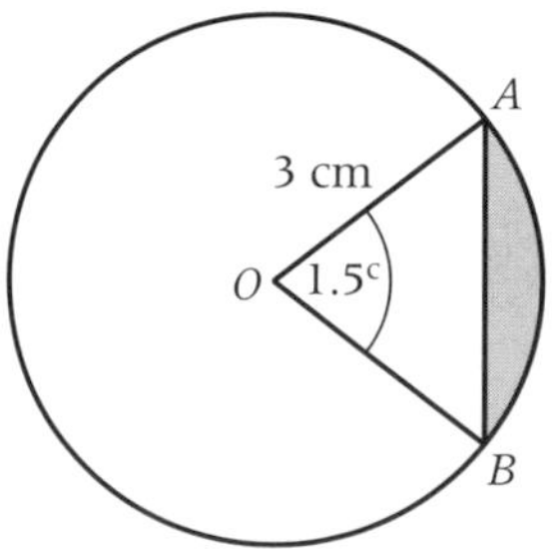

(a) Find the length of the minor arc AB.

(b) Find the area of the minor sector OAB.

(c) Show that the area of the shaded segment is approximately 2.3 cm^2. [A]

8 The diagram shows a sector of a circle, with centre O and radius 6 cm. The mid-point of the chord PQ is M, and the angle $POM = \frac{\pi}{6}$ radians.

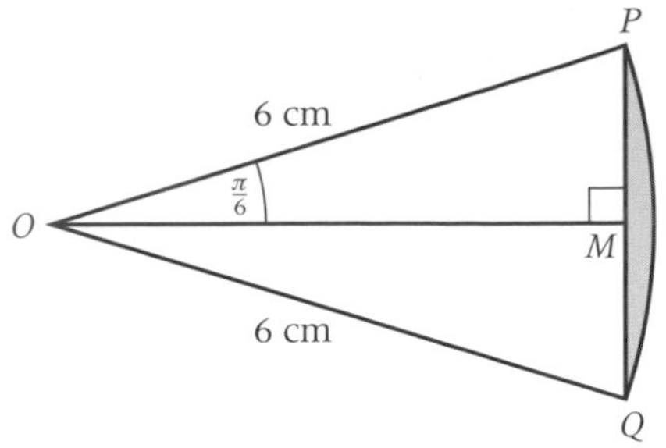

(a) Write down the exact values of:

(i) the lengths of PM and OM,

(ii) the length of the arc PQ,

(iii) the area of the sector POQ.

(b) Use the appropriate answers from **(a)** to show that the area of the shaded region is $m(2\pi - 3\sqrt{3})$ cm^2, for some integer m whose value is to be determined. [A]

9 A circle PQR has centre O and radius r. It is divided into three equal areas by the chords PQ and PR. Given that angle $QPR = \theta$ radians, find the area of

(a) the sector OQR,

(b) the triangle OPQ.

Deduce that $\sin\theta = \frac{\pi}{3} - \theta$. [A]

10 The diagram shows an arc ABC of a circle centre O and radius 10 cm.

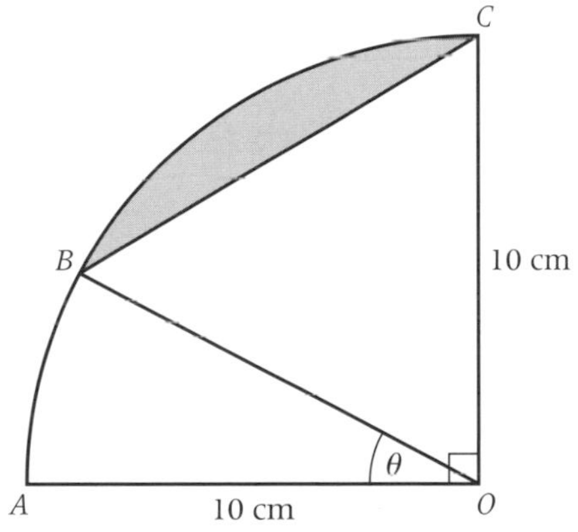

Angle $AOB = \theta$ radians and angle AOC is a right angle.

The shaded segment is bounded by the arc BC and the chord BC.

(a) Write down, in terms of θ, the area of the sector AOB.

(b) Show that the area of the shaded segment is $25(\pi - 2\theta - 2\cos\theta)$ cm^2.

(c) Given that the length of the arc BC is 4 times the length of the arc AB, show that $\theta = \frac{\pi}{10}$. [A]

9

Key point summary

1 One radian is the angle subtended at the centre of a circle by an arc whose length is equal to the radius of the circle. *p337*

1 radian is written as 1 rad or as 1^c

1 rad $\approx 57°$

2 Converting between degrees and radians use $360° = 2\pi$ rads. *p337*

Hence $x° = x \times \dfrac{2\pi}{360}$ rads.

3 The length l of an arc of a circle is given by $l = r\,\theta$, where r is the radius and θ is the angle, in radians, subtended by the arc at the centre of the circle. *p340*

4 The area A of a sector of a circle is given by $A = \dfrac{1}{2}r^2\theta$, where r is the radius and θ is the angle, in radians, subtended by the arc at the centre of the circle. *p343*

Test yourself

What to review

[*In this section give final numerical answers to 3 sf*]

	What to review
1 In triangle DEF, angle $D = 1.8$ radians, angle $E = \dfrac{\pi}{4}$ radians and angle $F = \theta$ radians. Find the value of $\cos\theta$.	*Section 9.2*
2 Points A and B lie on the circumference of a circle of radius 8 cm and centre O. Angle AOB, subtended by the minor arc AB, is 1.4 radians. **(a)** Find the area of the minor sector AOB. **(b)** Find the area of the minor segment. **(c)** Find the perimeter of the **major** sector.	*Section 9.3 and 9.4*
3 The perimeter of a sector of a circle of radius r cm is P cm. The area of the sector is A cm^2. Show that $A = \dfrac{1}{2}r(P - 2r)$.	*Sections 9.3 and 9.4*

Test yourself ANSWERS

1 0.849.

2 **(a)** 44.8 cm^2; **(b)** 13.3 cm^2; **(c)** 55.1 cm.

CHAPTER 10

C2: Further trigonometry with radians

Learning objectives

After studying this chapter, you should be able:

- to solve trigonometrical equations of the type considered in C2 chapters 4 and 6 but where answers are in radians.

10.1 Introduction

In this chapter you will apply the methods you used in chapters 4 and 6 to solve trigonometrical equations but you will be using radians instead of degrees. It is important that you remember that both 2π radians and $360°$ represent one complete turn. The following table summarises for you some of the conversions that are used most. (You used many of them in Exercise 9A).

θ in degrees	θ in radians
0	0
30	$\frac{\pi}{6}$
45	$\frac{\pi}{4}$
60	$\frac{\pi}{3}$
90	$\frac{\pi}{2}$
180	π
360	2π

In chapter 6, a common interval for θ given in questions, was $-180° \leqslant \theta \leqslant 180°$. In radians this would be shown as $-\pi \leqslant \theta \leqslant \pi$. Similarly, the interval $-90° \leqslant \theta < 360°$ would be shown in radians as $-\frac{\pi}{2} \leqslant \theta < 2\pi$.

If the interval for θ is given in terms of π you are to assume that θ is measured in radians. In such cases all solutions to the trigonometrical equation must be given in radians.

The decimal approximation for π is 3.141 59... so the question 'solve the equation $\sin\theta = 0.4$ in the interval $-\frac{\pi}{2} \leqslant \theta < 2\pi$' is asking you to find the values of θ, in radians, that lie between $-1.5707...^c$ and $6.283...^c$.

The graphs of $\sin x$, $\cos x$ and $\tan x$, where x is in radians, are the same shape as those in sections 4.5, 4.6 and 4.9, but values on the x-axis are shown in radians instead of degrees.

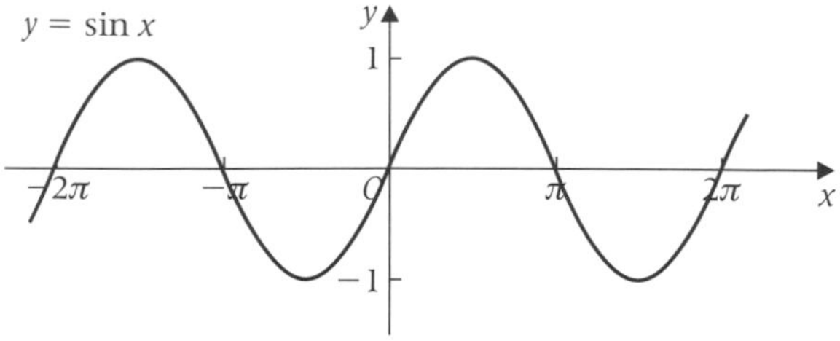

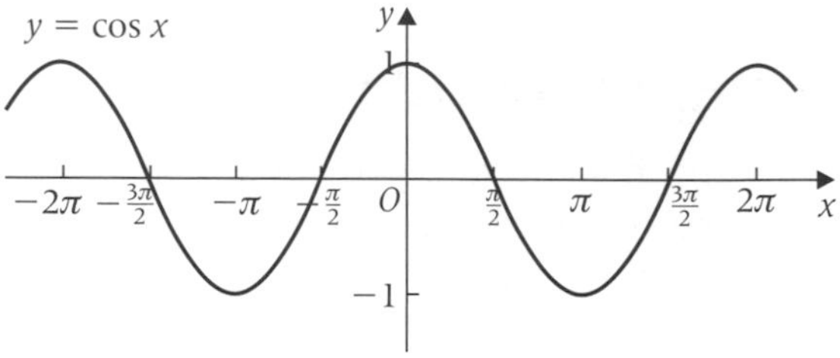

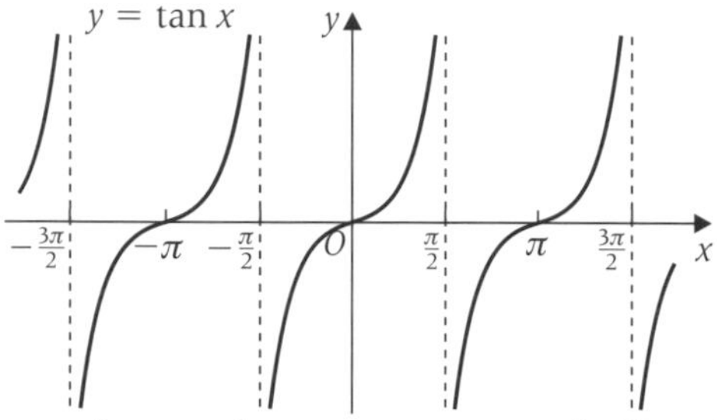

$y = \sin x$ and $y = \cos x$ are periodic with period 2π radians.
$y = \tan x$ is periodic with period π radians.

EXERCISE 10A

By considering the graphs above:

1 Write the following in terms of $\sin\theta$:

See section 4.5.

(a) $\sin(\pi - \theta)$, **(b)** $\sin(\pi + \theta)$,
(c) $\sin(2\pi + \theta)$, **(d)** $\sin(2\pi - \theta)$.

2 Write the following in terms of $\cos\theta$:

(a) $\cos(\pi - \theta)$, **(b)** $\cos(\pi + \theta)$,
(c) $\cos(2\pi + \theta)$, **(d)** $\cos(2\pi - \theta)$.

3 Write the following in terms of $\tan\theta$:

(a) $\tan(\pi - \theta)$, **(b)** $\tan(\pi + \theta)$,
(c) $\tan(2\pi + \theta)$, **(d)** $\tan(2\pi - \theta)$.

10.2 Solving simple trigonometrical equations in terms of radians

Worked example 10.1

Solve the equation $\sin x = -0.6$, in the interval $-\pi < x < \pi$. Give your answer to three significant figures.

Solution

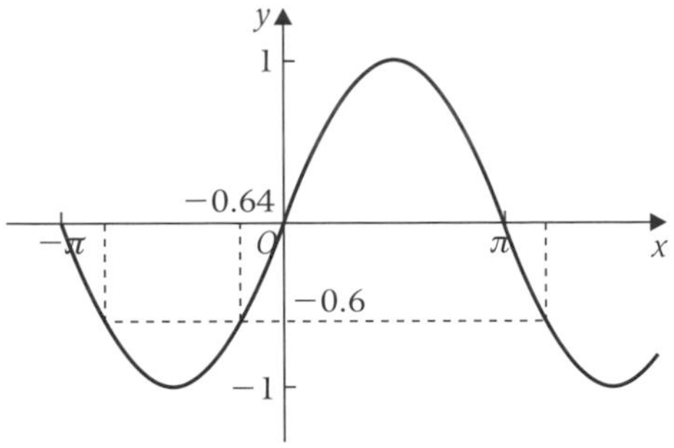

> The given interval is in terms of π so you must work in radians.

Using a calculator, $\sin^{-1}(-0.6) = -0.6435...^c$.

> Make sure your calculator is in radian mode.

By symmetry, the second solution in the interval $-\pi < x < \pi$ is $x = -\pi + 0.6435... = -2.498....$

> You could have found the 2nd solution by using $(\pi - \alpha) - 2\pi$, where α is $-0.6435....$

To three significant figures the solutions of $\sin x = -0.6$, in the interval $-\pi < x < \pi$ are $x = -2.50^c$ and -0.644^c.

Worked example 10.2

Solve the equation $\cos 4x = 0.3$, in the interval $0 \leqslant x \leqslant \pi$.
Give your answer to three significant figures.

Solution

Let $u = 4x \Rightarrow \cos u = 0.3$.
Finding the interval for u, $0 \leqslant x \leqslant \pi \Rightarrow 0 \leqslant 4x \leqslant 4\pi$, so the interval for u is $0 \leqslant u \leqslant 4\pi$.
You need to find all values for u in the interval $0 \leqslant u \leqslant 4\pi$ that satisfy the equation $\cos u = 0.3$.

> In two complete cycles (since $0 \leqslant u \leqslant 4\pi$) you would expect four solutions of $\cos u = 0.3$ (see section 4.10).

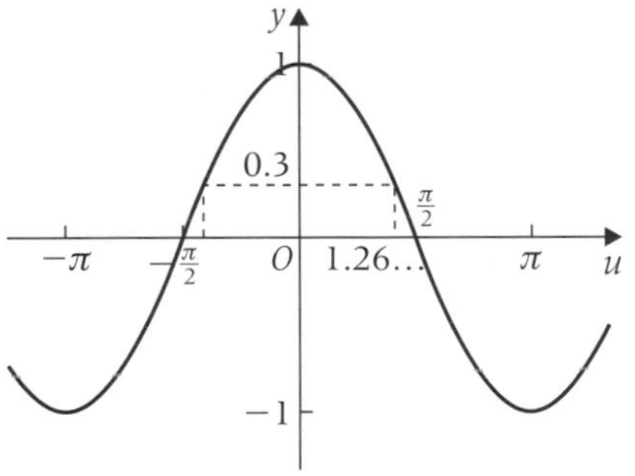

Using a calculator, $u = \cos^{-1}(0.3) = 1.2661...^c$.

> Make sure your calculator is in radian mode.

By symmetry, a second solution in a 2π cycle is $u = -1.2661...^c$.
Since $y = \cos u$ has period 2π radians, the solutions of $\cos u = 0.3$ are

$$u = 1.2661..., 1.2661... \pm 2\pi, 1.2661... \pm 4\pi, ...$$
$$u = -1.2661..., -1.2661... \pm 2\pi, -1.2661... \pm 4\pi, ...$$

Choosing those values of u that lie in the interval $0 \leqslant u \leqslant 4\pi$ give $u = 1.2661\ldots$, $1.2661\ldots + 2\pi$, $-1.2661\ldots + 2\pi$ and $-1.2661\ldots + 4\pi$.

You must find all relevant values for u before calculating x.

$4x = u = 1.2661\ldots, 7.5492\ldots, 5.017\,08\ldots, 11.3002\ldots$

$\Rightarrow \quad x = 0.317, 1.89, 1.25, 2.83$ (to 3 sf).

Worked example 10.3

Solve the equation $\tan(x + 0.4) = 0.9$, in the interval $0 < x < 2\pi$. Give your answer to three significant figures.

Solution

Let $u = x + 0.4 \Rightarrow \tan u = 0.9$.
$0 < x < 2\pi \Rightarrow 0 + 0.4 < x + 0.4 < 2\pi + 0.4 \Rightarrow 0.4 < u < 6.683\ldots$.
You need to find all values for u in the interval $0.4 < u < 6.683\ldots$ that satisfy the equation $\tan u = 0.9$.

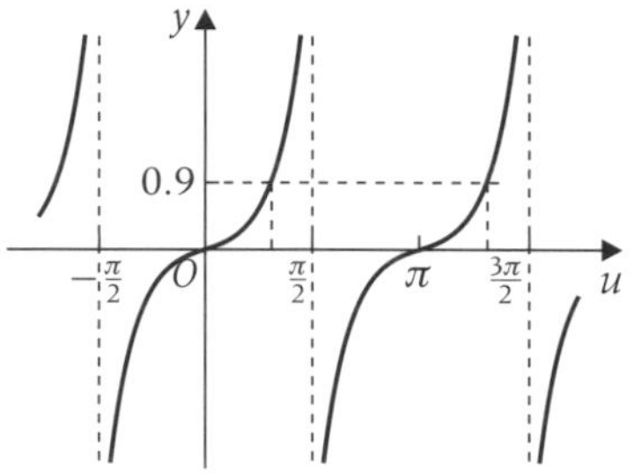

Using a calculator, $u = \tan^{-1}(0.9) = 0.7328\ldots^{c}$.

Make sure your calculator is in radian mode.

Since $y = \tan u$ has period π radians, the solutions of $\tan u = 0.9$ are $u = 0.7328\ldots^{c}$, $0.7328^{c} \pm \pi$, $0.7328\ldots^{c} \pm 2\pi, \ldots$.

Choosing those values of u that lie in the interval $0.4 < u < 6.683\ldots$ gives $u = 0.7328\ldots^{c}$, $0.7328\ldots^{c} + \pi$,

There are many other solutions but these are the only ones in the interval $0 < x < 2\pi$.

$x + 0.4 = u = 0.7328\ldots, 3.8744\ldots$

$\Rightarrow \quad x = 0.333, 3.47$ (to 3 sf).

Worked example 10.4

Solve the equation $\cos\left(x - \frac{\pi}{3}\right) = \cos\frac{\pi}{3}$, in the interval $-2\pi \leqslant x < 2\pi$. Leave your answer in terms of π.

Solution

Let $u = x - \frac{\pi}{3} \Rightarrow \cos u = \cos\frac{\pi}{3}$

The interval for u is $-\frac{7\pi}{3} \leqslant u < \frac{5\pi}{3}$.

Finding the interval for u,
$-2\pi \leqslant x < 2\pi$
$\Rightarrow -2\pi - \frac{\pi}{3} \leqslant x - \frac{\pi}{3} < 2\pi - \frac{\pi}{3}$.

You need to find all values for u in the interval $-\frac{7\pi}{3} \leqslant u < \frac{5\pi}{3}$ that satisfy the equation $\cos u = \cos\frac{\pi}{3}$.

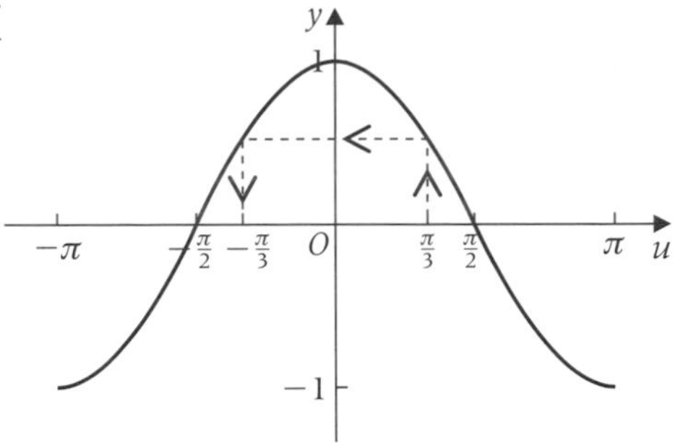

One solution is clearly, $u = \frac{\pi}{3}$.

By symmetry, a second solution in a 2π cycle is $u = -\frac{\pi}{3}$.

Since $y = \cos u$ has period 2π radians, the solutions of $\cos u = \cos\frac{\pi}{3}$ are

$$u = \frac{\pi}{3}, \frac{\pi}{3} \pm 2\pi, \frac{\pi}{3} \pm 4\pi, \ldots$$

and $u = -\frac{\pi}{3}, -\frac{\pi}{3} \pm 2\pi, -\frac{\pi}{3} \pm 4\pi, \ldots$.

Choosing those values of u that lie in the interval $-\frac{7\pi}{3} \leqslant u < \frac{5\pi}{3}$ gives

$$u = \frac{\pi}{3}, -\frac{5\pi}{3}, -\frac{\pi}{3} \text{ and } -\frac{7\pi}{3}.$$

$$x - \frac{\pi}{3} = \frac{\pi}{3}, -\frac{5\pi}{3}, -\frac{\pi}{3} \text{ and } -\frac{7\pi}{3}.$$

$$\Rightarrow \quad x = \frac{2\pi}{3}, -\frac{4\pi}{3}, 0 \text{ and } -2\pi.$$

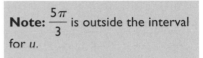

Note: $\frac{5\pi}{3}$ is outside the interval for u.

Do not attempt to use a decimal approximation for π.

EXERCISE 10B

1 For each of the following equations, find the solution that lies in the interval $0 < \theta < \frac{\pi}{2}$. In each case give your answer to three significant figures.

(a) $\sin\theta = 0.45$, **(b)** $\cos\theta = 0.75$,

(c) $\tan\theta = 1.6$, **(d)** $\sin\left(\theta - \frac{\pi}{4}\right) = 0.5$,

(e) $\cos(\theta + 0.1) = 0.48$, **(f)** $\tan\left(\theta + \frac{\pi}{6}\right) = 2$.

2 Solve the following equations in the interval $0 < x < 2\pi$. In each case give final answers in radians correct to three significant figures.

(a) $\sin x = 0.6$, **(b)** $\cos x = 0.4$, **(c)** $\tan x = 1.5$,

(d) $\sin x = -0.8$, **(e)** $\cos x = -0.7$, **(f)** $\tan x = -2$,

(g) $\sin x = \frac{3}{4}$, **(h)** $\cos x = -\frac{5}{8}$,

(i) $\tan 2x = -0.5$, **(j)** $\sin\frac{x}{2} = -\frac{1}{4}$,

(k) $\cos(x + 0.6) = 1$, **(l)** $\tan\left(x + \frac{1}{2}\right) = 3$.

3 Solve the equation $\cos\left(x + \frac{\pi}{2}\right) = -\frac{1}{2}$ in the interval $0 < x < 2\pi$. Give your answer to three significant figures.

4 Solve the following equations in the interval $-\pi < x < \pi$. Leave your answers in terms of π.

(a) $\sin x = \sin\frac{\pi}{6}$,
(b) $\cos x = \cos\frac{5\pi}{6}$,
(c) $\tan x = \tan\frac{\pi}{4}$,
(d) $\sin x = \sin\left(-\frac{\pi}{4}\right)$,
(e) $\cos 2x = \cos\frac{\pi}{4}$,
(f) $\tan 3x = \tan\frac{3\pi}{4}$,
(g) $\sin 7x = \sin \pi$,
(h) $\cos 3x = \cos\frac{2\pi}{3}$,
(i) $\tan 2x = \tan\left(-\frac{\pi}{3}\right)$.

5 (a) Write down the maximum value of $\cos\left(\theta + \frac{\pi}{6}\right)$.

(b) Find the value of θ in the interval $0 < \theta < 2\pi$, for which $\cos\left(\theta + \frac{\pi}{6}\right)$ is a maximum.

10.3 Using trigonometrical identities to solve equations where answers are given in radians

In chapter 6 you were shown the two identities

$$\cos^2\theta + \sin^2\theta \equiv 1 \quad \text{and} \quad \tan\theta \equiv \frac{\sin\theta}{\cos\theta}.$$

In this section we shall use these identities to reduce more complicated trigonometrical equations into the simpler forms $\sin k\theta = a$ or $\cos k\theta = b$ or $\tan k\theta = c$ and then proceed as in the previous section to give answers in radians.

In this section you will need to factorise quadratics. You must also remember to include a $\pm$ sign when taking square roots. You should also recall that $-1 \leqslant \cos\theta \leqslant 1$ and $-1 \leqslant \sin\theta \leqslant 1$.

Worked example 10.5

Solve the equation $3\cos^2 x + 1 = 6\sin^2 x$ in the interval $0 < x < 2\pi$. Give your answers to three significant figures.

Solution

Using the identity $\cos^2 x + \sin^2 x \equiv 1$,
$3\cos^2 x + 1 = 6\sin^2 x$ becomes $3(1 - \sin^2 x) + 1 = 6\sin^2 x$

$\Rightarrow 9\sin^2 x = 4 \Rightarrow \sin^2 x = \frac{4}{9}$

$\Rightarrow \sin x = \pm\frac{2}{3}$

You could eliminate the 1 to get $4\cos^2 x = 5\sin^2 x$. Applying the identity for $\tan x$ would then lead to $\tan^2 x = \frac{4}{5}$.

When taking square roots remember the $\pm$ sign. Both these values for $\sin x$ are possible.

When $\sin x = \frac{2}{3} \Rightarrow \quad x = 0.7297\ldots, \pi - 0.7297\ldots, 0.7297\ldots \pm 2\pi, \ldots.$

$\sin^{-1}\left(\frac{2}{3}\right) = 0.7297\ldots^c$

Choosing values in the interval $0 < x < 2\pi$ gives $x = 0.7297\ldots,$ $\pi - 0.7297\ldots.$

When $\sin x = -\frac{2}{3} \Rightarrow \quad x = -0.7297\ldots, \pi - (-0.7297\ldots),$ $-0.7297\ldots \pm 2\pi, \ldots.$

$\sin^{-1}\left(-\frac{2}{3}\right) = -0.7297\ldots^c$

Choosing values in the interval $0 < x < 2\pi$ gives $x = \pi + 0.7297\ldots, 2\pi - 0.7297\ldots.$

Combining these we have four solutions in the given interval. $x = 0.730, 2.41, 3.87, 5.55$ (correct to 3 sf).

Worked example 10.6

Solve the equation $\cos^2 x - \sin^2 x + 3\cos x = 1$ in the interval $-\pi < x < \pi$. Give your answer to three significant figures.

Solution

Using the identity $\cos^2 x + \sin^2 x \equiv 1$,
$\cos^2 x - \sin^2 x + 3\cos x = 1$ becomes
$\cos^2 x - (1 - \cos^2 x) + 3\cos x = 1$
$\Rightarrow 2\cos^2 x + 3\cos x - 2 = 0$
$\Rightarrow (\cos x + 2)(2\cos x - 1) = 0$
$\Rightarrow \cos x = -2$ or $\cos x = \frac{1}{2}$

Look at the non-squared trig term to decide what to keep. Since here it is cos *x*, we eliminate sin *x*.

Since cos *x* cannot be less than -1, the only solutions come from $\cos x = \frac{1}{2}$.

$-1 \leqslant \cos x \leqslant 1$

When $\cos x = \frac{1}{2}$,
$\Rightarrow x = 1.047\ldots, -1.047\ldots, 1.047\ldots \pm 2\pi, -1.047\ldots \pm 2\pi, \ldots.$

$\cos^{-1}\left(\frac{1}{2}\right) = 1.047\ldots.$

In the interval $-\pi < x < \pi$, the only solutions are $x = 1.05$, -1.05 (to 3 sf).

10

Worked example 10.7

Solve the equation $3\sin 2x = 2\cos 2x$ in the interval $0 < x < \pi$. Give your answers to three significant figures.

Solution

Since $3\sin 2x = 2\cos 2x$ we can deduce that $\cos 2x \neq 0$. Dividing both sides by $3\cos 2x$

$$\frac{3\sin 2x}{3\cos 2x} = \frac{2\cos 2x}{3\cos 2x} \Rightarrow \tan 2x = \frac{2}{3}.$$

Used the identity $\tan\theta = \frac{\sin\theta}{\cos\theta}$ with $\theta = 2x$.

The interval $0 < x < \pi \Rightarrow 0 < 2x < 2\pi$.

When $\tan 2x = \frac{2}{3}$,

$2x = 0.5880\ldots, 0.5880\ldots \pm \pi, 0.5880\ldots \pm 2\pi\ldots.$

In the interval $0 < 2x < 2\pi$,
$2x = 0.5880\ \ldots, 0.5880\ldots + \pi$
$\Rightarrow x = 0.294, 1.86$ (to 3 sf).

$\tan^{-1}\left(\frac{2}{3}\right) = 0.5880\ldots$

You must find all relevant values for $2x$ before calculating x.

EXERCISE 10C

1 Solve the following equations in the interval $0 < x < 2\pi$. In each case give final answers in radians correct to three significant figures.

(a) $\cos^2 x = 0.9$,

(b) $\sin^2 x = 0.4$,

(c) $\tan^2 x = 4$,

(d) $5\cos^2 x + \cos x - 4 = 0$,

(e) $3\sin^2 x + 2\sin x = 0$,

(f) $\tan^2 x + 2\tan x - 8 = 0$,

(g) $5\cos^2 x = 8\cos x - 5\sin^2 x$,

(h) $5\cos^2 x + 7\cos x - 6 = 0$,

(i) $3\sin^2 x + 7\sin x - 6 = 0$,

(j) $\tan^2 x + 2\tan x - 1 = 0$,

(k) $3\cos^2 x = \cos x$.

2 Solve the following equations in the interval $-\pi < x < \pi$. Give your answers to three significant figures.

(a) $\cos^2 x = 0.5$,

(b) $\sin^2 x = 0.25$,

(c) $\tan^2 x - 3 = 0$,

(d) $2\cos^2 x - 5\cos x - 3 = 0$,

(e) $2\sin^2 x + 5\sin x = 3$,

(f) $\tan^2 x + 2\tan x + 1 = 0$,

(g) $3\tan^2 x = 1$,

(h) $10\cos^2 x + 7\cos x = 6$,

(i) $4\cos^3 x = 3\cos x$.

3 Solve the following equations in the interval $0 < x < 2\pi$. In each case give final answers in radians correct to three significant figures.

(a) $\cos x = 3\sin x$,

(b) $5\cos x = 4\sin x$,

(c) $\sin^2 x = 4\cos^2 x$,

(d) $5\cos^2 x = \sin x + 1$,

(e) $3\sin^2 x - 2\cos x = 3$,

(f) $3\sin x = 2\tan x$,

(g) $2\cos x = 3\tan x$,

(h) $3\cos^2 x - 5\sin x\cos x - 2\sin^2 x = 0$
[Hint: $3p^2 - 5qp - 2q^2 = (3p + q)(p - 2q)$],

(i) $3\sin^2 x + 5\sin x\cos x = 2\cos^2 x$.

4 Solve the following equations in the interval $-\pi \leqslant x \leqslant \pi$. Give your answers to three significant figures.

(a) $\sin^2 2x = 0.5$,

(b) $\cos^2 3x = 1$,

(c) $\tan^2\left(x + \frac{\pi}{3}\right) = 3$,

(d) $2\cos^2 2x + \cos 2x - 1 = 0$,

(e) $2\sin^2 3x + 5\sin 3x = 3$,

(f) $\tan^3 2x = 3\tan 2x$,

(g) $3\tan^2\left(x + \frac{\pi}{4}\right) = 1$,

(h) $2\sin^2 3x = \cos 3x + 1$,

(i) $\cos^3 2x = 3\sin^2 2x \cos 2x$.

MIXED EXERCISE

1 Solve the equation $\cos\left(x + \frac{\pi}{6}\right) = -0.5$, in the interval $0 < x < 2\pi$, giving your answers to three significant figures. [A]

2 Solve the equation $\tan 3x = 1$, in the interval $0 < x < 2\pi$, giving your answers to three significant figures. [A]

3 Solve the equation $8\sin^2 x - 7 = 2\cos x$, giving all answers for x in radians, to three significant figures, in the interval $0 \leqslant x \leqslant 2\pi$. [A]

4 The angle θ radians, where $0 \leqslant \theta \leqslant 2\pi$, satisfies the equation $3\tan\theta = 2\cos\theta$.

(a) Show that $3\sin\theta = 2\cos^2\theta$.

(b) Hence use an appropriate identity to show that $2\sin^2\theta + 3\sin\theta - 2 = 0$.

(c) **(i)** Solve the quadratic equation in **(b)**. Hence explain why the only possible value of $\sin\theta$, which will satisfy it, is $\frac{1}{2}$.

(ii) Find the values of θ for which $\sin\theta = \frac{1}{2}$ and $0 \leqslant \theta \leqslant 2\pi$. [A]

5 It is given that x satisfies the equation $2\cos^2 x = 2 + \sin x$.

(a) Use an appropriate trigonometrical identity to show that $2\sin^2 x + \sin x = 0$.

(b) Solve this quadratic equation and hence find all the possible values of x in the interval $0 \leqslant x < 2\pi$. [A]

10

Key point summary

1 *p349*

θ in degrees	θ in radians
0	0
30	$\frac{\pi}{6}$
45	$\frac{\pi}{4}$
60	$\frac{\pi}{3}$
90	$\frac{\pi}{2}$
180	π
360	2π

2 If the interval for θ is given in terms of π you are to assume that θ is measured in radians. In such cases, all solutions to the trigonometrical equation must be given in radians. *p349*

Test yourself

Test yourself	What to review
1 Solve the equation $\tan x = 0.85$ in the interval $0 < x < 2\pi$. Give your answers in radians to three significant figures.	*Section 10.2*
2 Solve the equation $\sin 2x = 1$ in the interval $0 < x < 2\pi$. Give your answers in radians to three significant figures.	*Section 10.2*
3 Solve the equation $\cos\left(x + \frac{\pi}{4}\right) = \cos\frac{7\pi}{6}$ in the interval $-\pi < x < \pi$. Leave your answers in terms of π.	*Section 10.2*
4 Solve the equation $4\sin^2 x - 2\cos^2 x + 3\sin x = 1$ in the interval $0 < x < 2\pi$. Give your answers to three significant figures.	*Section 10.3*
5 Solve the equation $(3\sin x - 4\cos x)(4\sin^2 x - 3) = 0$ in the interval $0 < x < 2\pi$. Give your answers in radians to 3 sf.	*Section 10.3*

Test yourself ANSWERS

1 0.704, 3.85. **2** 0.785, 3.93. **3** $\frac{7\pi}{12}, \frac{11\pi}{12}$.

4 0.524, 2.62, 4.71.

5 $x = 0.927, 1.05, 2.09, 4.07, 4.19, 5.24$.

CHAPTER 11

C2: Exponentials and logarithms

Learning objectives

After studying this chapter, you should:

- be able to recognise and sketch graphs of exponential functions
- be familiar with the notation for logarithms
- be able to convert between exponentials and logarithms
- know that $\log_a a = 1$ and $\log_a 1 = 0$
- know and be able to use the three laws of logarithms
- be able to solve equations of the form $a^x = b$
- be able to evaluate logarithms to any base.

11.1 Exponential functions

Exponential functions occur naturally in formulae involving population growth.

> An **exponent** is another name for an index or power. Functions such as f where $f(x) = 2^x$ are called **exponential functions**, because the variable x appears in the exponent.

2^x, $(0.5)^{2x}$, 3^{t-2} are examples of exponential functions.

An exponential function increases very rapidly.

You may be able to use your calculator to show that $f(1) = 2$; $f(10) = 1024$; $f(100) \approx 1.27 \times 10^{30}$, but it will probably give you an error message if you try to calculate $f(1000)$.
Since $2^0 = 1$, $f(0) = 1$.

You can evaluate $f(x)$ when x is negative. For instance, $f(-1) = \frac{1}{2}$, $f(-3) = \frac{1}{8}$, $f(-10) = \frac{1}{1024}$, etc.

As x becomes large and negative, $f(x)$ approaches zero. Even so, you should find that $f(x)$ is always positive for all values of x.

The graph of $y = 2^x$ is sketched opposite.

In general, the graph of $y = a^x$, where a is a positive constant, would have a similar shape to the graph opposite.

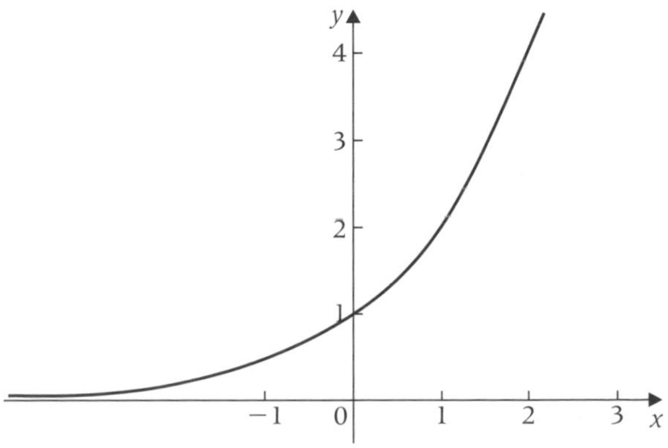

For all values of the positive constant a, the curve $y = a^x$ will always pass through the point (0,1).

The next Worked example shows how the sketch graphs of other exponential functions can be obtained from the sketch graph of $y = a^x$ by applying a single geometrical transformation.

Worked example 11.1

Sketch the graphs of **(a)** $y = 3^{-x}$, **(b)** $y = 5^{\frac{x}{2}}$.

Solution

(a) The graph of $y = f(-x)$ is obtained from the graph of $y = f(x)$ by a reflection in the y-axis. The graph of $y = 3^x$ is shown as a broken line. The required graph, namely $y = 3^{-x}$ is shown as a full line.

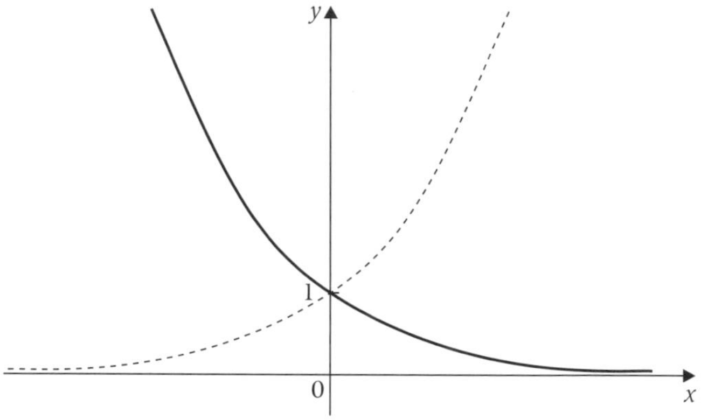

(b) The graph of $y = g(\frac{x}{2})$ is obtained from the graph of $y = g(x)$ by a one-way stretch of scale factor 2 parallel to the x-axis. The graph of $y = 5^x$ is shown below as a broken line. The required graph, namely $y = 5^{\frac{x}{2}}$, is shown as a full line.

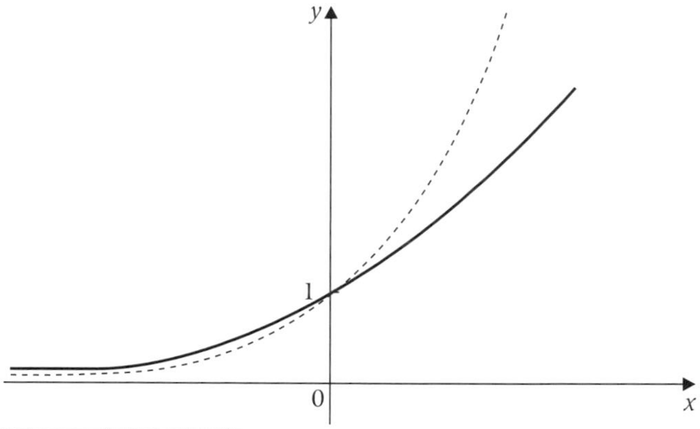

EXERCISE 11A

1 Sketch the graphs of:

(a) $y = 4^x$, **(b)** $y = 4^{-x}$,

(c) $y = 4^x - 4$, **(d)** $y = 2 + 4^{-x}$.

In each case, indicate any intercepts with the coordinate axes.

2 State a transformation which maps the graph of $y = 2^x$ onto

(a) $y = 2^{3x}$, **(b)** $y = 2^{x-3}$, **(c)** $y = 2^{\frac{x}{3}}$,

(d) $y = 2^{x+5}$, **(e)** $y = 2^{-x}$.

3 Find a transformation which maps the graph of $y = 3^x$ onto
   (a) $y = 3^{5x}$, **(b)** $y = 3^{x-4}$, **(c)** $y = 3^{\frac{2x}{5}}$,
   (d) $y = 2 \times 3^x$, **(e)** $y = 4 + 3^x$.

4 The graph of $y = 5^x$ passes through the points $(0, a)$, $(b, \sqrt{5})$ and $(-2, c)$. Find the value of a, b and c.

5 The graph of $y = 2^{x-3}$ passes through the points $(0, d)$, $(e, \sqrt{2})$ and $(f, 1)$. Find the value of d, e and f.

6 State the value of y for each of the following **(i)** when $x = 0$ and **(ii)** when $x = 1$, and sketch their graphs:
   (a) $y = 5^{-x}$, **(b)** $y = 3^{x+4}$, **(c)** $y = 2^{3x}$, **(d)** $y = 4^{1-x}$.

7 Sketch on the same axes the graphs of $y = 3^{x+1}$ and $y = 3^{2-x}$. Find the coordinates of the point of intersection of the two graphs.

11.2 Introduction to logarithms and the notation used

It is frequently useful to express a number N in exponential form,

$$N = a^x.$$

In this case the power x is called the **logarithm** of the number to the **base** a.

$100 = 10^2$; $2 = 4^{\frac{1}{2}}$; $\frac{1}{25} = 5^{-2}$.

The logarithm of N to base a is written as $\log_a N$.

So the statements $N = a^x$ and $x = \log_a N$ are identical and interchangeable. We write such a result as

$$N = a^x \Leftrightarrow x = \log_a N \qquad (*)$$

In words; the statement N equals a to the power x implies, and is implied by, the statement x equals the logarithm of N to base a.

Here are some numerical equivalents

$$100 = 10^2 \Leftrightarrow 2 = \log_{10} 100,$$
$$2 = 4^{\frac{1}{2}} \Leftrightarrow \frac{1}{2} = \log_4 2,$$
$$\frac{1}{25} = 5^{-2} \Leftrightarrow -2 = \log_5\left(\frac{1}{25}\right).$$

Putting $N = a$ in the result $(*)$ we can deduce

$$x = \log_a a \Rightarrow a^x = a = a^1 \Rightarrow x = 1$$
$$\text{so } \log_a a = 1.$$

11

Putting $N = 1$ in the result (*) we can deduce

$$x = \log_a 1 \Rightarrow a^x = 1 = a^0 \Rightarrow x = 0$$
$$\text{so } \log_a 1 = 0.$$

> The two results, $\log_a a = 1$ and $\log_a 1 = 0$, are frequently required in solutions to examination questions and should be memorised.

Worked example 11.2

Given that $\log_4 k = 2.5$, find the value of k.

Solution

$2.5 = \log_4 k \Rightarrow k = 4^{2.5} = 32$

$4^{2.5} = 4^2 \times 4^{\frac{1}{2}} = 16\sqrt{4} = 32$

Worked example 11.3

Evaluate:

(a) $\log_2 8$, **(b)** $\log_4 8$, **(c)** $\log_a \sqrt{a}$.

Solution

Used $x = \log_a N \Rightarrow N = a^x$

(a) Let $x = \log_2 8$ $\Rightarrow 8 = 2^x$
so $2^3 = 2^x \Rightarrow x = 3$ so $\log_2 8 = 3$;

(b) Let $x = \log_4 8$ $\Rightarrow 8 = 4^x$
so $2^3 = (2^2)^x = 2^{2x}$
$\Rightarrow 2x = 3 \Rightarrow x = 1.5$ so $\log_4 8 = 1.5$;

(c) Let $x = \log_a \sqrt{a}$ $\Rightarrow \sqrt{a} = a^x$
so $a^{\frac{1}{2}} = a^x \Rightarrow x = \frac{1}{2}$ so $\log_a \sqrt{a} = \frac{1}{2}$.

EXERCISE 11B

1 Find the value of k in each of the following:

(a) $\log_4 k = 2$, **(b)** $\log_4 k = 3$, **(c)** $\log_3 k = 1$,
(d) $\log_8 k = 0$, **(e)** $\log_{10} k = 5$, **(f)** $\log_9 k = 1.5$,
(g) $\log_4 k = 0.5$, **(h)** $\log_4 k = -2$, **(i)** $\log_4 k = -2.5$,
(j) $\log_k 4 = 2$, **(k)** $\log_k 4 = 0.5$, **(l)** $\log_k 27 = -1.5$.

2 Evaluate the following:

(a) $\log_2 32$, **(b)** $\log_2 2$, **(c)** $\log_3 27$,
(d) $\log_8 32$, **(e)** $\log_{10} 0.001$, **(f)** $\log_9 243$,
(g) $\log_4 32$, **(h)** $\log_4 0.125$, **(i)** $\log_4 0.5$,
(j) $\log_5 0.04$, **(k)** $\log_{16} 1$, **(l)** $7\log_{16} 16$,
(m) $\dfrac{\log_2 8}{\log_2 2}$, **(n)** $\dfrac{\log_4 8}{\log_2 4}$, **(o)** $\dfrac{\log_8 2}{\log_2 8}$.

3 Solve the equation $3 + \log_2 x = 0$.

11.3 Laws of logarithms

You have already used the laws of indices to simplify algebraic expressions.

Since logarithms are powers (or indices) there are corresponding laws of logarithms which we shall now derive.

Let $p = \log_a x \Rightarrow x = a^p$ and let $q = \log_a y \Rightarrow y = a^q$

So $xy = a^p \times a^q$

$\Rightarrow \quad xy = a^{p+q}$.

Used result (*) in Section 11.2.

$\Rightarrow p + q = \log_a xy$

and so $\log_a x + \log_a y = \log_a(xy)$.

Similarly $\dfrac{x}{y} = \dfrac{a^p}{a^q} \Rightarrow \dfrac{x}{y} = a^{p-q}$

$\Rightarrow p - q = \log_a\left(\dfrac{x}{y}\right)$

and so $\log_a x - \log_a y = \log_a\left(\dfrac{x}{y}\right)$.

A further law of logarithms can be found by using the index law $(a^p)^k = a^{pk.}$

Consider $x^k = (a^p)^k \Rightarrow x^k = a^{pk}$

$\Rightarrow pk = \log_a (x^k)$

and so $k \log_a x = \log_a (x^k)$

To apply the laws all logarithms must have the same base.

To summarise, the three laws of logarithms are

$\log_a x + \log_a y = \log_a(xy)$	[Law 1]
$\log_a x - \log_a y = \log_a\left(\dfrac{x}{y}\right)$	[Law 2]
$k \log_a x = \log_a(x^k)$	[Law 3]

For identification purposes only, we shall refer to them in this chapter as Laws 1, 2 and 3, respectively.

Worked example 11.4

Write $2 + \log_a 3$ as a single logarithm.

Solution

$2 + \log_a 3 = 2 \times \log_a a + \log_a 3$

Since the only logarithm in the question is to base a you write the non-logarithm term as a log to base a by using $1 = \log_a a$.

$= \log_a a^2 + \log_a 3$

Used Law 3.

$= \log_a(a^2 \times 3)$

Used Law 1.

Expressed as a single logarithm: $2 + \log_a 3 = \log_a(3a^2)$.

Worked example 11.5

Given that

$$3\log_a x - 4\log_a y = 1,$$

express a in terms of x and y.

Solution

$$3\log_a x - 4\log_a y = 1 \Rightarrow \log_a x^3 - \log_a y^4 = 1$$

Used Law 3 twice.

$$\Rightarrow \quad \log_a\left(\frac{x^3}{y^4}\right) = 1$$

Used Law 2.

$$\Rightarrow \quad \left(\frac{x^3}{y^4}\right) = a^1$$

Used result (*) in Section 11.2.

and so $a = \dfrac{x^3}{y^4}$.

Worked example 11.6

Express

$$\log_a \frac{x^2}{y^3\sqrt{z}} + \log_2 8$$

in terms of $\log_a x$, $\log_a y$ and $\log_a z$.

Solution

Since the bases of the two logarithms are different we must evaluate one of them or write them both in the same base.

Since $8 = 2^3 \Rightarrow 3 = \log_2 8$

$$\log_a \frac{x^2}{y^3\sqrt{z}} + \log_2 8 = \log_a \frac{x^2}{y^3\sqrt{z}} + 3$$

$$= \log_a x^2 - \log_a[y^3\sqrt{z}] + 3$$

Used Law 2.

$$= \log_a x^2 - [\log_a y^3 + \log_a z^{\frac{1}{2}}] + 3$$

Used Law 1.

$$= 2\log_a x - \left[3\log_a y + \frac{1}{2}\log_a z\right] + 3$$

Used Law 3.

and so $\log_a \dfrac{x^2}{y^3\sqrt{z}} + \log_2 8 = 2\log_a x - 3\log_a y - \dfrac{1}{2}\log_a z + 3$

EXERCISE 11C

1 Write as a single logarithm:

(a) $\log_a 32 - \log_a 4$, **(b)** $\log_a 6 + \log_a 4$,

(c) $2\log_a 6 - \log_a 4$, **(d)** $\log_a 6 + \log_a 8 - \log_a 4$,

(e) $\log_a 6 - \log_a 4 + \log_a a$, **(f)** $1 + \log_a 6$,

(g) $2 + \log_a 6 - \log_a 2$, **(h)** $2\log_a 3 - 3\log_a 6 + \log_a 8$.

2 Express in terms of $\log_a x$, $\log_a y$ and $\log_a z$:

(a) $\log_a(xyz)$, **(b)** $\log_a(axyz)$, **(c)** $\log_a(x^2y^3z^4)$,

(d) $\log_a \dfrac{ax}{yz}$, **(e)** $\log_a \dfrac{x^3}{y^5\sqrt{z}}$, **(f)** $\log_a \dfrac{x^2\sqrt{z}}{y^3}$,

(g) $\log_a \dfrac{\sqrt[3]{z}}{x^2y^4}$, **(h)** $\log_2 4 + \log_a \dfrac{\sqrt{axz^5}}{y^3}$.

3 Given that $\log_{24}12 = 0.782$, find the value of:

(a) $\log_{24}24$, **(b)** $\log_{24}1$, **(c)** $\log_{24}\sqrt{12}$, **(d)** $\log_{24}\left(\dfrac{1}{12}\right)$,

(e) $\log_{24}2$, **(f)** $\log_{24}6$, **(g)** $\log_{24}144$, **(h)** $\log_{24}3$.

4 Find the value of $2\log_a\left(\dfrac{3}{4}\right) - \log_a\left(7\dfrac{1}{2}\right) + \log_a\left(13\dfrac{1}{3}\right)$.

5 Solve the equations:

(a) $\log_2 x = 2\log_2 3 + \log_2 5$,

(b) $\log_2 x = 2\log_2 3 - 1$.

6 Find the two integer values of n which satisfy the equation

$$2\log_a n - \log_a(3n - 4) = \log_a 2.$$

7 Solve for x and y the simultaneous equations

$$\log_3 xy = 5$$
$$\log_3 x \log_3 y = 6.$$

Give your answers as simply as possible.

11.4 Solving equations of the form $a^x = b$

It is sometimes possible to find an exact solution of an equation of the form $a^x = b$. For example, the equation $2^x = 32$ has solution $x = 5$.
However, it is far more likely that equations of this type do not have exact solutions and they need to be solved using logarithms.

There are two numerical bases for logarithms that are used more than others.
They are 10 and e. If you look at your calculator you should be able to find two buttons which relate to logarithms: one will be log [or lg] which refers to base 10 and the other one is ln which relates to base e. You will learn about e in the next module.

In this section we will concentrate on the use of base 10.

11

$\log_{10}x$ is sometimes written as $\lg x$.

Check that your calculator gives $\lg 2 = 0.301\,029\,9\ldots$ and $\lg 0.4 = -0.3979\ldots$. Remember lg on your calculator may be log.

The next worked example shows you how to use logarithms to solve equations of the form $a^x = b$.

Worked example 11.7

Solve each of the following equations, giving your answers to three significant figures:

(a) $2^x = 30$,

(b) $7^{2y} = 3.5$.

Solution

(a) $2^x = 30$

Taking logarithms to base 10 of both sides gives

$$\log_{10}(2^x) = \log_{10} 30$$

Used Law 3.

$$\Rightarrow \quad x \log_{10} 2 = \log_{10} 30$$

$$\Rightarrow \quad x = \frac{\log_{10} 30}{\log_{10} 2} = \frac{1.477\,12\ldots}{0.301\,029\ldots} = 4.906\,89\ldots$$

Since $2^x = 32$ has the solution $x = 5$, you would expect the answer to $2^x = 30$ to be a little less than 5.

Therefore, $x = 4.91$ (to three significant figures).

(b) $7^{2y} = 3.5$

Taking logarithms to base 10 of both sides gives

$$\log_{10} 7^{2y} = \log_{10} 3.5$$

Used Law 3.

$$\Rightarrow \quad 2y \log_{10} 7 = \log_{10} 3.5$$

A common **mistake** is to write $\frac{\log_{10} 3.5}{\log_{10} 7} = \log_{10}\frac{1}{2}$.

$$\Rightarrow \quad 2y = \frac{\log_{10} 3.5}{\log_{10} 7} = \frac{0.544\,068\ldots}{0.845\,098\ldots} = 0.643\,79\ldots$$

$$\Rightarrow \quad y = 0.321\,89\ldots$$

Therefore $y = 0.322$ (to three significant figures).

Provided you are consistent you can take natural logarithms (logarithms to base e) of both sides and obtain the same answers. For example, in **(b)** you get $2y = \frac{\ln 3.5}{\ln 7} = \frac{1.252\,76\ldots}{1.945\,91\ldots} = 0.643\,79$, which gives $y = 0.322$ (as before).

The solution to the equation $a^x = b$ is $x = \frac{\log_{10} b}{\log_{10} a}$, where logarithms are taken to base 10.

Worked example 11.8

Solve the equation $3^{2-x} = 5^x$, giving your answer to three significant figures.

Solution

$3^{2-x} = 5^x$

Taking logarithms to base 10 of both sides gives

$$\log_{10} 3^{2-x} = \log_{10} 5^x$$
$$\Rightarrow \quad (2-x)\log_{10} 3 = x\log_{10} 5$$
$$\Rightarrow \quad 2\log_{10} 3 - x\log_{10} 3 = x\log_{10} 5$$
$$\Rightarrow \quad 2\log_{10} 3 = x(\log_{10} 5 + \log_{10} 3)$$
$$\Rightarrow \quad x = \frac{2 \times \log_{10} 3}{(\log_{10} 5 + \log_{10} 3)} = \frac{0.954\,24\ldots}{1.176\,09\ldots}$$
$$\Rightarrow \quad x = 0.811\,36\ldots$$

Used Law 3 twice.

You could use Law 3 and Law 1 to get $x = \dfrac{\log_{10} 3^2}{\log_{10}(5 \times 3)} = \dfrac{\log_{10} 9}{\log_{10} 15}$.

Therefore $x = 0.811$ (to three significant figures).

The next Worked example shows you how to apply the method for solving $a^x = b$ to find the value of a logarithm to any positive base.

Worked example 11.9

Find the value of $\log_2 3$, giving your answer to four significant figures.

Solution

Let $x - \log_2 3 \quad \Rightarrow \quad 2^x = 3$

Used key point (*) in section 11.2.

Taking logarithms to base 10 of both sides leads to

$$\log_{10} 2^x = \log_{10} 3$$
$$\Rightarrow \quad x\log_{10} 2 = \log_{10} 3$$
$$\Rightarrow \quad x = \frac{\log_{10} 3}{\log_{10} 2} = 1.584\,96\ldots$$

Therefore, $\log_2 3 = 1.585$ (to four significant figures).

In general, to find the value of $\log_a b$:

1 Let $x = \log_a b \Rightarrow a^x = b$.

2 Take logarithms to base 10 of both sides $\Rightarrow \log_{10} a^x = \log_{10} b$.

3 Use Law 3 $\Rightarrow x\log_{10} a = \log_{10} b$.

4 Rearrange $\Rightarrow x = \dfrac{\log_{10} b}{\log_{10} a}$ so $\log_a b = \dfrac{\log_{10} b}{\log_{10} a}$, then evaluate the right-hand side using a calculator.

EXERCISE 11D

1 Find the exact solutions to each of the following equations:

(a) $2^x = 64$, **(b)** $3^x = 2187$,

(c) $2^x = 0.125$, **(d)** $5^x = 3125$.

2 Find the solutions to each of the following, giving your answers to three significant figures. *You may find the answers to question 1 helpful in confirming your results.*

(a) $2^x = 62$, **(b)** $3^x = 2200$,

(c) $2^x = 0.13$, **(d)** $5^x = 3110$.

3 Solve each of the following, giving your answers to three significant figures.

(a) $7^x = 0.46$, **(b)** $4^x = 200$,

(c) $3^x = 19$, **(d)** $12^x = 58.9$.

4 Given that $3^{2x+5} = 17$, show that $x = \frac{1}{2}\left(\frac{\ln 17}{\ln 3} - 5\right)$.

5 Solve the following equations, leaving your answers in terms of natural logarithms:

(a) $2^{x+4} = 6$, **(b)** $3^{2x-1} = 17$,

(c) $2^{1-4x} = 5$, **(d)** $5^{3x+4} = 31$.

6 Solve each of the following, giving your answers to three significant figures:

(a) $3^{2x-5} = 7$, **(b)** $5^{7x-1} = 12$,

(c) $2^{5x-2} = 19$, **(d)** $7^{3-2x} = 19$.

7 Solve each of the following, giving your answers to four significant figures:

(a) $9^{x-7} = 17.5$, **(b)** $3^{4x-5} = 0.123$,

(c) $13^{x-5} = 1.345$, **(d)** $17^{5-4x} = 19.436$,

(e) $8^{6x-7} = 67.23$, **(f)** $6^{7x-4} = 23.89$.

8 Solve each of the following, giving your answers to three significant figures:

(a) $3^x = 7^{x+1}$, **(b)** $5^{x-1} = 4^{1-3x}$,

(c) $7^{1-2x} = 4^{x+3}$, **(d)** $7^{x+1} = 5^{1+3x}$.

9 Find, to three significant figures, the x-coordinate of the point where the line $y = 3$ intersects the curve:

(a) $y = 5^x$, **(b)** $y = 1 + 5^{-x}$.

10 Find, to three significant figures, the value of the following:

(a) $\log_3 5$, **(b)** $\log_3 2$, **(c)** $\log_5 3$, **(d)** $\log_2 \sqrt{3}$.

MIXED EXERCISE

1 **(a)** Write down the value of $\log_2 8$.

(b) Express $\log_2 9$ in the form $n \log_2 3$.

(c) Hence, show that $\log_2 72 = m + n \log_2 3$, where m and n are integers. [A]

2 (a) Show that $\log_2 32 = 5$.

(b) Find the value of: **(i)** $\log_2(32)^4$, **(ii)** $\log_2\left(\frac{1}{\sqrt{32}}\right)$.

3 (a) Given that $\log_a x = 2(\log_a k - \log_a 2)$, where a is a positive constant, show that $k^2 - 4x = 0$.

(b) (i) Find the exact value of $\log_9 27$.

(ii) Hence, given that $\log_3 y = \log_9 27$, show that $y = k \times \sqrt{k}$, where k is an integer to be found. [A]

4 (a) Given that $\log_a x = \log_a 5 + 2\log_a 3$, where a is a positive constant, show that $x = 45$.

(b) (i) Write down the value of $\log_4 4$.

(ii) Find the value of $\log_4 2$.

(iii) Given that $\log_2 y = \log_4 2$, find, in surd form, the value of y.

5 Use logarithms to solve the equation $2^x = 7$, giving your answer to three significant figures. [A]

6 By taking logarithms, solve the equation $0.6 = 2^{-x}$, giving your answer to three significant figures. [A]

7 (a) Sketch the graph of $y = 2^x$, stating the coordinates of any points where the graph meets the coordinate axes.

(b) Write down the value of:

(i) $\log_2 2$,

(ii) $\log_2 8$.

(c) Find the value of $\log_2 3 - \log_2 24$.

(d) The point $(k, 5)$ lies on the curve $y = 2^x$. Find the value of k, giving your answer to three significant figures. [A]

8 A curve, C, has equation $y = x - 2 - 2\log_{10} x$, where $x > 0$. The vertical lines $x = 6$ and $x = 7$ meet the curve at points R and S, respectively.

(a) Show that the y-coordinate of R is $4 - \log_{10} 36$.

(b) The area of the trapezium bounded by the lines RS, $x = 6$, $x = 7$ and the x-axis is A square units. Show that $A = \frac{p}{2} - \log_{10} q$, stating the values of the positive integers p and q. [A]

Key point summary

1 An **exponent** is another name for an index or power. Functions such as f where $f(x) = 2^x$ are called **exponential functions** because the variable x appears in the exponent. *p359*

2 For all values of the positive constant a, the curve $y = a^x$ will always pass through the point (0, 1). *p360*

3 The logarithm of N to base a is written as $\log_a N$. *p361*

4 $N = a^x \Leftrightarrow x = \log_a N$ *p361*

5 $\log_a a = 1$ and $\log_a 1 = 0$ *p362*

6 The laws of logarithms are *p363*

$$\log_a x + \log_a y = \log_a (xy); \quad \text{[Law 1]}$$

$$\log_a x - \log_a y = \log_a \left(\frac{x}{y}\right); \quad \text{[Law 2]}$$

$$k \log_a x = \log_a (x^k). \quad \text{[Law 3]}$$

7 $\log_{10} x$ is sometimes written as $\lg x$. *p366*

8 The solution to the equation $a^x = b$ is $x = \dfrac{\log_{10} b}{\log_{10} a}$, where logarithms are taken to base 10. *p366*

9 In general, to find the value of $\log_a b$, *p367*

1 Let $x = \log_a b \Rightarrow a^x = b$.

2 Take logarithms to base 10 of both sides $\Rightarrow \log_{10} a^x = \log_{10} b$.

3 Use Law 3 $\Rightarrow x \log_{10} a = \log_{10} b$.

4 Rearrange $\Rightarrow x = \dfrac{\log_{10} b}{\log_{10} a}$ so $\log_a b = \dfrac{\log_{10} b}{\log_{10} a}$, then evaluate the right-hand side using a calculator.

Test yourself

Test yourself	What to review
1 (a) Sketch the graphs of $y = 5^x$ and $y = 5^{-x}$ on the same axes. **(b)** State a geometrical transformation that maps the graph of $y = 5^x$ onto the graph of $y = 5^{-x}$.	*Section 11.1*
2 Evaluate: **(a)** $\log_8 64$, **(b)** $\log_9 3$, **(c)** $\log_8 64 + \log_{10} 0.01 + \log_9 27$.	*Section 11.2*
3 Solve the equation $\log_2 x = 2\log_2 5 + \log_2 1 - 2$.	*Section 11.3*
4 (a) Solve the equation $8^x = 12$, giving your answer to four significant figures. **(b)** Hence, or otherwise, find the value, to three significant figures, of $\log_8\left(\dfrac{12}{8}\right)$.	*Section 11.4*

Test yourself ANSWERS

1 (a)

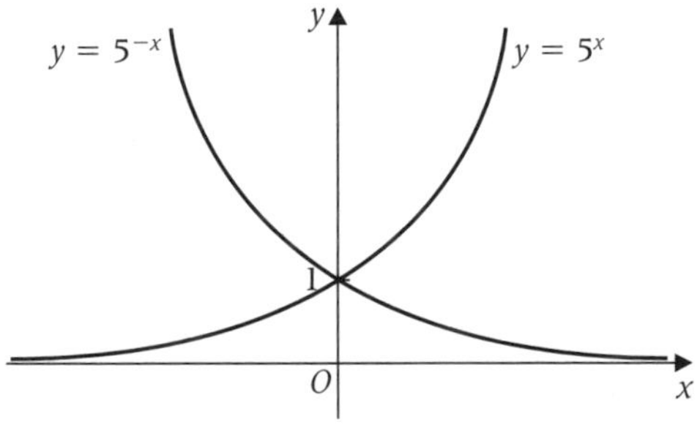

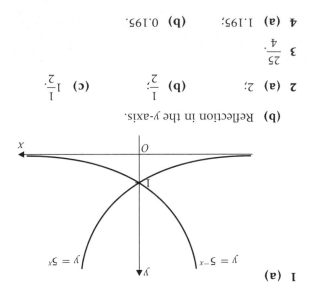

(b) Reflection in the y-axis.

2 (a) 2; **(b)** $\frac{1}{2}$; **(c)** $1\frac{1}{2}$.

3 $\frac{25}{4}$.

4 (a) 1.195; **(b)** 0.195.

11

CHAPTER 12

C2: Geometric series

Learning objectives

After studying this chapter, you should be able to:

- understand what is meant by a geometric series
- find the nth term of a geometric series
- find the sum of the first n terms of a geometric series
- know the condition for a geometric series to have a sum to infinity
- find the sum to infinity of certain geometric series.

12.1 Would you like to be a millionaire?

Imagine you have a chess board with 64 squares and someone promises to give you some pennies. She places one on the first square, two on the second, four on the third and so on, doubling the previous amount of money for each subsequent square.

1	2	4	8	?	?		
							?

How much money would you have when all the squares have been filled?

The amounts of money on each square form a pattern

$$1, 2, 4, 8, 16, \ldots$$

What is the 64th term in this sequence?

The numbers can be written as powers of 2, namely

$$1, 2, 2^2, 2^3, 2^4, \ldots$$

Since the third term is 2^2 and the fourth term is 2^3, etc. you can see that the 64th term will be 2^{63}.

The total sum of money on the chessboard is S and you can write

$$S = 1 + 2 + 2^2 + 2^3 + 2^4 + \ldots + 2^{62} + 2^{63}.$$

If you multiply each term of this series by 2, you get

$$2S = 2 + 2^2 + 2^3 + 2^4 + 2^5 \ldots + 2^{63} + 2^{64}.$$

Now, if you take the expression for $2S$ and subtract the expression for S you will notice that most of the terms cancel , leaving

$$2S - S = 2^{64} + (2^{63} - 2^{63}) + \ldots + (2^3 - 2^3) + (2^2 - 2^2) + (2 - 2) - 1 \Rightarrow S = 2^{64} - 1.$$

You can get an approximation to this number of pennies using your calculator. It comes to 1.84467×10^{19}.

So you would be worth more than £180 000 million million. Probably you would be the richest person in the universe!

12.2 Geometric series

The problem already considered involved a **geometric series**.

A sequence such as 3, 15, 75, 375, ... is a geometric sequence or geometric progression, because each term has been formed from the previous one by multiplication by the same number – in this case 5.

By looking at the ratio of one term to the term before it, you get

$$\frac{15}{3} = 5, \quad \frac{75}{15} = 5, \quad \frac{375}{75} = 5, \ldots$$

The number 5 is called the **common ratio**.

In general a geometric sequence is of the form

$$a, ar, ar^2, ar^3, \ldots$$

When terms are added together they form a series.

A geometric series is of the form

$$a + ar + ar^2 + ar^3 + \ldots$$

with first term a and common ratio r.

The nth term of a geometric series is given by ar^{n-1}. This formula is in the formulae booklet.

Worked example 12.1

A geometric series is given by $28 + 14 + 7 + \ldots$

(a) State the common ratio.

(b) Write down an expression for the 94th term.

Solution

(a) The common ratio is $\frac{1}{2}$.

Each term has been divided by 2 in order to get the next term. In other words, terms have been multiplied by $\frac{1}{2}$ in order to get the next term.

(b) The nth term is given by ar^{n-1}. Here $a = 28$ and $r = \frac{1}{2}$. Hence the 94th term is $28 \times (\frac{1}{2})^{93}$.

Worked example 12.2

The third term of a geometric series is 324 and the fifth term is 36. Find the first term and the two possible values of the common ratio.

Solution

Let the first term be a and the common ratio r.

Third term is $ar^2 = 324$. Fifth term is $ar^4 = 36$.

These equations can be solved simultaneously by dividing so that the letter a is eliminated.

$$\frac{ar^4}{ar^2} = \frac{36}{324} \Rightarrow r^2 = \frac{1}{9}$$

This equation has two solutions $r = \pm\frac{1}{3}$.

Since $ar^2 = 324$ and $r^2 = \frac{1}{9}$, $a = 9 \times 324 = 2916$.

The first term is 2916 and the common ratio could be either $\frac{1}{3}$ or $-\frac{1}{3}$.

EXERCISE 12A

1 Find **(i)** the common ratio, **(ii)** the next missing term, for each of the following geometric series.

(a) $180 + 60 + 20 + \dots$

(b) $4 + 16 + 64 + \dots$

(c) $3 - 6 + 12 - 24 + 48 \dots$

(d) $200 + 150 + 112\frac{1}{2} + \dots$

(e) $81 - 54 + 36 - 24 + \dots$

(f) $100 + 80 + 64 + \dots$

2 Find expressions for the 38th term in each of the geometric series in question **1**.

3 A geometric series has first term 120 and common ratio $\frac{2}{3}$. Find the fourth term.

4 A geometric series has second term 40 and common ratio $\frac{4}{9}$. Find the first term and the third term.

5 The fourth term of a geometric series is 6 and the fifth term is 4. Find the first term and the common ratio.

6 The third term of a geometric series is 80 and the sixth term is 10. Find the first term and the common ratio.

7 The fifth term of a geometric series is 72 and the seventh term is 32. Find the first term and the two possible values of the common ratio.

8 The first three terms of a geometric sequence are 2, q, 18. Find the possible values of q.

12.3 Sum of the first *n* terms of a geometric series

A formula for the first n terms of a geometric series can be found in a similar way to that used to find the number of pennies on the chessboard.

Let the sum of the first n terms be denoted by S.

$$S = a + ar + ar^2 + ar^3 + \dots + ar^{n-2} + ar^{n-1}$$

Multiplying each term of this series by r gives

$$rS = ar + ar^2 + ar^3 + \dots + ar^{n-1} + ar^n$$

Subtracting these results many of the terms cancel each other out.

$$S - rS = a + (ar - ar) + (ar^2 - ar^2) + \dots + (ar^{n-1} - ar^{n-1}) - ar^n$$

$$\Rightarrow S - rS = S(1 - r) = a - ar^n$$

$$\Rightarrow S = \frac{a - ar^n}{(1 - r)} = \frac{a(1 - r^n)}{(1 - r)}.$$

The sum of the first n terms is usually denoted by S_n.

The formula assumes $r \neq 1$. The case $r = 1$ would be the geometric series where all the terms are equal and the sum would then be n times the first term.

> The sum of the first n terms of a geometric series is
>
> $$S_n = \frac{a(1 - r^n)}{(1 - r)}.$$
>
> This formula is in the formulae booklet.

Worked example 12.3

A geometric series has first term 20 and common ratio $\frac{3}{4}$. Find the sum of the first ten terms of the series, giving your answer to three decimal places.

Solution

Using the formula $S_n = \dfrac{a(1 - r^n)}{(1 - r)}$ with $n = 10$.

Since $a = 20$ and $r = \frac{3}{4}$, $\Rightarrow S_{10} = \dfrac{20(1 - (\frac{3}{4})^{10})}{(1 - \frac{3}{4})} = 75.494\,918\,82 \dots$

Rounding to three decimal places the value is 75.495.

EXERCISE 12B

1 Find the sum of the first twenty terms of the following geometric series, giving your answers to three decimal places.

(a) $10 + 5 + 2.5 + \dots$ **(b)** $18 - 12 + 8 - \dots$

(c) $150 + 90 + 54 + \dots$ **(d)** $\frac{1}{4} + \frac{1}{2} + 1 + 2 + \dots$

(e) $1 - 3 + 9 - 27 + \dots$ **(f)** $3 - 3 + 3 - 3 + 3 - \dots$

(g) $5 + 5 + 5 + 5 + \dots$

12

2 The common ratio of a geometric series is 3 and the third term is 45. Find the sum of the first eight terms.

3 The fourth term of a geometric series is 20 and the third term is 16. Find **(a)** the common ratio, **(b)** the first term, **(c)** the sum of the first five terms of the series.

4 A geometric series has second term 24 and third term 36. Find the common ratio and the sum of the first ten terms.

5 A geometric series has third term 12 and fourth term 9. Find the sum of the terms from the fifth term to the fifteenth term.

6 A geometric series has fourth term equal to eight times the seventh term. The first term is 1024. Find the sum of the first fifteen terms of the series.

7 The sum of the first four terms of a geometric series is 5468.75 and the first term is 2000. Find the value of the common ratio if all the terms are positive.

8 The sum of the first n terms of the geometric series

$$3 - 6 + 12 - \ldots$$

is 129. Find the value of n.

12.4 Sum to infinity of a geometric series

The terms of the sequence $\frac{1}{2}$, $(\frac{1}{2})^2$, $(\frac{1}{2})^3$, $(\frac{1}{2})^4$, get closer to zero as the power increases and we say that $(\frac{1}{2})^n \to 0$ as $n \to \infty$.

$(\frac{1}{2})^n$ tends to zero as n tends to infinity.

On the other hand, the geometric sequence 2, 6, 18, 54, … diverges. The terms are getting larger in magnitude.

Provided $-1 < r < 1$, the value of r^n tends to zero as n tends to infinity. Using modulus notation, $-1 < r < 1$ is equivalent to $|r| < 1$.

A geometric series with common ratio r converges when $|r| < 1$. This formula is in the formulae booklet.

When the condition $-1 < r < 1$ is satisfied, the sum of the terms of the geometric series, S_n, tends to a limit.

This limit can be written as S_∞ and is called the sum to infinity of the geometric series.

If $r^n \to 0$, then $1 - r^n \to 1$.

The formula $S_n = \dfrac{a(1 - r^n)}{(1 - r)}$ becomes $S_\infty = \dfrac{a}{(1 - r)}$.

A convergent geometric series has sum to infinity,

$$S_\infty = \frac{a}{1-r}.$$

Worked example 12.4

Explain why the geometric series $24 - 16 + 10\frac{2}{3} - \ldots$ converges. Find its sum to infinity.

Solution

The common ratio, r is given by $-\dfrac{16}{24} = -\dfrac{2}{3}$.

Since $|r| = |-\frac{2}{3}| = \frac{2}{3} < 1$, the series converges.

The sum to infinity is given by $S_\infty = \dfrac{a}{(1-r)}$, with $a = 24$, $r = -\frac{2}{3}$

$$\Rightarrow S_\infty = \frac{24}{1-(-\frac{2}{3})} = \frac{24}{\frac{5}{3}} = \frac{72}{5} = 14.4$$

Worked example 12.5

Find the values of x for which the geometric series $1 + 4x + 16x^2 + \ldots$ converges.

Solution

The common ratio, r, of the geometric series is $4x$.

The condition for the series to converge is $|r| < 1$.

Hence $|4x| < 1$. This means $-1 < 4x < 1$

or $-\dfrac{1}{4} < x < \dfrac{1}{4}$.

Worked example 12.6

The sum to infinity of a convergent geometric series is 20. The first term is 16. Find the third term of the series.

Solution

You need to use the formula $S_\infty = \dfrac{a}{(1-r)}$.

You are given that $S_\infty = 20$ and that $a = 16$.
Therefore, you can find the common ratio, r.

$$20 = \frac{16}{1-r} \Rightarrow 1 - r = \frac{16}{20} \Rightarrow 1 - r = \frac{4}{5}$$

Hence $r = \dfrac{1}{5}$.

The third term of the series is given by $ar^2 = 16 \times \left(\dfrac{1}{5}\right)^2 = \dfrac{16}{25}$.

Worked example 12.7

Use the formula for the sum to infinity to find the fractional equivalent of the recurring decimal 0.121 212 12 …

Usually written as $0.\dot{1}\dot{2}$.

Solution

The number is equivalent to 0.12 + 0.0012 + 0.000 012 + … which is an infinite geometric series with first term 0.12 and common ratio 0.01. The sum to infinity of this series is

$$\frac{0.12}{1-0.01} = \frac{0.12}{0.99} = \frac{12}{99} = \frac{4}{33}.$$

Check it by typing 4 divided by 33 on your calculator.

The fractional equivalent is $\frac{4}{33}$.

EXERCISE 12C

1 For each of the following geometric series:

(i) state whether it converges or not,

(ii) find the sum to infinity if it is convergent.

(a) $64 + 32 + 16 + \ldots$

(b) $48 - 24 + 12 - 6 + \ldots$

(c) $2 - 2 + 2 - 2 + 2 - \ldots$

(d) $144 + 96 + 64 + \ldots$

(e) $144 - 96 + 64 - \ldots$

(f) $1000 + 1100 + 1210 + \ldots$

(g) $50 - 20 + 8 - \ldots$

2 A geometric series has third term 20 and common ratio $\frac{2}{3}$. Find the first term and the sum to infinity.

3 A geometric series has fourth term 27 and common ratio $\frac{3}{4}$. Find the first term and the sum to infinity.

4 The sum to infinity of a convergent geometric series is 60 and the first term is 48. Find the common ratio.

5 The sum to infinity of a convergent geometric series is 8 and the first term is 10. Find the fourth term of the series.

6 The seventh term of a geometric series is 64 and the twelfth term is -2.

(a) Find the first term and the common ratio of the series.

(b) Calculate the sum to infinity of the series.

7 A geometric series has first term 16 and its sum to infinity is 12. Find the fourth term of the series.

8 The second term of a geometric series is 300 and the third term is 75.

(a) Find the common ratio and the first term.

(b) Find the sum to infinity.

9 Find the values of x for which the following geometric series converge.

(a) $2 + 4x + 8x^2 + 16x^3 + \ldots$

(b) $1 + (3 - 2x) + (3 - 2x)^2 + (3 - 2x)^3 \ldots$

(c) $5 + 25(3x + 4) + 125(3x + 4)^2 + \ldots$

10 The first term of a geometric series is equal to the common ratio. The sum to infinity is 4. Find the value of the third term.

11 A pump is used to extract air from a bottle. The first operation of the pump extracts 56 cm^3 of air and subsequent extractions follow a geometric progression. The third operation of the pump extracts 31.5 cm^3 of air. Determine the common ratio of the geometric progression and calculate the total amount of air that could be extracted from the bottle, if the pump were to extract air indefinitely. [A]

12 Find the fractional equivalents of the following recurring decimals:

(a) 0.777 77... ;

(b) 0.123 123 123....

13 *Discussion question:*
Is the recurring decimal 0.9999 ... less than 1?

12.5 Use of logarithms to find number of terms

Suppose you wish to know how many terms you would need for the sum of the geometric series

$$1 + 2 + 4 + 8 + 16 + \ldots$$

to exceed a million.

The sum of the first n terms is given by $S_n = \dfrac{a(1 - r^n)}{(1 - r)}$.

You know that $a = 1$ and $r = 2$

$$\Rightarrow S_n = \frac{(1 - 2^n)}{(1 - 2)} = -1 \times (1 - 2^n) = 2^n - 1$$

You need to solve $2^n > 1\,000\,000$.

You could keep trying values on your calculator, but you can use logarithms to solve this directly.

You can take logarithms to base 10 of both sides.

$$\log_{10}(2^n) > \log_{10}(1\,000\,000)$$
$$\Rightarrow n \log_{10} 2 > 6$$
$$\Rightarrow \quad n > \frac{6}{\lg 2}$$
$$\Rightarrow \quad n > 19.93 \ldots$$

$\log_{10}(10^6) = 6$

You would need 20 terms before the sum was greater than a million.

Worked example 12.8

Find the least number of terms of the geometric series

$16 + 20 + 25 + \ldots$

required to give a sum greater than 25 000.

Solution

The common ratio, r, of the series is $\frac{5}{4}$.

The sum of the first n terms is $S_n = \frac{a(1 - r^n)}{(1 - r)}$, where $a = 16$.

When $r > 1$, it is often easier to use the formula where the top and bottom are multiplied by -1, namely $S_n = \frac{a(r^n - 1)}{(r - 1)}$.

Solving $S_n > 25\,000 \Rightarrow \frac{16(r^n - 1)}{(\frac{5}{4} - 1)} > 25\,000$

$$\Rightarrow \quad 64(r^n - 1) > 25\,000$$
$$\Rightarrow \quad (r^n - 1) > \frac{25\,000}{64}$$
$$\Rightarrow \quad (r^n - 1) > 390.625$$
$$\Rightarrow \quad r^n > 391.625$$

It is more convenient to use r^n rather than $\left(\frac{5}{4}\right)^n$ until you are ready to take logarithms.

Taking natural logarithms of both sides

$$\Rightarrow \quad n \ln(r) > \ln(391.625)$$
$$\Rightarrow \quad n \ln(1.25) > \ln(391.625)$$
$$\Rightarrow \quad n > \frac{\ln(391.625)}{\ln(1.25)}$$
$$\Rightarrow \quad n > 26.755 \ldots$$

The number of terms required is 27.

Notice that you could have used logarithms to base 10 since $\frac{\log_{10} 391.625}{\log_{10} 1.25} = 26.755 \ldots$

Worked examination question 12.9

A graduate signs a contract for a job with a salary of £20 000 in the first year and an annual increase of 5% of the previous year's salary.

(a) The amounts of annual salary form a geometric progression. Write down the second and third terms in pounds and state the common ratio.

(b) Calculate the amount, to the nearest pound, that the graduate will earn during the fifteenth year on this contract.

(c) Assuming the salary is paid monthly, determine the least length of time, in years and months, that the graduate will need to work in order for the total earnings to exceed a million pounds.

Solution

(a) Second term in pounds = 21 000.
Third term in pounds = 22 050.
Hence, the common ratio $r = 1.05$

(b) You need the fifteenth term of the geometric series which is of the form $ar^{14} = 20\,000 \times (1.05)^{14} = 39\,598.63 \ldots$
The amount earned (to the nearest pound) is £39 599

(c) In order to exceed £1 million, the sum of the first n terms must be greater than 1 million.

$$20\,000 \frac{(r^n - 1)}{(r - 1)} > 10^6$$

$$\Rightarrow 20\,000 \frac{(r^n - 1)}{0.05} > 10^6$$

$$\Rightarrow 400\,000\,(r^n - 1) > 10^6$$

$$\Rightarrow \quad (r^n - 1) > 2.5$$

$r = 1.05$

$$\Rightarrow \quad 1.05^n > 3.5$$

Taking logarithms of both sides

$$\Rightarrow \quad n \log 1.05 > \log 3.5$$

$$\Rightarrow \quad n > \frac{\log 3.5}{\log 1.05}$$

$$\Rightarrow \quad n > 25.6765 \ldots$$

You can use any base of logarithm – usually ln or lg since these are both on your calculator.

The time taken to have earned one million pounds is therefore 25 years and 9 months.

12

EXERCISE 12D

1 Find the least number of terms for which the sum of each of the following geometric series exceeds 25 million:

(a) $1 + 5 + 25 + 125 + \ldots,$

(b) $100 + 120 + 144 + \ldots,$

(c) $144 + 156 + 169 + \ldots.$

2 A geometric series has first term 12 and common ratio $\frac{1}{3}$. Find the least number of terms whose sum differs from the sum to infinity by less than 10^{-6}.

3 The third term of a geometric series is 81 and the sixth term is 24.

(a) Show that the common ratio of the series is $\dfrac{2}{3}$.

(b) Find the sum to infinity of the series. [A]

4 The first four terms of a geometric sequence are 10, 9, 8.1, 7.29.

(a) Show that the common ratio of the sequence is 0.9.

(b) Find the nth term.

(c) Show that the sum of the first 25 terms is approximately 92.8.

(d) Find the sum to infinity. [A]

5 Find the sum of the geometric series
$2 + 6 + 18 + \ldots + 2 \times 3^{n-1}$, giving your answer in the form $p^n - q$, where p and q are integers. [A]

6 A geometric series has common ratio r. The first term of the series is 2.

(a) (i) Write down the first four terms of the series in terms of r.

(ii) The sum of the first four terms of the series is $\dfrac{15}{4}$.
Show that $8r^3 + 8r^2 + 8r - 7 = 0$.

(b) (i) By multiplying out the brackets, show that $(2r - 1)(4r^2 + 6r + 7) = 8r^3 + 8r^2 + 8r - 7$.

(ii) Hence show that $8r^3 + 8r^2 + 8r - 7 = 0$ has only one real solution.

(c) Find the sum to infinity of the series. [A]

7 The sum to infinity of a convergent geometric series is $5a$, where a is the first term of the series.

(a) Show that the common ratio of the series is $\dfrac{4}{5}$.

(b) The third term of the series is 64.

(i) Find the first term of the series.

(ii) Find the **exact** decimal value for the sum of the first six terms. [A]

8 Evaluate $\sum_{n=1}^{\infty} (\frac{3}{4})^n$.

9 A geometric series has first term a and common ratio r, with $a > 0$ and $r > 0$.

(a) The second term of the series is 4 and the eighth term is $\frac{1}{2}$.

Show that $r = \dfrac{1}{\sqrt{2}}$ and find the value of a in surd form.

(b) Find the sum to infinity of the series in the form $k(\sqrt{2} + 1)$, and state the value of the integer k. [A]

10 An electrician cuts a piece of wire of length 27 metres from a long roll of wire. He then cuts off more pieces, each piece being two-thirds as long as the preceding piece.

(a) Calculate the length of the fourth piece cut off.

(b) Show that, no matter how many pieces he cuts off, their total length cannot exceed 81 metres.

(c) Show that the total length of the first eleven pieces cut off is more than 80 metres.

(d) The length of the eleventh piece of wire is u metres. Write down an expression for u. Hence show that

$$\log u = a \log 2 - b \log 3$$

where a and b are positive integers to be determined. [A]

11 The third term of a geometric sequence is 81 and the sixth term is 24.

(a) Show that the common ratio of the series is $\dfrac{2}{3}$.

(b) Find the sum to infinity of the series.

(c) **(i)** The first term of the series is a. Show that $\log_{10} a = 6 \log_{10} 3 - 2 \log_{10} 2$.

(ii) The hundredth term of the series is u. Write down an expression for u and hence show that $\log_{10} u = p \log_{10} 2 - q \log_{10} 3$, where p and q are positive integers to be found.

12 A ball is dropped from the top of a building and it takes 8 seconds to reach the ground. It bounces and it takes $\frac{4}{5}$ of its previous falling time to reach its greatest height. How long does it take altogether before coming to rest?

13 The first three terms of a geometric series are the fifth, ninth and twelfth terms of an arithmetic series that has common difference 2.

(a) Find the common ratio of the geometric series.

(b) Find the sum to infinity of the geometric series.

Key point summary

1 A geometric series is of the form *p373*

$$a + ar + ar^2 + ar^3 \dots$$

with first term a and common ratio r.

2 The nth term of a geometric series is given by ar^{n-1}. This formula is in the formulae booklet. *p373*

3 The sum of the first n terms of a geometric series is *p375*

$$S_n = \frac{a(1 - r^n)}{(1 - r)}.$$

This formula is in the formulae booklet.

4 A geometric series with common ratio r converges when $|r| < 1$. This formula is in the formulae booklet. *p376*

5 A convergent geometric series has sum to infinity *p377*

$$S_\infty = \frac{a}{(1 - r)}.$$

Test yourself	What to review
1 Find the common ratio of the geometric series $18 + 24 + 32 + \dots$ and state whether the series is convergent.	*Sections 12.2 and 12.4*
2 The second term of a geometric series is 81 and the fifth term is 3. Find the common ratio and the first term.	*Section 12.2*
3 Find the sum of the first ten terms of the geometric series whose eighth term is 28 and whose ninth term is 21. (*Give your answer to one decimal place.*)	*Section 12.3*
4 Find the set of values of x for which the geometric series $16 + 8(x - 1) + 4(x - 1)^2 + 2(x - 1)^3 + \dots$ converges. Determine the sum to infinity of the series in the case when $x = 0$.	*Section 12.4*
5 Determine the least number of terms needed for the sum of the geometric series $200 + 240 + 288 + \dots$ to exceed 35 million.	*Section 12.5*

Test yourself ANSWERS

1 $\frac{4}{3}$; no. **2** $\frac{1}{3}$; 243. **3** 791.8. **4** $-1 < x < 3$; $10\frac{2}{3}$. **5** 58.

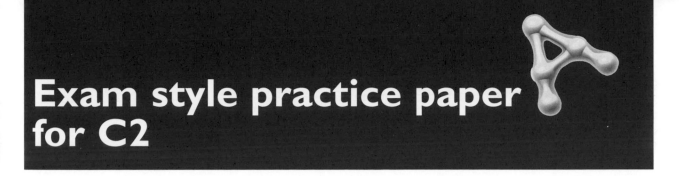

Exam style practice paper for C2

Time allowed 1 hour 30 minutes

Answer **all** questions

---

1 The diagram shows triangle ABC.

The lengths of AB and BC are 8 cm and 10 cm, respectively. The size of angle BAC is 80°.

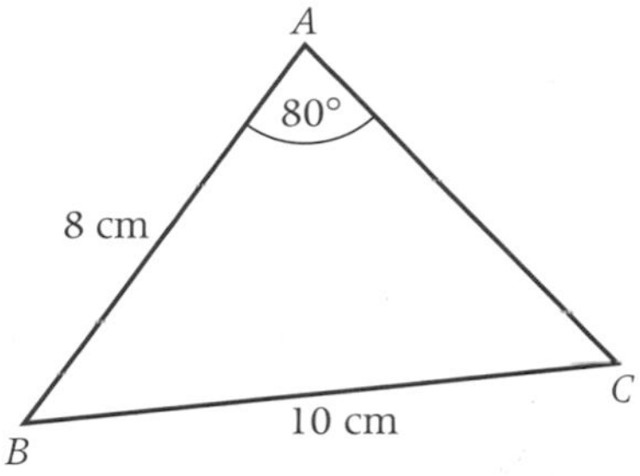

(a) Calculate the size of angle ACB giving your answer to the nearest 0.1°. (3 marks)

(b) Calculate the area of triangle ABC giving your answer to three significant figures. (3 marks)

2 The diagram shows a sector of a circle of radius 10 cm and angle θ radians.

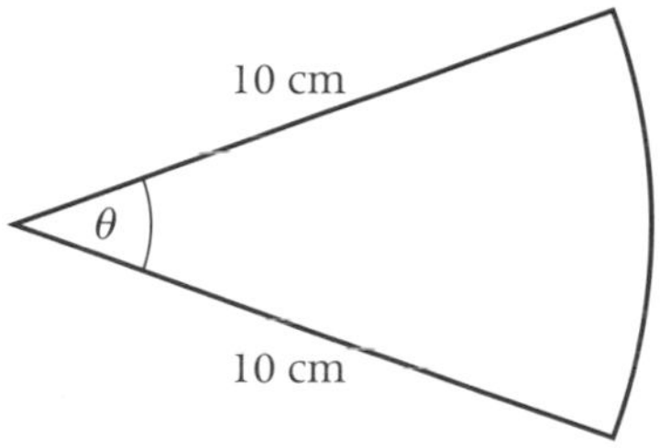

The perimeter of the sector is 28 cm.

(a) Show that $\theta = \frac{4}{5}$. (3 marks)

(b) Find the area of the sector. (2 marks)

3 An arithmetic series has sixth term 22 and tenth term 38.

(a) Find the first term and the common difference. (4 marks)

(b) Hence explain briefly why $\sum_{n=1}^{k} = u_n$, where u_n is the nth term of the arithmetic series, is even for all positive integer values of k. (2 marks)

4 **(a)** Write down the values of:

(i) $\log_5 5$, (1 mark)

(ii) $\log_5 1$. (1 mark)

(b) Given that

$$\log_a p^2 + \log_a q^2 = \log_a p - \log_a q,$$

where a, p and q are positive constants, show that $pq^3 = 1$. (4 marks)

5 **(a)** Use the binomial expansion to express $(1 + y)^3$ in the form $1 + py + qy^2 + y^3$, where p and q are integers. (3 marks)

(b) **(i)** Use your answer to **(a)** to write down the expansion of $(1 + \sqrt{x})^3$ and hence find $\int (1 + x^{\frac{1}{2}})^3 \, dx$. (4 marks)

(ii) Hence show that $\int_0^1 (1 + x^{\frac{1}{2}})^3 \, dx = 4.9$. (2 marks)

6 An infinite geometric series has common ratio r.
The second term of the series is -8.
The sum to infinity of the series is 18.

(a) Show that r satisfies the equation $9r^2 - 9r - 4 = 0$. (4 marks)

(b) Hence find the value of r. (3 marks)

7 The angle θ radians, where $0 \leqslant \theta \leqslant 2\pi$, satisfies the equation $4\cos^2\theta = 5\sin\theta - 2$.

(a) Show that $4y^2 + 5y - 6 = 0$, where $y = \sin\theta$. (3 marks)

(b) **(i)** Solve $4y^2 + 5y - 6 = 0$. (2 marks)

(ii) Hence find all values of θ for which $4\cos^2\theta = 5\sin\theta - 2$ and $0 \leqslant \theta \leqslant 2\pi$. Give your answers to three significant figures. (3 marks)

8 It is given that $y = 4x^2 + \dfrac{729}{x}$.

(a) **(i)** Find $\dfrac{dy}{dx}$. (3 marks)

(ii) Show that $\dfrac{dy}{dx} = 0$ when $x^3 = \dfrac{729}{8}$. (2 marks)

(iii) Find $\dfrac{d^2y}{dx^2}$. (2 marks)

(iv) Verify that $\dfrac{d^2y}{dx^2} = 24$ when $x^3 = \dfrac{729}{8}$. (2 marks)

(v) Find the value of x for which y has a stationary value and state whether this stationary value is a maximum or a minimum. (2 marks)

(b) A closed rectangular tin box contains 243 cm^3 of liquid when full. The box has length x cm and total external surface area A cm^2, where

$$A = 4x^2 + \frac{729}{x}.$$

Use your results from **(a)** to find the smallest possible value for the total external surface area of the box. (2 marks)

9 The diagram shows a sketch of the curve with equation $y = 16 - 2^x$.

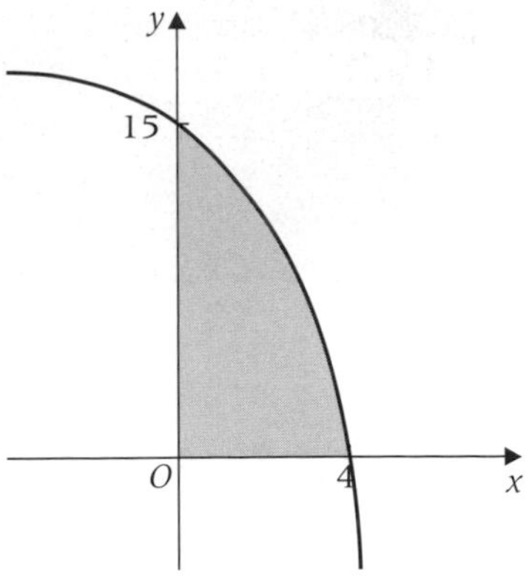

The curve intersects the x-axis at the point (4, 0) and intersects the y-axis at the point (0, 15). The region bounded by the curve and the coordinate axes is shaded.

(a) **(i)** Use the trapezium rule with five ordinates (four strips) to find an approximation for

$$\int_0^4 (16 - 2^x)\,dx. \qquad \text{(4 marks)}$$

(ii) By considering the graph of $y = 16 - 2^x$, explain with the aid of a diagram whether your approximation will be an overestimate or an underestimate of the true value for $\int_0^4 (16 - 2^x)\,dx$. (2 marks)

(b) The curve $y = -\frac{1}{5} \times 2^x$ intersects the curve $y = 16 - 2^x$ at the point P.

(i) Show that the x-coordinate of the point P is a root of the equation $2^x = 20$. (2 marks)

(ii) Solve this equation to find the x-coordinate of the point P. Give your answer to six significant figures. (3 marks)

(c) **(i)** Describe the single transformation by which the curve with equation $y = -2^x$ can be obtained from the curve with equation $y = 16 - 2^x$. (2 marks)

(ii) Describe the single transformation by which the curve with equation $y = -\frac{1}{5} \times 2^x$ can be obtained from the curve with equation $y = -2^x$. (2 marks)

Answers

Core maths 1

1 Algebra review

EXERCISE 1A

1 3.6.	**2** 4.5.	**3** $2\frac{2}{7}$.	**4** -13.	**5** $-8\frac{2}{3}$.
6 $6\frac{1}{5}$.	**7** 3.	**8** 6.	**9** 2.	**10** $1\frac{15}{29}$.
11 3.	**12** $1\frac{13}{21}$.	**13** 5.125 cm.	**14** 21.	**15** 43.

EXERCISE 1B

1 $x = 1, y = 3$.	**2** $x = 2, y = -1$.
3 $x = -2, y = -3$.	**4** $x = -4, y = 1$.
5 $x = 3\frac{7}{13}, y = -\frac{4}{13}$.	**6** $x = \frac{1}{17}, y = \frac{7}{17}$.
7 $x = -2\frac{3}{13}, y = \frac{11}{13}$.	**8** $x = -\frac{27}{35}, y = -\frac{11}{35}$.
9 $x = 2.5, y = 0$.	**10** $x = 2, y = -2$.

EXERCISE 1C

1 $x < 16$.	**2** $x \geqslant -\frac{7}{2}$.	**3** $x < 12$.	**4** $x > -\frac{10}{19}$.
5 $x > 53$.	**6** $x \geqslant 6\frac{1}{5}$.	**7** $x > 2\frac{15}{23}$.	**8** $x \leqslant 37$.

9 $5x - 4 > 0$ (can't have a negative length), $\therefore 5x > 4$, $\therefore x > \frac{4}{5}$, then $x < 3$.

10 $2(2x + 5) - 20 > 0$ which gives $x > 2.5$ and
$x + (2x + 5) + [2(2x + 5) - 20] < 100$ which gives $x < 15$.

EXERCISE 1D

1 **(a)** multiply the number by five and add two to the result;
(b) multiply the number by three and subtract four from the result;
(c) square the number and add seven to the result.

2 **(a)** 7;	**(b)** -5;	**(c)** -1;	**(d)** -13.
3 **(a)** 2;	**(b)** -1;	**(c)** 2;	**(d)** 11.
4 **(a)** 8;	**(b)** -1;	**(c)** 8;	**(d)** 35.

5 **(a)** **(i)** 9,	**(ii)** 1,	**(iii)** 0,	**(iv)** 4;
(b) **(i)** 9,	**(ii)** 1,	**(iii)** 0,	**(iv)** -26;
(c) **(i)** -8,	**(ii)** 4,	**(iii)** 1,	**(iv)** -23;
(d) **(i)** $\frac{1}{4}$,	**(ii)** $\frac{1}{2}$,	**(iii)** 1,	**(iv)** -1.

2 Surds

EXERCISE 2A

1 Irrational. 2 Rational. 3 Irrational. 4 Rational.
5 Rational. 6 Rational. 7 Irrational. 8 Irrational.
9 Rational. 10 Irrational.

EXERCISE 2B

1 $2\sqrt{2}$. 2 $2\sqrt{3}$. 3 $2\sqrt{5}$. 4 $5\sqrt{3}$.
5 $2\sqrt{13}$. 6 $2\sqrt{30}$. 7 $7\sqrt{5}$. 8 $6\sqrt{7}$.
9 $8\sqrt{3}$. 10 $10\sqrt{10}$. 11 $\sqrt{3}$. 12 $\sqrt{2}$.
13 5. 14 $2\sqrt{7}$. 15 $7\sqrt{5}$. 16 45.
17 $30\sqrt{2}$. 18 $\frac{\sqrt{6}}{3}$ or $\frac{\sqrt{2}}{\sqrt{3}}$.
19 (a) $4\sqrt{2}$; (b) $7\sqrt{7}$; (c) $-\sqrt{3}$.
20 $^3\sqrt{3}$, $^6\sqrt{10}$, $^4\sqrt{5}$.

EXERCISE 2C

1 $-2\sqrt{7}$. 2 $3\sqrt{3}$. 3 $-\sqrt{5}$. 4 $6\sqrt{2} + 2\sqrt{3}$.
5 $5\sqrt{2}$. 6 $5\sqrt{3}$. 7 $-2\sqrt{5}$. 8 $3\sqrt{6}$.
9 $33\sqrt{5}$. 10 $9\sqrt{7} + 42\sqrt{2}$. 11 $3\sqrt{5}$ m 12 $6\sqrt{7}$ cm.
13 (a) $\sqrt{41}$ cm; (b) $2\sqrt{3}$ cm

EXERCISE 2D

1 $30\sqrt{2}$. 2 $60\sqrt{5}$.
3 $\sqrt{30} - 2\sqrt{5}$. 4 12.
5 $6\sqrt{5} + 60$. 6 $21 - 3\sqrt{3} + 7\sqrt{2} - \sqrt{6}$.
7 2. 8 $4 + 5\sqrt{42}$.
9 6. 10 $132 + 8\sqrt{11} - 6\sqrt{33} - 4\sqrt{3}$.

EXERCISE 2E

1 $\frac{\sqrt{10}}{10}$. 2 $\frac{3\sqrt{2}}{2}$. 3 $\frac{\sqrt{35}}{10}$.
4 $\sqrt{2} + 1$. 5 $\frac{3 + \sqrt{21}}{4}$. 6 $\frac{2(\sqrt{5} + \sqrt{2})}{3}$.
7 $\frac{\sqrt{14} - 2}{2}$. 8 $2 - \sqrt{2}$. 9 $\frac{7 - \sqrt{21} + 5\sqrt{7} - 5\sqrt{3}}{4}$.
10 $\frac{23 + 3\sqrt{21}}{68}$. 11 $\frac{9 + \sqrt{65}}{4}$. 12 $\frac{20\sqrt{3} + 10}{11}$.
13 $\frac{37 + 8\sqrt{10}}{27}$. 14 $-(8 + 2\sqrt{14})$.

EXERCISE 2F

1 $-\frac{(24+56\sqrt{3})}{69}$. 2 $\frac{15-11\sqrt{5}}{19}$. 3 $\frac{29+3\sqrt{3}}{22}$.

4 $\frac{136-27\sqrt{7}}{59}$. 5 $\frac{8+2\sqrt{2}}{7}$. 6 $\frac{46-27\sqrt{3}}{71}$.

7 $x<-\frac{(5\sqrt{3}+3)}{33}$. 8 $x<\frac{13+\sqrt{5}}{41}$. 9 $x>\frac{82-31\sqrt{7}}{3}$.

10 $x<\frac{13\sqrt{2}-11}{7}$.

MIXED EXERCISE

1 **(a)** $5\sqrt{3}$; **(b)** $10\sqrt{3}$; **(c)** $8\sqrt{2}$; **(d)** 0;
(e) $11\sqrt{2}$; **(f)** $\sqrt{6}$.

2 **(a)** $5\sqrt{3}-6$; **(b)** $3\sqrt{5}+15$; **(c)** 4; **(d)** $5\sqrt{3}-11$;
(e) $-13-17\sqrt{5}$; **(f)** $21+12\sqrt{3}$.

3 **(a)** $\frac{7\sqrt{15}}{5}$; **(b)** $\frac{\sqrt{3}+1}{2}$; **(c)** $\frac{3+\sqrt{5}}{2}$; **(d)** $\frac{\sqrt{13}+\sqrt{11}}{2}$;
(e) $\frac{7-3\sqrt{3}}{2}$; **(f)** $\frac{33-17\sqrt{3}}{3}$; **(g)** $27-19\sqrt{2}$; **(h)** $\frac{2+\sqrt{2}}{2}$.

4 **(a)** $11-6\sqrt{2}$; **(b)** $\frac{11+6\sqrt{2}}{49}$.

5 **(a)** **(i)** Student's answer. A possible answer is $\frac{9}{2}$.
(ii) Student's answer. A possible answer is $4+\frac{\sqrt{2}}{2}$;
(b) 1, the statement is false.

6 **(a)** $-32+11\sqrt{7}$; **(b)** $11+4\sqrt{7}$.

7 **(a)** **(i)** $3\sqrt{5}$, **(ii)** $4\sqrt{5}$; **(b)** $7\sqrt{5}$.

8 **(a)** $8+2\sqrt{7}$; **(b)** $-\frac{2}{3}+\frac{4}{3}\sqrt{7}$.

9 **(a)** $4+\sqrt{3}$; **(b)** $8+2\sqrt{3}$.

10 **(a)** $2\sqrt{2}+3$; **(b)** $x<6+4\sqrt{2}$.

3 Coordinate geometry of straight lines

EXERCISE 3A

1 **(a)** 5; **(b)** $\sqrt{17}$; **(c)** $\sqrt{34}$; **(d)** $\sqrt{29}$;
(e) $\sqrt{5}$; **(f)** $3\sqrt{2}$; **(g)** $\sqrt{89}$; **(h)** $\sqrt{73}$;
(i) 5; **(j)** 2.5; **(k)** 6.5; **(l)** 12.5.

2 **(a)** $AB=6$, $BC=5$, $AC=5$, $\therefore$ not right-angled;
(b) $AB=2\sqrt{17}$, $AC=2\sqrt{34}$, $BC=2\sqrt{17}$,
$\therefore AC^2=AB^2+BC^2$, $\therefore$ it is right-angled;
(c) $AB=2\sqrt{2}$, $AC=\sqrt{26}$, $BC=3\sqrt{2}$,
$\therefore AC^2=AB^2+BC^2$, $\therefore$ it is right-angled.

4 $p = +4$ or $p = -4$.

5 **(a)** $PQ = 3\sqrt{2}$, $PR = 3\sqrt{2}$, $QR = 6$;
(c) $(1, 0)$;
(d) $(1, -3)$.

EXERCISE 3B

1 **(a)** $(5, 2)$; **(b)** $(1, 0.5)$; **(c)** $(3, 2)$; **(d)** $(-2, 4.5)$;
(e) $(3.5, 2)$; **(f)** $(-4.5, -0.5)$; **(g)** $(0, 2)$; **(h)** $(2, 1)$;
(i) $(1, 3)$; **(j)** $(0.25, 0.5)$.

2 **(a)** $(4, 6)$; **(b)** $(4, 5)$; **(c)** $(0, 7)$; **(d)** $(8, 5)$;
(e) $(4, 1)$; **(f)** $(6, -6)$.

3 **(a)** $(7, 3)$.

EXERCISE 3C

1 **(a)** $m_{AB} = 2$, $m_{CD} = 2$, $\therefore$ parallel;
(b) $m_{AB} = -\frac{1}{2}$, $m_{CD} = \frac{1}{2}$, $\therefore$ not parallel;
(c) $m_{AB} = 0$, $m_{CD} = 0$, $\therefore$ parallel;
(d) $m_{AB} = \frac{1}{7}$, $m_{CD} = \frac{1}{7}$, $\therefore$ parallel.

3 **(a)** $m_{AB} = \frac{1}{3}$, $m_{AC} = \frac{6}{11}$, $m_{BC} = \frac{4}{5}$;
(b) $m_{OP} = \frac{6}{11}$, $\therefore$ AC is parallel to OP.

EXERCISE 3D

1 **(a)** $-\frac{5}{2}$; **(b)** 3; **(c)** $-\frac{1}{4}$; **(d)** $\frac{2}{7}$; **(e)** $-\frac{4}{9}$.

2 **(a)** $-\frac{1}{3}$; **(b)** 3.

3 **(d)** 20.

EXERCISE 3E

1 **(a)** $2x - y - 3 = 0$; **(b)** $2x + 3y - 6 = 0$; **(c)** $x + 2y + 6 = 0$.

2 **(a)** gradient 3, intercept 2; **(b)** gradient 2, intercept $-\frac{5}{2}$;
(c) gradient $\frac{1}{2}$, intercept $\frac{7}{4}$; **(d)** gradient $-\frac{2}{3}$, intercept $\frac{8}{3}$;
(e) gradient $\frac{5}{4}$, intercept -2; **(f)** gradient 8, intercept -6;
(g) gradient $\frac{3}{5}$, intercept $-\frac{2}{5}$; **(h)** gradient $-\frac{3}{2}$, intercept 2;
(i) gradient 2, intercept $-\frac{6}{5}$; **(j)** gradient 0, intercept 2.

EXERCISE 3F

1 $y = 2x + 4$.

2 $y = -\frac{1}{3}x + 2$.

3 $y = \frac{1}{2}x + \frac{5}{2}$.

4 $y = \frac{3}{2}x + \frac{3}{2}$.

5 $y = -2x$.

6 -1.

7 $6y = 4x + 23$.

8 **(a)** $y = -\frac{1}{2}x + 2$; **(b)** $y = 2x - 3$.

9 **(b)** $(-1, -2)$; **(c)** $y = 5x + 3$.

10 40.

EXERCISE 3G

2 b, c and d.

3 **(a)** $(-4, 0)$; **(b)** $(3, 0)$; **(c)** $(-3, 0)$; **(d)** $(-4, 0)$.

4 $\frac{3}{4}$.

6 **(a)** $y = -x + 4$; **(b)** 1.

7 **(a)** $y = \frac{5}{2}x + 8$; **(b)** not on BD.

8 **(a)** $(-6, -5)$; **(b)** $(1, 2)$; **(c)** $(4, -2)$; **(d)** $(8, 64)$;
(e) $(34, -7)$ **(f)** $(3, 12)$; **(g)** $(1\frac{1}{3}, -1)$ **(h)** $(-7, -4)$.

9 $x = -3\frac{1}{6}, y = \frac{3}{5}$.

10 $x = -1.5, y = -1.5$.

MIXED EXERCISE

1 **(a)** $m_{OA} = \frac{3}{2}, \therefore y = \frac{3}{2}x$;

2 **(a)** **(i)** 4, **(ii)** $\frac{5}{2}$.
(b) $y = -\frac{2}{5}x - \frac{3}{5}$;
(c) $p = 5$.

3 **(b)** **(i)** $y = \frac{1}{2}x + \frac{23}{4}$, **(ii)** $(-\frac{23}{2}, 0)$.

4 **(a)** **(i)** -3, **(ii)** $3x + y = 4$;
(b) **(i)** $(2, -2)$, **(ii)** $\frac{1}{3}$, **(iii)** -1.

5 **(a)** **(i)** $\frac{3}{4}$, **(ii)** $y = \frac{3}{4}x - \frac{19}{4}$;
(b) **(ii)** $(5, -1)$.

6 **(a)** $\frac{5}{3}$;
(b) **(ii)** $3x + 5y = 2$;
(c) $(7, 3)$.

7 **(b)** $y = 2x - 5$;
(c) **(i)** $y - 6 = \frac{1}{3}(x - 3)$ or $3y = x + 15$,
(ii) $(6, 7)$.

8 **(a)** $y = -2x + 1$; **(b)** **(ii)** 0.25.

9 **(a)** $3x + 4y = 5$;
(b) $(3, -1)$;
(c) Side AD: $y = x + 3$, side CD: $3x + 4y = 19$;
(d) $(1, 4)$.

10 **(a)** AB: $y = -5x + 4$, BC: $5y = x - 6$;
(c) 52;
(d) $(0, 30)$.

4 Quadratic functions and their graphs

EXERCISE 4A

1 **(a)** $(x - 2)(x - 1)$; **(b)** $(x - 8)(x + 1)$; **(c)** $(x + 3)(x + 4)$;
(d) $(2x - 3)(x + 1)$; **(e)** $(x - 2)(3x - 1)$; **(f)** $(4x - 3)(x + 1)$;
(g) $(1 - 2x)(7 + x)$; **(h)** $(3x - 2)(2x + 3)$; **(i)** $(2 - x)(3 + 4x)$;

(j) $x(5 - 2x)$; **(k)** $(7 - x)(3 + 4x)$; **(l)** $(5 + 2x)(3 - 4x)$;
(m) $(2 + 3x)(6 - x)$; **(n)** $x(3 + 4x)$; **(o)** $(9 + 5x)(6 - 5x)$.

2 **(a)** $x = 1, x = 2$ and ∪ shape;
(b) $x = -1, x = 8$ and ∪ shape;
(c) $x = -4, x = -3$ and ∪ shape;
(d) $x = 1.5, x = -1$ and ∪ shape;
(e) $x = \frac{1}{3}, x = 2$ and ∪ shape;
(f) $x = \frac{3}{4}, x = -1$ and ∪ shape;
(g) $x = -7, x = \frac{1}{2}$ and ∩ shape;
(h) $x = -1.5, x = \frac{2}{3}$ and ∪ shape;
(i) $x = -\frac{3}{4}, x = 2$ and ∩ shape;
(j) $x = 0, x = 2.5$ and ∩ shape;
(k) $x = -\frac{3}{4}, x = 7$ and ∩ shape;
(l) $x = -2.5, x = \frac{3}{4}$ and ∩ shape;
(m) $x = -\frac{2}{3}, x = 6$ and ∩ shape;
(n) $x = -\frac{3}{4}, x = 0$ and ∪ shape;
(o) $x = -1.8, x = 1.2$ and ∩ shape.

3 **(a)** ∪ shape; **(b)** ∩ shape; **(c)** a straight line.

4 **(a)** $x = 1.5$; **(b)** $x = 3.5$; **(c)** $x = -3.5$; **(d)** $x = 0.25$;
(e) $x = \frac{7}{6}$; **(f)** $x = -\frac{1}{8}$; **(g)** $x = -3.25$; **(h)** $x = -\frac{5}{12}$;
(i) $x = \frac{5}{8}$; **(j)** $x = 1.25$; **(k)** $x = \frac{25}{8}$; **(l)** $x = -\frac{7}{8}$;
(m) $x = \frac{8}{3}$; **(n)** $x = -\frac{3}{8}$; **(o)** $x = -0.3$.

5 **(a)** $y = (x - 3)(x - 5)$, i.e. $y = x^2 - 8x + 15$;
(b) $y = (x - 3)(x + 2)$, i.e. $y = x^2 - x - 6$;
(c) $y = 2(x - 4)(x - 6)$, i.e. $y = 2x^2 - 20x + 48$;
(d) $y = (1 - x)(1 + x)$, i.e. $y = 1 - x^2$;
(e) $y = \frac{1}{2}(x - 2)(x - 3)$, i.e. $y = \frac{1}{2}x^2 - \frac{5}{2}x + 3$;
(f) $y = -3(x + 2)(x + 5)$, i.e. $y = -3x^2 - 21x - 30$.

EXERCISE 4B

1 **(a)** $(x + 4)^2 + 3$; **(b)** $(x + 2)^2 + 9$; **(c)** $(x + 5)^2 - 11$;
(d) $(x - 5)^2 + 5$; **(e)** $(x - 4)^2 - 13$; **(f)** $(x + \frac{3}{2})^2 + \frac{3}{4}$;
(g) $(x + \frac{1}{2})^2 + \frac{3}{4}$; **(h)** $(x - \frac{5}{2})^2 + \frac{3}{4}$; **(i)** $(x - \frac{1}{2})^2 + \frac{7}{4}$;
(j) $(x - \frac{7}{2})^2 - \frac{57}{4}$.

2 **(a)** $y = (x + 2)^2 + 8$, $(-2, 8)$, $x = -2$;
(b) $y = (x + 6)^2 + 4$, $(-6, 4)$, $x = -6$;
(c) $y = (x - 3)^2 - 7$, $(3, -7)$, $x = 3$;
(d) $y = (x + 4)^2 - 11$, $(-4, -11)$, $x = -4$;
(e) $y = (x - 1)^2 - 4$, $(1, -4)$, $x = 1$;
(f) $y = (x - 7)^2 - 17$, $(7, -17)$, $x = 7$;
(g) $y = (x + \frac{1}{2})^2 + \frac{11}{4}$, $(-\frac{1}{2}, \frac{11}{4})$, $x = -\frac{1}{2}$;
(h) $y = (x - \frac{3}{2})^2 - \frac{1}{4}$, $(\frac{3}{2}, -\frac{1}{4})$, $x = \frac{3}{2}$;
(i) $y = (x - \frac{5}{2})^2 - \frac{21}{4}$, $(\frac{5}{2}, -\frac{21}{4})$, $x = \frac{5}{2}$;
(j) $y = (x - \frac{9}{2})^2 - \frac{21}{4}$, $(\frac{9}{2}, -\frac{21}{4})$, $x = \frac{9}{2}$.

3 **(a)** $x = -2 \pm\sqrt{7}$; **(b)** $x = -3 \pm\sqrt{5}$; **(c)** $x = 4 \pm\sqrt{11}$;
(d) $x = 1 \pm\sqrt{5}$; **(e)** $x = 5 \pm 2\sqrt{7}$; **(f)** $x = 7 \pm 3\sqrt{5}$;
(g) $x = \frac{1}{2} \pm \frac{\sqrt{5}}{2}$; **(h)** $x = -\frac{3}{2} \pm \frac{\sqrt{29}}{2}$; **(i)** $x = \frac{3}{2} \pm \frac{\sqrt{5}}{2}$;
(j) $x = \frac{7}{2} \pm \frac{\sqrt{53}}{2}$; **(k)** $x = \frac{5}{2} \pm \frac{\sqrt{13}}{2}$; **(l)** $x = \frac{1}{2} \pm \frac{\sqrt{13}}{2}$.

EXERCISE 4C

1 **(a)** $3(x + 1)^2 - 5$; **(b)** $5(x + 4)^2 - 73$; **(c)** $2(x + 3)^2 - 19$;
(d) $4(x + 1)^2 - 15$; **(e)** $5(x + \frac{1}{2})^2 + \frac{39}{4}$; **(f)** $3(x + \frac{3}{2})^2 - \frac{59}{4}$;
(g) $2(x + \frac{5}{4})^2 - \frac{17}{8}$; **(h)** $3(x + \frac{2}{3})^2 + \frac{17}{3}$; **(i)** $4(x + \frac{7}{8})^2 - \frac{81}{16}$.

2 **(a)** $3(x - 2)^2 - 7$; **(b)** **(i)** -7, **(ii)** 0;
(c) **(i)** $x = 2$, **(ii)** $x = 2 \pm \sqrt{\left(\frac{7}{3}\right)}$.

3 **(a)** $a = 26, b = 5$ **(b)** 26, when $x = -\frac{5}{3}$.

4 $\left(-\frac{b}{2a}, \frac{4ac - b^2}{4a}\right)$, so $y = 0$ when $4ac - b^2 = 0$, i.e. when $b^2 = 4ac$.

5 **(a)** $3(x + 2)^2 + 5$;
(b) least value is 5 occurring when $x = -2$;
(c) greatest value is $\frac{1}{5}$.

6 **(a)** $3\sqrt{5}$; **(b)** $\sqrt{2}$; **(c)** $\sqrt{2}$.

7 $A = -(x - 5)^2 + 25$, maximum value 25 when $x = 5$.

8 $A = -2(x - 30)^2 + 1800$, maximum value is 1800 when $x = 30$.

9 $x + y = 25$, maximum area $= 350$.

EXERCISE 4D

1 **(a)** $\begin{bmatrix} 7 \\ 0 \end{bmatrix}$ **(b)** $\begin{bmatrix} -1 \\ 0 \end{bmatrix}$; **(c)** $\begin{bmatrix} 6 \\ 0 \end{bmatrix}$; **(d)** $\begin{bmatrix} -2 \\ 0 \end{bmatrix}$;
(e) $\begin{bmatrix} 1 \\ 5 \end{bmatrix}$; **(f)** $\begin{bmatrix} -2 \\ 3 \end{bmatrix}$; **(g)** $\begin{bmatrix} 1 \\ -7 \end{bmatrix}$; **(h)** $\begin{bmatrix} -8 \\ 3 \end{bmatrix}$.

2 **(a)** $y = (x - 4)^2$; **(b)** $y = x^2 + 3$; **(c)** $y = (x - 2)^2 + 5$;
(d) $y = (x + 1)^2 - 1$; **(e)** $y = (x - 3)^2 + 2$; **(f)** $y = (x + 1)^2 - 3$;
(g) $y = (x + 3)^2 - 4$; **(h)** $y = (x - 1)^2 - 5$; **(i)** $y = (x - 3)^2 + 4$.

3 **(a)** $(x - 2)^2 + 3$;
(b) **(i)** $\begin{bmatrix} 2 \\ 3 \end{bmatrix}$, **(ii)** $\begin{bmatrix} -2 \\ -3 \end{bmatrix}$; **(iii)** $\begin{bmatrix} 0 \\ 3 \end{bmatrix}$, **(iv)** $\begin{bmatrix} 2 \\ 0 \end{bmatrix}$, **(v)** $\begin{bmatrix} 1 \\ -2 \end{bmatrix}$.

EXERCISE 4E

1 **(a)** $\frac{3}{2} \pm \frac{\sqrt{13}}{2}$; **(b)** $-\frac{7}{4} \pm \frac{\sqrt{57}}{4}$; **(c)** no real roots;

(d) $\frac{5}{8} \pm \frac{\sqrt{57}}{8}$; **(e)** no real roots; **(f)** $-\frac{3}{10} \pm \frac{\sqrt{29}}{10}$;

(g) $\frac{5}{4} \pm \frac{\sqrt{17}}{4}$; **(h)** no real roots; **(i)** $1, -\frac{7}{4}$.

2 **(a)** $2 \pm \sqrt{5}$; **(b)** $-\frac{5}{4} \pm \frac{\sqrt{73}}{4}$; **(c)** $\frac{13}{8} \pm \frac{\sqrt{89}}{8}$;

(d) $\frac{3}{10} \pm \frac{\sqrt{69}}{10}$; **(e)** $-\frac{7}{8} \pm \frac{\sqrt{17}}{8}$; **(f)** $-\frac{7}{10} \pm \frac{\sqrt{69}}{10}$;

(g) $\frac{11}{4} \pm \frac{\sqrt{65}}{4}$; **(h)** $-\frac{5}{6} \pm \frac{\sqrt{13}}{6}$; **(i)** $-\frac{7}{8} \pm \frac{\sqrt{65}}{8}$.

EXERCISE 4F

1 **(a)** 8^2, rational; **(b)** 7^2, rational; **(c)** 37, irrational;
(d) 17, irrational; **(e)** 9^2, rational; **(f)** 1, rational;
(g) 3^2, rational; **(h)** 41, irrational; **(i)** 2436, irrational.

2 **(a)** $k > -1$; **(b)** $k < -2$; **(c)** $k > -\frac{25}{24}$;
(d) $k > -\frac{113}{32}$; **(e)** $k < \frac{109}{40}$; **(f)** $k < \frac{37}{36}$.

3 **(a)** $p = \pm\frac{5}{2}$; **(b)** $p = 2$; **(c)** $p = \pm\frac{4\sqrt{3}}{3}$;
(d) $p = 3$; **(e)** $p = 7$ or $p = -9$; **(f)** $p = -8$ or $p = 2$.

4 **(a)** two; **(b)** zero; **(c)** one; **(d)** two;
(e) one; **(f)** two; **(g)** two **(h)** two;
(i) two; **(j)** zero.

MIXED EXERCISE

1 **(a)** **(i)** $(x + 6)^2 - 25$, **(ii)** -25;
(b) $k = 4$, $k = \frac{4}{9}$.

2 **(a)** $2(x + 2)^2 - 1$;
(b) **(i)** -1, **(ii)** -2.

3 **(a)** $-2 + 3\sqrt{2}$, $-2 - 3\sqrt{2}$; **(b)** $(x + 2)^2 - 18$, -18.

4 **(a)** $(\frac{1}{2}, 0)$, $(-\frac{1}{2}, 0)$, $(0, 1)$;
(b) 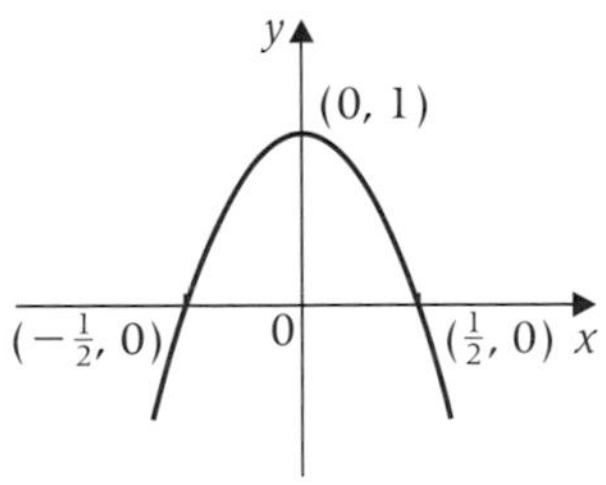

;

(c) $-4x^2 + 8x + 1$.

5 **(a)** **(i)** $f(x) = (x + 4)^2 - 3$, **(ii)** -3, when $x = -4$;
(b) **(i)** translation $\begin{bmatrix} 3 \\ 0 \end{bmatrix}$, **(ii)** $f(x - 3) = x^2 + 2x - 2$.

6 **(a)** $-8 + \frac{3}{2}\sqrt{2}$, $-8 - \frac{3}{2}\sqrt{2}$; **(b)** **(i)** $2(x + 8)^2 - 9$, **(ii)** -9.

5 Polynomials

EXERCISE 5A

1 (a) yes, 1; (b) no; (c) no; (d) yes, 5;
(e) yes, 4; (f) yes, 6; (g) yes, 0; (h) no.

2 $5x^3 + 8x^2 + x + 3$.

3 $7x^4 + 2x^3 - 10x^2 + x + 5$.

4 $4z^2 + 4z + 1$.

5 $3x^3 + 4x^2 - 2x + 12$.

6 (a) $4x^3 + x - 2$; (b) $10h^4 + 3h^3 - 5h^2 + 6$;
(c) $9x^6 + 8x^5 - 5x^3 - 3x - 1$; (d) $13x^2 + 11x - 9$;
(e) $8t^4 - 7t^2 - 14$; (f) $42x^2 - 9x + 48$.

7 (a) $3x^2 - x - 5$; (b) $-y^2 + 11y - 10$;
(c) $x^3 + 13x^2 - 10x - 10$; (d) $6q^3 - 9q^2 - 8q + 28$.

8 (a) $3x^4 + 13x^3 - 6x^2 + x + 12$; (b) $5x^3 - 10x^2 + x + 17$;
(c) $-3x^4 - 8x^3 + 14x^2 - 21x - 1$; (d) $3x^4 + 23x^3 - 26x^2 + 3x + 46$.

EXERCISE 5B

1 $2x^2 + 11x + 12$.

2 $3x^2 - 13x + 14$.

3 $x^3 + 4x^2 - x - 10$.

4 $3y^3 - 10y^2 + 7y - 12$.

5 $-4k^3 + 30k^2 - 19k + 35$.

6 $3x^4 + 13x^3 - 11x^2 - 39x + 30$.

7 (a) $2x^3 - 4x^2 - 28x + 48$;
(b) $-3x^3 + 33x^2 - 73x + 8$;
(c) $e^4 + e^3 - 26e^2 - 53e - 21$;
(d) $4x^5 + 11x^4 - 36x^3 - 19x^2 + 69x - 30$;
(e) $-18t^5 + 24t^4 - 69t^3 - 35t^2 - 18t - 55$;
(f) $x^3 + 4x^2 - 15x - 18$;
(g) $x^3 - 8x^2 - 28x + 80$.

8 (a) -28; (b) -23; (c) -13; (d) 4.

9 (a) -9; (b) -31; (c) 17; (d) 5.

10 (a) -14; (b) -5; (c) -7; (d) 6.

11 (a) $(x + 5)$; (b) $x^2 + 3x + 2$; (c) $x^2 + 5x - 3$;
(d) $2x^2 + 6x - 1$; (e) $4x^2 + 5x - 7$.

12 (a) $a = 1, a = -3, a = -7, a = \frac{1}{2}$; (b) $b = 2, c = 5$.

6 Factors, remainders and cubic graphs

EXERCISE 6A

4 $k = 3$. 5 $p = -3$. 6 (a) $k = 6$. 7 $k = 3$.

8 32. 9 $a = 2, b = 11$. 10 $a = -3, b = -22$.

EXERCISE 6B

1 $f(x) = (x - 1)(x + 1)(x + 2)$.

2 (b) $f(x) = (x + 5)(x + 2)(x - 4)$ $x = -5, x = -2, x = 4$.

3 (b) $P(x) = (x - 3)(x - 6)(x + 1)$ $x = 6, x = 3, x = -1$.

4 $P(x) = (x - 1)(x^2 + 2x - 4)$ $x = 1, x = -1 \pm \sqrt{5}$.

5 (a) $k = 3$; **(b)** $P(x) = (x + 2)(x + 4)(x - 3)$

6 $f(x) = (x - 4)(x - 1)(x + 2)$.

7 (a) $P(x) = (x - 3)(x^2 + 3x + 9)$;
(b) $(x - 2)(x^2 + 2x + 4)$;
(c) $(x - a)(x^2 + ax + a^2)$.

8 (a) $Q(x) = (x + 4)(x^2 - 4x + 16)$;
(b) $(x + 5)(x^2 - 5x + 25)$;
(c) $(x + b)(x^2 - bx + b^2)$.

9 (a) $c = 5, d = 2$; **(b)** $P(x) = (x + 4)(x + 2)(x - 1)$.

10 (b) (i) $P(x) = (x + 3)(x + 1)(x - 9)$,
(ii) $(y^2 + 3)(y^2 + 1)(y - 3)(y + 3)$.

11 (a) $P(x) = (x - 1)^2(x - 4)$; **(b)** $y = \pm 1, y = \pm 2$.

EXERCISE 6C

1 (a)
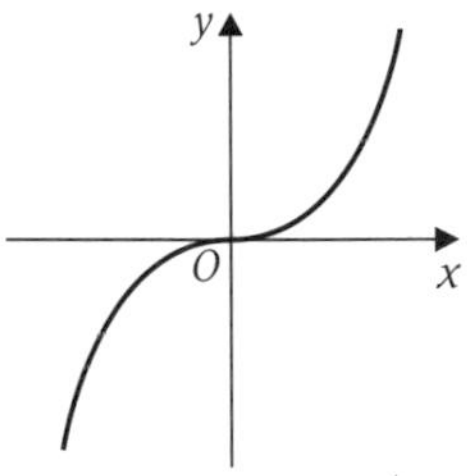

;

(b)
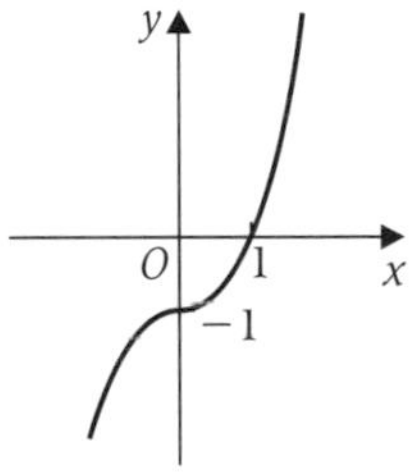

;

(c)
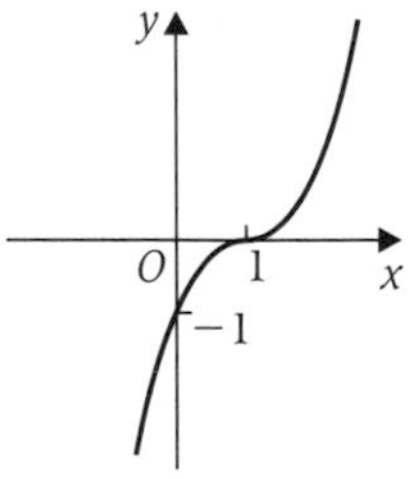

;

(d)
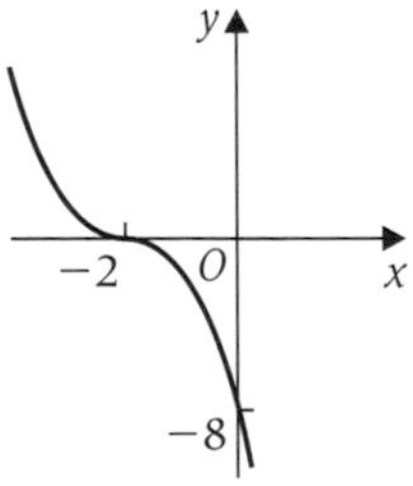

;

(e)
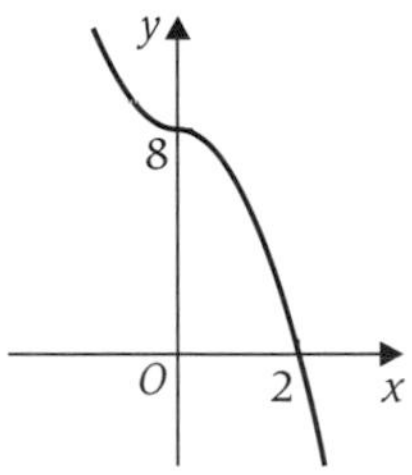

;

(f)
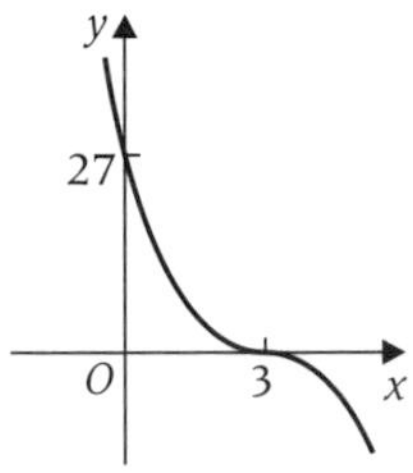

.

2 (a)

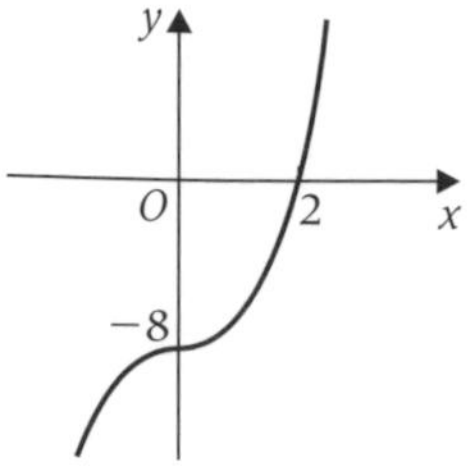

;

(b)

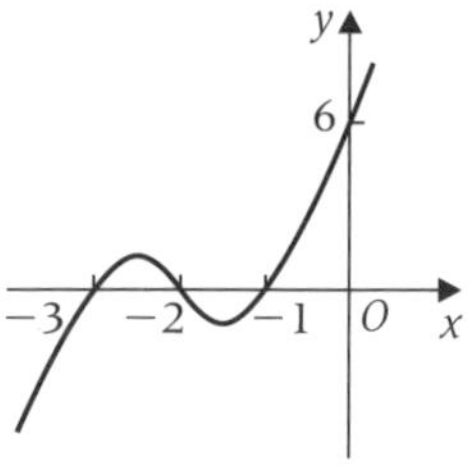

;

(c)

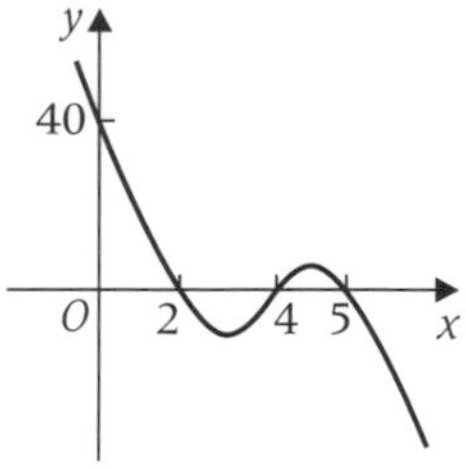

;

(d)

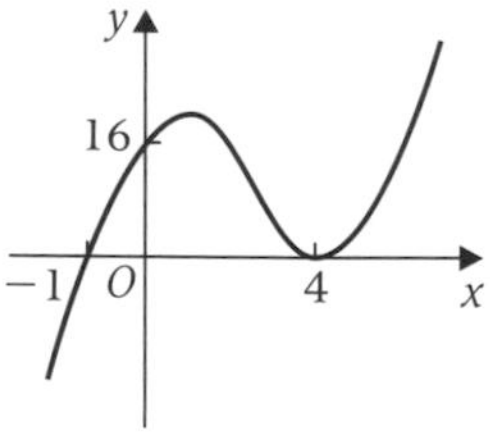

;

(e)

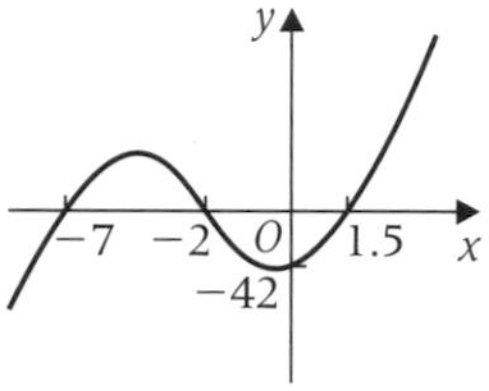

;

(f)

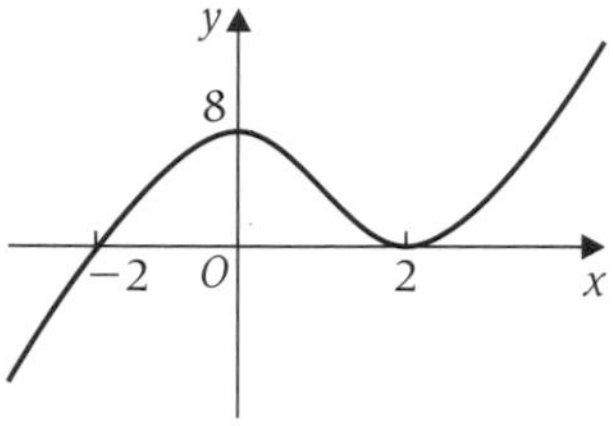

.

3 $P(x) = (x-2)(x-3)(x+2)$,

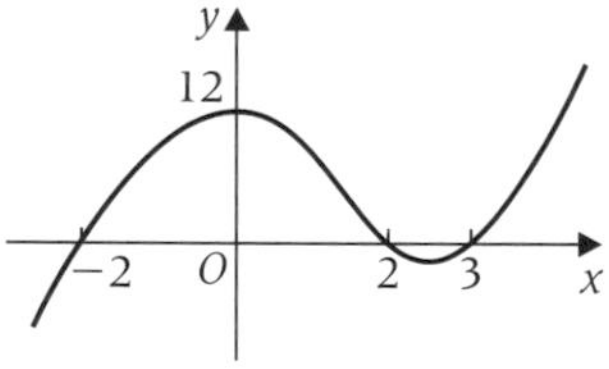

.

EXERCISE 6D

1 **(a)** $1+\dfrac{6}{x}$; **(b)** $1-\dfrac{1}{x+1}$; **(c)** $1+\dfrac{1}{x+2}$;

(d) $1-\dfrac{6}{x+2}$; **(e)** $x+\dfrac{1}{x+1}$; **(f)** $x+1-\dfrac{5}{x+1}$;

(g) $x+4-\dfrac{1}{x-1}$; **(h)** $x+5+\dfrac{17}{x-3}$; **(i)** $x-2-\dfrac{4}{x-4}$;

(j) $x+2-\dfrac{2}{x-2}$.

2 **(a)** $x^2+2x+3+\dfrac{2}{x-1}$; **(b)** $x^2+1-\dfrac{2}{x+1}$;

(c) $x^2-1+\dfrac{1}{x+2}$; **(d)** $x^2+3x+9+\dfrac{13}{x-2}$;

(e) $x^2+1-\dfrac{4}{x+3}$; **(f)** $x^2+7x+20+\dfrac{63}{x-3}$;

(g) $x^2 - 8x + 33 - \frac{127}{x+4}$; (h) $x^2 - x - 2 - \frac{15}{x-4}$;

(i) $x^2 - 5x + 1 + \frac{1}{x+5}$; (j) $x^2 + x + 9 + \frac{45}{x-5}$.

3 (a) (i) $x + 2$, (ii) -1 $P(1) = -1$ (i.e. the same as the remainder);

(b) (i) $x^2 + 4x + 5$, (ii) 2 $P(1) = 2$ (i.e. the same as the remainder);

(c) (i) $x^2 - x$, (ii) -3 $P(1) = -3$ (i.e. the same as the remainder);

4 (a) (i) $x + 3$, (ii) 3 $P(2) = 3$ (i.e. the same as the remainder);

(b) (i) $x^2 + 5x$, (ii) -3 $P(2) = -3$ (i.e. the same as the remainder);

(c) (i) $x^2 + 1$, (ii) -1 $P(2) = -1$ (i.e. the same as the remainder).

5 (a) $k = 3$; (b) $x^2 + 1$.

EXERCISE 6E

1 (a) -4; (b) -2; (c) -14; (d) 6; (e) -34; (f) -6.

2 (a) 2; (b) 1; (c) 8; (d) -7.

4 $k = 11$. 5 6. 6 44. 7 2.5.

8 (a) 5; (b) $P(2) = -2 \neq 0$, $\therefore (x - 2)$ is not a factor of $P(x)$.

9 (a) $a = -3, b = -4$; (c) $x = \pm 2, 3$.

10 $p = -3, R = -1$.

11 (a) $k = 4$; (b) -65; (c) $x = 2, x = 1 + \sqrt{3}, x = 1 - \sqrt{3}$.

7 Simultaneous equations and quadratic inequalities

EXERCISE 7A

1 (a) $\left.\begin{array}{r} x = -4 \\ y = -12 \end{array}\right\}$ or $\left.\begin{array}{r} x = 3 \\ y = -5 \end{array}\right\}$; (b) $\left.\begin{array}{r} x = 1 \\ y = 4 \end{array}\right\}$ or $\left.\begin{array}{r} x = 7 \\ y = -2 \end{array}\right\}$;

(c) $\left.\begin{array}{r} x = 1 \\ y = 2 \end{array}\right\}$ or $\left.\begin{array}{r} x = 2 \\ y = 1 \end{array}\right\}$; (d) $\left.\begin{array}{r} x = 3 \\ y = 1 \end{array}\right\}$ or $\left.\begin{array}{r} x = 1 \\ y = 3 \end{array}\right\}$.

2 (a) $x(2x + 1) = 3$; $\therefore 2x^2 + x - 3 = 0$;

(b) $\left.\begin{array}{r} x = 1 \\ y = 3 \end{array}\right\}$ or $\left.\begin{array}{r} x = -1.5 \\ y = -2 \end{array}\right\}$.

3 $\left.\begin{array}{r} x = 1 \\ y = 1 \end{array}\right\}$ or $\left.\begin{array}{r} x = 3 \\ y = -1 \end{array}\right\}$.

4 $\left.\begin{array}{r} x = 3 \\ y = 0 \end{array}\right\}$ or $\left.\begin{array}{r} x = 0.5 \\ y = 5 \end{array}\right\}$.

5 **(a)** $\left.\begin{array}{l}x=3\\y=1\end{array}\right\}$ or $\left.\begin{array}{l}x=-\frac{4}{3}\\y=\frac{19}{6}\end{array}\right\}$; **(b)** $\left.\begin{array}{l}x=\frac{4}{3}\\y=0\end{array}\right\}$ or $\left.\begin{array}{l}x=-4\\y=8\end{array}\right\}$;

(c) $\left.\begin{array}{l}x=3\\y=5\end{array}\right\}$ or $\left.\begin{array}{l}x=\frac{7}{3}\\y=\frac{7}{3}\end{array}\right\}$; **(d)** $x=2, y=-2$.

EXERCISE 7B

1 **(a)** $\left.\begin{array}{l}x=-2\\y=0\end{array}\right\}$ or $\left.\begin{array}{l}x=3\\y=5\end{array}\right\}$; **(b)** $\left.\begin{array}{l}x=1\\y=3\end{array}\right\}$ or $\left.\begin{array}{l}x=2\\y=2\end{array}\right\}$;

(c) $\left.\begin{array}{l}x=-\frac{1}{2}\\y=-\frac{17}{4}\end{array}\right\}$ or $\left.\begin{array}{l}x=7\\y=7\end{array}\right\}$; **(d)** $\left.\begin{array}{l}x=-\frac{10}{3}\\y=\frac{32}{3}\end{array}\right\}$ or $\left.\begin{array}{l}x=0\\y=-6\end{array}\right\}$.

(e) $\left.\begin{array}{l}x=-\frac{23}{15}\\y=\frac{106}{45}\end{array}\right\}$ or $\left.\begin{array}{l}x=2\\y=0\end{array}\right\}$.

2 **(a)** discriminant $= 0$, $\therefore$ 1 point;
(b) discriminant $= 68$, $\therefore$ 2 points;
(c) discriminant $= -3$, $\therefore$ 0 points;
(d) discriminant $= 33$, $\therefore$ 2 points;
(e) discriminant $= 97$, $\therefore$ 2 points.

3 $(\sqrt{3}, 2\sqrt{3}+5)$ and $(-\sqrt{3}, 5-2\sqrt{3})$.

4 **(a)** $2+2\sqrt{2}$ or $2-2\sqrt{2}$;
(b) $\frac{1}{4}(3+\sqrt{57})$ and $\frac{1}{4}(3-\sqrt{57})$;
(c) $\frac{1}{24}(25 \pm \sqrt{673})$.

5 Equation is $x^2-4x+4=0$, i.e. $(x-2)^2=0$, $\therefore$ repeated root of $x=2$. The line is a tangent to the curve.

6 $c=-\frac{21}{4}$.

7 **(b)** **(i)** $k=18$ or $k=-6$,
(ii) the line is a tangent to the curve.

8 $k<\frac{5}{3}$.

EXERCISE 7C

1 $-\frac{7}{2}<x<3$.
2 $x\leqslant 1$ or $x\geqslant 1.5$.
3 $x<-2$ or $x>3$.
4 $-1<x<6$.
5 $x\leqslant -5$ or $x\geqslant 2$.
6 $-4<x<-3$.
7 $x<-0.5$ or $x>1$.
8 $x<-4$ or $x>4$.
9 $-\frac{2}{3}\leqslant x\leqslant 2$.
10 $x<0.25$ or $x>1$.
11 $x<-1.5$ or $x>3$.
12 $-5\leqslant x\leqslant 5$.
13 $x\leqslant -1$ or $x\geqslant 3$.
14 $x<0.6$ or $x>1$.
15 $x<-1$ or $x>1.75$.
16 $0.375<x<1$.
17 $x<-3$ or $x>0.5$.
18 $-\frac{11}{12}<x<2$.
19 $x<\frac{3}{10}$ or $x>\frac{1}{2}$.
20 $x\leqslant\frac{1}{3}$ or $x\geqslant\frac{5}{3}$.

EXERCISE 7D

1 $3-\sqrt{13}<x<3+\sqrt{13}$.

2 $-4-\sqrt{13}\leqslant x\leqslant -4+\sqrt{13}$.

3 $x<2-\sqrt{5}$ or $x>2+\sqrt{5}$.

4 $1-\sqrt{6}<x<1+\sqrt{6}$.

5 $x\leqslant 5-\sqrt{22}$ or $x\geqslant 5+\sqrt{22}$.

6 $-\frac{3}{2}-\frac{\sqrt{5}}{2}<x<-\frac{3}{2}+\frac{\sqrt{5}}{2}$.

7 $x<1-\sqrt{\left(\frac{3}{2}\right)}$ or $x>1+\sqrt{\left(\frac{3}{2}\right)}$.

8 $x<1-\sqrt{\left(\frac{10}{3}\right)}$ or $x>1+\sqrt{\left(\frac{10}{3}\right)}$.

9 $2-\sqrt{\left(\frac{17}{3}\right)}\leqslant x\leqslant 2+\sqrt{\left(\frac{17}{3}\right)}$.

10 $x<\frac{3}{2}-\sqrt{\left(\frac{3}{2}\right)}$ or $x>\frac{3}{2}+\sqrt{\left(\frac{3}{2}\right)}$.

EXERCISE 7E

1 $-\frac{4}{7}\leqslant k\leqslant 4$.

2 $13k^2+40k+28\geqslant 0$.

3 $k<\frac{2}{9}$ or $k>2$.

4 $-1\frac{7}{8}<k<1$.

5 $7k^2-34k+39<0$.

6 $23k^2+62k+39\leqslant 0$.

MIXED EXERCISE

1 **(a)** **(i)** $(x+3)^2+2$, **(ii)** $x=-3$;
(b) $x<-4$ or $x>-2$;
(c) translation $\begin{bmatrix}-2\\0\end{bmatrix}$.

2 **(a)** $(x-3)^2-2$; **(b)** $3-\sqrt{2}<x<3+\sqrt{2}$.

3 **(a)** $(x+2)^2-9$; **(b)** $x<-5$ or $x>1$.

4 $c<-7.5$.

5 **(b)** **(i)** $k^2-24k+80=0$, **(ii)** $k^2-24k+80>0$.

6 **(a)** $x=-2-\frac{\sqrt{2}}{2}$ or $x=-2+\frac{\sqrt{2}}{2}$;
(b) $x<-2-\frac{\sqrt{2}}{2}$ or $x>-2+\frac{\sqrt{2}}{2}$.

7 $k^2-4k+3\leqslant 0$.

8 $k(x+1)^2+k-3$, always positive if $k>3$.

8 Coordinate geometry of circles

EXERCISE 8A

1 **(a)** $x^2+y^2=49$; **(b)** $(x-2)^2+y^2=4$;
(c) $x^2+(y-2)^2=25$; **(d)** $(x-2)^2+(y-3)^2=16$;
(e) $(x+3)^2+(y+2)^2=9$; **(f)** $(x-2)^2+(y+1)^2=3$.

2 **(a)** (2, 0), 2; **(b)** (0, −1), 1; **(c)** (1, −2), $\sqrt{5}$;
(d) (3, −2), 4; **(e)** (6, −2), $5\sqrt{2}$; **(f)** (−4, 5), 6;
(g) (−2, −1), $\sqrt{11}$; **(h)** (4, −1), $\frac{1}{2}\sqrt{70}$.

3 **(a)** $(x + 3)^2 + (y - 4)^2 = 25$; **(b)** (−6, 0), (0, 0), (0, 8).

4 **(a)** 9π; **(b)** 8π.

EXERCISE 8B

1 **(a)** translation $\begin{bmatrix} 2 \\ 0 \end{bmatrix}$; **(b)** translation $\begin{bmatrix} 0 \\ -3 \end{bmatrix}$; **(c)** translation $\begin{bmatrix} 2 \\ -3 \end{bmatrix}$;

(d) translation $\begin{bmatrix} -1 \\ 4 \end{bmatrix}$; **(e)** translation $\begin{bmatrix} 2 \\ -3 \end{bmatrix}$; **(f)** translation $\begin{bmatrix} -5 \\ 2 \end{bmatrix}$.

2 **(a)** $x^2 + (y - 3)^2 = 49$; **(b)** $(x + 2)^2 + y^2 = 1$;
(c) $(x - 2)^2 + (y - 5)^2 = 16$ **(d)** $(x + 4)^2 + (y - 1)^2 - 81 = 0$;
(e) $(x - 1)^2 + y^2 = 9$; **(f)** $(x + 8)^2 + (y - 2)^2 = 25$;
(g) $x^2 + (y - 1)^2 = 10$; **(h)** $(x - 1)^2 + (y + 7)^2 = 8$;
(i) $x^2 + y^2 + 2x - 2y = 5$; **(j)** $x^2 + y^2 - 2x - 2y = 7$.

3 $x^2 + y^2 = 9$, translation $\begin{bmatrix} 2 \\ 0 \end{bmatrix}$.

4 $x^2 + y^2 = 25$, translation $\begin{bmatrix} -3 \\ 2 \end{bmatrix}$.

5 **(a)**

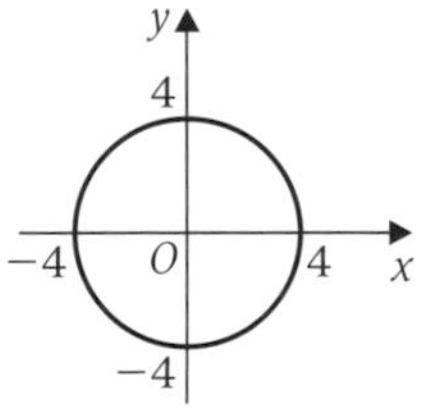

(b)

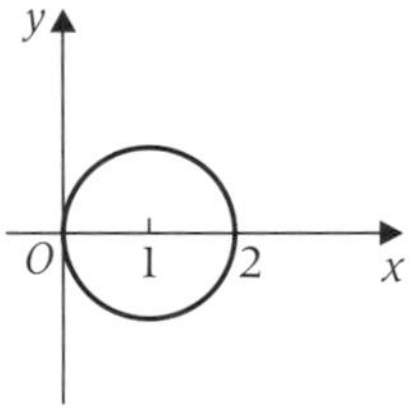

(c)

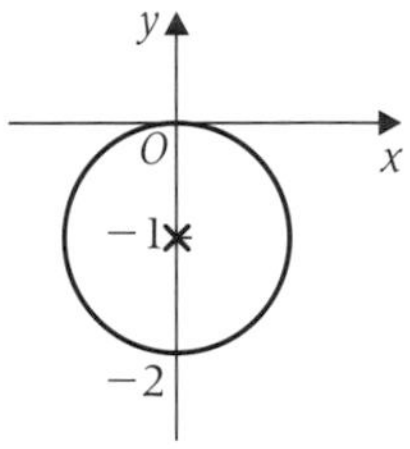

(d)

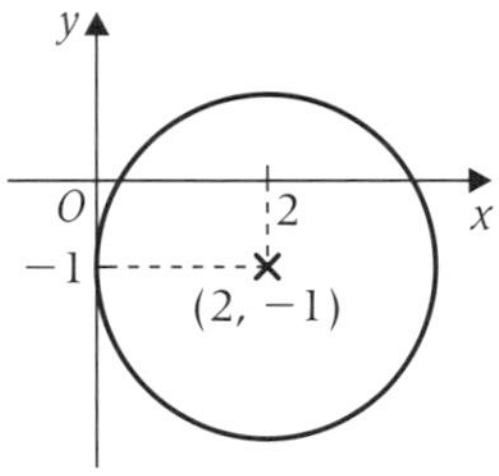

(e)

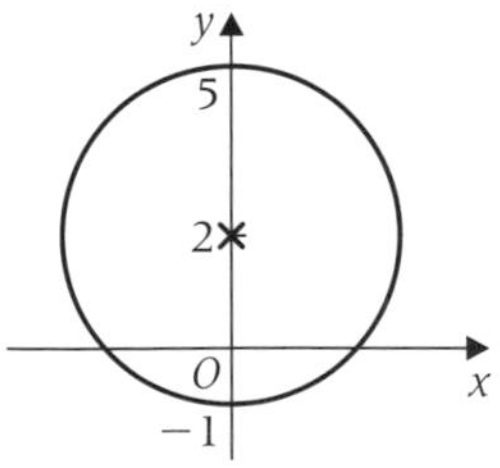

(f)

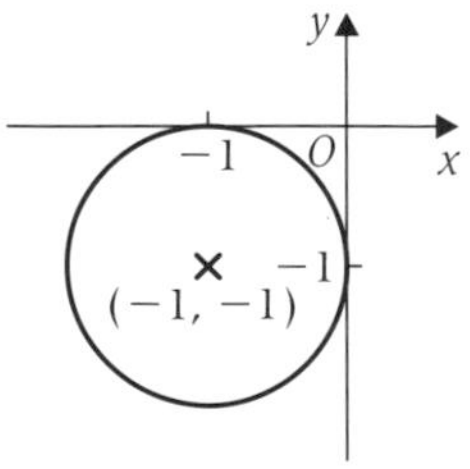

(g)

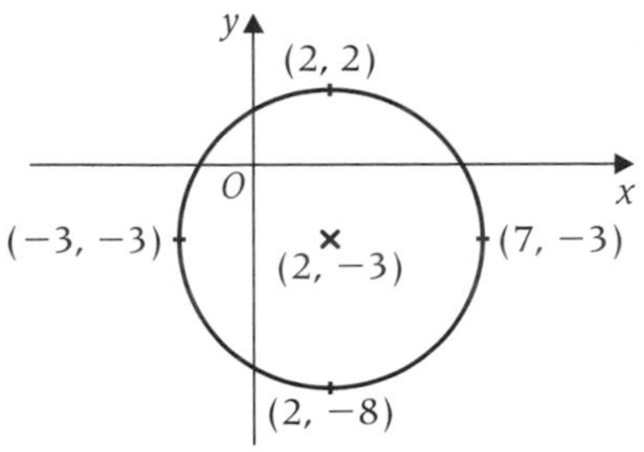

(h)

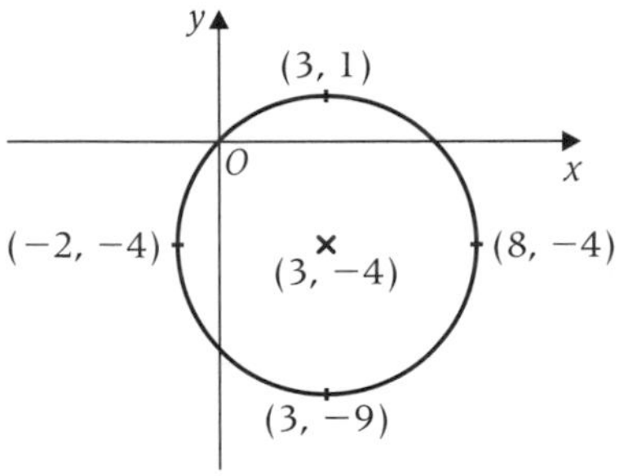

6 translation $\begin{bmatrix} -4 \\ 2 \end{bmatrix}$.

EXERCISE 8C

1 **(a)** (2, 2); **(b)** $2\sqrt{2}$.

2 **(b)** $(x + \frac{9}{2})^2 + (y - \frac{3}{2})^2 = \frac{25}{2}$.

3 **(a)** $(x - 4)^2 + y^2 = 4$;
(b) $x^2 + (y - 2)^2 = 9$;
(c) $x^2 + y^2 - 8x - 5y + 16 = 0$;
(d) $x^2 + y^2 - 10x - 2y + 6 = 0$;
(e) $x^2 + y^2 - 4x + y - 18 = 0$;
(f) $x^2 + y^2 - x + y - 12 = 0$.

4 (−1, 4).

5 (4, −2).

6 **(a)** $x = 2, y = \frac{5}{2} - \frac{1}{2}x$; **(b)** $(2, \frac{3}{2})$; **(c)** $(x - 2)^2 + (y - \frac{3}{2})^2 = \frac{25}{4}$.

7 $(x - 3)^2 + (y - 2)^2 = 10$.

EXERCISE 8D

1 (5, 0) and (−3, 4).

3 (−2, 1).

4 (−1, −2).

6 (2, 1) and (8, −5).

7 **(a)** $m = \pm\frac{4}{3}$; **(b)** 4.8.

EXERCISE 8E

1 $\sqrt{21}$.

2 $4\sqrt{2}$.

3 ±5.

EXERCISE 8F

1 **(a)** (1, 2); **(b)** $y = 3 - x$; **(c)** $y = x - 3$.

2 **(a)** $x = -5, x = 5$; **(b)** $3y = 4x - 25$.

3 $y = 8 - 2x$.

4 $y = 20 - 3x$.

5 **(a)** $\frac{4}{3}$; **(b)** $3x + 4y = 18$.

6 **(a)** $x + y = 6$; **(b)** $y = x$; **(c)** $(-3, -3)$.

7 $y = 2x - 7$.

8 **(a)** $(3, -4)$, 5; **(b)** $(7, -1)$, $(7, -7)$;
(c) $4x + 3y = 25$, $3y + 49 = 4x$.

9 **(a)** -1; **(b)** $y = x - 3$; **(c)** $(x + 1)^2 + (y - 4)^2 = 32$.

10 **(b)** $3y = 4x$; **(d)** $(x - 6)^2 + (y - 8)^2 = 25$.

MIXED EXERCISE

1 **(a)** $(x - 3)^2 + (y + 2)^2 = (\sqrt{20})^2$; **(b)** $(3, -2)$, $2\sqrt{5}$;
(c) $-1, +7$; **(d)** $(+3, 0)$, 2.

2 **(a)** $x^2 + y^2 = 29$, $(x - 7)^2 + (y - 4)^2 = 26$;
(b) $2x + 5y = 29$, $y = 5x - 5$.

3 **(a)** translation $\begin{bmatrix} 1 \\ -3 \end{bmatrix}$; **(b)** $y = 1, y = -7$.

4 **(a)** $(x + 3)^2 + (y - 2)^2 = 25$; **(b)** $(0, -2)$, $(0, 6)$;
(c) $-\frac{3}{4}, \frac{3}{4}$; **(d)** $(5\frac{1}{3}, 2)$.

5 **(a)** $(6, -3)$, $\sqrt{65}$; **(b)** **(ii)** $3\sqrt{5}$, **(iii)** $4\sqrt{5}$.

6 **(a)** $x = 4, y = 1$;
(b) The line is a tangent to the circle at the point (4, 1).

7 **(a)** 5.

8 **(b)** $x = 3 - \sqrt{5}$, $x = 3 + \sqrt{5}$.

9 **(a)** 5; **(b)** $x^2 + y^2 + 2x - 18y + 57 = 0$; **(c)** $4x + 3y = 23$.

10 **(a)** $3\sqrt{2}$, $(1, 0)$; **(b)** $\begin{bmatrix} 1 \\ 0 \end{bmatrix}$; **(c)** **(i)** $x + y = 1$.

9 Introduction to differentiation

EXERCISE 9A

1 7. **2** 19. **3** 3. **4** -4.75.

EXERCISE 9B

1 4.1.

2 4.641.

3 **(a)** 6.1051; **(b)** 5.101 005 01; **(c)** 5.010 010 005; 5.

4 12.

EXERCISE 9C

1 $5(2+h)$ or $10+5h$, 10.

2 **(a)** $4+6h+4h^2+h^3$, 4; **(b)** $32+24h+8h^2+h^3$, 32;
(c) $4a^3+6a^2h+4ah^2+h^3$, $4a^3$.

EXERCISE 9D

1 **(a)** 0; **(b)** $9x^2-14x$; **(c)** $6x^5-8x+5$;
(d) $6x^5+3x^2$; **(e)** $-2+10x$; **(f)** $21x^2$;
(g) $8x-27x^2$; **(h)** $80x^9-42x^5$; **(i)** $16x+4$;
(j) $12x-4$; **(k)** $7x^6-6x+2$; **(l)** $4x^3-9x^2-1$.

2 **(a)** x^2; **(b)** $30x^4-16x+\frac{x}{3}$; **(c)** $14x-\frac{1}{5}$;
(d) $4x^5-6x^7$; **(e)** $\frac{15x^4}{4}-\frac{10x}{9}$; **(f)** $8x^{11}-\frac{7}{2}x^2+\frac{5}{4}$.

3 **(a)** $\frac{dp}{dq}=18q^2-7$; **(b)** $\frac{dy}{dt}=27t^8-0.4$;
(c) $\frac{dm}{dn}=\frac{8n}{3}-\frac{4}{7}$; **(d)** $\frac{dr}{ds}=21s^2-\frac{1}{4}$;
(e) $\frac{dt}{dw}=\frac{3w}{2}-4w^6$; **(f)** $\frac{dz}{dp}=15p^2-2p^7+2$.

EXERCISE 9E

1 32. 2 8. 3 51. 4 3. 5 23.

6 **(a)** 3; **(b)** -9; **(c)** -2; **(d)** 128; **(e)** -13.

7 **(a)** 32; **(b)** 3; **(c)** -24; **(d)** 40.

8 **(a)** -2; **(b)** -10, 10.

9 0, tangent is horizontal, ∪ or ∩ shape or  .

EXERCISE 9F

1 (4, 4). 2 (2, 5). 3 $(-2, 9)$. 4 $2\frac{1}{3}$, 1.

5 $(3, 21\frac{1}{2})$ and $(-2, -17\frac{2}{3})$.

6 **(a)** $(3.5, -2.25)$; **(b)** $(2, -9)$.

7 (2, 17) and (4, 13).

EXERCISE 9G

1 -2. 2 -13. 3 3.

4 **(a)** -10; **(b)** 6 and -6.

5 (3, 0). 6 $(1, -6)$. 7 $(2, -21)$ and $(1\frac{1}{3}, -17\frac{14}{27})$.

8 **(a)** (3, 5); **(b)** $(1\frac{2}{3}, -\frac{1}{3})$.

10 Applications of differentiation

EXERCISE 10A

1 **(b)** 5; **(c)** $y=5x+1$.

2 **(a)** $y=8x-11$; **(b)** $y=9-9x$; **(c)** $y=19x+13$.

3 **(a)** $y = 30x - 55$; **(b)** $y = 21 - 12x$; **(c)** $y = 5x - 2$;
(d) $y = -6x - 3$; **(e)** $y = 1 - x$.

4 **(a)** $y = 4x - 4$; **(b)** $y = 24 - 3x$; **(c)** $y = 14x - 4$;
(d) $y = 9 - 10x$.

5 **(b)** $y = 7x - 21$.

6 **(a)** -12; **(b)** $(0, 24)$.

8 $y = 6x + 6$, $y = -6x + 30$.

9 $y = x + 8$, $y = 4x - 4$, $(4, 12)$.

10 **(a)** $y + 6 = 11x$; **(b)** $1\frac{7}{11}$.

11 **(a)** $y = 3 - x$; **(b)** $y = x - 1$.

EXERCISE 10B

1 **(a)** **(i)** 36, **(ii)** $-\frac{1}{36}$; **(b)** $36y + x = 218$.

2 **(a)** $3y + x + 5 = 0$; **(b)** $6y + x = 13$;
(c) $26y - x + 106 = 0$; **(d)** $11y + x + 34 = 0$.

3 **(a)** $15y + x = 47$; **(b)** $2y + 5 = x$;
(c) $6y - 18 = x$; **(d)** $10y = x + 118$;
(e) $8y + 503 = x$.

4 **(a)** $5y = x + 9$; **(b)** 8.1.

5 **(a)** $k = 11$; **(c)** $4y + x = 28$.

6 $y = 4 - x$, $4y = 1 - x$, $(5, -1)$.

7 **(a)** $2y + x = 5$; **(b)** $7y + x + 14 = 0$; **(c)** $(12\frac{3}{5}, -3\frac{4}{5})$.

EXERCISE 10C

1 **(a)** $90x^2 - 40x^3$; **(b)** $-3 - 8x + 3x^2$.

2 -25.

3 **(a)** 4 m; **(b)** 54 m min^{-1}.

4 $16\pi \text{ cm}^3 \text{ s}^{-1}$.

5 **(a)** $94 \text{ m}^3 \text{ hour}^{-1}$ (increasing); **(b)** $-81 \text{ m}^3 \text{ hour}^{-1}$ (decreasing);
(c) $-84.25 \text{ m}^3 \text{ hour}^{-1}$ (decreasing); **(d)** $166 \text{ m}^3 \text{ hour}^{-1}$ (increasing).

6 10 m s^{-1}.

7 **(a)** -14 (decreasing); **(b)** -6 (decreasing); **(c)** 34 (increasing).

EXERCISE 10D

2 **(a)** $2x - 4$; **(b)** $2x - 4 > 0, \rightarrow 2x > 4, \rightarrow x > 2$.

3 **(a)** $-3 - 3x^2$; **(b)** decreasing since always negative.

4 $x < -3$.

5 $-1\frac{1}{3} < x < 2$.

6 **(a)** $3x^2 - 12x + 13$; **(b)** $3(x - 2)^2 + 1$, always greater than or equal to $+1$, $\therefore$ increasing.

11 Maximum and minimum points

EXERCISE 11A

1 **(a)** (2, 1); **(b)** (1, 4); **(c)** (−5, 33); **(d)** (2, −46).

2 (2.5, 18.5) maximum.

3 **(a)** (5, −71) minimum;
(b) (5, 109) maximum;
(c) (1, 3) minimum, (−1, 7) maximum;
(d) (4, −128) minimum, (−4, 128) maximum.

4 **(a)** (3, 155) maximum;
(b) (0, 0) maximum, (4, −160) minimum;
(c) (1, −8) minimum, (−1, 8) maximum.

5 (0, −19), ($\frac{1}{2}$, $-19\frac{1}{4}$).

7 **(a)** $x = \frac{1}{2}$; **(b)** **(i)** $6x^2 - 2x + 8$.

8 $A(-1, 9)$, $B(3, -23)$.

9 $14\frac{1}{16}$ m.

EXERCISE 11B

1 **(a)** **(i)** $6x^2 + 12x - 9$, **(ii)** $12x + 12$;
(b) **(i)** $60x^5 - 27x^2 + 4x - 7$, **(ii)** $300x^4 - 54x + 4$;
(c) **(i)** $20x^4 - 18x$, **(ii)** $80x^3 - 18$.

2 **(a)** $40x^3$; **(b)** $12x^2 + 24$; **(c)** $6x - 2$.

3 **(a)** $2x - 8$, (4, −11); **(b)** $\frac{d^2y}{dx^2} = 2 > 0$, ∴ minimum.

4 (−2, 38), $\frac{d^2y}{dx^2} = -18$ maximum

(4, −70), $\frac{d^2y}{dx^2} = 18$ maximum

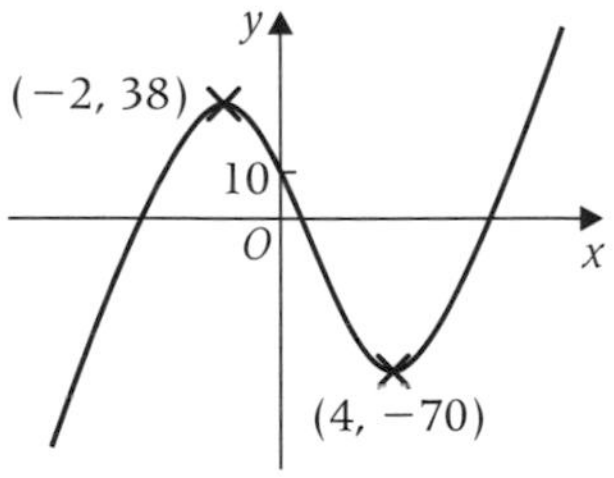

5 (1, 0), $\frac{d^2y}{dx^2} = -6$ maximum

(2, −1), $\frac{d^2y}{dx^2} = 6$ minimum

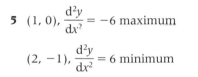

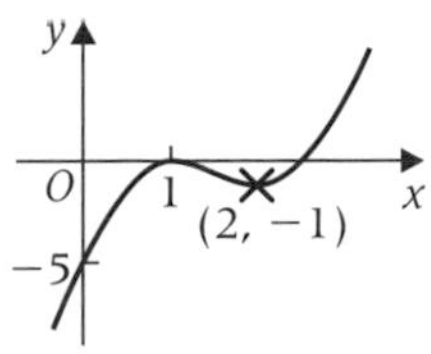

6 **(a)** **(i)** (2.5, −6.25), **(ii)** 2 minimum, **(iii)**

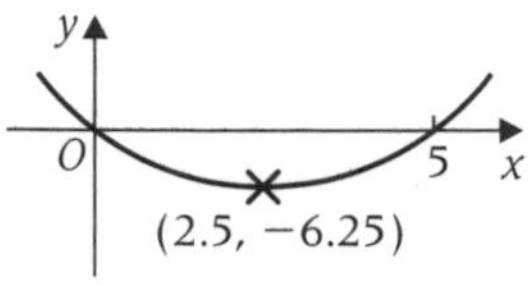

(b) **(i)** (2.5, 0), **(ii)** 8 minimum, **(iii)**

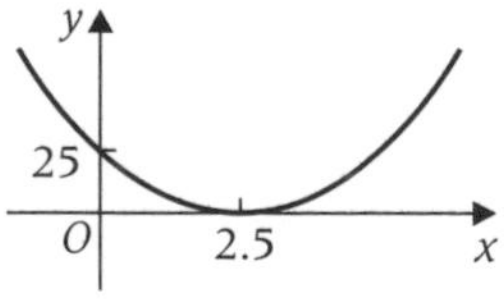

(c) **(i)** (3, -3), **(ii)** 108 minimum, **(iii)**

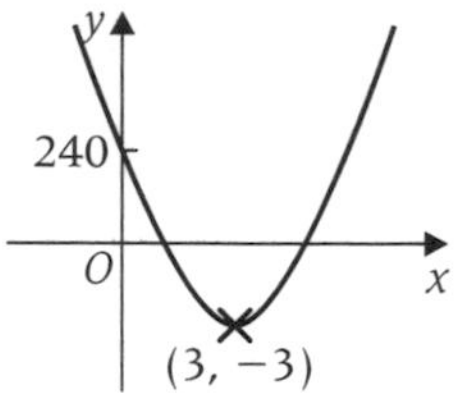

(d) **(i)** (1, 4), (3, 0),

(ii) (1, 4), $\frac{d^2y}{dx^2} = -6$ maximum

(3, 0), $\frac{d^2y}{dx^2} = 6$ minimum, **(iii)**

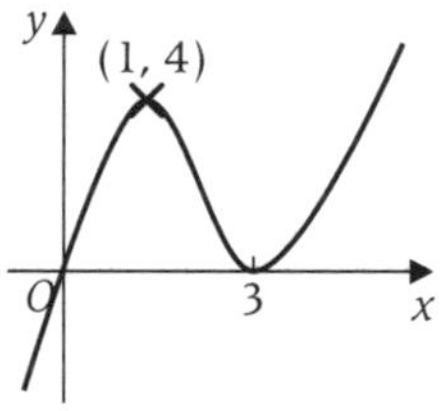

7 (0, 0) maximum, (2, −12) minimum,

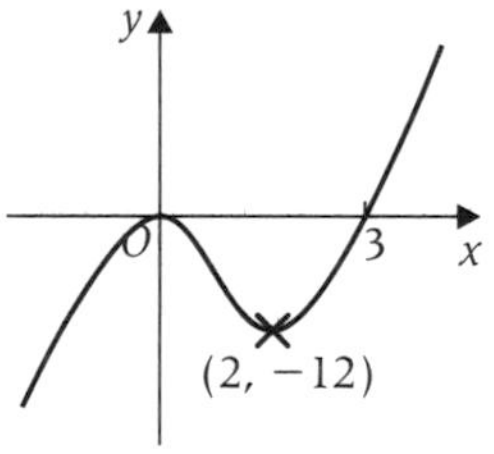

8 **(a)** $4x^3 - 24x^2 + 32x$; **(b)** 4, 2, 0;

(c) $\frac{d^2y}{dx^2} = -16$, i.e. maximum; **(d)** 8.24 a.m.

9 **(a)** $117 + 114t - 3t^2$;

(b) **(i)** $t = 39$, **(ii)** $\frac{d^2p}{dt^2} = -120$, ∴ maximum;

(c) 2009.

EXERCISE 11C

1 121 cm^2, each side 11 cm.

2 7 and 14.

3 16 and 24.

4 **(b)** 15 m, $\dfrac{d^2A}{dx^2} = -4$; **(c)** 450 m^2.

5 **(c)** $x = 12$, 180 m^2.

6 **(b)** 6 and $1\frac{1}{3}$, 6 is impossible since this can't be cut out to make the box!

(c) $\dfrac{d^2V}{dx^2} = -56 < 0$;

(d) $59\frac{7}{27}$ cm^3.

7 **(a)** $0 < 2x < 5$ cm (considering the smallest side), $\therefore 0 < x < 2.5$ cm;

(c) $x = 1$;

(d) 18 cm^3.

8 **(a)** $V = hx^2$, $A = x^2 + 4hx$;

(b) **(ii)** $10\sqrt{10}$, **(iii)** $5000\sqrt{10}$.

12 Integration

EXERCISE 12A

1 **(a)** $y = 2x^6 + c$; **(b)** $y = \dfrac{x^4}{2} + c$;

(c) $y = \dfrac{x^8}{2} + c$; **(d)** $y = 12x + c$;

(e) $y = x^3 - x^2 + 5x + c$; **(f)** $y = x^6 - 6x^3 - \dfrac{x^2}{2} + c$;

(g) $y = \dfrac{x^8}{2} - 2x^6 - 4x + c$; **(h)** $y = \dfrac{x^{12}}{2} + c$;

(i) $y = \dfrac{5x^4}{4} + c$; **(j)** $y = \dfrac{x^9}{3} - 2x + c$;

(k) $y = 2x - \dfrac{3x^8}{4} + c$; **(l)** $y = \dfrac{x^6}{2} - 2x^2 + c$.

2 **(a)** $f(x) = 2x^3 - 2x^2 + 5x + c$; **(b)** $f(x) = x^3 + 5x^2 + 4x + c$;

(c) $f(x) = 2x^5 - \dfrac{x^6}{2} + 7x + c$; **(d)** $f(x) = x^{10} - 2x^4 + x + c$;

(e) $f(x) = x^2 - \dfrac{x^6}{2} + c$; **(f)** $f(x) = 2x^5 - 3x^4 + \dfrac{7x^2}{2} + c$;

(g) $f(x) = \dfrac{x^5}{5} + x^3 - 5x + c$; **(h)** $f(x) = \dfrac{x^4}{2} - \dfrac{3x^8}{2} + c$;

(i) $f(x) = 2x^7 - x^5 + \dfrac{3x^2}{2} + c$; **(j)** $f(x) = 2x^4 + \dfrac{3x^6}{2} - 4x + c$.

3 $f(x) = 4x^3 + 4x^2 - 3x - 8$.

4 $y = 2x^4 - 4x^3 + 3x - 1$.

5 $y = x^5 - 4x^2 + 3x - 7$.

6 $n = -1$.

EXERCISE 12B

1 **(a)** $y = x^3 - x^2 + c$; **(b)** $y = 2x^5 + \frac{3x^4}{2} + c$;

(c) $y = x^8 + 2x^6 + c$; **(d)** $y = \frac{x^6}{2} + x^5 + c$;

(e) $y = -2x + \frac{5x^2}{2} - x^3 + c$; **(f)** $y = \frac{x^4}{4} + \frac{x^3}{3} - 3x^2 + c$;

(g) $y = \frac{x^5}{5} + \frac{3x^4}{4} + \frac{2x^3}{3} + c$; **(h)** $y = \frac{x^4}{4} + \frac{8x^3}{3} + \frac{15x^2}{2} + c$.

2 **(a)** $f(x) = 8x^2 - \frac{4x^3}{3} - 15x + c$; **(b)** $f(x) = 2x^3 - 2x^2 + c$;

(c) $f(x) = 2x^4 - 4x^2 + c$; **(d)** $f(x) = 3x^4 - 2x^2 + c$;

(e) $f(x) = \frac{x^5}{5} - x^3 + c$; **(f)** $f(x) = 2x^6 - 3x^4 + c$;

(g) $f(x) = \frac{x^7}{7} + \frac{x^6}{6} - \frac{2x^5}{5} + c$;

(h) $f(x) = 4x^7 - \frac{14x^6}{3} + \frac{112x^5}{5} - 28x^4 + c$.

3 $y = 5x^5 - 4x^4 + 12$.

4 $y = 2x^3 + \frac{x^2}{2} - 15x + 7$.

5 $y = 9x^4 - 16x^3 - 90x^2 + 99$.

EXERCISE 12C

1 $2x^6 + c$. **2** $4x^5 + c$.

3 $x^2 + c$. **4** $x^4 + 2x^3 + c$.

5 $8x^{11} + c$. **6** $5x^4 + c$.

7 $x^3 + c$. **8** $\frac{5x^4}{4} + x^6 + c$.

9 $x^5 - \frac{3x^2}{2} + c$. **10** $\frac{3x^5}{5} - x^2 + x + c$.

11 $\frac{x^4}{4} + \frac{x^3}{3} - \frac{x^2}{2} - x + c$. **12** $\frac{x^4}{4} - \frac{3x^2}{2} + x + c$.

13 $x^3 + 4x^2 + 5x + c$. **14** $2x^3 - \frac{11x^2}{2} + 5x + c$.

15 $9x - \frac{2x^3}{3} - \frac{3x^2}{2} + c$. **16** $\frac{9x^2}{2} - \frac{x^4}{2} - x^3 + c$.

17 $9x + 6x^2 + \frac{4x^3}{3} + c$. **18** $4x - 6x^2 + 3x^3 + c$.

19 $x^5 - \frac{x^4}{4} + 5x^2 - 2x + c$. **20** $\frac{x^7}{7} - \frac{x^4}{2} + x$.

21 $25x - 30x^2 + 12x^3 + c$. **22** $\frac{x^8}{8} - \frac{2x^5}{5} + \frac{x^2}{2} + c$.

EXERCISE 12D

1 3.	**2** 484.	**3** −2.	**4** 127.5.
5 −12.	**6** $1\frac{1}{3}$.	**7** −3.5.	**8** −42.
9 $-5\frac{1}{6}$.	**10** $-2\frac{5}{12}$.	**11** 64.	**12** 81.
13 9.25.	**14** 61.	**15** 33.	**16** 18.

EXERCISE 12E

1

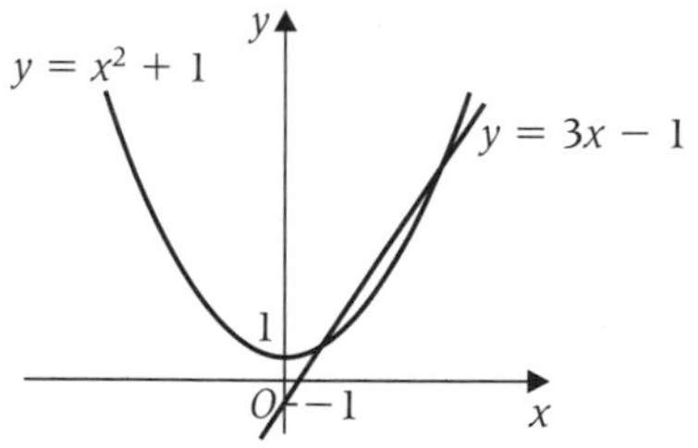

intersect at (1, 2) and at (2, 5)
area = $\frac{1}{6}$.

2 (0, 18), (6, 0), area = 36.

3 (3, 6), (−2, 11), area = $20\frac{5}{6}$.

4

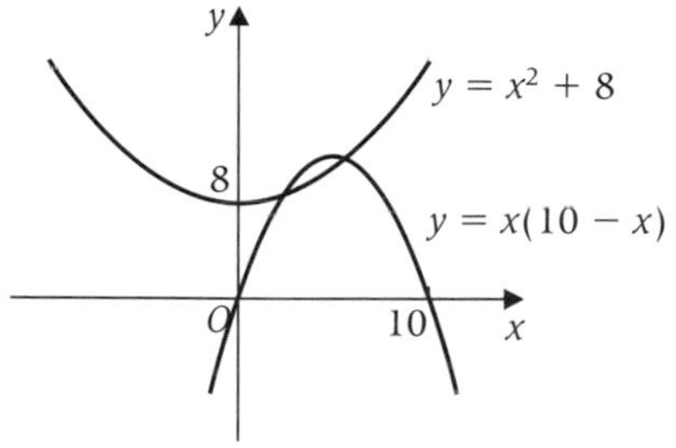

intersect at (1, 9), (4, 24), area = 9.

5

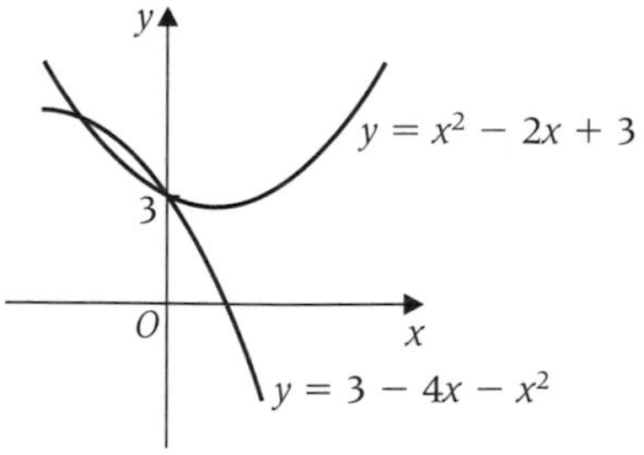

intersect at (−1, 6) and (0, 3), area = $\frac{1}{3}$.

6 (1, 0), (2, 0), area = $\frac{1}{6}$.

7 (2, 0), (−3, 0), area = $20\frac{5}{6}$.

8 $20\frac{5}{6}$.

9 4.5

10

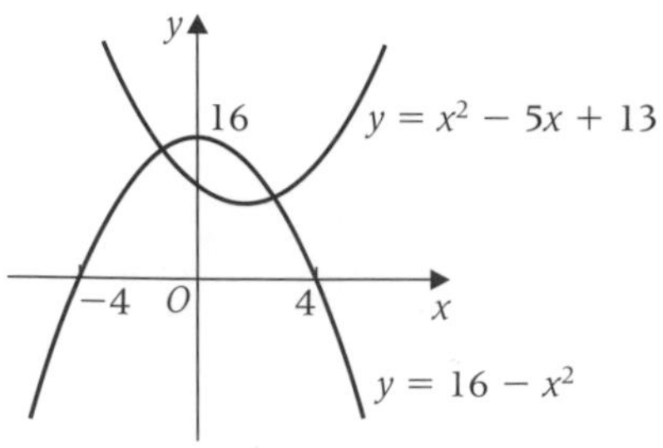

intersect at $(3, 7)$ and $(-\frac{1}{2}, 15\frac{3}{4})$, area $= 14\frac{7}{24}$.

11 $3\frac{1}{12}$, it would give you the difference in magnitude between the shaded areas: the upper area is positive, the lower negative.

12 **(a)** $A(-2, 0)$, $B(2, 0)$, $C(-1, -3)$;
(b) 4.5.

CALCULUS REVISION EXERCISE

1 **(a)** **(i)** $\frac{dy}{dx} = 1 - 8x^3$, **(iii)** $\frac{3}{8}$;
(b) **(i)** $\frac{x^2}{2} - \frac{2x^5}{5} + c$, **(ii)** $\frac{9}{80}$.

2 **(a)** **(i)** $\frac{dy}{dx} = 3x^2 - 12x + 9$, **(ii)** $x = 1, x = 3$;
(b) **(i)** 47.25, **(ii)** 20.25.

3 6.75

4 **(a)** $x = -2$;
(b) 2.25
(c) **(i)** $-\frac{2x^3}{3} + \frac{x^2}{2} + 6x + c$, **(ii)** $5\frac{5}{8}$.

5 **(a)** $f(3) = -2$, $f(4) = 0$;
(b) $f(x) = (x - 2)(x - 4)(x - 1)$;
(c) **(i)** $\frac{dy}{dx} = 3x^2 - 14x + 14$, **(ii)** $\frac{dy}{dx} = -1$, ∴ decreasing,
(iii) $\frac{x^4}{4} - \frac{7x^3}{3} + 7x^2 - 8x + c$, **(iv)** $\frac{5}{12}$.

6 **(b)** $y = 4x - 10$; **(c)** $20\frac{5}{6}$.

7 **(a)** $Q(\sqrt{3}, 2\sqrt{3} + 5)$, $P(-\sqrt{3}, 5 - 2\sqrt{3})$; **(b)** $4\sqrt{3}$.

8 **(a)** 16
(b) **(i)** -12, **(ii)** 24;
(c) 8.

9 **(a)** 6, $6x - y - 6 = 0$; **(b)** 0.25.

10 **(b)** $(-2, 0)$.

Exam style practice paper for C1

1 **(a)** $y < 3.5$; **(b)** $x \leqslant -2.5, x \geqslant 2$.

2 **(a)** $6 + 4\sqrt{3}$; **(b)** $10 + 2\sqrt{3}$.

3 **(a)** $\frac{dy}{dx} = 6x^2 - 18x + 12$, $\frac{d^2y}{dx^2} = 12x - 18$;
(b) $x = 1$ and $x = 2$;
(c) -6 and 6, min $(2, 4)$.

4 **(a)** **(i)** 2, **(ii)** $y = 2x + 2$, **(iii)** $3\sqrt{5}$;
(b) $(1.5, 5)$.

5 **(a)** -6;
(c) $x^2 + 5x + 6$;
(d) $(x - 1)(x + 2)(x + 3)$;
(e)

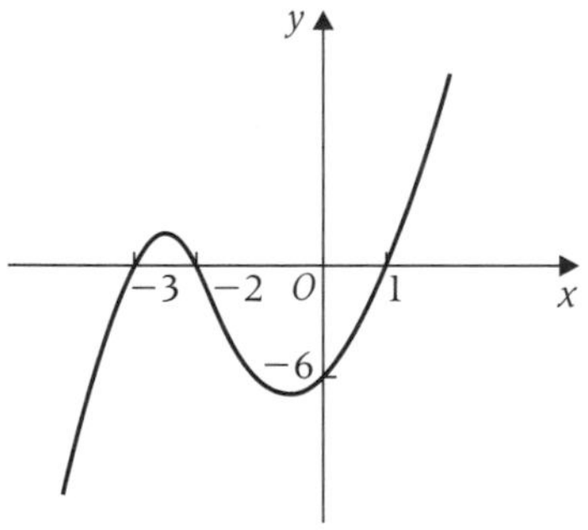

6 **(a)** $(x + 1)^2 + (y - 3)^2 = 25$; **(b)** $(-1, 3)$;
(c) Translation $\begin{bmatrix} -1 \\ 2 \end{bmatrix}$; **(d)** **(ii)** $3x = 4y + 10$.

7 **(a)** $3x^2 - 3$; **(b)** $y = \frac{1}{3}x + 2$;
(c) **(i)** $\frac{1}{4}x^4 - \frac{3}{2}x^2 + 2x + c$.

8 **(a)**

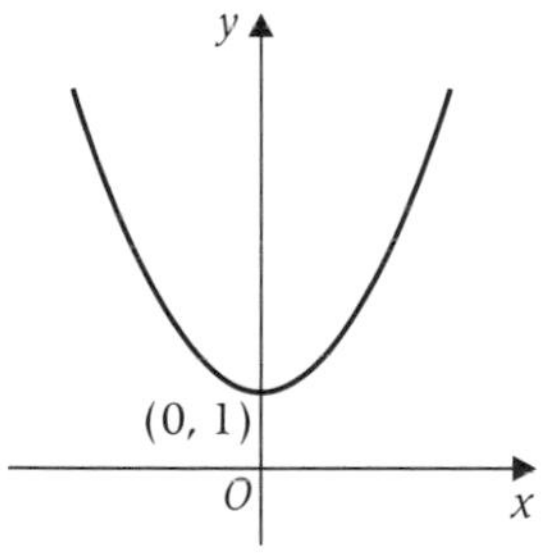

(c) $m = \pm 16$.

Core maths 2

1 Indices

EXERCISE 1A

1	$6m^5$	**2**	$2b^3$	**3**	$24p^3q^2$	**4**	r^9
5	$56t^8$	**6**	$36a^3b^6$	**7**	$54a^{12}$	**8**	$24a^{12}b^5$
9	$60p^5q^4$	**10**	$10p^{10}q^8$	**11**	$36a^4p^3q^4$	**12**	m^{15}
13	m^{15}	**14**	$a^{10}b^{10}$	**15**	$8x^6$	**16**	a^6b^{12}
17	$3p^9q^8$	**18**	$10m^6n^7$	**19**	$4r^{11}t^7$	**20**	$20p^{14}q^{11}$

EXERCISE 1B

1 a^3

2 $4p^2$

3 $5t$

4 10 *or* $10h^0$

5 x^3

6 $3a^3$

7 $\frac{1}{3}pq$ or $\frac{pq}{3}$

8 $\frac{1}{2}a^8$

9 $2pq^3$

10 $\frac{5}{6}b$

11 $10t^4$

12 $10a^8$

13 $2p^4q^3$

14 $x^{11}y^{-1}$ or $\frac{x^{11}}{y}$

15 mn^{-2} or $\frac{m}{n^2}$

16 $\frac{1}{3}a^0c^{-5}$ or $\frac{1}{3c^5}$

17 $\frac{3}{4}m^{-1}n^3$ or $\frac{3n^3}{4m}$

18 x^0y^0 or 1

19 $30xyz^{-1}$ or $\frac{30xy}{z}$

20 $2a^2b^5c^{-5}$ or $\frac{2a^2b^5}{c^5}$

EXERCISE 1C

1 **(a)** 2^4; **(b)** 3^3; **(c)** 5^3; **(d)** (any prime)^0; **(e)** 7^2; **(f)** 2^{-3}; **(g)** 3^{-4}; **(h)** 3^{-1}; **(i)** 11^{-2}; **(j)** 17^{-2}; **(k)** 2^{-7}.

2 **(a)** $\frac{1}{2}$; **(b)** 1; **(c)** 625; **(d)** $\frac{1}{49}$; **(e)** 1; **(f)** $-\frac{1}{125}$; **(g)** $\frac{1}{12}$; **(h)** $\frac{1}{9}$; **(i)** $\frac{1}{25}$; **(j)** 9.

3 **(a)** $\frac{1}{2}$; **(b)** $\frac{1}{81}$; **(c)** 3; **(d)** $\frac{1}{25}$; **(e)** $\frac{4}{9}$; **(f)** $\frac{1}{64}$.

4 **(a)** x^{-10} or $\frac{1}{x^{10}}$; **(b)** a^{21}; **(c)** 1; **(d)** $c^{-15}d^{20}$ or $\frac{d^{20}}{c^{15}}$, **(e)** x^{-12} or $\frac{1}{x^{12}}$; **(f)** a^{-2} or $\frac{1}{a^2}$; **(g)** p^{-11} or $\frac{1}{p^{11}}$; **(h)** $5p^{12}$; **(i)** $4a^2$; **(j)** $13a^{-4}$ or $\frac{13}{a^4}$; **(k)** $7a^{-6}$ or $\frac{7}{a^6}$; **(l)** 2; **(m)** $4y^{12}$; **(n)** $7x^2$; **(o)** $3p^{-3}q^{-4}$ or $\frac{3}{p^3q^4}$; **(p)** $3a^7b^{-6}c^4$ or $\frac{3a^7c^4}{b^6}$; **(q)** $3a^{-10}b^2$ or $\frac{3b^2}{a^{10}}$; **(r)** $32a^{-4}b^0c^{-5}$ or $\frac{32}{a^4c^5}$; **(s)** $\frac{3}{2}a^{-2}bc^3$ or $\frac{3bc^3}{2a^2}$; **(t)** $6x^{-8}y^{-1}z^{-2}$ or $\frac{6}{x^8yz^2}$.

5 (a) x^5; (b) y^{-3} or $\frac{1}{y^3}$; (c) p^{-1} or $\frac{1}{p}$;
(d) r^{-6} or $\frac{1}{r^6}$; (e) $t^{-8}u^2$ or $\frac{u^2}{t^8}$; (f) p^3q^{-1} or $\frac{p^3}{q}$;
(g) $a^{-1}b^{-1}$ or $\frac{1}{ab}$.

EXERCISE 1D

1 (a) 8; (b) 100 000; (c) 9;
(d) 10 000; (e) 32; (f) 4.

2 (a) $\frac{1}{4}$; (b) $\frac{1}{8}$; (c) $\frac{1}{625}$;
(d) $\frac{7}{2}$; (e) $\frac{25}{4}$; (f) $\frac{4}{3}$;
(g) $\frac{5}{4}$; (h) $\frac{9}{16}$; (i) $\frac{13}{9}$.

3 (a) x^2; (b) a^5; (c) x^5;
(d) y; (e) $a^{\frac{7}{12}}$; (f) 2.

4 (a) $x = -1$; (b) $x = \frac{4}{5}$; (c) $x = \frac{1}{3}$; (d) $x = \frac{11}{6}$.

5 $x = \frac{13}{6}$.

6 (a) (i) 3^{-3}, (ii) 3^{2x}; (b) $x = -4$.

7 (a) $2^{\frac{1}{2}}$; (b) $2^{\frac{5}{2}}$; (c) $x = -\frac{1}{2}$.

8 (b) $q = 2^{\frac{3}{2}}$; (c) 2^3.

9 (a) (i) $3^{\frac{1}{2}}$, (ii) $3^{x-\frac{1}{2}}$;
(b) $x = -\frac{1}{2}$.

10 (a) (i) $5^{\frac{3}{2}}$, (ii) 5^{-2}; (b) $x = -\frac{1}{2}$.

11 (b) $y = 4$ or $y = 1$; (c) $x = 2$ or $x = 0$.

12 (a) $x = 0$ or $x = 1$; (b) $x = -2$ or $x = 3$;
(c) $x = 0$ is only real solution.

2 Further differentiation

EXERCISE 2A

1 (a) $9x^2 + 2x^{-2}$; (b) $5 + \frac{8}{3}x^{-3}$;
(c) $-\frac{4}{7}x^{-2} - \frac{6}{5}x^{-3} - 16x$; (d) $-35x^{-8} + \frac{2}{5}x \quad 27$;
(e) $-25x^{-6} - 6x^{-7} - 7$; (f) $-\frac{8}{5}x^{-3} + \frac{2}{3}x^{-2} + \frac{8}{3}$;
(g) $-\frac{3}{2}x^{-2} - 4x^{-4} - 6x^2$; (h) $-40x^{-5} + \frac{x^6}{2}$.

2 (a) $\frac{5}{2}x^{-\frac{1}{2}}$; (b) $\frac{1}{3}x^{-\frac{2}{3}}$;
(c) $x^{-\frac{2}{3}} + \frac{2}{5}x^{-\frac{4}{5}}$; (d) $21x^2 + \frac{1}{2}x^{-\frac{3}{2}}$;
(e) $-\frac{24}{5}x^{-5} - \frac{3x^2}{5}$; (f) $-x^{-\frac{4}{3}} + 4x^{-\frac{7}{5}}$;
(g) $12x - \frac{3}{2}x^{\frac{1}{2}} + 4x^{-3}$; (h) $-\frac{4}{5}x^{-\frac{6}{5}} - \frac{9}{2}x^{\frac{1}{2}}$.

3 (a) $-\frac{18}{5}x^{-4} + 12x^2$; (b) $3x^{-\frac{1}{2}} - 3$;
(c) $-4 + 6x^{-3}$; (d) $-10x^{-3} - \frac{15}{2}x^{-6}$;
(e) $\frac{3}{4}x^{-\frac{1}{4}} + \frac{5}{4}x^{\frac{1}{4}}$; (f) $12x^3 + \frac{3}{5}x^{-\frac{6}{5}} - \frac{8}{5}x^{-3}$;
(g) $-\frac{5}{2}x^{-\frac{1}{2}} + 4x^{-3}$; (h) $-4x^{-\frac{5}{3}} + 2x^{-\frac{4}{5}}$.

4 (a) $\frac{dz}{dx} = 6x + 8x^{-5}$; (b) $\frac{dv}{dt} = 3 + 7t^{-3} + \frac{3}{2}t^{-\frac{1}{2}}$;
(c) $\frac{dp}{db} = -12b^{-4} - 9b^{-13} - 56b^6$; (d) $\frac{dy}{dz} = -10z^{-3} + \frac{3}{25}z^{-\frac{2}{5}} - 27$;
(e) $\frac{ds}{dt} = -72t^{-7} - 2t^{-3}$; (f) $\frac{dh}{dc} = -\frac{8}{5}c^{-3} + \frac{1}{3}c^{-\frac{3}{2}} + \frac{4}{3}c^{-\frac{1}{2}}$.

5 (a) $-24x^{-4}$ or $-\frac{24}{x^4}$; (b) $-3x^{-\frac{3}{2}} - 3x^{-\frac{1}{2}}$;
(c) $4x^{-3} + x^{-\frac{3}{2}}$; (d) $12x^{-3} - 12x^{-5}$.

6 (a) $64x - 4x^{-2}$; (b) $64 + 8x^{-3}$.

EXERCISE 2B

1 (a) $6x^{\frac{1}{2}} - 12$; (b) $5x^{\frac{3}{2}} - 6x$; (c) $12 + \frac{19}{2}x^{-\frac{1}{2}}$;
(d) $1 - 3x^{-\frac{1}{2}}$; (e) $\frac{3\sqrt{x}}{2} + \frac{2}{\sqrt{x}}$; (f) $1 - 3\sqrt{x}$.

2 (a) $12x^3 + 16x$; (b) $4 + 6x^{-3}$; (c) $9x^2 + 4 - \frac{6}{x^3}$;
(d) $-3x^{-3} + 18x^{-4}$; (e) $1 + \frac{12}{x^2}$; (f) $\frac{3}{2}x^{-2} - 2x^{-5}$;
(g) $-\frac{1}{2}x^{-\frac{7}{2}} - 3x^4$; (h) $\frac{18}{x^4} - \frac{1}{x^2} - \frac{2}{x^3}$.

3 (a) $\frac{13}{2}\sqrt{3}$; (b) -12; (c) 2.5.

4 (a) $-8x^{-2} - 3x^2$; (b) $\frac{-18}{x^{\frac{5}{2}}} - \frac{1}{2x^{\frac{3}{2}}}$;
(c) $3\sqrt{x} + 9 - \frac{7}{3x^{\frac{2}{3}}}$; (d) $\frac{20x^{\frac{2}{3}}}{3} - \frac{7}{6x^{\frac{5}{6}}} + \frac{3}{x^2}$.

EXERCISE 2C

1 2.

2 1.

3 (a) -7; (b) $1\frac{2}{3}$.

4 $1\frac{3}{8}$, $3\frac{2}{3}$.

5 $y = 34x - 56$.

6 $y = -\frac{2}{19}x - \frac{36}{19}$.

7 (a) (i) $y = -10x + 9$, (ii) $y = \frac{1}{10}x - \frac{11}{10}$;
(b) (i) $y = 6$, (ii) $x = 4$;
(c) (i) $y = -x + 10$, (ii) $y = x + 4$.

EXERCISE 2D

1 **(a)** $x = -3$;

(b) At (3, 6), $\frac{d^2y}{dx^2} = \frac{2}{3}$, minimum;

At (−3, −6), $\frac{d^2y}{dx^2} = -\frac{2}{3}$, maximum.

2 $(\frac{1}{4}, 8)$, minimum; $(-\frac{1}{4}, -8)$ maximum.

3 (4, 8), maximum.

4 **(a)** (0.09, 1.45), maximum;

(b) (4, 0), minimum;

(c) $(\frac{5}{2}, 20)$, minimum; $(-\frac{5}{2}, -20)$, maximum.

5 **(a)** $x^3 - \frac{64}{x^3}$; **(b)** −63; **(d)** $x = 2$ or $x = -2$.

6 **(a)** $2 - \frac{54}{x^3}$; **(b)** $x = 3$;

(c) $\frac{d^2y}{dx^2} = 2$ is positive, ∴ M is minimum.

7 **(a)** $2 - \frac{3}{2}x^{\frac{1}{2}}$.

8 **(a)** $(\frac{1}{4}, -\frac{11}{8})$; **(b)** $3\frac{1}{2}$ is positive, ∴ M is a minimum.

9 **(a)** $(3\frac{1}{3}, -37\frac{1}{27})$; **(b)** (1, 2) and (−1, −2); **(c)** (3, 54).

10 **(a)** **(i)** (2.5, −6.25), **(ii)** $\frac{d^2y}{dx^2} = 2$, minimum;

(b) **(i)** (2.5, 0), **(ii)** $\frac{d^2y}{dx^2} = 8$, minimum;

(c) **(i)** (2, 3), **(ii)** $\frac{d^2y}{dx^2} = 1.5$, minimum;

(d) **(i)** (4, −4), **(ii)** $\frac{d^2y}{dx^2} = \frac{3}{8}$, minimum.

11 (2, 0) and (−2, 0).
At both these points $\frac{d^2y}{dx^2} = -8$ is negative so both are maximum points. There are no further solutions of $\frac{dy}{dx} = 0$, ∴ no minimum points.

EXERCISE 2E

1 **(b)** $x = 9$;

(d) minimum area is 972 cm² when dimensions are 9 cm, 18 cm and 12 cm.

2 **(b)** $r = 2$, $\frac{d^2s}{dr^2} = 12\pi$; **(c)** 24π cm².

3 **(a)** **(i)** $2hr + \frac{\pi r^2}{2} = 500$;

(b) **(i)** $r = 11.8$ cm (3 sf),

(ii) $\frac{d^2p}{dr^2} = 0.604$ (3 sf), minimum value.

4 (a) (i) $2x - \dfrac{16}{x^2}$, **(ii)** $x = 2$,

(iii) 6, **(iv)** minimum;

(b) (i) 1.41 cm, **(ii)** 10.9 cm^2.

3 Further integration and the trapezium rule

EXERCISE 3A

1 (a) $y = -\frac{1}{4}x^{-4} + c$; **(b)** $y = -\dfrac{x^{-3}}{3} + c$;

(c) $y = -2x^{-2} + c$; **(d)** $y = -6x^{-1} + c$;

(e) $y = 4x^{\frac{3}{4}} + c$; **(f)** $y = \frac{3}{4}x^{\frac{4}{3}} + c$;

(g) $y = 4x^{\frac{3}{2}} + c$; **(h)** $y = 4\sqrt{x} + c$;

(i) $y = \frac{2}{5}x^{\frac{5}{2}} + \dfrac{1}{x} + c$; **(j)** $y = -\dfrac{1}{x^3} + \dfrac{1}{x} + c$.

2 (a) $f(x) = -\dfrac{2}{x} + \dfrac{1}{x^3} + c$; **(b)** $f(x) = \dfrac{2}{5x^5} - \dfrac{7}{x} + c$;

(c) $f(x) = -\dfrac{1}{x^3} - \dfrac{2}{x^5} + c$; **(d)** $f(x) = -\dfrac{1}{x^2} + \dfrac{3}{x^4} + c$;

(e) $f(x) = -\dfrac{2}{x^9} + \dfrac{2}{x^7} + c$; **(f)** $f(x) = 4x\sqrt{x} - \dfrac{6}{x\sqrt{x}} + c$.

3 $y = -\dfrac{4}{x} - 4x^2 + 5x + 31$.

4 $y = 2x\sqrt{x} + 4x^2\sqrt{x} - 133$.

5 $y = 6x\sqrt[3]{x} - 3(\sqrt[3]{x})^2 - 455$.

EXERCISE 3B

1 (a) $y = -\dfrac{3}{2x^2} + \dfrac{2}{3x^3} + c$; **(b)** $y = -\dfrac{10}{x} - \dfrac{3}{x^2} + c$;

(c) $y = \frac{4}{7}x^3\sqrt{x} + 2x\sqrt{x} + c$; **(d)** $y = -x^{-3} - 2x^{-4} + c$;

(e) $y = 2x^2\sqrt{x} + 2x^3\sqrt{x} + c$; **(f)** $y = \frac{2}{7}x^3\sqrt{x} + \frac{2}{5}x^2\sqrt{x} - 4x\sqrt{x} + c$;

(g) $y = 6x^{\frac{8}{3}} + 3x^{\frac{5}{3}} + c$; **(h)** $y = c - \dfrac{1}{3x^3} - \dfrac{2}{x^4} - \dfrac{3}{x^5}$;

(i) $y = c - \dfrac{2}{x} - \dfrac{11}{2x^2}, - \dfrac{4}{x^3}$; **(j)** $\dfrac{6x^{\frac{5}{2}}}{5} + \dfrac{4x^{\frac{3}{2}}}{3} - 16x^{\frac{1}{2}} + c$.

2 (a) $y = -\dfrac{3}{x^4} + \dfrac{1}{x^3} + c$; **(b)** $f(x) = 2x\sqrt{x}(x - 1) + c$;

(c) $f(x) = 2x^6 - 3x^4 + c$; **(d)** $f(x) = 4x\sqrt{x} - x^2 + c$;

(e) $f(x) = \dfrac{x^3}{3} - 3x - \dfrac{7}{x} + c$; **(f)** $f(x) = \dfrac{2}{5}x^{\frac{5}{2}} - 2x^{\frac{3}{2}} + 10x^{\frac{1}{2}} + c$;

(g) $f(x) = \dfrac{12}{5}x^{\frac{5}{3}} - \dfrac{15}{2}x^{\frac{2}{3}} + c$; **(h)** $f(x) = x - \dfrac{8}{3}x^{\frac{3}{2}} + 12x^{\frac{1}{2}} + c$.

3 $y = \dfrac{8}{x^2} - \dfrac{25}{x} - 30$.

4 $y = \dfrac{x^4}{4} - 3x - \dfrac{2}{x} + 1$.

5 $y = \frac{2}{x} - \frac{11}{2x^2} + \frac{5}{x^3} + \frac{1}{2}$.

6 $y = \frac{2}{5}x^{\frac{5}{2}} - 2x^{\frac{3}{2}} + 14x^{\frac{1}{2}} + \frac{1}{5}$.

7 $y = \frac{3}{4}x^{\frac{8}{3}} - 3x^{\frac{5}{3}} - \frac{9}{2}x^{\frac{2}{3}} + \frac{7}{4}$.

EXERCISE 3C

1 $8x^{\frac{3}{2}} + c$.

2 $8x^{\frac{5}{2}} + c$.

3 $\frac{-1}{x^2} + c$.

4 $-\frac{4}{5x^5} - \frac{1}{x^6} + c$.

5 $-\frac{8}{x^{11}} + c$.

6 $\frac{-20}{3x^3} + c$.

7 $-3x^{-1} + c$.

8 $-4x^{-1} + x^6 + c$.

9 $-\frac{2}{x^3} + c$.

10 $2x^{\frac{3}{2}} + c$.

11 $\frac{-1}{2x^2} + c$.

12 $-\frac{4}{x} - \frac{3}{x^2} + c$.

13 $\frac{6}{5}x^{\frac{5}{2}} + \frac{10}{3}x^{\frac{3}{2}} + c$.

14 $4x^{\frac{3}{2}} - 10x^{\frac{1}{2}} + c$.

15 $\frac{2}{7}x^{\frac{7}{2}} + c$.

16 $\frac{2}{5}x^{\frac{5}{2}} - 2x^{\frac{1}{2}} + c$.

17 $-3x^{-3} - 6x^{-2} - 4x^{-1} + c$.

18 $8x^{\frac{1}{2}} - 8x^{\frac{3}{2}} + \frac{18}{5}x^{\frac{5}{2}} + c$.

19 $\frac{x^2}{2} + 4x\sqrt{x} + 9x + c$.

20 $\frac{x^2}{2} - \frac{4}{3}x^{\frac{3}{2}} + x + c$.

21 $6x^{\frac{3}{2}} - \frac{2}{7}x^{\frac{7}{2}} + c$.

22 $\frac{3}{13}x^{\frac{13}{3}} - \frac{3}{4}x^{\frac{4}{3}} + c$.

23 $25x - 30x^2 + 12x^3 + c$.

24 $x^2 - \frac{1}{x^2} + c$.

25 $4x^4\sqrt{x} + c$.

26 $\frac{2}{7}x^{\frac{7}{2}} + 6x^{-\frac{1}{2}} + c$.

27 $\frac{2}{15}x^{\frac{15}{2}} - \frac{4}{9}x^{\frac{9}{2}} + \frac{2}{3}x^{\frac{3}{2}} + c$.

28 $\frac{2}{5}x^{\frac{5}{2}} - 2x^{-\frac{5}{2}} + c$.

29 $\frac{x^3}{3} - \frac{12x^2\sqrt{x}}{5} + \frac{9x^2}{2} + c$.

EXERCISE 3D

1 3.

2 6.

3 48.

4 62.

5 -31.

6 $\frac{1}{2}$.

7 2.

8 $13\frac{17}{32}$.

9 $\frac{2}{5}$.

10 $\frac{2}{3}$.

11 $-\frac{1}{4}$.

12 **(a)** $x^3 + x^{-2}$; **(b)** $4\frac{1}{4}$.

13 18.

14 $\frac{23}{6}$.

15 $12\frac{4}{5}$.

16 23.

17 $3\frac{1}{5}$.

18 **(a)** $x^2 - \frac{27}{x} - 7x + c$; **(b)** $9\frac{1}{2}$.

19 (4, 2), $1\frac{1}{3}$.

20 **(a)** $x = 1, x = 2$; **(b)** $\frac{2}{3}$.

EXERCISE 3E

1 1.218. **2** 8.712.

3 3.268. **4** 1.356.

5 1.911. **6** 0.0781% (3 sf), use more intervals.

7

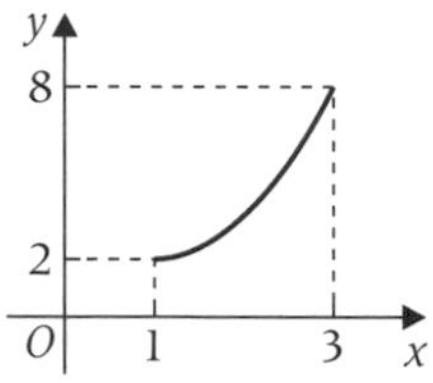

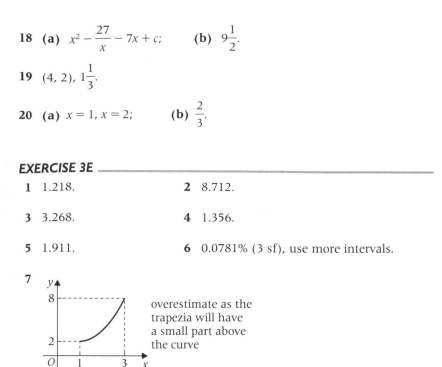

overestimate as the trapezia will have a small part above the curve

8 1.356 (4 sf), both are the same correct to 4 sf. If more significant figures are used then six strips should give a better approximation since there are less 'errors' from the curve in total.

9 **(a)** underestimate; **(b)** underestimate; **(c)** overestimate.

10

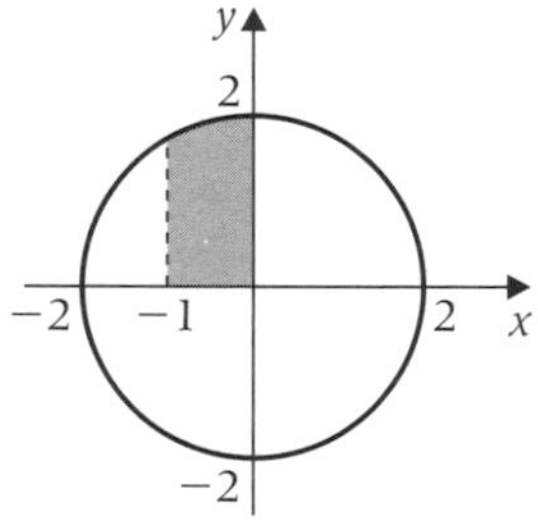

4 Basic trigonometry

EXERCISE 4A

1 **(a)** **(i)** sine positive, **(ii)** cosine positive;
(b) **(i)** sine positive, **(ii)** cosine negative;
(c) **(i)** sine negative, **(ii)** cosine positive;
(d) **(i)** sine negative, **(ii)** cosine negative;
(e) **(i)** sine negative, **(ii)** cosine positive;
(f) **(i)** sine negative, **(ii)** cosine negative.

2 **(b)** **(i)** $\theta = -340°, -200°, 20°, 160°$, **(ii)** $-230°, 50°, 130°$.

3 **(b)** **(i)** $\theta = -410°, -50°, 50°, 310°$, **(ii)** $\theta = 140°, 220°, -220°$

EXERCISE 4B

1 **(a)** $\theta = 50^\circ, -310^\circ, -230^\circ, 130^\circ$; **(b)** $\theta = -20^\circ, -160^\circ$;
(c) $\theta = 90^\circ, -270^\circ$; **(d)** $\theta = 0^\circ, -180^\circ, 180^\circ$;
(e) $\theta = 170^\circ, -350^\circ, -190^\circ, 10^\circ$; **(f)** $\theta = -130^\circ, -50^\circ, 230^\circ, 310^\circ$.

2 **(a)** $\theta = -40^\circ, 40^\circ$; **(b)** $\theta = 70^\circ, 290^\circ$; **(c)** $\theta = 110^\circ, 250^\circ$;
(d) $\theta = 0^\circ, -360^\circ$; **(e)** $\theta = 90^\circ, 270^\circ$; **(f)** $\theta = 20^\circ, -20^\circ$.

3 **(a)** $\theta = 35^\circ$; **(b)** $\theta = 20^\circ$; **(c)** $\theta = 180^\circ$;
(d) $\theta = 225^\circ$; **(e)** $\theta = 230^\circ$; **(f)** $\theta = 135^\circ$.

EXERCISE 4C

1 **(a)** $-130^\circ, 50^\circ, 230^\circ$; **(b)** $160^\circ, 340^\circ$;
(c) $-240^\circ, -60^\circ, 120^\circ$; **(d)** $-340^\circ, -160^\circ$;
(e) $-130^\circ, 50^\circ, 230^\circ$; **(f)** $-240^\circ, -60^\circ, 120^\circ$.

2 **(a)** 145°;
(b) 20°;
(c) 80°;
(d) 135°;
(e) 130°;
(f) 135°.

EXERCISE 4D

1 **(a)** $\theta = 14.5^\circ, 165.5^\circ$;
(b) $x = -323.1^\circ, -36.9^\circ, 36.9^\circ$;
(c) $x = -336.4^\circ, -203.6^\circ, 23.6^\circ, 156.4^\circ$;
(d) $\theta = -225.6^\circ, -134.4^\circ, 134.4^\circ, 225.6^\circ$;
(e) $x = -287.5^\circ, -72.5^\circ$;
(f) $\theta = 210^\circ, 330^\circ$;
(g) $x = 180^\circ$;
(h) $\theta = -360^\circ, 0^\circ, 360^\circ$.

2 **(a)** $\theta = 30.0^\circ, 150.0^\circ, 390.0^\circ, 510.0^\circ$;
(b) $x = 214.8^\circ, 325.2^\circ, 574.8^\circ, 685.2^\circ$;
(c) $x = 113.6^\circ, 246.4^\circ, 473.6^\circ, 606.4^\circ$;
(d) $\theta = 191.5^\circ, 348.5^\circ, 551.5^\circ, 708.5^\circ$;
(e) $\theta = 48.2^\circ, 311.8^\circ, 408.2^\circ, 671.8^\circ$;
(f) $x = 41.8^\circ, 138.2^\circ, 401.8^\circ, 498.2^\circ$.

3 **(a)** $x = 26.6^\circ, 206.6^\circ$;
(b) $\theta = 45^\circ, 225^\circ$;
(c) $x = 0^\circ, 180^\circ, 360^\circ$;
(d) $x = 9.5^\circ$;
(e) $\theta = -16.7^\circ, 163.3^\circ$.

4 **(a)** $56.3^\circ, 236.3^\circ$;
(b) $-21.8^\circ, 158.2^\circ, 338.2^\circ, 518.2^\circ, 698.2^\circ$;
(c) $498.8^\circ, 678.8^\circ$;
(d) $-863.1^\circ, -683{,}1^\circ, -503.1^\circ, -323.1^\circ, -143.1^\circ$.

EXERCISE 4E

1 **(a)** $\theta = 50$, $x = 7.88$, $R = 4$ cm;
(b) $\theta = 80.1$, $x = 7.54$, $R = 5.08$ cm;
(c) $\theta = 12.5$, $x = 10.2$, $R = 6.93$ cm.

2 **(a)** 7.55 cm; **(b)** 14 cm; **(c)** 8.45 cm.

3 **(a)** 21.2 cm^2; **(b)** 22.0 cm^2; **(c)** 34.2 cm^2.

4 **(a)** 90°; **(b)** 82.8°; **(c)** 133°.

5 **(a)** 31.4 cm; **(b)** 19.0 cm; **(c)** 51.9 cm.

6 **(a)** $B = 37.2°$, $b = 7.26$ cm, $c = 9.20$ cm;
(b) **either** $A = 58.5°$, $B = 78.5°$, $b = 5.75$ cm;
or $A = 122°$, $B = 15.5°$, $b = 1.57$ cm.
[*Note* A = 121.5° 1 dp];
(c) $A = 37.4°$, $B = 90.6°$, $b = 7.74$ cm;
(d) **either** $B = 65.8°$, $C = 63.2°$, $b = 3.17$ cm;
or $B = 12.2°$, $C = 117°$, $b = 0.732$ cm.
[*Note* C = 116.8° 1 dp].

7 **(a)** $AB = 18.1$ cm; **(b)** Area = 136 cm^2.

8 **(a)** $R = 10.1$ cm; **(b)** Area = 320 cm^2.

9 319 cm^2.

10 61.4 cm^2.

5 Transformations of graphs

EXERCISE 5A

1 **(a)** $\begin{bmatrix}0\\-5\end{bmatrix}$; **(b)** $\begin{bmatrix}3\\0\end{bmatrix}$; **(c)** $\begin{bmatrix}6\\0\end{bmatrix}$; **(d)** $\begin{bmatrix}0\\2\end{bmatrix}$;
(e) $\begin{bmatrix}1\\0\end{bmatrix}$; **(f)** $\begin{bmatrix}-2\\3\end{bmatrix}$; **(g)** $\begin{bmatrix}0\\6\end{bmatrix}$; **(h)** $\begin{bmatrix}-\frac{1}{2}\\0\end{bmatrix}$.

2 **(a)** $y = 3(x-4)$; **(b)** $y = x^6 + 3$; **(c)** $y = (x-2)^2 + 5$;
(d) $y = 5x^2 - 1$; **(e)** $y = 2(x-3)^3$; **(f)** $y = (x+2)^5$;
(g) $y = (x+3)^7$; **(h)** $y = (x-5)^3 - 5$; **(i)** $y = \frac{2}{x-2}$.

3 **(a)** $\begin{bmatrix}0\\4\end{bmatrix}$ translation; **(b)** $\begin{bmatrix}10°\\0\end{bmatrix}$ translation;
(c) $\begin{bmatrix}-70°\\0\end{bmatrix}$ translation; **(d)** $\begin{bmatrix}30°\\2\end{bmatrix}$ translation.

4 **(a)** $\begin{bmatrix}0\\1\end{bmatrix}$ translation; **(b)** $\begin{bmatrix}-120°\\0\end{bmatrix}$ translation;
(c) $\begin{bmatrix}50°\\0\end{bmatrix}$ translation; **(d)** $\begin{bmatrix}80°\\-3\end{bmatrix}$ translation.

5 **(a)**

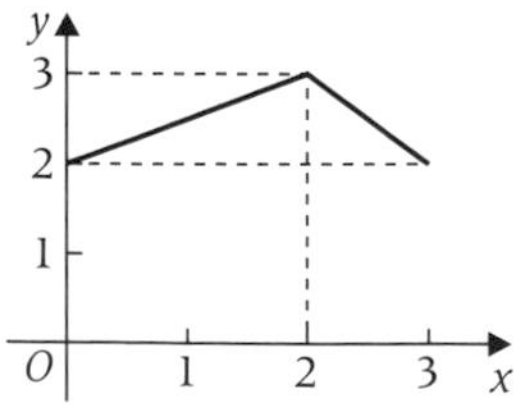

(b)

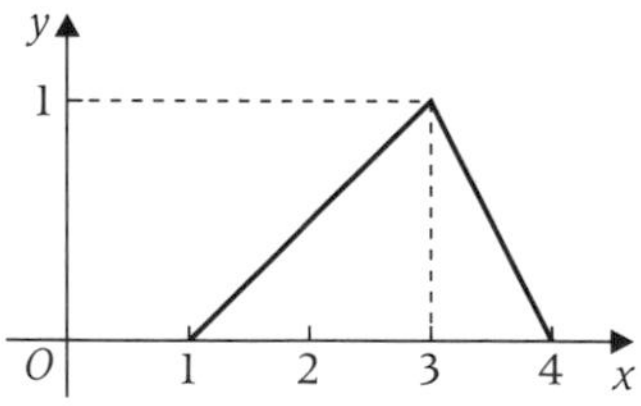

(c)

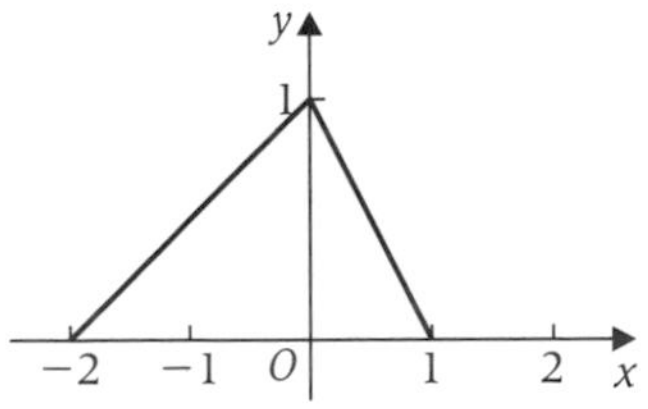

(d)

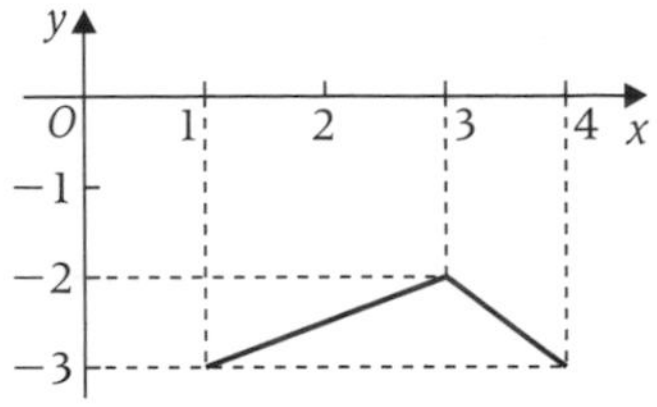

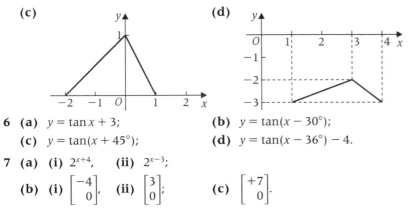

6 (a) $y = \tan x + 3$; (b) $y = \tan(x - 30°)$;
(c) $y = \tan(x + 45°)$; (d) $y = \tan(x - 36°) - 4$.

7 (a) (i) 2^{x+4}, (ii) 2^{x-3};
(b) (i) $\begin{bmatrix} -4 \\ 0 \end{bmatrix}$, (ii) $\begin{bmatrix} 3 \\ 0 \end{bmatrix}$; (c) $\begin{bmatrix} +7 \\ 0 \end{bmatrix}$.

8 270°, 630°, 990°; translation of 270°, etc., in x-direction maps $y = \sin x$ onto $y = \cos x$.

EXERCISE 5B

1 (a) reflection in the line $y = 0$ (x-axis);
(b) one-way stretch, scale factor 5, in the y-direction;
(c) translation $\begin{bmatrix} 3 \\ 0 \end{bmatrix}$;
(d) one-way stretch, scale factor 3, in the x-direction;
(e) one-way stretch, scale factor 3, in the y-direction;
(f) translation $\begin{bmatrix} -2 \\ -4 \end{bmatrix}$;
(g) one-way stretch, scale factor 2, in the x-direction;
(h) reflection in the line $x = 0$ (y-axis).

2 (a) $y = \left(\frac{x}{5}\right)^3$; (b) $y = 4 - x^6$;
(c) $y = 1 - 3x - x^2$; (d) $y = 27x^2 + 12$;
(e) $y = 2\left(-\frac{x}{3}\right)^3 - \left(-\frac{x}{3}\right)^2 + 7$; (f) $y = (3x)^2$;
(g) $y = -2x^7$; (h) $y = (2x)^3 - 5$;
(i) $y = \frac{3}{(-x + 2)}$.

3 (a) one-way stretch, scale factor 3, in y-direction;
(b) one-way stretch, scale factor $\frac{1}{2}$, in x-direction;
(c) one-way stretch, scale factor 5, in x-direction.

4 (a) reflection in line $y = 0$ (x-axis);
(b) one-way stretch, scale factor $\frac{1}{3}$, in x-direction;
(c) one-way stretch, scale factor 5, in y-direction.

5 (a)

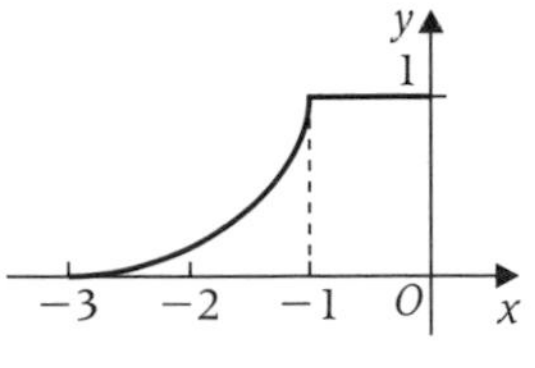

(b)

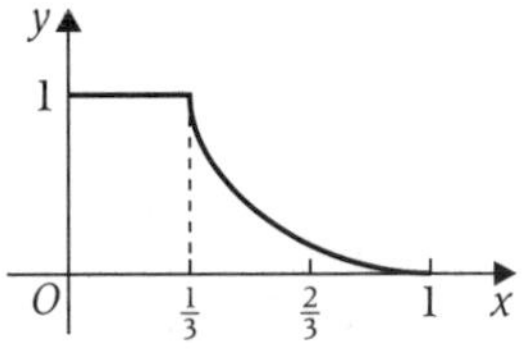

(c)

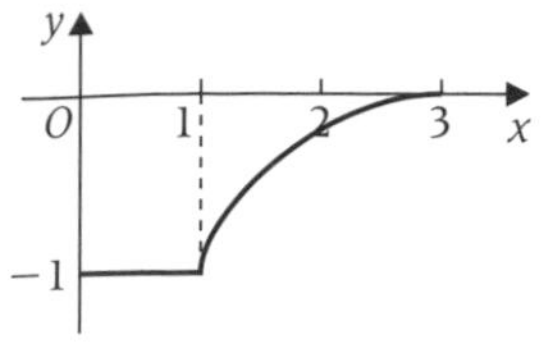

(d)

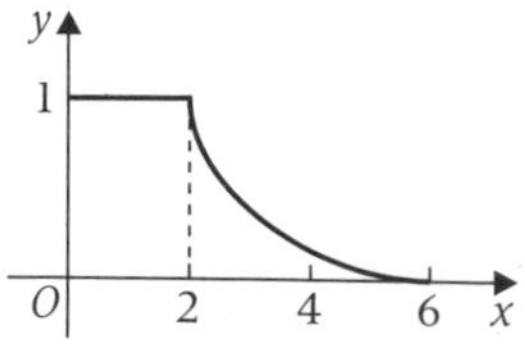

6 **(a)** $y = \sin(2x)$; **(b)** $y = 2\sin(6x)$; **(c)** $y = \sin(-6x)$.

7 **(a)** one-way stretch, scale factor $\frac{1}{32}$, in y-direction;

(b) $\begin{bmatrix} 5 \\ 0 \end{bmatrix}$.

6 Solving trigonometrical equations

EXERCISE 6A

1 **(a)** $-720° \leqslant 2x \leqslant 720°$; **(b)** $-1440° \leqslant 4x \leqslant 1440°$;

(c) $-1080° \leqslant 3x \leqslant 1080$; **(d)** $-180° \leqslant \dfrac{x}{2} \leqslant 180°$;

(e) $-90° \leqslant \dfrac{x}{4} \leqslant 90°$.

2 **(a)** $x = -165°, -105°, -45°, 15°, 75°, 135°$;

(b) $x = 6°, 30°, 78°$;

(c) $x = 12.4°, 77.6°, 192.4°, 257.6°$;

(d) $\theta = -137.7°, -102.3°, -17.7°, 17.7°$;

(e) $\theta = -172.8°, -97.2°, 7.2°, 82.8°$;

(f) $\theta = 32.4°, 92.4°, 152.4°$;

(g) $x = 22.5°, 112.5°$;

(h) $\theta = \pm 43.9°, \pm 76.1°$;

(i) $\theta = -151.8°, -61.8°, 28.2°, 118.2°$;

(j) $\theta = \pm 180°$;

(k) $\theta = 34.9°, 325.1°$;

(l) $x = 7.9°, 52.1°, 127.9°, 172.1°$;

(m) $x = 135°, 315°$.

EXERCISE 6B

1 **(a)** $-450° \leqslant x - 90° \leqslant 270°$; **(b)** $-300° \leqslant x + 60° \leqslant 420°$;

(c) $-660° \leqslant x - 300° \leqslant 60°$; **(d)** $-130° \leqslant x + 230° \leqslant 590°$.

2 **(a)** $x = -15°, 105°, 345°$;

(b) $x = -345°, -165°, 15°, 195°$;

(c) $x = -400°, -340°, -40°, 20°, 320°$;

(d) $x = -155.6°, -64.4°, 204.4°$;

(e) $x = -140°, 40°, 220°$;

(f) $36.9°, 143.1°$.

EXERCISE 6C

1 **(a)** $x = 0°, \pm 180°$; **(b)** $x = \pm 60°, \pm 120°$;

(c) $x = \pm 30, \pm 150°$; **(d)** $x = \pm 30°, \pm 150°$.

2 **(a)** $\theta = 0°, 120°, 240°$; **(b)** $\theta = 0°, 19.5°, 160.5°, 180°$;
(c) $\theta = 0°, 180°$; **(d)** $\theta = 90°, 101.5°, 270°, 258.5°$;
(e) $\theta = 30°, 150°, 199.5°, 340.5°$; **(f)** $\theta = 75.5°, 284.5°$;
(g) $\theta = 70.5°, 120°, 240°, 289.5°$; **(h)** $\theta = 14.5°, 165.5°, 270°$.

3 **(a)** $x = 51.8°, 308.2°$; **(b)** $x = 189.9°, 350.1°$;
(c) $x = 47.1°, 132.9°$; **(d)** $x = 73.1°, 286.9°$;
(e) $x = 35.5°, 144.5°$; **(f)** $x = 145.1°, 214.9°$.

4 **(a)** $\theta = -153.4°, -26.6°, 153.4°, 26.6°, 206.6°$;
(b) $x = -84.3°, -76.0°$;
(c) $\theta = -116.6°, -45°, 63.4°, 135°$;
(d) $\theta = 17.4°, 72.6°, 197.4°, 252.6°$;
(e) $x = -51.6°, -165.2°, 14.8°, 128.4°$.

EXERCISE 6D

2 **(a)** $\tan\theta = -\frac{4}{3}$; **(b)** $\tan x = \frac{12}{5}$; **(c)** $\sin\theta = \frac{5}{13}$;
(d) $\sin x = -\frac{4}{5}$; **(e)** $\tan x = -\frac{7}{24}$; **(f)** $\cos\theta = \frac{7}{25}$;
(g) $\cos x = -\frac{4}{5}$.

4 **(a)** $x = 133.3°, 226.7°$;
(b) $\theta = 67.5°, 292.5°$;
(c) $x = -343.0°, -197.0°, 17.0°, 163.0°$;
(d) $x = -141.8°, -38.2°$;
(e) $\theta = 45.0°, 135.0°\ 225.0°\ 315.0°$.

5 **(a)** 2.5; **(b)** $\frac{1}{6}$; **(c)** 12; **(d)** $\frac{4}{3}$.

6 **(a)** $-315°, -135°, 45°, 225°$;
(b) $63.4°, 243.4°$;
(c) $-32.0°, 148.0°$.

7 **(a)** $18.2°, 78.2°, 138.2°$; **(b)** $55.9°, 145.9°$.

8 **(a)** $3 + \sin\theta$; **(b)** $x = 105°, 165°, 285°, 345°$.

9 **(b)** $7.9°, 52.1°, 127.9°, 172.1°, 247.9°, 292.1°$.

10 **(b)** **(i)** $x = -\frac{1}{3}, x = -1$, **(ii)** $199°, 341°, 270°$.

11 **(b)** $19.5°, 160.5°, 270°$.

12 **(a)** $\frac{3}{4}$; **(b)** $7.4°, 43.4°, 79.4°$.

13 **(a)**

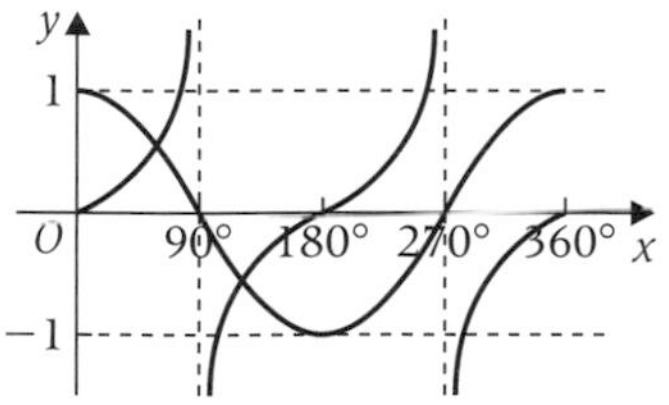

(c) $38.2°, 141.8°$.

14 $210°, 330°$.

7 Factorials and binomial expansions

EXERCISE 7A

1 **(a)** 6; **(b)** 720; **(c)** 40 320; **(d)** 576; **(e)** 1.

2 **(a)** 20; **(b)** 210; **(c)** 14;
(d) 19; **(e)** 336; **(f)** 24.

3 **(a)** 140; **(b)** 2520; **(c)** 1980; **(d)** 36 720; **(e)** 285.

4 **(a)** 10; **(b)** 35; **(c)** 1; **(d)** 1; **(e)** 28;
(f) 28; **(g)** 36; **(h)** 25; **(i)** 1; **(j)** 286;
(k) 680; **(l)** 100; **(m)** 3160.

5 15 504.

6 27 405.

EXERCISE 7B

1

1
1 1
1 2 1
1 3 3 1
1 4 6 4 1
1 5 10 10 5 1
1 6 15 20 15 6 1
1 7 21 35 35 21 7 1
1 8 28 56 70 56 28 8 1
1 9 36 84 126 126 84 36 9 1

2 1 16 120 560 1820
1 17 136 680 2380

3 1 21 210 1330 5985
1 22 231 1540 7315

4 **(a)** $a^3 + 3a^2b + 3ab^2 + b^3$;
(b) $c^5 + 5c^4d + 10c^3d^2 + 10c^2d^3 + 5cd^4 + d^5$;
(c) $x^7 + 7x^6y + 21x^5y^2 + 35x^4y^3 + 35^3y^4 + 21x^2y^5 + 7xy^6 + y^7$;
(d) $r^8 + 8r^7s + 28r^6s^2 + 56r^5s^3 + 70r^4s^4 + 56r^3s^5 + 28r^2s^6 + 8rs^7 + s^8$.

5 **(a)** $1 + 5x + 10x^2 + 10x^3 + 5x^4 + x^5$;
(b) $1 + 6y + 12y^2 + 8y^3$;
(c) $1 + 2x + \frac{5}{3}x^2 + \frac{20}{27}x^3 + \frac{5}{27}x^4 + \frac{2}{81}x^5 + \frac{1}{729}x^6$;
(d) $1 + 2p + \frac{3}{2}p^2 + \frac{1}{2}p^3 + \frac{1}{16}p^4$.

6 **(a)** $1 - 3a + 3a^2 - a^3$;
(b) $1 - 12b + 54b^2 - 108b^3 + 81b^4$;
(c) $1 - \frac{5}{2}x + \frac{5}{2}x^2 - \frac{5}{4}x^3 + \frac{5}{16}x^4 - \frac{1}{32}x^5$;
(d) $1 - 2x + \frac{4}{3}x^2 - \frac{8}{27}x^3$.

7 **(a)** $243 + 405p + 270p^2 + 90p^3 + 15p^4 + p^5$;
(b) $625 + 500x + 150x^2 + 20x^3 + x^4$;
(c) $8 - 12m + 6m^2 - m^3$;
(d) $729 - 2916t + 4860t^2 - 4320t^3 + 2160t^4 - 576t^5 + 64t^6$.

8 **(a)** $1 - 5x + 10x^2 - 10x^3$; **(b)** $1 + 12x + 54x^2 + 108x^3$;
(c) $1 - 2x + \frac{5}{3}x^2 - \frac{20}{27}x^3$; **(d)** $1 + 2x + \frac{8}{5}x^2 + \frac{16}{25}x^3$.

9 **(a)** $y^5 + 10y^4 + 40y^3 + 80y^2$;
(b) $64y^6 - 192y^5 + 240y^4 - 160y^3$;
(c) $256y^4 + 768y^3 + 864y^2 + 432y$;
(d) $2187y^7 - 5103y^6 + 5103y^5 - 2835y^4$.

10 **(a)** $216x^3 + 24x$;
(b) $480y^4 + 2160y^2 + 486$.

11 $\binom{4}{0}\binom{4}{1}\binom{4}{2}\binom{4}{3}\binom{4}{4}$; $\binom{5}{0}\binom{5}{1}\binom{5}{2}\binom{5}{3}\binom{5}{4}\binom{5}{5}$.

EXERCISE 7C

1 **(a)** $1 + 11x + 55x^2 + 165x^3$; **(b)** $1 + 15x + 105x^2 + 455x^3$;
(c) $1 - 8x + 28x^2 - 56x^3$; **(d)** $1 + 14x + 84x^2 + 280x^3$;
(e) $1 - 12x + 66x^2 - 220x^3$; **(f)** $1 + 3x + 4x^2 + \frac{28}{9}x^3$;
(g) $1 - 5x + \frac{45}{4}x^2 - 15x^3$;
(h) $512 - 2304x + 4608x^2 - 5376x^3$.

2 **(a)** $x^{12} + 12x^{11} + 66x^{10} + 220x^9$;
(b) $x^{17} + 34x^{16} + 544x^{15} + 5440x^{14}$;
(c) $x^{10} - 30x^9 + 405x^8 - 3240x^7$;
(d) $128x^7 - 448x^6 + 672x^5 - 560x^4$.

3 **(a)** $21\,840x^4$; **(b)** $1224x^2$;
(c) $-18\,304x^3$; **(d)** $364x^{12}$;
(c) $\frac{880}{9}x^4$; **(f)** $247\,860x^{14}$.

4 **(a)** -330; **(b)** $11\,440$;
(c) $-20\,127\,744$; **(d)** $-21\,840$;
(e) $\frac{14}{9}$; **(f)** -3400.

5 $1 + 4x + 7x^2 + 7x^3$.

6 $1 + 7x + 21x^2 + 35x^3$, 1.000 070 002 100 035.

7 15 120

8 **(a)** $1 + 8x + 28x^2 + 56x^3$; **(b)** 112.

9 **(a)** -63; **(b)** $x^4 - 9x^3 + 21x^2 - 35x + 15$.

10 **(a)** $1 + 10x + 45x^2 + 120x^3$; **(b)** 810.

11 $x^5 + 4x^4 + 14x^3 + 4x^2 + 9x$.

12 $1 - 33x + 495x^2 - 4455x^3$, -5445.

13 **(a)** $h^4 + 10h^3 + 40h^2 + 80h + 80$; **(b)** 80.

14 **(a)** $5 + 176x + 2826x^2 + 27\,324x^3$; **(b)** $2 - 5x + 3x^2 + 5x^3$.

8 Sequences and series

EXERCISE 8A

1 1, 4, 9.

2 2, 26, 242.

3 **(a)** 3, 5, 7, 9, 11; **(b)** 0, 3, 8, 15, 24;
(c) 0, 2, 6, 12, 20; **(d)** 2, 9, 28, 65, 126;
(e) 2, 4, 8, 16, 32; **(f)** 2, 8, 26, 80, 242.

4 **(a)** 4, 7, 10, 13, 16; **(b)** 20, 18, 16, 14, 12;
(c) 3, 7, 16, 32, 57; **(d)** 2, 6, 13, 23, 36;
(e) 1, 21, 66, 146, 271; **(f)** 100, 92, 74, 42, -8.

5 **(a)** 3; **(b)** 5; **(c)** -2; **(d)** $2n-1$;
(e) $2n-1$; **(f)** $6n-3$; **(g)** $6n-2$.

6 **(a)** $2n-1$; **(b)** 2^n; **(c)** $7-n$; **(d)** $\dfrac{n}{n+1}$.

7 **(a)** 5, 7, 9, 11; $2n+3$; **(b)** 40, 37, 34, 31; $43-3n$;
(c) 4, 7, 12, 19; n^2+3, **(d)** 2, 4, 7, 11; $1+\dfrac{n}{2}(n+1)$.

8 **(a)** 4.5, 3.875, 4.031 25, 3.992 187 5; **(b)** 4.

9 **(a)** 2, 1.4, 1.28, 1.256; **(b)** 1.25.

10 **(a)** 5.5, 3.25, 4.375, 3.8125; **(b)** 4.

EXERCISE 8B

1 **(a)** 55; **(b)** 210; **(c)** 2485.

2 **(a)** 4860; **(b)** 495 550; **(c)** 60 100; **(d)** 240 694.

3 **(a)** 27; **(b)** 47; **(c)** 137.

4 **(a)** 3; **(b)** 84; **(c)** 183.

5 54.

6 $d=3$.

7 -4.

8 $a=-2, d=3$.

9 $a=32.5, d=2.5$.

10 $a=-6, d=6$.

11 78.

12 95.

EXERCISE 8C

1 200, $\frac{14}{19}$.

2 **(a)** 285; **(b)** 1070.

3 $a=-30$.

4 23.

5 **(a)** 575; **(b)** 22 422; **(c)** 4692; **(d)** 231; **(e)** -777.

6 3875

7 **(a)** $a=8, d=4$; **(b)** 5300.

8 **(a)** 16; **(b)** 890.

9 **(a)** $d=1.5$, 65; **(b)** 75.

10 **(a)** 4; **(b)** 5250; **(c)** **(ii)** 250.

11 **(b)** $a=-13, d=3$.

EXERCISE 8D

1 **(a)** 25; **(b)** 46; **(c)** 55;
(d) 34; **(e)** 99; **(f)** 20.

2 **(a)** 4972; **(b)** 10 000; **(c)** 44 795;
(d) 14 895; **(e)** 14 850; **(f)** 500 500.

3 **(a)** $3n - 2, \sum_{n=1}^{31} 3n - 2$; **(b)** $9n + 16, \sum_{n=1}^{100} 9n + 16$;

(c) $79 - 4n, \sum_{n=1}^{18} 79 - 4n$; **(d)** $13 - 2n, \sum_{n=1}^{22} 13 - 2n$.

4 **(a)** −18, −13; **(b)** +5; **(c)** 3180.

5 **(a)** 17, 14; **(b)** −3; **(c)** −1660.

EXERCISE 8E

1 $a = 3, d = 1.75$.

2 **(a)** $a = 5, d = 0.25$; **(b)** 5975.

3 $a = -3, d = 2.5$.

4 **(a)** $d = 1.5$; **(b)** 345.

5 173.

6 $d = 44$.

7 75150

8 **(a)** 10.5, 11; **(b)** +0.5; **(c)** 30; **(d)** 532.5.

9 **(a)** 392; **(b)** **(i)** 47, 44, 41, 38.

10 **(a)** $3n + 12$; **(c)** 16 miles.

9 Radian measure

EXERCISE 9A

1 **(a)** 90°; **(b)** 720°; **(c)** 60°; **(d)** 45°;
(e) 150°; **(f)** 135°; **(g)** 120°; **(h)** 330°;
(i) 315°; **(j)** 450°.

2 **(a)** π; **(b)** $\frac{2\pi}{3}$; **(c)** $\frac{\pi}{5}$; **(d)** $\frac{2\pi}{15}$;
(e) $\frac{3\pi}{5}$; **(f)** 3π; **(g)** $\frac{4\pi}{9}$; **(h)** $\frac{5\pi}{4}$;
(i) $\frac{9\pi}{4}$; **(j)** $\frac{\pi}{12}$.

3 **(a)** 115°; **(b)** 28.6°; **(c)** 103°; **(d)** 172°;
(e) 17.2°; **(f)** 132°; **(g)** 73.3°; **(h)** 91.7°;
(i) 115°; **(j)** 77.1°.

4 **(a)** 1.05^c; **(b)** 2.62^c; **(c)** 0.436^c; **(d)** 5.32^c;
(e) 1.68^c; **(f)** 1.36^c; **(g)** 0.244^c; **(h)** 1.43^c;
(i) 0.663^c; **(j)** 8.73^c.

5 **(a)** **(i)** 0.389, **(ii)** 0.921, **(iii)** 0.423;
(b) **(i)** 0.932, **(ii)** 0.362, **(iii)** 2.57;
(c) **(i)** 0.909, **(ii)** −0.416, **(iii)** −2.19;
(d) **(i)** 0.924, **(ii)** −0.383, **(iii)** −2.41

6 (a) 4π; (b) 48π.

7 (a) $\frac{2\pi}{3}$; (b) $\frac{7\pi}{12}$; (c) $\frac{7\pi}{8}$.

EXERCISE 9B

1 (a) 27 cm; (b) 50 cm; (c) 2.1 cm; (d) 7.7 cm.

2 (a) 122 cm; (b) 116 cm; (c) 55.9 cm; (d) 37.9 cm (to 3 sf).

3 (a) 2.1 rads; (b) 2.5 rads; (c) 0.625 rads; (d) $2\frac{2}{3}$ rads; (e) 1 rad; (f) 0.5 rads; (g) $2\frac{1}{6}$ rads; (h) 1.6 rads.

4 (a) 5 cm; (b) 3.2 cm; (c) 1.6 cm (d) $\frac{108}{\pi}$(= 34.4 cm to 3 sf); (e) $\frac{45}{\pi}$(= 14.3 cm to 3 sf); (f) 32 cm.

7 (a) 16.8 cm; (b) 7.7 cm; (c) 16.4 cm; (d) 29.9 cm.

EXERCISE 9C

1 (a) 16 cm^2; (b) 27 cm^2; (c) 0.844 cm^2; (d) 0.524 cm^2.

2 (a) 4 cm; (b) 10 cm; (c) 6 cm; (d) 4.37 cm (to 3 sf).

3 (a) 288 cm^2; (b) 94.2 cm^2; (c) 12.3 cm^2; (d) 50.9 cm^2.

4 (a) 40.3 cm; (b) 31.7 cm; (c) 54.4 cm; (d) 70.4 cm (to 3 sf).

5 (a) 15.7 cm^2; (b) 10.5 cm^2; (c) 20 cm^2; (d) 20.9 cm^2.

7 55.5 cm^2.

MIXED EXERCISE

1 15 cm.

2 (a) $\frac{\pi}{6}$ rads; (b) $(\frac{40}{3}\pi + 8)$ cm.

3 (a) 375 cm^2; (b) 1880 cm^2.

4 (b) 99 metres.

5 5.7 cm.

6 (a) $\frac{\pi}{3}$; (b) 2π cm.

7 (a) 4.5 cm; (b) 6.75 cm^2.

8 (a) (i) $PM = 3$ cm, $OM = 3\sqrt{3}$ cm, (ii) 2π cm, (iii) 6π cm^2; (b) $m = 3$.

9 (a) $r^2\theta$; (b) $\frac{1}{2}r^2 \sin\theta$.

10 (a) 50θ cm^2.

10 Further trigonometry with radians

EXERCISE 10A

1 (a) $\sin\theta$; (b) $-\sin\theta$; (c) $\sin\theta$; (d) $-\sin\theta$.

2 (a) $-\cos\theta$; (b) $-\cos\theta$; (c) $\cos\theta$; (d) $\cos\theta$.

3 (a) $-\tan\theta$; (b) $\tan\theta$; (c) $\tan\theta$; (d) $-\tan\theta$.

EXERCISE 10B

1 (a) 0.467; (b) 0.723; (c) 1.01;
(d) 1.31; (e) 0.970; (f) 0.584.

2 (a) 0.644, 2.50; (b) 1.16, 5.12;
(c) 0.983, 4.12; (d) 4.07, 5.36;
(e) 2.35, 3.94; (f) 2.03, 5.18;
(g) 0.848, 2.29; (h) 2.25, 4.04;
(i) 1.34, 2.91, 4.48, 6.05; (j) no solutions in given interval;
(k) 5.68; (l) 0.749, 3.89.

3 0.524, 2.62

4 (a) $\frac{\pi}{6}, \frac{5\pi}{6}$; (b) $\pm\frac{5\pi}{6}$;
(c) $-\frac{3\pi}{4}, \frac{\pi}{4}$; (d) $-\frac{3\pi}{4}, -\frac{\pi}{4}$;
(e) $\pm\frac{\pi}{8}, \pm\frac{7\pi}{8}$; (f) $-\frac{3\pi}{4}, -\frac{5\pi}{12}, -\frac{\pi}{12}, \frac{\pi}{4}, \frac{7\pi}{12}, \frac{11\pi}{12}$;
(g) $0, \pm\frac{\pi}{7}, \pm\frac{2\pi}{7}, \pm\frac{3\pi}{7}, \pm\frac{4\pi}{7}, \pm\frac{5\pi}{7}, \pm\frac{6\pi}{7}$;
(h) $\pm\frac{2\pi}{9}, \pm\frac{4\pi}{9}, \pm\frac{8\pi}{9}$; (i) $-\frac{2\pi}{3}, -\frac{\pi}{6}, \frac{\pi}{3}, \frac{5\pi}{6}$.

5 (a) 1; (b) $\frac{11\pi}{6}$.

EXERCISE 10C

1 (a) 0.322, 2.82, 3.46, 5.96; (b) 0.685, 2.46, 3.83, 5.60;
(c) 1.11, 2.03, 4.25, 5.18; (d) 0.644, 3.14, 5.64;
(e) 3.14, 3.87, 5.55; (f) 1.11, 1.82, 4.25, 4.96;
(g) 0.896, 5.39; (h) 0.927, 5.36;
(i) 0.730, 2.41; (j) 0.393, 1.96, 3.53, 5.11;
(k) 1.23, 1.57, 4.71, 5.05.

2 (a) ±0.785, ±2.36; (b) ±0.524, ±2.62;
(c) ±1.05, ±2.09; (d) ±2.09;
(e) 0.524, 2.62; (f) −0.785, 2.36;
(g) ±0.524, ±2.62 (h) ±1.05;
(i) ±0.524, ±1.57, ±2.62.

3 (a) 0.322, 3.46; (b) 0.896, 4.04;
(c) 1.11, 2.03, 4.25, 5.18; (d) 0.927, 2.21, 4.71;

(e) 1.57, 2.30, 3.98, 4.71; (f) 0.841, 3.14, 5.44;
(g) 0.524, 2.62; (h) 0.464, 1.89, 3.61, 5.03;
(i) 0.322, 2.03, 3.46, 5.18.

4 (a) ±0.393, ±1.18, ±1.96, ±2.75;
(b) 0, ±2.09;
(c) ±2.09, 0, ±3.14, 1.05;
(d) ±0.524, ±1.57, ±2.62;
(e) −1.92, −1.22, 0.175, 0.873, 2.27, 2.97;
(f) 0, ±0.524, ±1.05, ±1.57, ±2.09, ±2.62, ±3.14;
(g) −1.31, −0.262, 1.83, 2.88;
(h) ±0.349, ±1.05, ±1.75, ±2.44, ±3.14;
(i) ±0.262, ±0.785, ±1.31, ±1.83, ±2.36, ±2.88.

MIXED EXERCISE

1 1.57, 3.67.

2 0.262, 1.31, 2.36, 3.40, 4.45, 5.50.

3 1.32, 2.09, 4.19, 4.97.

4 (c) (i) $\frac{1}{2}$, −2, $-1 \leqslant \sin\theta \leqslant 1$, (ii) 0.524, 2.62 (to 3 sf).

5 (b) 0, 3.14, 3.67, 5.76 (to 3 sf).

11 Exponentials and logarithms

EXERCISE 11A

1 (a)
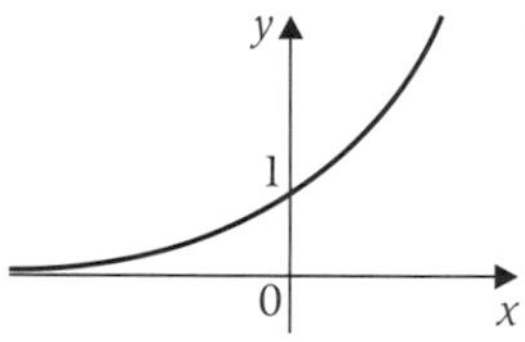

;

(b)
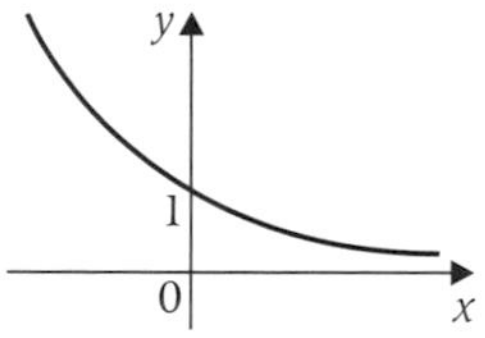

;

(c)
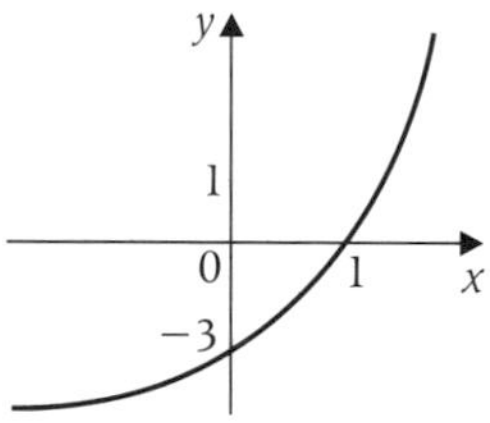

;

(d)
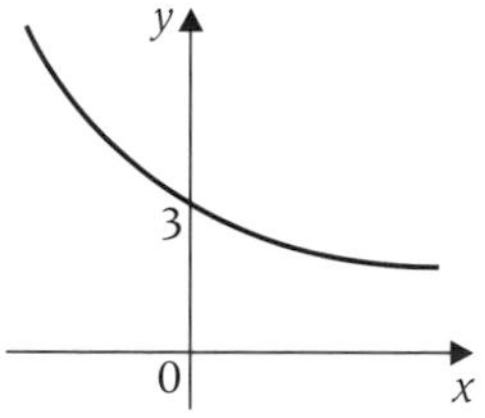

.

2 (a) One way stretch SF $\frac{1}{3}$ parallel to x-axis;
(b) Translation $\begin{bmatrix} 3 \\ 0 \end{bmatrix}$;
(c) One way stretch SF 3 parallel to x-axis;
(d) Translation $\begin{bmatrix} -5 \\ 0 \end{bmatrix}$;
(e) Reflection in y-axis.

3 **(a)** One way stretch SF $\frac{1}{5}$ parallel to x-axis;

(b) Translation $\begin{bmatrix} 4 \\ 0 \end{bmatrix}$;

(c) One way stretch SF $\frac{5}{2}$ parallel to x-axis;

(d) One way stretch SF 2 parallel to y-axis;

(e) Translation $\begin{bmatrix} 0 \\ 4 \end{bmatrix}$.

4 $a = 1$, $b = \frac{1}{2}$, $c = \frac{1}{25}$.

5 $d = \frac{1}{8}$, $e = 3\frac{1}{2}$, $f = 3$.

6 **(a)** $(0, 1)$, $(1, \frac{1}{5})$

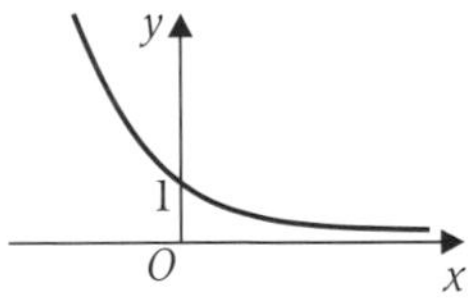

(b) $(0, 81)$, $(1, 243)$

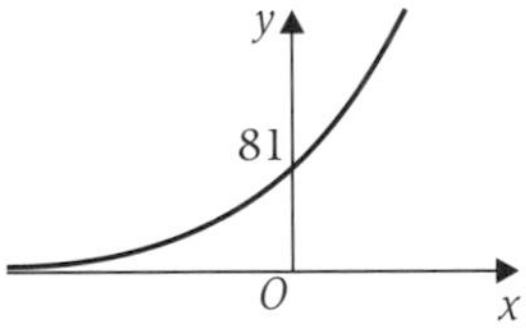

(c) $(0, 1)$, $(1, 8)$

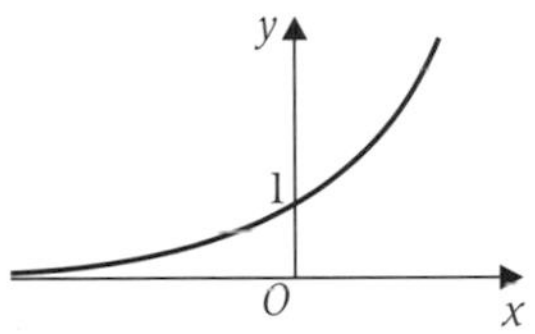

(d) $(0, 4)$, $(1, 1)$

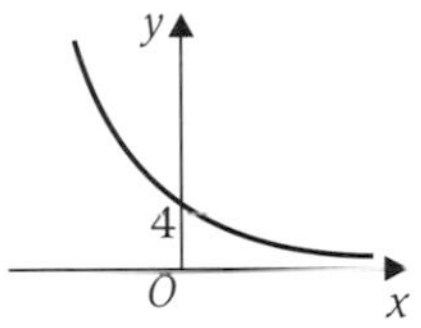

7

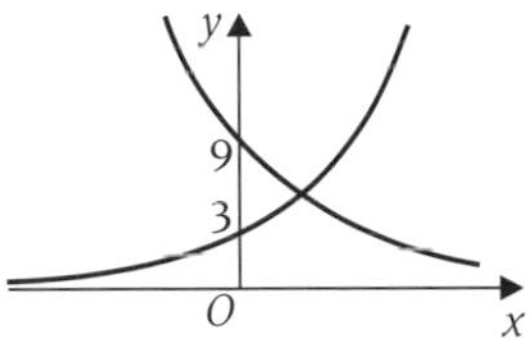

$(\frac{1}{2}, 3\sqrt{3})$.

EXERCISE 11B

1 **(a)** 16; **(b)** 64; **(c)** 3; **(d)** 1;
(e) 100 000; **(f)** 27; **(g)** 2; **(h)** 0.0625;
(i) 0.03125; **(j)** 2; **(k)** 16; **(l)** $\frac{1}{9}$.

2 **(a)** 5; **(b)** 1; **(c)** 3; **(d)** $1\frac{2}{3}$;
(e) -3; **(f)** 2.5; **(g)** 2.5; **(h)** -1.5;
(i) -0.5; **(j)** -2; **(k)** 0; **(l)** 7;
(m) 3; **(n)** 0.75; **(o)** $\frac{1}{9}$.

3 $x = 0.125$.

EXERCISE 11C

1 **(a)** $\log_a 8$; **(b)** $\log_a 24$; **(c)** $\log_a 9$; **(d)** $\log_a 12$;
(e) $\log_a \frac{3a}{2}$; **(f)** $\log_a (6a)$; **(g)** $\log_a (3a^2)$; **(h)** $\log_a \left(\frac{1}{3}\right)$.

2 **(a)** $\log_a x + \log_a y + \log_a z$; **(b)** $1 + \log_a x + \log_a y + \log_a z$;
(c) $2\log_a x + 3\log_a y + 4\log_a z$; **(d)** $1 + \log_a x - \log_a y - \log_a z$;

(e) $3\log_a x - 5\log_a y - \frac{1}{2}\log_a z$; (f) $2\log_a x - 3\log_a y + \frac{1}{2}\log_a z$;

(g) $-2\log_a x - 4\log_a y + \frac{1}{3}\log_a z$; (h) $\frac{5}{2} + \frac{1}{2}\log_a x - 3\log_a y + \frac{5}{2}\log_a z$.

3 (a) 1; (b) 0; (c) 0.391; (d) −0.782;
(e) 0.218; (f) 0.564; (g) 1.564; (h) 0.346.

4 0.

5 (a) 45; (b) $x = 4.5$.

6 $n = 2, n = 4$.

7 $x = 9, y = 27$ or $x = 27, y = 9$.

EXERCISE 11D

1 (a) 6; (b) 7; (c) −3; (d) 5.

2 (a) 5.95; (b) 7.01; (c) −2.94; (d) 5.00.

3 (a) −0.399; (b) 3.82; (c) 2.68; (d) 1.64.

5 (a) $\left(\frac{\ln 6}{\ln 2} - 4\right)$; (b) $\frac{1}{3}\left(\frac{\ln 17}{\ln 3} + 1\right)$;
(c) $\frac{1}{4}\left(1 - \frac{\ln 5}{\ln 2}\right)$; (d) $\frac{1}{3}\left(\frac{\ln 31}{\ln 5} - 4\right)$.

6 (a) 3.39; (b) 0.363; (c) 1.25; (d) 0.743.

7 (a) 8.303; (b) 0.7731; (c) 5.116;
(d) 0.9882; (e) 1.504; (f) 0.8244.

8 (a) −2.30; (b) 0.519; (c) −0.419; (d) 0.117.

9 (a) 0.683; (b) −0.431.

10 (a) 1.46; (b) 0.631; (c) 0.683; (d) 0.792.

MIXED EXERCISE

1 (a) 3; (b) $2\log_2 3$; (c) $m = 3, n = 2$.

2 (b) (i) 20, (ii) −2.5.

3 (b) (i) 1.5, (ii) $k = 3$.

4 (b) (i) 1, (ii) $\frac{1}{2}$, (iii) $\sqrt{2}$.

5 2.81.

6 0.737.

7 (a)

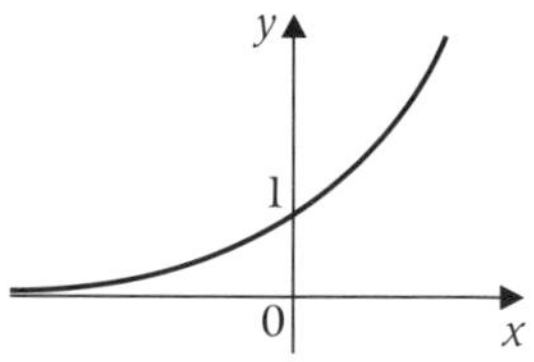

;

(b) (i) 1, (ii) 3;
(c) −3;
(d) 2.32.

8 (b) $p = 9, q = 42$.

12 Geometric series

EXERCISE 12A

1 **(a)** **(i)** $\frac{1}{3}$, **(ii)** $6\frac{2}{3}$; **(b)** **(i)** 4, **(ii)** 256;
(c) **(i)** -2, **(ii)** -96; **(d)** **(i)** $\frac{3}{4}$, **(ii)** $84\frac{3}{8}$;
(e) **(i)** $-\frac{2}{3}$, **(ii)** 16; **(f)** **(i)** $\frac{4}{5}$, **(ii)** $51\frac{1}{5}$.

2 **(a)** $180(\frac{1}{3})^{37}$; **(b)** 4^{38}; **(c)** -3×2^{37};
(d) $200(\frac{3}{4})^{37}$; **(e)** $-81(\frac{2}{3})^{37}$; **(f)** $100(\frac{4}{5})^{37}$.

3 $35\frac{5}{9}$.

4 90, $17\frac{7}{9}$.

5 $20\frac{1}{4}$, $\frac{2}{3}$.

6 320, $\frac{1}{2}$.

7 First term $= 364\frac{1}{2}$, common ratio $= \pm\frac{2}{3}$.

8 ± 6.

EXERCISE 12B

1 **(a)** 20.000; **(b)** 10.797; **(c)** 374.986;
(d) 262 143.750; **(e)** $-871\,696\,100.000$; **(f)** 0;
(g) 100.
Note that answers to (f) and (g) are exact whereas in parts (a) to (e) the answers have been rounded to three decimal places.

2 16 400.

3 **(a)** 1.25; **(b)** 10.24; **(c)** 84.04.

4 $r = \frac{3}{2}$, $S_{10} = 1813.281\,25$.

5 25.859 65 ...

6 2047.9375.

7 $\frac{3}{4}$.

8 $n = 7$.

EXERCISE 12C

1 **(a)** **(i)** converges, **(ii)** 128; **(b)** **(i)** converges, **(ii)** 32;
(c) not; **(d)** **(i)** converges, **(ii)** 432;
(e) **(i)** converges, **(ii)** 86.4; **(f)** not;
(g) **(i)** converges, **(ii)** $35\frac{5}{7}$.

2 $a = 45$, $S_\infty = 135$.

3 $a = 64$, $S_\infty = 256$.

4 0.2.

5 $-\frac{5}{32}$.

6 **(a)** $a = 4096$, $r = -\frac{1}{2}$; **(b)** $2730\frac{2}{3}$.

7 $-\frac{16}{27}$.

8 **(a)** $\frac{1}{4}$, 1200; **(b)** 1600.

9 **(a)** $-\frac{1}{2} < x < \frac{1}{2}$; **(b)** $1 < x < 2$; **(c)** $-\frac{7}{5} < x < -\frac{19}{15}$.

10 $\frac{64}{125}$.

11 $\frac{3}{4}$, 224 cm^3.

12 **(a)** $\frac{7}{9}$; **(b)** $\frac{41}{333}$.

EXERCISE 12D

1 **(a)** 12; **(b)** 60; **(c)** 120.

2 16.

3 **(b)** $546\frac{3}{4}$.

4 **(b)** $10 \times (0.9)^{n-1}$; **(d)** 100.

5 $3^n - 1$.

6 **(a)** **(i)** $2, 2r, 2r^2, 2r^3$; **(c)** 4.

7 **(b)** **(i)** $a = 100$, **(ii)** 368.928.

8 3.

9 **(a)** $4\sqrt{2}$; **(b)** $k = 8$.

10 **(a)** 8 metres; **(d)** $a = 10$, $b = 7$.

11 **(b)** 546.75; **(c)** **(ii)** $p = 97$, $q = 93$.

12 72 seconds.

13 **(a)** $\frac{3}{4}$; **(d)** -128.

Answers to practice paper for C2

1 **(a)** 52.0°; **(b)** 29.7 cm^2.

2 **(b)** 40 cm^2.

3 **(a)** 1st term = 2, common difference = 4;

(b) e.g. since the 1st term and common difference are both even, all terms in the series are even so the sum of the first k terms, for any positive integer k will always be even.

4 **(a)** **(i)** 1, **(ii)** 0.

5 **(a)** $1 + 3y + 3y^2 + y^3$;

(b) **(i)** $1 + 3\sqrt{x} + 3x + x\sqrt{x}$, $x + 2x\sqrt{x} + \frac{3}{2}x^2 + \frac{2}{5}x^2\sqrt{x} + c$.

6 **(b)** $-\frac{1}{3}$.

7 **(b)** **(i)** $y = -2, \frac{3}{4}$, **(ii)** 0.848^c, 2.29^c.

8 **(a)** **(i)** $8x - \frac{729}{x^2}$, **(iii)** $8 + \frac{1458}{x^3}$, **(v)** 4.5, minimum;

(b) 243 cm^2.

9 **(a)** **(i)** 41.5, **(ii)** underestimate;

(b) **(ii)** 4.321 93;

(c) **(i)** translation $\begin{bmatrix} 0 \\ -16 \end{bmatrix}$,

(ii) stretch with scale factor $\frac{1}{5}$ in the y-direction.

Index